HERBIVORES:
BETWEEN PLANTS
AND PREDATORS

D1757193

BATH SPA UNIVERSITY COLLEGE
NEWTON PARK LIBRARY

Please return or renew not later than
the latest date stamped below

Herbivores: Between Plants and Predators

THE 38TH SYMPOSIUM OF THE BRITISH ECOLOGICAL SOCIETY
IN COOPERATION WITH THE NETHERLANDS ECOLOGICAL SOCIETY
HELD AT THE WAGENINGEN AGRICULTURAL UNIVERSITY
THE NETHERLANDS, 1997

EDITED BY

H. Olff
Nature Conservation and Plant Ecology Group,
Department of Environmental Sciences,
Wageningen Agricultural University,
Wageningen, The Netherlands

V.K. Brown
CABI Bioscience (Environment)
UK Centre
Ascot, UK

R.H. Drent
Zoological Laboratory,
University of Groningen,
Haren, The Netherlands

Blackwell
Science

DISTRIBUTORS

Marston Book Services Ltd
PO Box 269
Abingdon, Oxon OX14 4YN
(Orders: Tel: 01235 465500
         Fax: 01235 465555)

USA
Blackwell Science, Inc.
Commerce Place
350 Main Street
Malden, MA 02148 5018
(Orders: Tel: 800 759 6102
              781 388 8250
         Fax: 781 388 8255)

Canada
Login Brothers Book Company
324 Saulteaux Crescent
Winnipeg, Manitoba R3J 3T2
(Orders: Tel: 204 837-2987)

Australia
Blackwell Science Pty Ltd
54 University Street
Carlton, Victoria 3053
(Orders: Tel: 3 9347 0300
         Fax: 3 9347 5001)

A catalogue record for this title
is available from the British Library

ISBN 0-632-05155-8 (hardback)
     0-632-05204-X (paperback)

Library of Congress
Cataloging-in-publication Data

British Ecological Society. Symposium (38th: 1997)
    Herbivores: Between Plants and Predators:
    the 38th symposium of the British
    Ecological Society 1997/edited by H. Olff, V.K.
    Brown, R.H. Drent.
       p.   cm.
    ISBN 0-632-05155-8 (hardback). — ISBN
    0-632-05204-X (pbk.)
    1. Animal–plant relationships — Congresses.
    2. Herbivores — Ecology — Congresses.
    3. Herbivores — Food — Congresses. 4. Plant
    defenses — Congresses. I. Olff, H. II. Brown,
    V.K. III. Drent, R.H.
    QH549.5.B75   1997
    591.5′4 — dc21   98-8024
                                        CIP

Contents

Contents

Preface

Herbivores play a central role in almost every ecosystem as consumers of primary production, and as food for predators. However, plants are not defenceless against herbivores — many physiological and morphological traits have evolved leading to avoidance or tolerance of herbivory, while traits in herbivores have evolved to cope with these defence strategies. Herbivores also need plants for other reasons than food, e.g. for protection against predation. Predators and parasitoids of herbivores possess specific features enabling them to eat 'plant eaters', involving specific search strategies and even responses to signals sent out by the plants themselves. Taking herbivory as a central focus enables us to integrate knowledge from different fields of interest and disciplines, such as plant chemistry, nutrient cycling, food webs, vegetation succession, animal foraging strategies, community structure and population dynamics. Several of these aspects have been covered by previous symposia of the British Ecological Society, however, not from the central perspective of herbivory. It is for this reason that the focus on the herbivore has become the central theme for the 1997 BES symposium.

Plant–herbivore–predator interactions have only been studied intensively from the early 1960s, starting mainly with work on the trophic structure of aquatic ecosystems (lumping all herbivores in one guild) or on plant chemical defence. New fields include evolutionary studies, research on interactions including more than two trophic levels, analysis of below-ground food webs, and studies on the interplay of plant community processes with herbivore community processes. In assembling this symposium, we tried to cover these new fields, and restricted ourselves to terrestrial ecosystems. We think the meeting was a great success, due to the enthusiasm of the participants and the speakers, and that it has helped to indicate promising directions for the future.

This volume is intended as an up-to-date overview of the current knowledge on the most important aspects of herbivory in terrestrial ecosystems. It contains many new ideas that still need to be tested, and hence may stimulate new and collaborative research in the near future. We hope this leads to more insight into the evolution, functioning and organization of natural populations, communities and ecosystems.

The meeting was co-sponsored by the Netherlands Ecological Society (NEVECOL) and we wish to acknowledge the unstinting financial assistance of the Stortenbeker Fund of that society for helping to make the meeting truly international in scope (participants from 15 countries attended the sessions in Wageningen). The post-symposium excursion to the island of Schiermonnikoog, graced by perfect weather, attracted 30 members who could see at first hand the outcome of the herbivore exclusion experiments on the marsh and we thank Professor J.P. Bakker for masterminding this.

H. Olff
V.K. Brown
R.H. Drent
FEBRUARY 1998

Part 1

Plants shaping the evolution of insect herbivores

Introductory remarks

P.M. Brakefield[1] and C. Wiklund[2]

The focus of much recent research in population biology has concerned the interactions between plants and their insect herbivores. This is an especially exciting topic, because it is multidisciplinary in nature with many of the most important advances coming from the combined application of ecological research with modelling or molecular tools. It is timely to assess what is known about the ways in which plants have influenced or shaped the evolution of insects in relation to both processes of adaptation and of speciation. In addition, these relationships can be viewed from the opposing perspective: how have insect herbivores shaped the evolution of plants? The following chapters not only review previous work, but also highlight some of the unresolved problems and the likely directions of future research.

The opening chapter by John Thompson describes the major ideas about the ecological, genetic and evolutionary processes involved in the coevolution of plants and insect herbivores. As covered in more detail in his recent book (*The Coevolutionary Process*, 1994, University of Chicago Press), attention is given to spatial patterns of variation in the tightness and nature of coevolutionary relationships. It is clear that subdivided population structures can have profound effects on how coevolution manifests itself in the population biology of the organisms involved. The extent of geographical variation in the relationships between plants and insect herbivores is only now beginning to become clear, both from the ecological and genetic perspectives. As Thompson shows, a challenge for the future will be to take this into account in the further development of evolutionary models of coevolution.

Many of the case studies cited in Thompson's chapter involve species of Lepidoptera. The following three chapters focus even more strongly on these insects, which might suggest a lack of representation of other insect

1 Institute of Evolutionary and Ecological Sciences, University of Leiden, PO Box 9516, 2300 RA Leiden, The Netherlands
2 Department of Zoology, Stockholm University, S-10691 Stockholm, Sweden

groups. We make no apology for this apparent bias, because we believe that this group of insects provides many excellent model systems, which have yielded insights about pattern and process of general relevance. The phylogenetic aspects are especially well represented in studies of the interactions between butterflies and their food plants. Sören Nylin and Niklas Janz review research on one model group of Nymphalid butterflies, for which the data are particularly rich. Their own recent efforts have demonstrated the power of combining ecological experiments on life history evolution and host plant choice, with robust descriptions of phylogenies from the application of modern molecular tools to both the plants and insects. This approach promises many exciting new insights over the coming decade.

An entirely different approach is taken by Patsy Haccou and colleagues in their case study of the interaction between the cinnabar moth, *Tyria jacobaeae*, and its patchily distributed host plant, *Senecio jacobaea*. They use a combination of ecological data and mathematical modelling of patch-leaving strategies and their payoff relationships to derive novel insights about how a particular interaction may evolve. Their study refers especially to circumstances in which the host plant is patchily distributed, in which an individual plant may be too small for completion of host development and where dispersal of pre-adult stages among plant hosts can occur. The future incorporation of the insights from such approaches into understanding metapopulation dynamics may provide a productive link to the geographical patterning of coevolutionary relationships discussed in Thompson's chapter.

May Berenbaum and Art Zangerl concentrate on their own studies of wild parsnips and their herbivore community in North America to describe the ways in which plant chemical defences may evolve. As with the phylogenetic studies of Nylin and Janz, research on plant defences has received a boost in recent years through the application of molecular biology. As is made clear by Berenbaum and Zangerl, plant chemical defence is beginning to be understood in terms of molecular mechanisms and gene regulation. However, there is much still to learn, especially in view of the complex cocktails of plant secondary compounds involved in most examples of their defence, and the finely tuned relationships between the presence of targeted herbivorous insects and plant genotype, phenotype and fitness.

The final chapter from Maurice Sabelis and colleagues effectively adds an additional level of organization and complexity to plant–insect interactions.

The authors show that such interactions may not only be influenced by the type of direct aspects of plant defence discussed by Berenbaum and Zangerl, but also by important indirect aspects involving predators of the herbivores. The latter aspects may range from plants providing protection or food for potential predators, to different ways in which they can 'call for help'. While some readers might consider that we know rather little about the consequences of direct plant defence, without including indirect defence, Sabelis and colleagues demonstrate the additional accuracy of modelling plant–herbivore dynamics, which may be attained by including the latter, at least for certain communities. Again, they suggest how understanding tritrophic interactions may make an important contribution to the theory of metapopulation dynamics. The field of tritrophic interactions is likely to represent a further area of expanding research involving interactions between plants and insects over the coming decade. As with other such areas of expanding research, the most exciting insights are likely to continue to result from the combination of different approaches and from integrated studies which involve different levels of biological organization. An increasing understanding of the genetic basis of variation in the relevant plant and insect traits, and of the ways in which genes regulate the phenotypes which are influenced by selection, is likely to be especially influential in the future.

Chapter 1

What we know and do not know about coevolution: insect herbivores and plants as a test case

J.N. Thompson[1]

Summary

The hard task of coevolutionary studies is to sort out which aspects of the great diversity of interspecific interactions are most important in shaping the coevolutionary process and the ecological patterns that result. A major part of that task is to understand how different spatial and temporal scales shape the coevolutionary process. Here I evaluate what we currently know and do not know about the coevolutionary process at the level of local populations, geographic mosaics of populations, and phylogenetic lineages of interacting taxa, using the interactions between plants and insect herbivores as a test case. Studies on the coevolutionary process have now indicated that the conditions necessary for coevolutionary change are met at least sometimes at each level in the hierarchy of interacting natural populations and lineages. There are, however, only a few studies that have been designed specifically to evaluate reciprocal evolutionary change in particular interactions at the level of local populations, geographic mosaics of differentiated populations, or lineages of species. Most studies are asymmetric, showing much observational and experimental work on one side of the interaction and only inferred effects on the other side. I end with a series of corrections of common misunderstandings and unfounded expectations about the coevolutionary process.

1 Departments of Botany and Zoology, Washington State University, Pullman, WA 99164, USA

Introduction

The study of coevolution—reciprocal evolution in interacting species driven by natural selection—between phytophagous insects and plants was one of the major initial stimuli for the blossoming of coevolutionary studies over the past 30 years. During these years, our perception of the importance of coevolution in organizing interactions within and among communities has swung wildly back and forth. This debate over pattern, process and the interpretation of individual studies has, in retrospect, been immensely useful. It has cast aside some naive views of the coevolutionary process and refined the questions we need to answer.

One of the casualties of the debate, however, has been the close ties between ecological and evolutionary approaches to the study of inter-specific interactions, including insect–plant relationships. Coevolutionary studies are continuing apace, but they are often divorced from ecological context as they focus on long time scales. Meanwhile, ecological studies of interspecific interactions and the forces shaping biodiversity often do not even consider the potential role of coevolution as a major organizing process in community ecology. Many current ecological and evolutionary studies of interspecific interactions seem to be proceeding more in parallel than as interwoven strands, even though some of the major advances in community ecology and coevolutionary theory continue to come from combining evolutionary and ecological approaches.

The gap between coevolutionary studies and many ecological approaches is now so wide that only one (Ricklefs & Schluter 1993) of four recent important edited volumes on biodiversity and interspecific interactions published between 1993 and 1996 included any discussion of coevolution. The other three volumes—two on biodiversity (Gaston 1996; Hochberg et al. 1996) and the other on interspecific interactions (Kareiva et al. 1993)—included contributions by more than 100 researchers. Authors in all three books discussed evolutionary issues such as adaptation of populations to changing conditions and the population structure of genetic diversity, but no chapter focused on the coevolutionary process or included it as a sufficiently major point to make it into the index.

Some of these same authors have discussed coevolution elsewhere. So, the point is not that coevolution is regarded as unimportant. Rather, it is more that coevolutionary studies have become uncoupled from many aspects of current ecological thought, developing in parallel with a focus on different time scales and different questions. As Pickett et al. (1994) have argued in their analysis of ecological theories in general, the gap is partially a result of differences in what is considered to be the domain of these different areas of research.

Much of the current gap between ecological and evolutionary approaches to the study of interspecific interactions seems to have arisen because we have often not been clear about what we actually know and do not know. This blurring of what is known and what is unknown developed from four problems:

1 an initial lack of coevolutionary hypotheses that made specific ecological predictions for anything other than the simplest pairwise interactions;

2 misinterpretation of the hypotheses that were available, especially the Ehrlich and Raven (1964) hypothesis of escape-and-radiate coevolution;

3 a view of the coevolutionary process that did not confront the geographic structuring of populations; and

4 the long time span required to gather critical data on both (or all) sides of interactions.

Each of these problems has led repeatedly to interpretations of the co-evolutionary process that have not followed from what we actually know or what has actually been hypothesized. During the past decade, we have resolved or clarified many of these points, but it is taking some additional time for the news of that progress to become generally known.

Here I evaluate what we know and do not know at the three major spatial and temporal levels in the hierarchy of the coevolutionary process, using plants and insect herbivores as the model group of interactions. Those three levels are local populations, geographic mosaics of interacting species, and phylogenetic lineages of interacting taxa. In cases where we lack studies from insect–plant interactions, I point to related studies from other interactions that can be used as guides for future work. In addition, I note the major misinterpretations and unfounded expectations of the coevolutionary process that have been partially responsible for generating the current gap between ecological and coevolutionary studies of evolving interactions. By necessity, this evaluation leaves out much. Parallel analyses could be written on what we know about patterns resulting from the process, how specific hypotheses on pattern and process compare with one another, and which interactions currently provide the most detailed understanding of specific patterns of adaptation and counteradaptation.

Coevolution within local communities

The results of several decades of work in evolutionary population genetics have shown beyond all doubt that species are usually collections of genetically differentiated populations. Because population structure is hierarchical, we cannot assume that the study of coevolution between two interacting local populations reflects accurately the coevolutionary dynamics of those two species. Populations adapt and counteradapt to one another at the level of local communities, but local interactions and communities are highly dynamic and transient. Similarly, we cannot assume that comparisons of cladograms of interacting taxa capture the dynamics of reciprocal change in any pair of species. Analyses of cladograms capture mostly patterns of change that have led to fixed traits, and most of the dynamics of coevolutionary change may not lead to fixed traits in the interacting species.

Nevertheless, the results of coevolution at the level of local populations are the raw material for the processes and patterns that result at higher levels in the hierarchy of the coevolutionary process. There are two crucial questions on coevolution at the level of local communities:

1 Is there reciprocal heritable variation for traits that affect interspecific interactions?

2 Is there evidence for reciprocal natural selection on interacting species?

What we know

There is now extensive evidence for heritable variation in traits important to plant–herbivore interactions, and good evidence of selective effects of plants and insects on each other. Studies showing these results are now numerous enough that I have listed in Table 1.1 only a summary of the kinds of results available and several example studies or reviews. Other major reviews on these questions have been published in recent years (e.g. Berenbaum & Zangerl 1992b; Fritz & Simms 1992; Futuyma & Keese 1992; Marquis 1992; Karban & Baldwin 1997).

Five points stand out in our current understanding.

1 Populations and species differ in the extent and kinds of genetic variation available to shape interactions (e.g. Futuyma *et al.* 1995; Bergelson & Purrington 1996), which creates an opportunity for evaluating how the pattern of genetic variation within and among populations shapes the coevolutionary process.

2 Because natural selection imposed by herbivores may act on any plant stage, the potential for selective effects of herbivores on plants is greater

Table 1.1 Studies demonstrating or indicating preconditions for coevolution between plants and phytophagous insects within local populations.

Some preconditions for local coevolution	Recent primary references or reviews
Genetic variation	
Plant resistance to insects and/or performance of insects on different plant genotypes	Berenbaum & Zangerl 1992b; Fritz & Simms 1992; Simms & Rausher 1993; Strong *et al*. 1993; Mitchell-Olds *et al*. 1996; van der Meijden 1996; Mopper 1996, Vrieling *et al*. 1996
Performance of insects on different plant species	Via 1994; Futuyma *et al*. 1995; Thompson 1996
Oviposition preference for different plant species	Fox 1993; Singer *et al*. 1993; Thompson 1993
Selective effects	
Decreased plant survival or reproduction following insect attack in natural populations	Root 1996
Selective attack on plant genotypes or phenotypes in natural populations	Marquis 1992; Zangerl & Berenbaum 1993

than has been evaluated by any one study. As Marquis (1992) noted, and is still true, no study has yet evaluated selective herbivory on all stages of the life cycle for any plant population.

3 Most studies evaluate how a few traits in one species influence selection on another species, or how an interaction affects one or several components of fitness. But increasingly sophisticated genetic (e.g. quantitative trait locus mapping), chemical and mathematical techniques are unravelling the complexity of the defence systems of plants (e.g. Cates 1996; Stamp & Yang 1996; Young 1996) and the genetic complexity of the determinants of insect performance on plants (e.g. Thompson *et al*. 1990; Via 1994; Keese 1996; Thompson 1996). These analyses indicate that studies using only a few index variables have a high potential for underestimating selective effects in insect–plant interactions.

4 Hence, determination of the costs of defence in plants and the effects of plants on insect performance in experimental studies is much more difficult than we once thought (e.g. Joshi & Thompson 1995; Mitchell-Olds *et al*. 1996; Bergelson & Purrington 1996; Nylin *et al*. 1996; Sheck & Gould 1996).

5 Finally, selection imposed by plants and herbivores on each other within natural populations can be erratic. Root's (1996) 6-year study on insect attack on goldenrod populations is particularly important. It shows that herbivore pressure on plant populations can be highly variable

among years, and little can be inferred from short-term studies (e.g. a few years in a single environment) demonstrating lack of selection.

What we do not know

We lack studies on heritable variation on both sides of insect–plant inter-actions within single communities, and we lack evidence for reciprocal selection within local communities. The best current data on reciprocal heritable variation in local insect–plant interactions are from Berenbaum and Zangerl's (e.g. Zangerl & Berenbaum 1990; Berenbaum & Zangerl 1992a; Zangerl & Rutledge 1996; Chapter 3) ongoing studies on genetic variation in plant chemistry in wild parsnip and P450-mediated metab-olism of furanocoumarins in the parsnip webworm. We need additional data of this sort, and we need related data of the kind that Henter and Via (Henter 1995; Henter & Via 1995) have gathered for pea aphids and their parasitoids, showing distributions of traits and outcomes of interactions within populations. These distributions are the local raw material for the coevolutionary process (Thompson 1988), and analysis of the shapes of those distributions has the potential to be used as a tool for understanding the trajectory of coevolutionary change.

From what I can determine, we do not yet have even a single study demonstrating actual reciprocal selection on herbivores and plants and its dynamics over multiple generations within a local community. This statement, however, should not be used as evidence against local coevolution between herbivores and plants for two reasons. Firstly, there have been few experimental research programmes designed specifically to monitor the dynamics of local reciprocal selection in these interactions. Studies consistently evaluate selection on either the animal side of the interaction or the plant side, and most studies have been very short term. Secondly, the current state of our understanding of reciprocal selection between herbivores and plants is not as dire as it may seem. In interactions between monophagous insects and plants, we can reasonably infer that the plant affects insect fitness. (The alternative would be to assert that all local genotypes of the insects have equal fitness on all local genotypes of the plant.) Any study demonstrating that a monophagous insect species directly affects local fitness of its host therefore suggests, as a first guess for future work, local reciprocal selection pressures.

There are data on local reciprocal selection for a few other, non-plant–herbivore interactions, and the results have come from studies specifically designed to evaluate reciprocal selection. For example, Burdon's long-term results on gene-for-gene coevolution between Australian wild flax and flax rust have linked natural selection with particular host and pathogen genotypes (Jarosz & Burdon 1991; Burdon 1994; Burdon & Thompson 1995). These studies have demonstrated that each rust genotype is able to attack plants of only particular genotypes, and attack by the rust has significant effects on the fitness of those susceptible plants within natural populations. Hence, there is selective survival of each rust genotype that depends upon its match with particular plant genotypes, and there is selective mortality on plant genotypes that depends upon rust genotypes. These studies have also indicated that, despite reciprocal local effects on fitness of genotypes, most of the dynamics of coevolution occur at scales above the level of local populations (Thompson & Burdon 1992; Burdon 1994).

The geographic mosaic of coevolution

What connects the local populations of interacting species to the fixed traits distributed among interacting taxa is the rich geographic mosaic of interactions among genetically differentiated populations. Novel adaptations may initially appear in one population and move by gene flow to other populations. Each adaptation will differ in the number of populations in which it is favoured by natural selection. Counteradaptations evolving in other species may similarly be effective in some populations but not others. This differential selection among populations, coupled with gene flow, random genetic drift and extinction of local populations creates a geographic mosaic of reciprocal evolutionary change in which local mismatches of traits may be common. In what I have called the geographic mosaic theory of coevolution, I have argued that much of the dynamics of coevolutionary change may occur at this intermediate level in the hierarchy of coevolution (Thompson 1994, 1997).

The geographic mosaic theory uses what we now know about the geographic structure of populations and interactions to emphasize genetic and ecological aspects of the coevolutionary process that are not evident from analyses carried out only within local populations or using only cladograms. Stated as a three-part hypothesis on process, it suggests that:

1 there is a selection mosaic among populations, favouring different evolutionary trajectories to interactions in different populations;

2 there are coevolutionary hotspots, which are the subset of communities in which much of the coevolutionary change occurs; and

3 there is a continual population remixing of the range of coevolving traits, resulting from the selection mosaic, coevolutionary hotspots, gene flow, random genetic drift and local extinction of populations.

This process, in turn, suggests three predictions on ecological patterns: populations will differ in the specific traits shaped by an interaction; traits of interacting species will be well matched in some communities and mismatched in others; and there will be few species-level coevolved traits, because few traits will be favoured in all populations (Thompson 1994, 1997).

The study of insect–plant interactions has been a major impetus to development of the geographic mosaic view of coevolution. Analyses of these associations have been at the forefront in demonstrating that there are aspects of evolving interactions that are not captured at the levels of local populations or the fixed traits of aspects. These ecological results include the following.

1 Plant and insect species are groups of genetically differentiated populations (e.g. more than 20 papers in *Evolution* and *Molecular Ecology* in 1996 alone).

2 Some species and interactions may persist through a metapopulation structure or a source–sink structure in which local extinctions and recolonizations are common (e.g. Harrison *et al.* 1988; Hanski *et al.* 1995).

3 Interacting species differ in their geographic ranges (e.g. Pellmyr 1992).

4 The number of species in an interaction varies geographically (e.g. Lawton *et al.* 1993).

5 The outcomes of interactions and the potential for local coevolution can differ among populations due to differences in abiotic conditions, the local genetic structure of populations, and the community context in which the interaction takes place (Thompson 1988).

6 The geographic population structure of insects and their host plants may differ (e.g. Michalakis *et al.* 1993).

These results form the ecological basis for a geographic mosaic view of the coevolutionary process, but they do not in themselves demonstrate that the geographic mosaic of interactions is important to the coevolutionary process. We must, in addition, be able to show that the hypothesis as formulated and its predictions hold for natural interspecific interactions. What we have currently are partial data sets supporting different components of the geographic mosaic view.

What we know and do not know

An increasing number of studies have shown differentiation among populations in traits important to the evolution of insect–plant interactions (Table 1.2). In contrast, there are still few data showing for both sides of an interaction the geographic pattern of traits known to affect fitness in the interacting species and reciprocal selection on those traits. Among studies of plants and insect herbivores, Nielsen's (1996, 1997) recent results on the geographic pattern of attack on *Barbarea vulgaris* populations by the flea beetle *Phyllotreta nemorum* hold promise, as do studies of the geographic mosaic of interactions between goldenrods and their gall-forming flies (Abrahamson & Weis 1997) and between the prodoxid moth *Greya politella* and its host plants (Thompson 1997). In the case of *Barbarea* and *Phyllotreta*, populations of the plant species are generally unsuitable for larval development in most *P. nemorum* populations within Denmark. At least one *P. nemorum* population, however, has evolved the ability to attack *B. vulgaris*, and Nielsen has worked out the genetic bases for this newly evolved ability.

Beginning with the pioneering efforts of Whitham (1989) to evaluate how interspecific hybridization in plants may affect their interactions with herbivores, we now have strong data suggesting that plant hybridization affects the geographic structure of insect–plant interactions (e.g. Whitham *et al.* 1994; Fritz *et al.* 1996). But we have few data on how these hybrid zones can influence the geographic mosaic of coevolution. We also have

Table 1.2 Studies demonstrating preconditions for a geographic mosaic view of coevolution between plants and phytophagous insects.

Some preconditions for a geographic mosaic view of coevolution	Recent primary references or reviews
Genetic differentiation among plant populations for traits affecting interactions with insect herbivores	Mattson *et al.* 1993; Sork *et al.* 1993; Adler *et al.* 1995; Mithen *et al.* 1995; Mithen & Campos 1996
Differential attack by insects across plant populations due to differences in plant genotypes	Strauss 1994; Whitham *et al.* 1994
Genetic differentiation among insect populations in host plant preference or specificity	Thompson 1993; Brown *et al.* 1996; Nielsen 1996; Scriber 1996, Singer & Thomas 1996
Genetic differentiation among insect populations in performance on different plant species	Fox *et al.* 1994; Bossart & Scriber 1995; Veenstra *et al.* 1995; Nielsen 1997
Geographically restricted utility of particular traits involved in attack or defence	Carroll & Boyd 1992; Nielsen 1996

only one set of results on how the evolution of polyploidy in plants can shape the geographic structure of interactions between plants and insects (Thompson *et al.* 1997). Since half of all plant species are thought to have originated through polyploidy, the effect of polyploidy on insect–plant coevolution is one of the most important major gaps in our understanding of the coevolutionary process between these taxa.

For other interactions, two of the best data sets showing geographic patterns on both sides of the interaction are Burdon's results on gene-for-gene coevolution between Australian wild flax and flax rust discussed earlier, and Kraaijeveld and van Alphen's results on encapsulation and resistance to encapsulation in *Drosophila melanogaster* and its parasitoids. Kraaijeveld and van Alphen's (1994, 1995) studies have shown a very complex geographic mosaic in the proportion of *D. melanogaster* individuals capable of encapsulating two parasitoid species, and a similar geographic mosaic in the ability of these parasitoids to escape encapsulation. Moreover, they have shown that the ability to encapsulate one parasitoid species is independent of the ability to encapsulate the other.

Coevolution and the phylogeny of interacting taxa

When Ehrlich and Raven (1964) published their landmark paper on coevolution between butterflies and plants, it sparked hundreds of studies that began to use the phylogenies of interacting taxa to evaluate the evolution of interactions at the level of species and higher taxa. The Ehrlich and Raven hypothesis, in its formal form now called escape-and-radiate coevolution, is a specific kind of reciprocal evolution that involves both adaptation and speciation. It is one of a variety of ways in which coevolution and speciation may be interrelated (Thompson 1994).

At the other extreme from escape-and-radiate coevolution is Jermy's (1976, 1984) hypothesis of sequential evolution, which argues that insect herbivores are generally so uncommon that they are unlikely to impose selection pressures on plant populations strong enough to drive plant diversification. In this view, phytophagous insects have diversified by following plant diversification, but they have not contributed to plant diversification or major patterns of adaptation. Tibor Jermy (personal communication, 1996), however, allows for the possibility that insects mutualistic with plants may impose important selection pressures on plants. Jermy developed his view initially as a counterpoint to the coevolution-is-everywhere overextrapolation that began to develop following publication of Ehrlich and Raven's paper. He was concerned primarily with whether selection pressures exerted by insects could be responsible for plant diversification. Although he argued against the importance of insects as selection pressures on plants, the focus of his hypothesis was not on patterns of coadaptation between insects and plants. Plant diversification driven by insects was the major issue for him.

Unfortunately, dozens of studies of the phylogeny of interactions have treated escape-and-radiate coevolution and sequential evolution as if they were the sole, and mutually exclusive, possible routes for diversification of interacting lineages. Failure to meet the expected patterns of diversification predicted by escape-and-radiate coevolution has often been treated as *ipso facto* evidence for sequential evolution and perhaps even evidence for a lack of any kind of coevolution between the taxa. There is, however, no necessary connection between coevolution and particular patterns of speciation. Coevolving taxa may have reciprocal effects on speciation that do not follow directly or solely from the specific process of escape-and-radiate coevolution. In other interactions, coevolution may be important in shaping the adaptation of populations without being a major force in reciprocal speciation.

What we know and do not know

Attempts at testing some of the components of escape-and-radiate coevolution have been made, especially using insects and plants (reviews in Futuyma & Keese 1992; Farrell & Mitter 1993, 1994; Thompson 1994; Miller & Wenzel 1995; Menken 1996). But we have not been able to test the overall hypothesis for several reasons. Until recently we have lacked robust data on the comparative phylogenies of interacting lineages. Often the data have been much more robust for one side of the interaction than the other. Moreover, only now are algorithms being developed for statistically comparing interacting lineages. And, just as importantly, most studies have searched for a pattern that is not predicted by the hypothesis.

The Ehrlich and Raven hypothesis is one of the most consistently misinterpreted papers in all of evolutionary biology. Although the steps of escape-and-radiate coevolution have been outlined repeatedly in publications over more than 30 years, these steps and their implications for the pattern of colonization of hosts have commonly been ignored in most studies purporting to test the hypothesis. Ehrlich and Raven developed the idea as a general view of how reciprocal evolution may have shaped adaptation and speciation in insects and plants, not as a lock-step

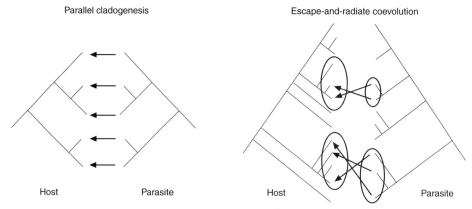

Parallel cladogenesis   Escape-and-radiate coevolution

Host   Parasite   Host   Parasite

Figure 1.1 Escape-and-radiate coevolution as compared with parallel cladogenesis. The hypothetical figure for escape-and-radiate coevolution shows two bouts of the escape-and-radiate process, creating starbursts of speciation in the host and, subsequently, in the parasite. Each oval encircles only the initial starburst of speciation in the host lineage resulting from new defences or the initial starburst of speciation in the parasite lineage resulting from new counterdefences. (Few species per starburst are shown to keep the figure simple to interpret.) Subsequent speciation events following each initial starburst in the host and parasite will add complexity to the phylogenetic patterns. Although not shown, each new defence and counterdefence may occur in all or some descendent taxa throughout the lineage (i.e. each oval encircles only the initial starburst of speciation, not the full distribution of a new defence and counterdefence in all descendent species). Parasite species not connected by an arrow to the host lineage represent species that have colonized hosts outside the host lineage shown here.

view in which every speciation event is matched one-on-one. Most importantly, *Ehrlich and Raven's hypothesis of escape-and-radiate coevolution does not predict parallel cladogenesis at the level of matched species. It predicts repeated starbursts of reciprocal speciation in mutant host and parasite lineages* (Fig. 1.1).

Consider the hypothesis step by step. It begins with a new mutant in the host lineage, which allows it to avoid attack by its parasites. This creates a new adaptive zone for the mutant plant population, allowing the population to spread and diversify in the absence of the interaction with its enemies. Note that host diversification happens in the absence of the interaction. Subsequently, a mutant parasite overcomes these new defences and then radiates in species onto the now diversified mutant host lineage.

Here is the most important point. Nothing about the hypothesis of escape-and-radiate coevolution predicts that the new mutant parasite will colonize first the most ancestral plant in the mutant host lineage and then radiate in lock-step fashion up through the more derived host species. Because the new plant lineage underwent diversification during the absence of the interaction, there is no reason to expect that there will be any escalating pattern of defences that make more derived hosts more resistant than more ancestral hosts *within* the new host clade. A mutant parasite should therefore be just as likely to colonize a derived species as it would an ancestral species within the mutant host lineage (Fig. 1.1). Any pattern of escalation consistent with escape-and-radiate coevolution will occur only at higher taxonomic levels. The patterns at these levels reflect repeated escapes and radiations in hosts and subsequent colonizations and radiations in parasites.

I suggest that any test of escape-and-radiate coevolution must show three patterns, outlined in Table 1.3. These patterns follow directly from the process as it operates within and between clades within larger lineages. There are two caveats to these patterns. The new mutant parasite lineage may colonize the mutant host lineage before the host lineage has fully undergone diversification in species. That is, initial diversification of the host clade could occur during the absence of the parasite, but some later diversification of the host clade may occur after the mutant parasite has colonized it. This situation may appear as a pattern in which the new parasite lineage does not colonize the basal host species in an ancestral–descendent sequence, but it does colonize the more derived host species (i.e. those appearing after parasite colonization) more or less in sequence. This is further complicated by the possibility that, as each new parasite species develops within the new lineage, some of these parasites may colonize more basal rather than more derived within the host lineage.

The second caveat is that Ehrlich and Raven were not concerned exclusively with repeated starbursts of speciation between two lineages. That version, which I have outlined here, is the most strictly coevolutionary version now formalized as escape-and-radiate coevolution. Their initial view, however, allowed for even broader, general patterns of escape and radiation in which subsequent insect colonizers may not even be part of the insect group from which the plants initially escaped. Nevertheless, neither the more strictly coevolutionary view nor the more general view predicts parallel cladogenesis at the species level.

19

Table 1.3 Phylogenetic predictions that follow from the process of escape-and-radiate coevolution, the version of the Ehrlich–Raven hypothesis formalized for reciprocal evolutionary change between specific taxa.

---

1 Novel defences and counterdefences occur *among* clades within lineages
- That is, novel defences and counterdefences are evident in each new bout of escape and species radiation in the host lineage and subsequent colonization and species radiation in the parasite lineage
- A host clade resulting from the escape-and-radiate process should be evident as a starburst of speciation associated with evolution of a new defence; a parasite clade resulting from this process should be evident as a subsequent starburst of speciation associated with evolution of a new counterdefence

2 Major novel defences and counterdefences among species do not occur *within* each host clade
- That is, to be attributed to the escape-and-radiate process diversification of parasites and hosts must result from starbursts of speciation associated with new defences and counterdefences rather than from species-by-species escalating evolutionary arms races among host and parasite species within clades

3 The parasite clade has not colonized each new species within a host clade in a systematic fashion, beginning with the most ancestral host species and proceeding in a lock-step way through the more derived species

(Two caveats are noted in the text.)

The recent work of Farrell and Mitter (1993, 1994) has begun to confront directly some of the components of escape-and-radiate coevolution and also the more general relationships between coevolution and speciation in general. Their work has focused on the ways in which specific evolutionary innovations shape subsequent diversification of interacting lineages. At the same time, other phylogenetic approaches are now being used increasingly to evaluate:

1 if an interaction between two lineages has multiple origins (Chapela *et al.* 1994; Hinkle *et al.* 1994);

2 which traits are unique to particular species or lineages and how those unique traits might have come about (Pellmyr *et al.* 1996; Armbruster *et al.* 1997); and

3 which traits involved in interactions are most phylogenetically conserved (e.g. Jordano 1995).

Together, these approaches should help to unravel the relationships between adaptation and speciation during coevolution.

Correction of misinterpretations and unfounded expectations of the coevolutionary process

In evaluating what we know and do not know about the coevolutionary process, I found several misinterpretations and unfounded expectations that keep arising in the introductions and discussion sections of papers. Some of these I have discussed in earlier sections, but I have collected them here by stating them as corrections to some commonly held views.

1 There is no reason to think that plants, as a rule, cannot adapt simultaneously to multiple herbivores (or any species to multiple other species). Adaptation to one herbivore need not necessarily come at the expense of adaptation to other herbivores. No one, for example, would argue that adaptation to herbivores comes necessarily at the expense of adaptation to ultraviolet light, pollinators or fruit dispersers. We assume that these adaptations can evolve at least partially independently of one another, although there is evidence that some traits and evolutionary responses are correlated with one another (e.g. Pilson 1996). Adaptation to two herbivores need not be any more linked genetically than adaptation to any other set of two or more environmental problems faced by plants. Some sets of problems are solved independently, others partially independently and yet others with a single solution. The results for plant resistance to insects vary from showing independence in plant response (Hougen-Eitzman & Rausher 1994) to correlated responses (Pilson 1996), and the results showing correlated responses are mostly for one population tested in one environment. The problem to solve is how such genetic correlations in plant response to two herbivores shape coevolution across environments as the selection pressures and expression of genes change.

Other studies of defence have shown independence or partial independence in response to enemies by studying multiple natural populations subjected to different combinations of enemies and evaluating how they have diverged from one another. None of the studies of which I am aware has been performed using insect–plant interactions. Two of the best studies for other interactions, showing two different approaches, are those indicating independence of evolutionary responses of *Daphnia* populations that differ in the suites of predator species that attack them (Dodson 1988), and those showing independence (or, at least, no correlation) in defences of wild *Drosophila melanogaster* populations against two parasitoids species (Kraaijeveld & van Alphen 1995). Both studies require follow-up work on the genetics of defence, but both show the utility of using selection experiments to explore the evolutionary independence of defences — either inferred, with some risk of misinterpretation, from natural populations or shaped by artificial selection. Mitchell-Olds *et al.* (1996) have emphasized the power of selection experiments over breeding experiments in testing for potential response to selection and correlations among traits.

2 Low mean herbivore loads do not, in themselves, imply low selection pressures. Root's (1996) studies, and the results for other interactions such as those involving Galápagos finches (Grant & Grant 1995) demonstrate the importance of erratic selection within natural populations.

3 Mismatches in adaptation and counteradaptation within local communities are not, in themselves, evidence against coevolution. If the geographic mosaic theory of coevolution holds, then mismatches should be common at the level of local communities, although there should be evidence of reciprocal change when many populations of the interacting species are evaluated together.

4 Coevolution does not require that interacting species have completely coincident geographic ranges. Such a view ignores the geographic mosaic of the coevolutionary process.

5 Escape-and-radiate coevolution does not predict parallel cladogenesis at the level of matched species. It predicts repeated starbursts of speciation in mutant host and parasite lineages.

6 Demonstration that a lineage arose and underwent some diversification before the evolutionary appearance of another lineage with which it interacts is not evidence against coevolution. It is only evidence against parallel cladogenesis.

7 Comparison of branching patterns of interacting lineages cannot, by itself, tell us whether two or more lineages have coevolved (i.e. whether reciprocal evolutionary change has shaped them). It can only tell us whether they speciated in parallel.

8 Demonstrating that only one or a few traits of interacting species have coevolved is not evidence for the lack of importance of coevolution. Even if a phylogenetic analysis showed that 99% of the traits involved in an interaction are accounted for at the genus level or higher, that would not by itself constitute evidence that ongoing coevolution is unimportant in shaping these relationships. The remaining few traits may be the most important focus of current natural selection on those species. It is not the number or percentage of coevolving traits that matters. It is the selection intensities on those traits. Arguing that coevolution is unimportant because only a few traits are coevolved is analogous (and partially homologous) to arguing that evolution in general is unimportant because species are polymorphic at only a small percentage of their loci or because only a small percentage of DNA codes for proteins.

Conclusions

The point of coevolutionary studies should not be solely to demonstrate reciprocal evolutionary change. The potential value of the study of co-evolution is in understanding how the coevolutionary process shapes ecological, genetic and phylogenetic patterns. To achieve that goal we need more detailed attention to the development of specific testable hypotheses on how coevolution proceeds for different forms of interaction at different spatial and temporal scales. Examples of such hypotheses are the developing theory of virulence and resistance between hosts and parasites (including the evolution of sex), only some of which is yet truly coevolutionary in approach (review in Ebert & Hamilton 1996), metapopulation hypotheses on gene-for-gene coevolution (Thompson & Burdon 1992), geographic hypotheses on the coevolutionary dynamics of ecological character displacement (Roughgarden 1995), the local and geographic dynamics of coevolutionary alternation between parasites and multiple hosts (Davies & Brooke 1989) or predators and multiple prey, and the development of a conceptual framework for determining when the dynamics of coevolution proceed mostly at geographic rather than local scales (Thompson 1994, 1997). The formation of these hypotheses in recent years is allowing for the further refinement of the theory of coevolution and tests on which aspects of the entangled bank of interspecific interactions are most important in driving the coevolutionary process.

Acknowledgements

I thank Steve Compton, Paul Ehrlich, Allen Herre, Tibor Jermy, Pedro Jordano, Olle Pellmyr and Arpad Szentesi for discussions on what we know about coevolution during preparation of the manuscript, and David Althoff, Bradley Cunningham, Scott Nuismer and Kari Segraves for comments on the manuscript. This work was supported by NSF grants DEB 93-17424 and DEB 97-07781 and a Sabbatical Fellowship at the National Center for Ecological Analysis and Synthesis (NSF Grant DEB 94-21535).

References

Abrahamson, W.G. & Weis, A.E. (1997). *Evolutionary Ecology across Three Trophic Levels*. Princeton University Press, Princeton.

Adler, L.S., Schmitt, J. & Bowers, M.D. (1995). Genetic variation in defensive chemistry in *Plantago lanceolata* (Plantaginaceae) and its effect on the specialist herbivore *Junonia coenia* (Nymphalidae). *Oecologia*, **101**, 75–85.

Armbruster, W.S., Howard, J.J., Clausen, T.P., Debevec, E.M., Loquvam, J.C., Matsuki, M., Cerendolo, B. & Andel, F. (1997). Do biochemical exaptations link evolution of plant defense and pollination systems? Historical hypotheses and experimental tests with *Dalechampia* vines. *American Naturalist*, **149**, 461–484.

Berenbaum, M.R. & Zangerl, A.R. (1992a). Genetics of physiological and behavioral resistance to host furanocoumarins in the parsnip webworm. *Evolution*, **46**, 1373–1384.

Berenbaum, M.R. & Zangerl, A.R. (1992b). Genetics of secondary metabolism and herbivore resistance in plants. II. Ecological and evolutionary processes. In *Herbivores: Their Interactions with Secondary Plant Metabolites* (Ed. by G.A. Rosenthal & M.R. Berenbaum), pp. 415–438. Academic Press, San Diego.

Bergelson, J. & Purrington, C.B. (1996). Surveying patterns in the cost of resistance in plants. *American Naturalist*, **148**, 536–558.

Bossart, J.L. & Scriber, J.M. (1995). Maintenance of ecologically significant genetic variation in the Tiger Swallowtail butterfly through differential selection and gene flow. *Evolution*, **49**, 1163–1171.

Brown, J.M., Abrahamson, W.G. & Way, P.A. (1996). Mitochondrial DNA phylogeography of host races of the goldenrod ball gallmaker, *Eurosta solidaginis* (Diptera: Tephritidae). *Evolution*, **50**, 777–786.

Burdon, J.J. (1994). The distribution and origin of genes for race-specific resistance to *Melampsora lini* in *Linum marginale*. *Evolution*, **48**, 1564–1575.

Burdon, J.J. & Thompson, J.N. (1995). Changed patterns of resistance in a population of *Linum marginale* attacked by the rust pathogen *Melampsora lini*. *Journal of Ecology*, **83**, 199–206.

Carroll, S.P. & Boyd, C. (1992). Host race radiation in the soapberry bug: natural history with the history. *Evolution*, **46**, 1052–1069.

Cates, R.G. (1996). The role of mixtures and variation in the production of terpenoids in conifer–insect–pathogen interactions. *Recent Advances in Phytochemistry*, **30**, 179–216.

Chapela, I.H., Rehner, S.A., Schultz, T.R. & Mueller, U.G. (1994). Evolutionary history of the symbiosis between fungus-growing ants and their fungi. *Science*, **266**, 1691–1694.

Davies, N.B. & Brooke, M. de L. (1989). An experimental study of co-evolution between the cuckoo, *Cuculus canorus*, and its hosts. II. Host egg markings, chick discrimination and general discussion. *Journal of Animal Ecology*, **58**, 225–236.

Dodson, S. (1988). The ecological role of chemical stimuli for the zooplankton: predator-avoidance behavior in *Daphnia*. *Limnology and Oceanography*, **33**, 1431–1439.

Ebert, D. & Hamilton, W.D. (1996). Sex against virulence: the coevolution of parasitic diseases. *Trends in Ecology and Evolution*, **11**, 79–82.

Ehrlich, P.R. & Raven, P.H. (1964). Butterflies and plants: a study in coevolution. *Evolution*, **18**, 586–608.

Farrell, B.D. & Mitter, C. (1993). Phylogenetic determinants of insect/plant community diversity. In *Species Diversity in Ecological Communities* (Ed. by R.E. Ricklefs & D. Schluter), pp. 253–266. University of Chicago Press, Chicago.

Farrell, B.D. & Mitter, C. (1994). Adaptive radiation in insects and plants: time and opportunity. *American Zoologist*, **34**, 57–69.

Fox, C.W. (1993). A quantitative genetic analysis of oviposition preference and larval performance on two hosts in the bruchid beetle, *Callosobruchus maculatus*. *Evolution*, **47**, 166–175.

Fox, C.W., Waddell, K.J. & Mousseau, T.A. (1994). Host-associated fitness variation in a seed beetle (Coleoptera: Bruchidae): evidence for local adaptation to a poor quality host. *Oecologia*, **99**, 329–336.

Fritz, R.S. & Simms, E.L. (Eds) (1992). *Plant Resistance to Herbivores and Pathogens: Ecology, Evolution, and Genetics*. University of Chicago Press, Chicago.

Fritz, R.S., Roche, B.M., Brunsfeld, S.J. & Orians, C.M. (1996) Interspecific and temporal variation in herbivore responses to hybrid willows. *Oecologia*, **108**, 121–129.

Futuyma, D.J. & Keese, M.C. (1992). Evolution and coevolution of plants and phytophagous arthropods. In *Herbivores: Their Interactions with Secondary Plant Metabolites*, Vol. II. *Ecological and Evolutionary Processes* (Ed. by G.A. Rosenthal & M.R. Berenbaum), pp. 439–475. Academic Press, San Diego.

Futuyma, D.J., Keese, M.C. & Funk, D.J. (1995). Genetic constraints on macroevolution: the evolution of host affiliation in the leaf beetle genus *Ophraella*. *Evolution*, **49**, 797–809.

Gaston, K.J. (Ed.) (1996). *Biodiversity: A Biology of Numbers and Difference*. Blackwell Science Ltd, Oxford.

Grant, P.R. & Grant, B.R. (1995). Predicting microevolutionary responses to directional selection on heritable variation. *Evolution*, **49**, 241–251.

Hanski, I., Pakkala, T., Kuussaari, M. & Lei, G. (1995). Metapopulation persistence of an endangered butterfly in a fragmented landscape. *Oikos*, **72**, 21–28.

Harrison, S., Murphy, D.D. & Ehrlich, P.R. (1988). Distribution of the bay checkerspot butterfly, *Euphydryas editha bayensis*: evidence for a metapopulation model. *American Naturalist*, **132**, 360–382.

Henter, H.J. (1995). The potential for coevolution in a host-parasitoid system. II. Genetic variation within a population of wasps in the ability to parasitize an aphid host. *Evolution*, **49**, 439–445.

Henter, H.J. & Via, S. (1995). The potential for coevolution in a host-parasitoid system. I. Genetic variation within an aphid population in susceptibility to a parasitic wasp. *Evolution*, **49**, 427–438.

Hinkle, G., Wetterer, J.K., Schultz, T.R. & Sogin, M.L. (1994). Phylogeny of the attine ant fungi based on analysis of small subunit ribosomal RNA gene sequences. *Science*, **266**, 1695–1697.

Hochberg, M.E., Clobert, J. & Barbault, R. (Eds) (1996). *Aspects of the Genesis and Maintenance of Biological Diversity*. Oxford University Press, Oxford.

Hougen-Eitzman, D. & Rausher, M.D. (1994). Interactions between herbivorous insects and plant–insect coevolution. *American Naturalist*, **143**, 677–697.

Jarosz, A.M. & Burdon, J.J. (1991). Host–pathogen interactions in natural populations of *Linum marginale* and *Melampsora lini*. II. Local and regional variation in pattern of resistance and racial structure. *Evolution*, **45**, 1618–1627.

Jermy, T. (1976). Insect–host–plant relationship — co-evolution or sequential evolution? In *The Host–Plant in Relation to Insect Behavior and Reproduction* (Ed. by T. Jermy), pp. 109–113. Plenum, New York.

Jermy, T. (1984). Evolution of insect/host plant relationships. *American Naturalist*, **124**, 609–630.

Jordano, P. (1995). Angiosperm fleshy fruits and seed dispersers: a comparative analysis of adaptation and constraints in plant–animal interactions. *American Naturalist*, **145**, 163–191.

Joshi, A. & Thompson, J.N. (1995). Trade-offs and the evolution of host specialization. *Evolutionary Ecology*, **9**, 82–92.

Karban, R. & Baldwin, I.T. (1997). *Induced Responses to Herbivory*. University of Chicago Press, Chicago.

Kareiva, P.M., Kingsolver, J.G. & Huey, R.B. (Eds) (1993). *Biotic Interactions and Global Change*. Sinauer Associates, Sunderland, MA.

Keese, M.C. (1996). Feeding responses of hybrids and the inheritance of host-use traits in leaf feeding beetles (Coleoptera: Chrysomelidae). *Heredity*, **76**, 36–42.

Kraaijeveld, A.R. & van Alphen, J.J.M. (1994). Geographical variation in resistance of the parasitoid *Asobara tabida* against encapsulation by

Drosophila melanogaster larvae: the mechanism explored. *Physiological Entomology*, **19**, 9–14.

Kraaijeveld, A.R. & van Alphen, J.J.M. (1995). Geographical variation in encapsulation ability of *Drosophila melanogaster* larvae and evidence for parasitoid-specific components. *Evolutionary Ecology*, **9**, 10–17.

Lawton, J.H., Lewinsohn, T.M. & Compton, S.G. (1993). Patterns of diversity for the insect herbivores on bracken. In *Species Diversity in Ecological Communities* (Ed. by R.E. Ricklefs & D. Schluter), pp. 178–184. University of Chicago Press, Chicago.

Marquis, R.J. (1992). Selective impact of herbivores. In *Plant Resistance to Herbivores and Pathogens: Ecology, Evolution, and Genetics* (Ed. by R.S. Fritz & E.L. Simms), pp. 301–325. University of Chicago Press, Chicago.

Mattson, W.J., Birr, B.A. & Lawrence, R.K. (1993). Variation in the susceptibility of North American white spruce populations to the gall-forming adelgid, *Adelges abeitis* (Homoptera: Adelgidae). In *Department of Agriculture General Technical Report NC-174, The Ecology and Evolution of Gall-Forming Insects* (Ed. by P.W. Price, W.J. Mattson & Y.N. Baranchikov), pp. 135–147. Department of Agriculture, Washington, DC, United States.

Menken, S.B.J. (1996). Pattern and process in the evolution of insect–plant associations: *Yponomeuta* as an example. *Entomologia Experimentalis et Applicata*, **80**, 297–305.

Michalakis, Y., Sheppard, A.W., Noël, V. & Olivieri, I. (1993). Population structure of a herbivorous insect and its host plant on a microgeographic scale. *Evolution*, **47**, 1611–1616.

Miller, J.S. & Wenzel, J.W. (1995). Ecological characters and phylogeny. *Annual Review of Entomology*, **40**, 389–415.

Mitchell-Olds, T., Siemens, D. & Pedersen, D. (1996). Physiology and costs of resistance to herbivory and disease in *Brassica. Entomologia Experimentalis et Applicata*, **80**, 231–237.

Mithen, R. & Campos, H. (1996). Genetic variation of aliphatic glucosinolates in *Arabidopsis thaliana* and prospects for map based gene cloning. *Entomologia Experimentalis et Applicata*, **80**, 202–205.

Mithen, R., Raybould, A.F. & Giamoustaris, A. (1995). Divergent selection for secondary metabolites between wild populations of *Brassica oleracea* and its implications for plant–herbivore interactions. *Heredity*, **75**, 472–484.

Mopper, S. (1996). Adaptive genetic structure in phytophagous insect populations. *Trends in Ecology and Evolution*, **11**, 235–238.

Nielsen, J.K. (1996). Intraspecific variability in adult flea beetle behaviour and larval performance on an atypical host plant. *Entomologia Experimentalis et Applicata*, **80**, 160–162.

Nielsen, J.K. (1997). Genetics of the ability of *Phyllotreta nemorum* to survive in an atypical host plant, *Barbarea vulgaris* ssp. *arcuata*. *Entomologia Experimentalis et Applicata*, **82**, 37–44.

Nylin, S., Janz, N. & Wedell, N. (1996). Oviposition preference and offspring performance in the comma butterfly: correlations and conflicts. *Entomologia Experimentalis et Applicata*, **80**, 141–144.

Pellmyr, O. (1992). The phylogeny of a mutualism: evolution and coadaptation between *Trollius europaeus* and its pollinating parasites. *Biological Journal of the Linnean Society*, **47**, 337–365.

Pellmyr, O., Thompson, J.N., Brown, J.M. & Harrison, R.G. (1996). Evolution of pollination and mutualism in the yucca moth lineage. *American Naturalist*, **148**, 827–847.

Pickett, S.T.A., Kolasa, J. & Jones, C.G. (1994). *Ecological Understanding: The Nature of Theory and the Theory of Nature*. Academic Press, San Diego.

Pilson, D. (1996). Two herbivores and constraints on selection for resistance in *Brassica rapa. Evolution*, **50**, 1492–1500.

Ricklefs, R.E. & Schluter, D. (Eds) (1993). *Species Diversity in Ecological Communities*. University of Chicago Press, Chicago.

Root, R.B. (1996). Herbivore pressure on goldenrods (*Solidago altissima*): its variation and cumulative effects. *Ecology*, **77**, 1074–1087.

Roughgarden, J. (1995). *Anolis Lizards of the Caribbean: Ecology, Evolution, and Plate Tectonics*. Oxford University Press, Oxford.

Scriber, J.M. (1996). A new "cold pocket" hypothesis to explain local host preference shifts in *Papilio canadensis. Entomologia Experimentalis et Applicata*, **80**, 315–319.

Sheck, A.L. & Gould, F. (1996). The genetic basis of differences in growth and behavior of specialist and generalist herbivore species: selection on hybrids of *Heliothis virescens* and *Heliothis subflexa* (Lepidoptera). *Evolution*, **50**, 831–841.

Simms, E.L. & Rausher, M.D. (1993). Patterns of selection on phytophage resistance in *Ipomoea purpurea. Evolution*, **47**, 970–976.

Singer, M.C. & Thomas, C.D. (1996). Evolutionary responses of a butterfly metapopulation to human- and climate-caused environmental variation. *American Naturalist*, **148**, S9–S39.

Singer, M.C., Thomas, C.D. & Parmesan, C. (1993). Rapid human-induced evolution of insect–host associations. *Nature*, **366**, 681–683.

Sork, V.L., Stowe, K.A. & Hochwender, C. (1993). Evidence for local adaptation in closely adjacent subpopulations of northern red oak (*Quercus rubra* L.) expressed as resistance to leaf herbivores. *American Naturalist*, **142**, 928–936.

Stamp, N.E. & Yang, Y. (1996). Response of insect herbivores to multiple allelochemicals under different thermal regimes. *Ecology*, **77**, 1088–1102.

Strauss, S.Y. (1994). Levels of herbivory and parasitism in host hybrid zones. *Trends in Ecology and Evolution*, **9**, 209–214.

Strong, D.R., Larsson, S. & Gullberg, U. (1993). Heritability of host–plant resistance to herbivory changes with gallmidge density during an outbreak on willow. *Evolution*, **47**, 291–300.

Thompson, J.N. (1988). Variation in interspecific interactions. *Annual Review of Ecology and Systematics*, **19**, 65–87.

Thompson, J.N. (1993). Preference hierarchies and the origin of geographic specialization in host use in swallowtail butterflies. *Evolution*, **47**, 1585–1594.

Thompson, J.N. (1994). *The Coevolutionary Process.* University of Chicago Press, Chicago.

Thompson, J.N. (1996). Trade-offs in larval performance on normal and novel hosts. *Entomologia Experimentalis et Applicata*, **80**, 133–139.

Thompson, J.N. (1997). Evaluating the dynamics of coevolution among geographically-structured population. *Ecology*, **78**, 1619–1623.

Thompson, J.N. & Burdon, J.J. (1992). Gene-for-gene coevolution between plants and parasites. *Nature*, **360**, 121–125.

Thompson, J.N., Wehling, W. & Podolsky, R. (1990). Evolution genetics of host use in swallowtail butterflies. *Nature*, **344**, 148–150.

Thompson, J.N., Cunningham, B.M., Segraves, K.A., Althoff, D.M. & Wagner, D. (1997). Plant polyploidy and insect/plant interactions. *American Naturalist*, **150**, 730–743.

van der Meijden, E. (1996). Plant defence, an evolutionary dilemma: contrasting effects of (specialist and generalist) herbivores and natural enemies. *Entomologia Experimentalis et Applicata*, **80**, 307–310.

Veenstra, K.H., Pashley, D.P. & Ottea, J.A. (1995). Host–plant adaptation in fall armyworm host strains: comparison of food consumption, utilization, and detoxication enzyme activities. *Annals of the Entomological Society of America*, **88**, 80–91.

Via, S. (1994). Population structure and local adaptation in a clonal herbivore. In *Ecological Genetics* (Ed. by L.A. Real), pp. 58–85. Princeton University Press, Princeton.

Vrieling, K., de Jong, T.J., Klinkhamer, P.G.L., van der Meijden, E. & van der Veen-van Wijk, C.A.M. (1996). Testing trade-offs among growth, regrowth and anti-herbivore defences in *Senecio jacobaea. Entomologia Experimentalis et Applicata*, **80**, 189–192.

Whitham, T.G. (1989). Plant hybrid zones as sinks for pests. *Science*, **244**, 1490–1493.

Whitham, T.G., Morrow, P.A. & Potts, B.M. (1994). Plant hybrid zones as centers of biodiversity—the herbivore community of two endemic Tasmanian eucalypts. *Oecologia*, **97**, 481–490.

Young, N.D. (1996). QTL mapping and quantitative disease resistance in plants. *Annual Review of Phytopathology*, **34**, 479–501.

Zangerl, A.R. & Berenbaum, M.R. (1990). Furanocoumarin induction in wild parsnip: genetic and population variation. *Ecology*, **71**, 1933–1940.

Zangerl, A.R. & Berenbaum, M.R. (1993). Plant chemistry, insect adaptations to plant chemistry, and host plant utilization patterns. *Ecology*, **74**, 47–54.

Zangerl, A.R. & Rutledge, C.E. (1996). The probability of attack and patterns of constitutive and induced defense: a test of optimal defense theory. *American Naturalist*, **147**, 599–608.

Chapter 2

The ecology and evolution of host plant range: butterflies as a model group

S. Nylin[1] and N. Janz[1]

Summary

Historically, much of the impetus in the study of insect–plant interactions has come from the model group of butterflies and their hosts. Such studies raise some general questions: Why is there such a high degree of specialization on particular plants by egg-laying females? Do these plants represent the best, or even the only, acceptable resources, in terms of offspring performance? Have insects and plants coevolved? We outline results of research performed with a focus on the unusually polyphagous comma butterfly *Polygonia c-album*. Results are reported and synthesized from the following.

1 A number of investigations on *P. c-album* itself. We stress the importance of varying outcomes of trade-offs between fitness parameters for the maintenance of polyphagy in this species, and the evolutionarily transient nature of such a mechanism.

2 Comparisons between species in the tribe Nymphalini, illustrating that constraints on the gathering and processing of information are likely to be factors of general importance in the evolution of specialization.

3 A phylogenetic study of Nymphalini. All taxa near the root of this tribe are specialists on the plant family Urticaceae and relatives. The host plant range in the lineage leading to *P. c-album* was evidently later broadened to include other plant families such as Salicaceae, Betulaceae and Grossulariaceae, and this historical sequence of events seems to have consequences for the host plant preferences observed today.

1 Department of Zoology, Stockholm University, S-10691 Stockholm, Sweden

4 A phylogenetic study covering the butterflies as a whole. We discuss the possibilities and limits for generalization to other insects.

Butterfly–plant model groups

The butterflies form one of the most widely used model groups for insect studies in evolutionary biology, rivalling the position of the birds among the vertebrates, and probably for the same reasons. They are conspicuous and have great aesthetic appeal, attracting the interest of non-scientists who have assembled a great deal of useful knowledge. It was not by chance that Ehrlich and Raven (1964) based their scenario of coevolution between plants and phytophagous insects on butterfly–plant interactions. A comparative host plant database would have been very hard to acquire for another insect taxon. Ehrlich and Raven's seminal paper prompted intense research on butterfly–plant and other insect–plant interactions, but before and after its publication the butterflies featured in research on a range of subjects such as aposematic colouration, mimicry, sexual selection, metapopulation dynamics and mutualism, most of which also show connections to host plant studies via plant chemistry, phenology and habitat requirements.

It seems reasonable that host plant colonization and switching, in an evolutionary sense, should be relatively easily accomplished in butterflies. Many other insects have a more intimate relationship with their host plants, and the association is retained throughout their life cycle. Yet most butterflies are highly specialized on a few closely related host plants, and there are clear phylogenetic patterns in host plant utilization among species (Ehrlich & Raven 1964; Janz & Nylin 1998). Why specialization, and why conservative plant utilization? These questions are two of those raised most prominently by butterfly–plant studies. Another is the question of whether coevolution is really probable between butterflies and plants. This chapter will address these questions. Using studies on butterfly model groups, and one such in particular, we will discuss the possibilities for generalization to other insect–plant interactions.

Introducing the comma butterfly

The holarctic genus *Polygonia* is a member of the tribe Nymphalini (*sensu* Harvey 1991), together with genera such as *Nymphalis* and *Vanessa*. The only representative of the genus in northern Europe is the comma butterfly,

P. c-album, which occurs widely across the whole of the Palearctic. A closely related species with similar host plant habits, *P. faunus*, occurs in the Nearctic.

P. c-album flies mostly near forest edges, and in open deciduous and mixed forests. Adults occur in two seasonal forms, one of which (the spring form, with dark wing undersides) hibernates before reproduction in the spring. The summer form, which has a lighter colour on the wing undersides, develops directly to reproduction. The life cycle is mainly regulated by photoperiod (Nylin 1989, 1992), but sex, temperature and host plant quality also affect the probability that adults will develop and reproduce directly (Wedell *et al.* 1997). Females are strongly polyandrous and males transfer large, nutritious, spermatophores at mating (Svärd & Wiklund 1989). They are important in sperm competition but also as a paternal investment in offspring (Wedell 1996).

For a butterfly, the host plant range of *P. c-album* is unusually large, and this is the reason for our interest in the species. To understand the evolution of the specialized feeding habits seen in most butterflies, and indeed in most phytophagous insects, it may be a good strategy to study an exception to the rule. In addition, a wider range of host plants means more scope to study adaptive variation in diet width within species. The generations of *P. c-album* differ in host plant preferences, the summer form being more prone to specialize on high-quality hosts (Nylin 1988; Nylin & Janz 1996). There are also differences among populations in this respect (Nylin 1988; Janz & Nylin 1997).

Larvae feed on herbs, vines, bushes and trees, utilizing distantly related plants in several orders: Urticales—stinging nettle *Urtica dioica* (Urticaceae), elm *Ulmus glabra* (Ulmaceae), hop *Humulus lupulus* (Cannabidaceae); Salicales—sallow *Salix caprea* (Salicaceae); Rosales—currants *Ribes* spp. (Grossulariaceae); Betulales—birch *Betula pubescens* (Betulaceae), hazel *Corylus avellana* (Corylaceae) (Ackery 1988; Nylin 1988; Ebert 1993). This list is roughly in descending order of female preference and larval performance (Fig. 2.1) (Nylin 1988; Janz *et al.* 1994). Larvae can initiate and complete development on all of these hosts (Nylin 1988), and females of the Swedish population seem to oviposit naturally on all of them (S. Nylin & C. Wiklund, personal communications, 1985–1998).

At the same time, it should be recognized that *P. c-album* is not a true generalist, in the sense of some relatively non-discriminating moths or grasshoppers. We have tried establishing newly hatched larvae on many unrelated plants occurring in the same habitats as the host plants, without

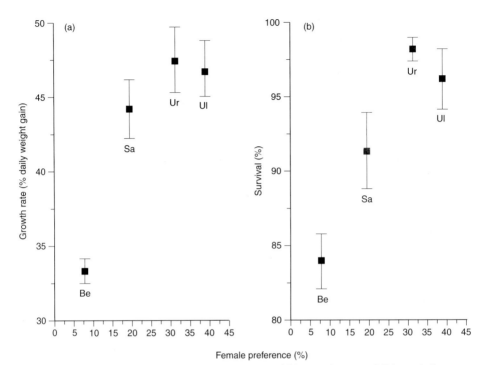

Figure 2.1 Larval performance in terms of (a) growth rate and (b) survival as a function of female preference for host plants in *Polygonia c-album*. Be, *Betula pubescens*; Sa, *Salix caprea*; Ur, *Urtica dioica*; Ul, *Ulmus glabra*. Data from Janz *et al.* (1994).

success (S. Nylin & N. Janz, unpublished results). Larvae did not even initiate feeding. Also, many plants that are relatives of the host plants, i.e. members of Urticaceae, Salicaceae and Betulaceae, were not suitable. For instance, although *B. pubescens* is a reasonably good host plant, the other common birch in Sweden, *Betula pendula*, is lethal for larvae and avoided by females (Nylin & Janz 1993). In this sense, *P. c-album* conforms to the norm of specialized feeding habits in butterflies, and it may be more correct to call it a specialist on several plant taxa than a generalist.

Preference and performance

Explanations for the high degree of specialization and conservatism in many phytophagous insects, and for other aspects of the evolution of

34

host plant range and insect–plant coevolution, can be sought at the level of adult female preference for different host plants, or at the level of offspring performance on different hosts. In addition, correlations or absence of correlations between preference and performance may give important clues (Thompson 1988).

Female preference

In butterflies, females typically search actively for suitable host plants. Larvae lack the adaptations for dispersal seen in more non-discriminating Lepidoptera (Tammaru *et al.* 1995). In most cases the eggs are laid directly on the host plants, but there are exceptions; when the plant is super-abundant in the habitat, or in species overwintering in the egg stage. In the latter case, eggs are often laid on a more stable structure near host plants (Wiklund 1984).

The decision of females to accept or reject the plant as a host has been modelled by Courtney *et al.* (1989). Most phytophagous insects are not strictly monophagous, but instead exploit at least a few related species. When given a choice among plants in the laboratory, females display a preference hierarchy, and some plants receive more eggs than others. Often there are fewer plants acceptable to females than can be eaten by larvae (Wiklund 1975; Smiley 1978), and individual females vary in the degree of acceptance of plants low in the hierarchy but not so much in the ranking of hosts (Wiklund 1974, 1981; Thompson 1993; Janz *et al.* 1994). The 'hierarchy-threshold' decision model suggests that a threshold exists which determines how low in the hierarchy a female will go, and this threshold can be modified by the state of the female or by natural selection (Courtney *et al.* 1989). Clearly, however, although the preference rank is relatively conservative in evolution, it does sometimes change (Singer *et al.* 1992).

Offspring performance

Ideally, offspring performance should be assessed as total offspring fitness, summed over the whole life cycle (Nylin *et al.* 1996). In practice, it may be necessary to choose a few life history traits that are probable fitness correlates, but these should be chosen with care and be based on knowledge of the organism. The most universally important aspects of offspring performance for the purposes of comparing hosts and investigating preference–performance correlations are probably larval survival and larval growth rate. A high growth rate combines the possible fitness advantages of large size and short development time (Nylin & Janz 1993; Janz *et al.* 1994).

Other aspects of offspring performance may or may not be important in a given organism. In some cases, especially in polyandrous species such as *P. c-album*, the male component of fitness cannot be ignored if the goal is a full understanding of fitness variation among hosts. In such cases male weight may correlate more strongly with reproductive success than female weight (Wiklund & Kaitala 1995). Male quality in other respects can also be affected by larval host plant and be better correlated with fitness (Wedell 1996).

In *P. c-album* we have investigated host plant effects on larval survival in the laboratory, larval development time and growth rate, pupal and adult weight, female fecundity, male spermatophore protein content and effects of male larval host plant on female fecundity and longevity. Not all of these traits correlated positively with each other. Larval survival was high on all hosts tested except *C. avellana* (Nylin 1988). In a more detailed test using only *U. dioica*, *U. glabra*, *S. caprea* and *B. pubescens*, survival was highest on the first two hosts (Fig. 2.1; Janz *et al.* 1994). Larval growth rates were also highest, and development time shortest, on these hosts (Fig. 2.1) (Nylin 1988; Janz *et al.* 1994; Nylin *et al.* 1996).

On the other hand, pupal weight and female fecundity was highest on *S. caprea* (Janz *et al.* 1994; Nylin *et al.* 1996). Despite the fact that males were also larger and heavier when reared on *S. caprea*, they produced spermatophores more rich in protein when reared on *U. dioica* (N. Wedell, S. Nylin & N. Janz, unpublished results). Radioactive label experiments showed that this protein is used in egg production. Females who were mated with males reared on *U. dioica* invested more in reproduction and lived longer (Wedell 1996). Lifetime fecundity did not differ in this experiment, probably because females were not allowed to remate (cf. Wiklund & Kaitala 1995). A pilot study suggested that females prefer to mate with males reared on *U. dioica* (Wedell 1996).

Preference–performance correlations, and lack of correlations

The subject of preference–performance correlations was first reviewed by Thompson (1988), who noted their central importance for studies on insect–plant interactions. A priori, we may expect a strong positive correlation between female preference and offspring performance, since it makes sense that the mother should choose good host plants. After all, the fitness of the ovipositing female should be largely determined by offspring fitness. If found, such a correlation implies that the research has revealed the most important fitness correlates determining the choice of host plant. Often, however, weak or even negative correlations are seen, and there are several types of explanations, some involving constraints on evolution and some not (Thompson 1988; Nylin & Janz 1996; Nylin *et al.* 1996).

Firstly, there is a possibility that trade-offs between the quantity and quality of offspring can weaken and even reverse correlations (Nylin & Janz 1996). The fitness of the ovipositing female is in fact not determined solely by offspring performance but also by the number of eggs laid before death, i.e. by oviposition rate and longevity. We can expect females to include not only the very best hosts in the diet if this increases oviposition rate. They may even prefer abundant, relatively good hosts over rare, very good hosts. We can also expect them to avoid plants associated with high mortality risks during search and oviposition. When we gave newly hatched larvae of *P. c-album* the same choice of host plants as we gave their mothers, parents and offspring agreed closely in the ranking of host plants. However, mothers laid many eggs on the low-ranked *S. caprea* and *B. pubescens*, plants that larvae seldom or never took as their first choice (Nylin & Janz 1996). We concluded that no strong trade-offs act between fitness components acting on mothers and their offspring, but that selection in favour of high oviposition rates can be part of the explanation for the polyphagy of the species.

Next, females may actually behave suboptimally, for instance if insects or plants are new to the area (Chew 1977). They may also be behaving in an adaptive manner, but be basing their choices on aspects of offspring performance that were not measured. Mortality risks for the eggs and larvae, associated with plants in the field, may be of importance (Singer 1971; Bernays 1989). Such extrinsic factors are not easily incorporated into laboratory studies. There can also be intrinsic effects of rearing larvae on different plants that were not investigated. For instance, effects of host plants on male reproductive success are seldom measured (Delisle & Bouchard 1995).

As can be seen from the above (e.g. Fig. 2.1) in *P. c-album*, preference and performance in terms of larval survival, development time, growth rate and male reproductive success were well correlated at the population level. In other words, females of the species seem to be basing their choices on intrinsic qualities of the plants, and we are probably justified in not pursuing investigations in the field in order to explain the ranking of the host plants.

S. caprea was a better host than *U. dioica* in terms of adult size and fecundity, and we have suggested that trade-offs between the importance of a short development time and large adult size may be another part of the explanation of the observed polyphagy. Comparisons between populations of *P. c-album* support this interpretation. It can be predicted that females of populations in transition areas, where it is only barely possible to fit an additional generation into the season, should be more prone to select host plants supporting fast larval growth than females elsewhere (Nylin 1988; Scriber & Lederhouse 1992). Thus, in *P. c-album*, females of the partially bivoltine English population would be expected to be more specialized on Urticales hosts than females of the Central Swedish population (in the centre of the univoltine area) and this is indeed the case (Nylin 1988; Janz & Nylin 1997).

Host plant range

Why are most butterflies, and indeed most phytophagous insects, relatively specialized? There are two major classes of explanations (Futuyma 1991). Either specialist species outperform generalists for reasons of superior physiological adaptations to host plant chemistry; it is the intrinsic qualities of plants as hosts that enforce specialization. Or there is selection for specialization *per se*, driven by ecological (extrinsic) factors such as competition, predation (Bernays 1989) or mate finding. The former process is given credence by cases where there is phylogenetic conservatism in host plant utilization (as in many butterfly groups), the latter is suggested more by the many insect taxa where related species specialize on distantly related hosts (Futuyma 1991).

Traditionally, explanations for observed patterns of insect–plant interactions have been sought primarily in trade-offs in offspring performance among potential hosts, especially physiological trade-offs necessitated by variation in plant chemistry (Ehrlich & Raven 1964; Berenbaum 1983). When investigated, measures of performance on different hosts have in fact most often turned out to be uncorrelated or positively correlated (Futuyma & Wasserman 1981; Fox & Caldwell 1994, and references therein). Hence, it is possible that physiological trade-offs are not the main reason why specialization evolves in the first place, but there can be little doubt that such trade-offs are of major importance in constraining many phytophagous insects to feed on chemically similar plants (Feeny 1992; Joshi & Thompson 1995).

In itself, the observed pattern that females often are more restricted than larvae in their acceptance of plants as hosts might suggest that explanations for specialization should be sought primarily in female oviposition behaviour, i.e. in a type of selection for specialization *per se* (Janz & Nylin 1997). One type of explanation deals with host plant finding. It stems from the fact that the task of recognizing host plants is a complicated process of crucial importance, and it may not be possible for females to simultaneously possess good selection rules for many plants, especially if they are very different (Courtney 1983; Futuyma 1983b; Fox & Lalonde 1993). Recent versions of this hypothesis are formulated in terms of constraints on information processing (Bernays & Wcislo 1994; Kotler & Mitchell 1995; Bernays 1996).

The other side of the coin is that in certain situations constraints on information processing mean that oviposition 'mistakes' should be accepted even by individuals behaving optimally (within these constraints), possibly facilitating polyphagy and host plant shifts (Larsson & Ekbom 1995). This area of research is also related to the trade-off between quantity and quality of offspring mentioned above. Ovipositing females often are under time constraints (Underwood 1994). To achieve a reasonably high realized fecundity it may be necessary to oviposit on 'suboptimal' host individuals and species, especially if it is not trivial to distinguish among them.

In a series of experiments designed to test predictions from theory on information processing, we gave females a choice between high- and low-quality specimens of *U. dioica* (Janz & Nylin 1997). Several species and populations of butterflies in the tribe Nymphalini were tested. *U. dioica* is a good host for all of them, but they vary in degree of specificity. We predicted that species and populations that include several plant species in their normal repertoire, and are able to correctly rank them in terms of offspring performance, should instead be constrained in their ability to assess quality within plant species. The prediction was borne out in all cases: comparing the polyphagous *P. c-album* with the *U. dioica* specialist *Polygonia satyrus*, comparing the polyphagous *Cynthia cardui* with the specialists *Vanessa indica* and *Inachis io*, and even comparing the Swedish population of *P. c-album* with the more specialized English population.

In the preceding section, we argued that trade-offs between fitness components among potential host plants is a probable factor that helps maintain the polyphagy of *P. c-album*. In northern populations, and in years when the second generation fails to reproduce successfully, *S. caprea* may be superior to *U. dioica* as host. Several other plants may be of at least equal quality. Gene flow between populations could contribute to the maintenance of a broad diet. A similar explanation was given by Krainacker *et al.* (1987) for the generalist food habits of the Mediterranean fruit fly. They reared larvae on 24 different hosts, assessing development time, survival, pupal size and fecundity, and found that these aspects of performance did not correlate well with each other. Moreover, a measure of fitness, the intrinsic rate of population increase (r), was calculated from these measurements. A relatively high r was achieved on all hosts despite great variation in the contributing parameters. The authors also were of the opinion that the unpredictability of host plant availability has been a major factor selecting for the generalized host plant choice (also cf. Novotny 1994).

What should be stressed at this point is the transient nature of these explanations for polyphagy. Trade-offs among fitness components, aided by gene flow and variability in space and time, may conceivably maintain polyphagy for many generations. However, it does seem likely that, sooner or later, one host plant will be the best one overall for a sufficient number of consecutive generations to select in favour of specialization on this host.

Tracing the evolution of polyphagy and specialization

It is often illuminating to consider the historical context of adaptation, in order to identify evolutionary events and sequences of events (Ronquist & Nylin 1990; Wanntorp *et al.* 1990). For this reason, a phylogenetic analysis of the tribe Nymphalini (*sensu* Harvey 1991) is being performed (S. Nylin, K. Nyblom, F. Ronquist, N. Janz, J. Belicek & M. Källersjö, unpublished data). A preliminary phylogeny, based mainly on morpho-logical data and the NDI sequence of mitochondrial DNA, is used here to illustrate some of the main patterns.

There is strong support for a monophyletic group which includes the genera *Aglais*, *Inachis*, *Nymphalis* and *Polygonia* (in the wide sense, including '*Kaniska*' *canace* and the controversial *P. l-album* or '*Nymphalis*' *vau-album*). With the exception of *Cynthia*, all genera in Nymphalini outside of this clade are specialists on the order Urticales. The family Urticaceae is the main host in all cases but some also feed on Ulmaceae or Cannabidaceae (Table 2.1). In addition, *Polygonia* and *Nymphalis* are most likely sister taxa, and *Aglais* feed exclusively on Urticaceae whereas *Inachis* utilizes Urticaceae and the related Cannabidaceae as hosts. In all

Table 2.1 Host plant families used by genera in the tribe Nymphalini. Based primarily on Scott (1986) and Ackery (1988).

Genus	Distribution	Host plants
Mynes	Australia, N.G.	Urticaceae
Symbrenthia	South-East Asia	Urticaceae
Araschnia	Palearctic	Urticaceae
Antanartia	Africa	Urticaceae
Hypanartia	Neotropical	Urticaceae, Ulmaceae
Cynthia	Cosmopolitan	Urticaceae, Asteraceae, Malvaceae etc.
Vanessa	Cosmopolitan	Urticaceae, Cannabidaceae
Bassaris	Australia, New Zealand	Urticaceae
Inachis	Palearctic	Urticaceae, Cannabidaceae
Aglais	Holarctic	Urticaceae
Nymphalis	Holarctic	Ulmaceae, Rosaceae, Salicaceae, Betulaceae, Rhamnaceae, Cannabidaceae?
Polygonia	Holarctic	Urticaceae, Ulmaceae, Cannabidaceae, Salicaceae, Betulaceae, Grossulariaceae, Ericaceae, Corylaceae, Fagaceae, Smilacacea

probability, then, the ancestral species of Nymphalini was a specialist on Urticaceae, and there have been at least two evolutionary events when the host plant range was extended, once in *Cynthia* and once in *Polygonia* + *Nymphalis* (Fig. 2.2).

To visualize what may have happened closer to *P. c-album*, we have optimized the utilization of plant families observed today onto the phylogeny, in an attempt to reconstruct colonization events (Fig. 2.3). There are around 10 plant families that are the main hosts of species in *Polygonia* + *Nymphalis* (Table 2.1). Fully half of them, Urticaceae, Ulmaceae, Cannabidaceae, Salicaceae and Betulaceae, are used by species in both genera. The most parsimonious interpretation would seem to be that the ancestor of this clade was a polyphagous species capable of feeding on all of these families. Several species have subsequently specialized on one or two of these families, and only *P. c-album* retains all of them as hosts.

The reconstruction of events seen in Fig. 2.3 suggests a much more dynamic picture of colonization and exclusion of plant families, with five colonizations of Ulmaceae, three of Cannabidaceae and Salicaceae, and two of Betulaceae. This is at least partly due to the fact that the reconstruction is based on the utilization of plants by ovipositing females rather than the capacities of larvae to feed on them. A colonization of a plant family at a single basal point in the phylogeny, followed by diversification of the

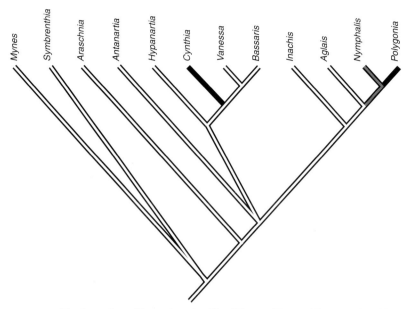

Figure 2.2 Total number of host plant families (□, 1–2; ▨, 3–6; ■, >6) utilized by genera in the Nymphalini *sensu* Harvey (1991). Data mainly from Ackery (1988), but aberrant records omitted if not supported elsewhere. Lineages shown in white feed only on the family Urticaceae and relatives (cf. Table 2.1). Phylogeny by S. Nylin *et al.* (unpublished data). The phylogeny is preliminary and partially unresolved, shown here only for the purpose of illustration.

butterfly lineage and a large enough number of host switching events, will eventually transform the single colonization event into independent colonizations in the most parsimonious reconstruction. For instance, all species in the tribe that we have tested so far (except *Polygonia gracilis*) have shown some ability to feed on *U. dioica*, regardless of female plant utilization. Still, the reconstruction suggests a loss and recolonization of Urticaceae. Work is in progress to determine the feeding capacities of larvae, but if the capacity to feed on ancestral hosts are frequently lost, it will often be impossible to distinguish with certainty between one and several colonizations. Also, in a sense, the complex changes in host plant utilization by females is the real pattern to be explained, since they are the ones who select the hosts and probably drive the evolution of host plant range.

The three colonizations of Grossulariaceae (*Ribes*) and two of Ericaceae (*Rhododendron, Vaccinium*) within *Polygonia* are probably not wholly

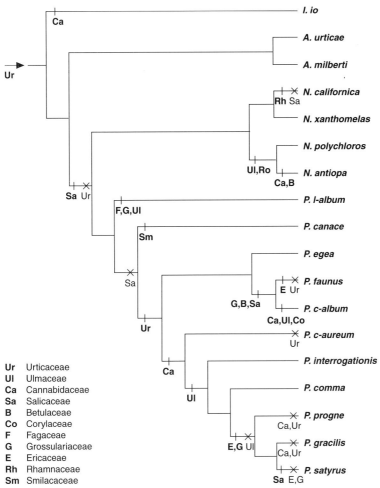

Figure 2.3 Host plant colonizations (bars, bold family labels) and exclusions from the host plant range (crosses, plain family labels) in the genera *Inachis*, *Aglais*, *Nymphalis* and *Polygonia*. Plant family labels according to legend. Dubious species and species without host plant data omitted. Host plant data mainly from Scott (1986), Ackery (1988) and Ebert (1993). The figure shows the most parsimonious reconstruction explaining the host plant ranges of recent species. When reconstructions were equivocal, the number of colonization events was minimized. Preliminary phylogeny by S. Nylin *et al.* (unpublished data). The phylogenetic relationships are uncertain and shown only for the purpose of illustration.

independent events. These families are of particular importance for Nearctic *Polygonia*, one of which (*P. gracilis*) more or less exclusively feeds on them. As far as is known, *P. c-album* does not utilize Ericaceae, but its sister species does. This led us to test *Vaccinium myrtillus* as a larval host plant, and we found that newly hatched larvae could be established on this plant and successfully reared to adulthood. Utilization of some plant families are more clearly the result of isolated events. One family, Rosaceae, is restricted to two species of *Nymphalis*. The family Rhamnaceae is restricted to *N. californica*, a specialist on *Ceanothus*. In other words, this species colonized Rhamnaceae and lost the ancestral host families. Similarly, *P. canace* has made an unexpected full switch to the monocotyledon family Smilacaceae.

Our analysis suggests that the preference hierarchy observed in *P. c-album* in part has a historical explanation. The overall preference for hosts in Urticales, and the strong performance of offspring on these hosts, should be seen in the light of the probability that over millions of years members of Nymphalini were specialists on the family Urticaceae, sometimes also including related families in the Urticales in the diet. This pattern suggests that colonization of such chemically related plants has been more likely than colonization of more distantly related families. It is quite possible that the families in Urticales, and perhaps also other families, were colonized repeatedly during this long time span, but that respecialization on Urticaceae followed because they remained the best host plants overall for the taxon. It is another limitation of phylogenetic optimization techniques that such events, if they are more rapid than cladogenesis, will necessarily go undetected (Frumhoff & Reeve 1994).

Later the host plant range was broadened, perhaps already in the ancestor of *Nymphalis* + *Polygonia*, to include several more distantly related families. Even closer to present (perhaps near the ancestor of *Polygonia*), Grossulariaceae and Ericaceae also were included. *P. c-album* (and probably its sister species *P. faunus*) has kept the ability to feed on all of these families, whereas other *Polygonia* feed on only a subset. Species of *Nymphalis* also have more narrow host ranges. In other words, following the broadening of the host plant range there have been several specialization events, in most cases involving a plant family with which the lineage has had ties for a very long time. At least six of the 15 species of *Nymphalis* + *Polygonia* included in our analysis are today relatively specialized on families in the Urticales. A seventh, *P. c-album* itself, strongly prefers these families.

To summarize, there is some evidence both of rapid widening of the host plant range and of stepwise addition of host plants to the repertoire, and there is evidence both of conservative use of the same plant taxa and of colonization events followed by rapid specialization. Moreover, the reconstruction suggests that the latter type of events frequently involves respecialization on the ancestral host, or specialization on a closely related host. Stepwise addition of plants is of interest because this suggests that incorporation of a new host does not require loss of association with a former host (Ronquist & Nylin 1990; Futuyma 1991).

Butterflies and plants: a phylogenetic study

To investigate the generality of the patterns seen in the clade mentioned in the preceding section, we have also performed a phylogenetic analysis of patterns of host plant utilization in the butterflies as a whole (Janz & Nylin 1998). The higher levels of traditional plant taxonomy (orders, subclasses) probably contain many para- and polyphyletic groups, and for this reason we used the clades found by the molecular study of seed plants by Chase *et al.* (1993) to delimit plant taxa. This phylogeny is not uncontroversial, but we made little use of the actual phylogenetic hierarchy, and the major clades are probably an improvement on traditional taxonomy.

We optimized the utilization of each plant clade onto the butterfly phylogeny. There was only one plant taxon that came close to being drawn back to the root of the phylogeny, i.e. to be a putative ancestral host. This was the clade that we have called Rosid 1B, including Fabaceae, Rosaceae, the families in Urticales and a few minor plant families (Janz & Nylin 1998). Since butterfly phylogeny is in some flux (de Jong *et al.* 1996) this result should be treated with caution, although there is no strong alternative candidate. Furthermore, results of phylogenetic reconstruction techniques should always be seen as hypotheses regarding history and not as empirical observations. It is possible that the ancestral butterfly fed on several plant taxa and rapidly lost all of them except Rosid 1B as hosts, but it seems more parsimonious to hypothesize that the ancestral butterfly was a 'specialist' on plants in this taxon or at least preferred them as hosts.

We also tested the prediction that colonizations and host shifts should most frequently involve closely related taxa. This prediction follows from the often observed pattern that insects feed on several plant species in the same family, or on several related families (Ehrlich & Raven 1964; Jermy 1984), often seen as evidence for the importance of plant chemistry in constraining insects to feed on related (or at least chemically similar) plants

(Feeny 1992). Our analyses were, however, performed at a much higher level of plant taxonomy, to test the generality of this often cited 'rule' for insect–plant relationships, and we found support for it also at this level (Janz & Nylin 1998). Evidently, then, phylogenetic relationships matter for the probability of colonizations and host shifts in butterflies. It is reasonable to assume that this is at least partly due to shared chemistry in related plant lineages.

Use of trees or shrubs is widespread among the butterfly taxa (73%), and colonizations of distantly related plants were more common in these lineages than in herb-feeding taxa. We see this as evidence in support of the importance of convergent defensive chemistry in trees or bushes as compared to herbs, and the role of such plants as 'bridges' for shifts between distantly related hosts (Feeny 1976, 1991).

Coevolution?

The question of whether coevolutionary processes have been important in shaping insect–plant relationships is treated in more detail in Chapter 1. Here, we only discuss the relevance of the studies mentioned above for this issue.

Firstly, the meaning of the term coevolution must be clarified. We take it to mean reciprocal adaptations of interacting lineages, following Janzen (1980) and Futuyma (1983a). These authors also pointed out that although Ehrlich and Raven (1964) popularized the term coevolution whilst referring to butterfly–plant interactions, the series of adaptive radiations of butterflies and plants that they envisioned did not involve coevolution in the now accepted sense. Rather, they described a diversity of butterflies elaborated against an already existing background of chemically diverse plants, presumably adapted for defence against insects and other herbivores existing at the time of seed plant diversification. This scenario is in part compatible with the 'sequential evolution' of Jermy (1984) (see also Ronquist & Nylin 1990; Feeny 1991).

The large-scale butterfly–plant interactions analysed in the preceding section likewise do not address the question of coevolution between these lineages, because the terminal plant taxa of Chase *et al.* (1993) were in existence long before the origin of the butterflies (Janz & Nylin 1998). It is still quite possible that there have been coevolutionary interactions between butterflies and plants, but if so they cannot be detected at this level of study. Rather, we must go up in scale towards phylogenetic analysis of the associations between these plant clades and the Lepidoptera as a whole, or down towards phylogenetic and ecological studies of butterflies and plants at the level of species and populations.

Even slight effects of herbivory should be selected against in plants, and could set off a reciprocal evolutionary interaction, but a better focus may still be studies of species that specialize (at least locally) on one plant species and that are important herbivores on this plant. There is little direct evidence of coevolution from such studies on butterfly–plant interactions (Futuyma 1983a; Vasconcellos Neto 1991; Underwood 1994); however, very few studies have been performed with this specific aim. *P. c-album* would not be the natural place to look for coevolutionary interactions, since this species lays single eggs on large plants, mostly trees and shrubs. Nevertheless, *P. c-album* is only one component of the herbivore fauna on its host plants, and plant defences against other more damaging insects (in the case of *U. dioica*, for instance, the gregarious relatives. *A. urticae* and *I. io*) are likely to often be effective against *P. c-album* as well. Such 'diffuse' coevolution is probably much more common than specific interactions, and equally relevant to study (Thompson 1994).

Conclusions

Taken together, the various studies on butterfly–plant interactions reported in this chapter suggest a general evolutionary scenario. The ancestral butterfly probably had strong ties with plants in the clade Rosid 1B, especially Fabaceae and relatives, i.e. it was a relatively specialized species. Since that time there have been many colonizations and complete host shifts, most often to other rosids. More drastic shifts have also taken place, most frequently in butterflies whose larvae feed on trees or bushes, perhaps due to the less diverse defence chemistry of such plants compared with herbs. Colonizations have not usually led to the immediate loss of the ancestral host plant but rather to some degree of polyphagy. This state has, however, been lost over time due to successive specialization, and only when it has been carried over several speciation events can we see the result in the form of related species specializing on different (but often related) plants. If the process was rapid enough we would, in a phylogenetic reconstruction, see either a seemingly instantaneous shift or nothing at all, if respecialization on the ancestral host occurred.

There are probably multiple reasons for the specialization events (cf. Strong 1988). There is little direct evidence of the importance of extrinsic factors such as predation, competition and mate finding in selecting for specialization in butterflies, but caterpillars of specialist butterflies seem less vulnerable to insect predation than generalist caterpillars of other Lepidoptera (Bernays 1989). There is also evidence that females avoid oviposition on plants associated with a high risk of parasitoid attack (Ohsaki & Sato 1994) and lycaenids may prefer plants on which they are protected by ants (Pierce & Elgar 1985). Such preferences could lead to monophagy, and eventually to chemical dependence on the preferred host (Smiley 1978).

Any taxon has its idiosyncracies and we should be aware that, as a model for insect–plant interactions, the butterflies have their own. The fact that adult butterflies are comparatively large and mobile insects has consequences. The unpredictability of the local environment is less likely to be a factor maintaining polyphagy than in more stationary species such as fruit flies (Krainacker et al. 1987) or leafhoppers (Novotny 1994). The choice of larval host plant can most often be made by the mother, who can search for suitable plants. There is little larval dispersal, which means that we expect reasonably good preference–performance correlations. The site of oviposition will matter more than in many other insects, where the larvae or nymphs disperse in search of potential hosts.

In butterflies, offspring leave the plant, often before pupation, emerge in the neighbourhood and fly elsewhere. The degree of intimacy with the plant is not as great as in taxa where offspring feed on the same plant as adults, and we have less reason to expect monophagy because of selection from extrinsic, ecological factors than in insect taxa where matings take place on the host plant or where adults gain from being cryptic on the host. 'Hopkins' host selection principle' (Hopkins 1917), that females should be conditioned to prefer the type of plant upon which they fed as juveniles, has (perhaps predictably) not received any support from butterfly studies (Wiklund 1974; Tabashnik et al. 1981; Williams 1983). The degree of intimacy is also less than in, for example, leaf-mining species, seed predators, which pollinate their hosts, or galling species, which need to manipulate host plant physiology. Clear and specific coevolutionary interactions are perhaps more likely for these insect taxa than in butterfly–plant interactions.

Despite the relatively low degree of intimacy with the plants, most butterflies are highly specialized on a few closely related host plants and evolutionarily conservative in their host plant utilization. This can probably be largely explained by two other butterfly characteristics: the parasite lifestyle mentioned above (most butterfly individuals develop to adulthood on a single plant species) and the fact that adults of most species take little nutrition as adults, except sugars for energy. Thus, nutrition from plants taken in the larval stage is of overwhelming importance later in life, and the mother's choice of larval host plant is very important for total fitness.

A common pattern in butterflies is specialization, not on a single plant species but on a set of related hosts. Similarly, evolutionary conservatism is evident at high levels of plant taxonomy but less evident at lower levels; there are frequent colonizations of related species, genera and families. We believe that this is best explained by a combination of selection for specialization *per se*, which causes the narrow host plant range, and physiological trade-offs among plants (due to diversity in host chemistry), which maintains it. Variation in host plant chemistry is also one of the most important factors involved in causing a performance hierarchy between plants, a situation selecting for specialization also in the absence of physiological trade-offs. Even though there may be advantages associated with polyphagy, such as faster oviposition and risk spreading, and even though trade-offs between fitness components may help maintain polyphagy, it is likely that in the long run some host plants will be better hosts than others. Constraints on information processing can be one factor further speeding up the process of specialization, if a large host range means a poorer ability to select good-quality plants within species.

Although butterflies and other phytophagous insects often are very specialized in local populations and subspecies, the species or genus as a whole can have a considerably wider diet. This suggests a process of rapid colonizations and specializations (Fox & Morrow 1981; Janz *et al.* 1994; Thompson 1994). Related plants are likely to be similar in many respects, which can simplify colonization, and shared secondary chemistry is one of them. Thus, larvae may be prevented from feeding on distantly related plants, but 'pre-adapted' to feed on new but related hosts. Females may also be predisposed to utilize them, because of shared chemical and visual cues triggering oviposition (Chew 1977). Larvae as well as females will also be pre-adapted (now literally) to recolonize related plants used in the past. Moreover, it seems likely that a related host can more easily be incorporated into the already existing decision rules for selecting among plant individuals of different quality (Janz & Nylin 1997).

Specialization has often been viewed as a 'dead end' in evolution (cf. Thompson 1994), and it may be surprising to see that in the Nymphalini specialization seems to be the ancestral state and polyphagy a derived condition. Polyphagous butterflies like *P. c-album* are not true generalists, however, but specialists on several sets of related plants. They are often excluded from feeding even on some close relatives of the utilized plants. The butterflies are also a highly derived group within the Lepidoptera, and there may exist a higher taxonomic level where the traditional scenario of generalists giving rise to specialists is given support. Furthermore, this is exactly what is seen at lower levels, if we are correct in believing that polyphagy in butterflies is a transient state and that polyphagous butterflies tend to respecialize or break up into specialized species. Nevertheless, the evolution of host plant range in phytophagous insects is clearly a much more dynamic and interesting process than the traditional scenario would lead us to believe.

References

Ackery, P.R. (1988). Hostplants and classification: a review of nymphalid butterflies. *Biological Journal of the Linnean Society*, **33**, 95–203.

Berenbaum, M.R. (1983). Coumarins and caterpillars: a case for coevolution. *Evolution*, **37**, 163–179.

Bernays, E.A. (1989). Host range in phytophagous insects: the potential role of generalist predators. *Evolutionary Ecology*, **3**, 299–311.

Bernays, E.A. (1996). Selective attention and host-plant specialization. *Entomologia Experimentalis et Applicata*, **80**, 125–131.

Bernays, E.A. & Wcislo, W.T. (1994). Sensory capabilities, information processing, and resource specialization. *Quarterly Review of Biology*, **69**, 187–204.

Chase, M.W., Soltis, D.E., Olmstead, R.G., Morgan, D., Les, D.H., Mishler, B.D. et al. (1993). Phylogenetics of seed plants — an analysis of nucleotide sequences from the plastid gene rbcL. *Annals of the Missouri Botanical Garden*, **80**, 528–580.

Chew, F.S. (1977). Coevolution of pierid butterflies and their cruciferous foodplants II. The distribution of eggs on potential foodplants. *Evolution*, **31**, 568–579.

Courtney, S.P. (1983). Models of host plant location by butterflies: the effect of search images and searching efficiency. *Oecologia*, **59**, 317–321.

Courtney, S.P., Chen, G.K. & Gardner, A. (1989). A general model for individual host selection. *Oikos*, **55**, 55–65.

de Jong, R., Vane-Wright, R.I. & Ackery, P.R. (1996). The higher classification of butterflies (Lepidoptera): problems and prospects. *Entomologica Scandinavica*, **27**, 65–101.

Delisle, J. & Bouchard, A. (1995). Male larval nutrition in *Choristoneura rosaceana* (Lepidoptera: Tortricidae): an important factor in reproductive success. *Oecologia*, **104**, 508–517.

Ebert, G. (1993). *Die Schmetterlinge Baden-Württembergs*, Vols 1 & 2. Verlag Eugen Ulmer, Stuttgart.

Ehrlich, P.R. & Raven, P.H. (1964). Butterflies and plants: a study in coevolution. *Evolution*, **18**, 586–608.

Feeny, P. (1976). Plant apparency and chemical defence. *Recent Advances in Phytochemistry*, **10**, 1–40.

Feeny, P. (1991). Chemical constraints on the evolution of swallowtail butterflies. In *Plant–Animal Interactions: Evolutionary Ecology in Tropical and Temperate Regions* (Ed. by P.W. Price, T.M. Lewinsohn, G.W. Fernandes & W.W. Benson), pp. 315–340. John Wiley & Sons, New York.

Feeny, P. (1992). The evolution of chemical ecology: contributions from the study of herbivorous insects. In *Herbivores: Their Interactions with Secondary Plant Metabolites* (Ed. by G.A. Rosenthal & M.R. Berenbaum), pp. 1–44. Academic Press, New York.

Fox, C.W. & Caldwell, R.L. (1994). Host-associated fitness trade-offs do not limit the evolution of diet breadth in the small milkweed bug *Lygaeus kalmii* (Hemiptera, Lygaeidae). *Oecologia*, **97**, 382–389.

Fox, C.W. & Lalonde, R.G. (1993). Host confusion and the evolution of insect diet breadths. *Oikos*, **67**, 577–581.

Fox, L.R. & Morrow, P.A. (1981). Specialization: species property or local phenomenon? *Science*, **211**, 887–893.

Frumhoff, P.C. & Reeve, H.K. (1994). Using phylogenies to test hypotheses of adaptation: a critique of some current proposals. *Evolution*, **48**, 172–180.

Futuyma, D.J. (1983a). Evolutionary interactions among herbivorous insects and plants. In *Coevolution* (Ed. by D.J. Futuyma & M. Slatkin), pp. 207–231. Sinauer Associates, Sunderland, MA.

Futuyma, D.J. (1983b). Selective factors in the evolution of host choice by phytophagous insects. In *Herbivorous Insects: Host-Seeking Behavior and Mechanisms* (Ed. by S. Ahmad), pp. 227–244. Academic Press, New York.

Futuyma, D.J. (1991). Evolution of host specificity in herbivorous insects: genetic, ecological, and phylogenetic aspects. In *Plant–Animal Interactions: Evolutionary Ecology in Tropical and Temperate Regions* (Ed. by P.W. Price, T.M. Lewinsohn, G.W. Fernandes & W.W. Benson), pp. 431–454. John Wiley & Sons, New York.

Futuyma, D.J. & Wasserman, S.S. (1981). Food plant specialization and feeding efficiency in the tent caterpillars *Malacosoma disstria* (Hübner) and *M. americanum* (Fabricius). *Entomologia Experimentalis et Applicata*, **30**, 106–110.

Harvey, D.J. (1991). Higher classification of the Nymphalidae. In *The Development and Evolution of Butterfly Wing Patterns* (Ed. by H.F. Nijhout), pp. 255–276. Smithsonian University Press, Washington, DC.

Hopkins, A.D. (1917). A discussion of C G Hewitt's paper on "Insect behavior". *Journal of Economical Entomology*, **10**, 92–93.

Janz, N. & Nylin, S. (1997). The role of female search behaviour in determining host plant range in plant feeding insects: a test of the information processing hypothesis. *Proceedings of the Royal Society of London B*, **264**, 701–707.

Janz, N. & Nylin, S. (1998) Butterflies and plants: a phylogenetic study. *Evolution*, **52**, 486–502.

Janz, N., Nylin, S. & Wedell, N. (1994). Host plant utilization in the comma butterfly: sources of variation and evolutionary implications. *Oecologia*, **99**, 132–140.

Janzen, D.H. (1980). When is it coevolution? *Evolution*, **34**, 611–612.

Jermy, T. (1984). Evolution of insect/host plant relationships. *American Naturalist*, **124**, 609–630.

Joshi, A. & Thompson, J.N. (1995). Trade-offs and the evolution of host specialization. *Evolutionary Ecology*, **9**, 82–92.

Kotler, B.P. & Mitchell, W.A. (1995). The effect of costly information in diet choice. *Evolutionary Ecology*, **9**, 18–29.

Krainacker, D.A., Carey, J.R. & Vargas, R.I. (1987). Effect of larval host on life history traits of the Mediterranean fruit fly, *Ceratitis capitata*. *Oecologia*, **73**, 583–590.

Larsson, S. & Ekbom, B. (1995). Oviposition mistakes in herbivorous insects: confusion or a step towards a new host plant? *Oikos*, **72**, 155–160.

Novotny, V. (1994). Association of polyphagy in leafhoppers (Auchenorrhyncha, Hemiptera) with unpredictable environments. *Oikos*, **70**, 223–232.

Nylin, S. (1988). Host plant specialization and seasonality in a polyphagous butterfly, *Polygonia c-album* (Nymphalidae). *Oikos*, **53**, 381–386.

Nylin, S. (1989). Effects of changing photoperiods in the life cycle regulation of the comma butterfly, *Polygonia c-album* (Nymphalidae). *Ecological Entomology*, **14**, 209–218.

Nylin, S. (1992). Seasonal plasticity in life history traits: growth and development in *Polygonia c-album* (Lepidoptera: Nymphalidae). *Biological Journal of the Linnean Society*, **47**, 301–323.

Nylin, S. & Janz, N. (1993). Oviposition preference and larval performance in *Polygonia c-album* (Lepidoptera: Nymphalidae) — the choice between bad and worse. *Ecological Entomology*, **18**, 394–398.

Nylin, S. & Janz, N. (1996). Host plant preferences in the comma butterfly

(*Polygonia c-album*): do parents and offspring agree? *Ecoscience*, **3**, 285–289.

Nylin, S., Janz, N. & Wedell, N. (1996). Oviposition plant preference and offspring performance in the comma butterfly: correlations and conflicts. *Entomologia Experimentalis et Applicata*, **80**, 141–144.

Ohsaki, N. & Sato, Y. (1994). Food plant choice of *Pieris* butterflies as a trade-off between parasitoid avoidance and quality of plants. *Ecology*, **75**, 59–68.

Pierce, N.E. & Elgar, M.A. (1985). The influence of ants on host plant selection by *Jalmenus evagoras*, a myrmecophilus lycaenid butterfly. *Behavioral Ecology and Sociobiology*, **16**, 209–222.

Ronquist, F. & Nylin, S. (1990). Process and pattern in the evolution of species associations. *Systematic Zoology*, **39**, 323–344.

Scott, J.A. (1986). *The Butterflies of North America*. Stanford University Press, Stanford, CA.

Scriber, J.M. & Lederhouse, R.C. (1992). The thermal environment as a resource dictating geographic patterns of feeding specialization of insect herbivores. In *Effect of Resource Distribution on Animal–Plant Interactions* (Ed. by M.R. Hunter, T. Ohgushi & P.W. Price), pp. 429–466. Academic Press, New York.

Singer, M.C. (1971). Evolution of food-plant preference in the butterfly *Euphydryas editha*. *Evolution*, **25**, 383–389.

Singer, M.C., Vasco, D., Parmesan, C., Thomas, C.D. & Ng, D. (1992). Distinguishing between 'preference' and 'motivation' in food choice: an example from insect oviposition. *Animal Behaviour*, **44**, 463–471.

Smiley, J. (1978). Plant chemistry and the evolution of host specificity: new evidence from *Heliconius* and *Passiflora*. *Science*, **201**, 745–747.

Strong, D.R. (Ed.) (1988). Special feature. Insect host range. *Ecology*, **69**, 885–915.

Svärd, L. & Wiklund, C. (1989). Mass and production rate of ejaculates in relation to monandry/protandry in butterflies. *Behavioural Ecology and Sociobiology*, **24**, 395–402.

Tabashnik, B.E., Wheelock, H., Rainbolt, J.D. & Watt, W.B. (1981). Individual variation in oviposition preference in the butterfly, *Colias eurytheme*. *Oecologia*, **50**, 225–230.

Tammaru, T., Kaitaniemi, P. & Ruohomaki, K. (1995). Oviposition choices of *Epirrita autumnata* (Lepidoptera: Geometridae) in relation to its eruptive population dynamics. *Oikos*, **74**, 296–304.

Thompson, J.N. (1988). Evolutionary ecology of the relationship between oviposition preference and performance of offspring in phytophagous insects. *Entomologia Experimentalis et Applicata*, **47**, 3–14.

Thompson, J.N. (1993). Preference hierarchies and the origin of geographic specialization in host use in swallowtail butterflies. *Evolution*, 47, 1585–1594.

Thompson, J.N. (1994). *The Coevolutionary Process*. Chicago University Press, Chicago.

Underwood, D.L.A. (1994). Intraspecific variability in host plant quality and ovipositional preferences in *Eucheira socialis* (Lepidoptera, Pieridae). *Ecological Entomology*, 19, 245–256.

Vasconcellos Neto, J. (1991). Interactions between Ithomiine butterflies and Solanaceae: feeding and reproductive strategies. In *Plant–Animal Interactions: Evolutionary Ecology in Tropical and Temperate Regions* (Ed. by P.W. Price, T.M. Lewinsohn, G.W. Fernandes & W.W. Benson), pp. 291–313. John Wiley & Sons, New York.

Wanntorp, H.-E., Brooks, D.R., Nilsson, T., Nylin, S., Ronquist, F., Stearns, S.C. & Wedell, N. (1990). Phylogenetic approaches in ecology. *Oikos*, 57, 119–132.

Wedell, N. (1996). Mate quality affects reproductive effort in a paternally investing species. *American Naturalist*, 148, 1075–1088.

Wedell, N., Nylin, S. & Janz, N. (1997). Effects of larval host plant and sex on the propensity to enter diapause in the comma butterfly. *Oikos*, 78, 569–575.

Wiklund, C. (1974). Oviposition preferences in *Papilio machaon* in relation to the host plants of the larvae. *Entomologia Experimentalis et Applicata*, 17, 189–198.

Wiklund, C. (1975). The evolutionary relationship between adult oviposition preferences and larval host plant range in *Papilio machaon*. *Oecologia*, 18, 185–197.

Wiklund, C. (1981). Generalist vs. specialist oviposition behaviour in *Papilio machaon* and functional aspects on the hierarchy of oviposition preferences. *Oikos*, 36, 163–170.

Wiklund, C. (1984). Egg-laying patterns in butterflies in relation to their phenology and the visual apparency and abundance of their host plants. *Oecologia*, 63, 23–29.

Wiklund, C. & Kaitala, A. (1995). Sexual selection for large male size in a polyandrous butterfly: the effect of body size on male versus female reproductive success in *Pieris napi*. *Behavioral Ecology*, 6, 6–13.

Williams, K.S. (1983). The coevolution of *Euphydryas chacedona* and their larval host plants. III. Oviposition behavior and host plant quality. *Oecologia*, 56, 336–340.

Chapter 3

Genetic variation in cytochrome P450-based resistance to plant allelochemicals and insecticides

M. Berenbaum[1] and A. Zangerl[1]

Summary

Despite some overlap in chemical structure, plant allelochemicals differ dramatically from synthetic organic insecticides in their modes of action, selectivity, environmental persistence, spatial distribution and efficacy as selective agents. In this review, we examine genetic variation in the enzymatic metabolism of xenobiotics by insects in the context of the selection pressure maintaining that variation, specifically contrasting the ways in which insects cope with these two major types of xenobiotics: plant secondary compounds and synthetic organic insecticides. The focus of the review is on cytochrome P450 mono-oxygenases, haeme-containing enzymes that are responsible for a wide range of detoxification reactions in a diversity of insects. The existence of genetic variation in resistance to both allelochemicals and synthetic organic insecticides suggests that forces exist that constrain the evolution of resistance. External forces include variation in the likelihood of encountering toxins. Among the internal forces include pleiotropic reductions in fitness associated with resistance alleles in the absence of the toxin (genetic costs), allocation costs associated with investment into detoxification at the expense of growth, and ecological costs, in the form of bioactivation and analogue synergism. The extraordinary rapidity with which insects overcome synthetic organic insecticides contrasts markedly with the durability of plant chemical

1 Department of Entomology, 320 Morrill Hall, University of Illinois, Urbana, IL 61801-3795, USA

defences characterizing most plant–insect interactions. Gaining insight into the ways in which plants maintain efficacy of allelochemical defences over evolutionary time may be useful in designing durable chemically based management programmes for phytophagous pests.

Introduction: the world as a chemically hostile place

From the perspective of most contemporary insects, the world is a chemically hostile place. Toxic substances are routinely encountered in food, in the air and in the substrates in which they make their homes. The number of toxins encountered, even over the briefest of insect life-spans, varies with the lifestyle, but (irrespective of diet choices) virtually all insects must cope with chemicals produced by organisms attempting to avoid being eaten. Herbivores must contend with plant secondary metabolites that can variously slow growth, interfere with moulting, disrupt reproduction or simply cause instantaneous death (Rosenthal & Berenbaum 1991). Insect carnivores are likely to encounter sprays, froths or toxin-charged blood in their arthropod prey, and in some circumstances must deal with reprocessed plant secondary compounds sequestered by their prey for defensive purposes (Blum 1981). Even detritivores in most circumstances encounter toxins of microbial origin in decomposing matter (Janzen 1977). Even if insect diets do exist that are entirely free of natural toxins, human innovations in pest control have increased the likelihood that the insects consuming those diets will encounter lethal substances. For the past five decades, the insecticidal properties of halogenated hydrocarbons have been exploited widely, and air, water and soil have become contaminated with their residues. These residues continue to pose challenges to target and non-target arthropods alike.

Insects are not without defences against these toxicological insults. Generally speaking, arthropods possess a variety of genetically based mechanisms for coping with both natural and synthetic toxins. Resistance can be:

1 structurally based, in that certain attributes of morphological structure prevent a toxin from interacting with the target site (Berenbaum 1986);

2 behaviourally based, in that certain behaviours minimize exposure to a toxin (Tallamy 1986); or

3 biochemically based, in that toxins are rendered ineffective by enzyme-catalysed transformations (Brattsten 1979, 1992).

Across taxa, the importance of each mechanism varies; within a taxon, even a single type of mechanism can display extraordinary variability.

The fact that genetic variation exists for resistance against toxins is in a sense puzzling; presumably, any agent capable of causing mortality should select for increased resistance. The presence of genetic variation suggests that forces exist that oppose the evolution of resistance. Such forces have been identified and can be external or internal in origin. External forces include variation in the likelihood of encountering toxins; internal forces include pleiotropic reductions in fitness associated with resistance alleles in the absence of the toxin.

In this review, we examine the nature of genetic variation in the enzymatic metabolism of xenobiotics by insects, specifically contrasting the ways in which insects cope with two major types of foreign substances: plant secondary compounds, which occur in the diets of virtually all herbivorous insects; and synthetic organic insecticides, which by dint of ubiquity are likely to affect the lives of many herbivorous insects. The genetic basis for many forms of biochemical resistance has been clearly defined and the literature associated with this type of resistance is extensive (although by no means complete), allowing for the formation of robust inferences.

Similarities between resistance to insecticides and resistance to allelochemicals

A toxin is a chemical that can interfere with basic life processes to cause death; its origin, whether on a laboratory bench or in a plant stem, is in some ways irrelevant. The synonymy of chemical control of insects with use of synthetic organic and inorganic insecticides is of relatively recent vintage. Among the earliest chemical insecticides to be used by humans were plant products; indeed, the use of plant extracts for insect control dates back thousands of years (Smith & Secoy 1975). Gordon (1961) was perhaps the first person to recognize that the facility with which insects develop resistance to synthetic organic insecticides may be a consequence of the ability to develop resistance to natural toxins encountered in the diet; pesticide resistance may result from modification of extant xenobiotic-metabolizing enzyme systems already active in detoxification of plant secondary metabolites. In support of this hypothesis, Krieger *et al.* (1971) demonstrated that the ability to epoxidize aldrin, a cyclodiene insecticide, was correlated with diet breadth in herbivorous Lepidoptera. These authors argued that, because they consume a wider range of hosts, polyphagous insects naturally encounter a broader array of plant secondary compounds and thus are better equipped for metabolizing xenobiotics in general (but see Rose 1985). Consistent with this notion that prior evolutionary exposure to plant toxins is conducive to the evolution of resistance to synthetic chemicals was the conspicuous absence of metabolically based insecticide resistance in non-feeding stages of herbivorous insects such as Lepidoptera (Wilkinson & Brattsten 1972), despite repeated efforts to document such resistance.

Subsequent correlative evidence in support of the pre-adaptation hypothesis was drawn from comparisons of different guilds of herbivores, which vary in the frequency with which secondary metabolites are encountered in their food. Sucking insects, for example, consume vascular sap generally lacking in large concentrations of secondary compounds; these species generally possess lower levels of xenobiotic-metabolizing enzymes than do foliage-feeding species (Mullin 1986; Hung *et al*. 1995) and are significantly less likely to evolve resistance to insecticides than are foliage-feeding species (Rosenheim *et al*. 1996). Similarly, developmental stages or species that consume pollen and nectar, which tend to be low in allelochemical content, may be expected to have lower levels of xenobiotic-metabolizing enzyme activity than developmental stages or species that consume allelochemical-rich foliage. Mullin (1985) demonstrated that herbivores generally have higher levels of activities than do adult stages of their natural enemies, which feed on pollen and nectar. Brattsten (1987a) compared midgut allelochemical metabolism in *Anthonomus grandis* (boll weevil), an insect that has remained insecticide-susceptible for decades despite intensive insecticide use, with those in a resistance-prone species, *Heliothis virescens*, and found that even susceptible budworms had higher levels of xenobiotic-metabolizing activity than did adult boll weevils. Brattsten argued that 'good metabolic defences may facilitate insecticide resistance development in the tobacco budworm'. Brattsten (1987b) subsequently showed that xenobiotic-metabolizing enzyme inducibility was also higher in the resistance-prone species than in the insecticide-susceptible species and suggested that the pollen-feeding habits of the adult boll weevil leave it metabolically ill-equipped for the evolution of insecticide resistance.

That plant secondary compounds such as terpenes and glucosinolates induce the enzymes responsible for model substrate and insecticide metabolism (Brattsten *et al*. 1977; Yu *et al*. 1979) strengthened the conceptual linkage between metabolism of synthetic toxins and adaptation to a naturally chemically rich diet, as did experimental evidence that selection for survival of a phytophagous mite on a toxic plant simultaneously increased survival in the presence of some insecticides (Gould *et al.* 1982). Particular plant chemicals were associated with host plant-specific enhancement of insecticide resistance; leaves of wild *Lycopersicon hirsutum* F. *glabratum* increased diazinon tolerance in tobacco budworms fourfold, probably due to elevated levels of the trichome constituent 2-tridecanone (Riskallah *et al.* 1986). Biochemical characterization of the enzymes involved in insecticide and allelochemical metabolism revealed considerable similarities. These studies revealed that one group of enzymes in particular, cytochrome P450 mono-oxygenases, are of considerable importance in metabolic transformations of both plant-derived and synthetic toxins in insects. Cytochrome P450 mono-oxygenases (P450s) are haeme-containing enzymes that catalyse NADPH-dependent reductive cleavage of molecular oxygen to produce an altered organic substrate and a molecule of water. In insects, P450s metabolize both endogenous and exogenous substrates (Hodgson 1985). Among the endogenous substrates transformed by P450s are intermediates in the biosynthesis of pheromones and steroid hormones; among the exogenous substrates, synthetic organic insecticides such as pyrethroids (Tomita & Scott 1995) and cyclodienes (Feyereisen *et al.* 1995) as well as host plant allelochemicals such as furanocoumarins (Ma *et al.* 1994).

There is currently substantial uncertainty as to how rapidly detoxicative enzymes, such as P450, evolve in response to selection in insects. Most studies of genetic variation in P450s have focused on drug-metabolizing isozymes in humans (Bertilsson 1995). The high level of allelic diversity documented to exist in these P450s has been attributed to the removal over time of plant toxins, the putative original substrates for many of these enzymes, from the human diet; as the selective regime, in terms of dietary toxins, is relaxed, particular P450s acquire mutations that alter function (Kimura *et al.* 1989) or even disappear altogether (Gaedigk *et al.* 1991). The scenario is potentially more complex in insects. On one hand, P450s are known to respond rapidly to intensive insecticide selection, indicating remarkable lability in this group of detoxification enzymes. On the other hand, host plant utilization preferences remain remarkably stable throughout lineages (e.g. Ehrlich & Raven 1964), indicating evolutionary conservatism in those P450s responsible for allelochemical metabolism. To date, the majority of studies of insect P450 evolution have focused on the evolution of insecticide resistance, generally in non-phytophagous species

(*Musca domestica, Drosophila melanogaster*). Only in rare instances is any information available on fine-scale genetic variation in these enzyme systems. Conclusions drawn from such studies may not be descriptive of how xenobiotic-metabolizing P450s evolve in a wide range of insects, particularly in response to selection pressures exerted by host plant toxins.

Cytochrome P450s have long been known for their broad but overlapping substrate specificities (Gonzalez & Nebert 1990). This characteristic has made reconstructing the evolution of particular P450s challenging. Enzymes with nearly identical structure can have dramatically different catalytic activities, and enzymes with little sequence identity can overlap substantially in substrate specificity. Thus, the long-standing assumption that allelochemical-metabolizing P450s are closely related (or identical) to insecticide-metabolizing P450s is not unreasonable, but it is certainly difficult to test, given the limited amount of available sequence data. Fewer than two dozen insect P450s have been characterized to date at the molecular level and, of these, substrate specificities are all but unknown for the vast majority (Table 3.1). What little is known, however, suggests that insecticide-metabolizing P450s are not necessarily closely related to allelochemical-metabolizing P450s. Cytochrome P450s are divided into families, based on levels of sequence identity (Nelson *et al.* 1995). Several of the putative xenobiotic-metabolizing insect P450s that have been characterized have been placed in family CYP6 (Feyereisen *et al.* 1989; Cohen *et al.* 1992; Wang & Hobbs 1995; Hung 1996; Saner *et al.* 1996). The cyclodiene-metabolizing CYP6A1 of *Musca domestica* shares less than 40% identity with the furanocoumarin-metabolizing CYP6B1 of the black swallowtail *Papilio polyxenes*, the putative furanocoumarin-metabolizing CYP6B3 of *P. polyxenes* and the furanocoumarin-metabolizing CYP6B4 of *Papilio glaucus*. Similarly, CYP6A2, from *Drosophila melanogaster*, which metabolizes aflatoxin B1, 7,12-dimethylbenz-anthracene, and 3-amino-1-methyl-5H-pyrido [4,3-b] indole (Saner *et al.* 1996), shares little sequence similarity with CYP6B1, CYP6B3 or CYP6B4.

A phylogenetic analysis of all insect P450 amino-acid sequences known at the time revealed that the xenobiotic-metabolizing P450s do not share a recent common ancestor (Berenbaum *et al.* 1996). The best predictor of patterns of relationship in this group of P450s is common membership in an order rather than putative substrate identity, e.g. CYP6D from the dipteran *Musca domestica*, which is thought to metabolize pyrethroid insecticides (Wheelock & Scott 1992), is more closely related to other house fly P450s that do not metabolize pyrethroids than it is to CYP6B2,

Table 3.1 Insect P450s. Modified from Hung (1996).

Species	Gene/cDNA[1]	Substrates	Reference
Blaberus discoidalis	CYP4C1 cDNA	Hypertrehalosemic hormone?	Bradfield *et al.* (1991)
Musca domestica	CYP6A1 cDNA CYP6A3 CYP6A4 CYP6A5 CYP6C1 CYP6C2	Aldrin, heptachlor	Feyereisen *et al.* (1989)
	CYP6D1 cDNA	Pyrethroids?	Tomita & Scott (1995)
Drosophila melanogaster	CYP6A2 cDNA	Aflatoxin B1, 7,12-dimethylbenz-anthracene, 3-amino-1-methyl-5H-pyrido [4,3-b] indole	Saner *et al.* (1996)
	CYP18 cDNA CYP4E1 CYP4D1 CYP4D2 cDNA		
Manduca sexta	CYP4L1 cDNA CYP4L2 cDNA		Snyder *et al.* (1995)
	CYP4M1 cDNA	Tridecanone? Undecanone?	
	CYP4M2 cDNA		
	CYP4M3 cDNA	Tridecanone? Undecanone?	
	CYP9A1 cDNA		
Helicoverpa armigera	CYP6B2 cDNA	Pyrethroids?	Wang & Hobbs (1995)
Papilio polyxenes	CYP6B1 cDNA	Furanocoumarins	Cohen *et al.* (1992)
	CYP6B3 cDNA	Furanocoumarins?	Hung (1996)
	CYP6B3v2	Furanocoumarins?	Hung (1996)
Papilio glaucus	CYP6B4v1 cDNA	Furanocoumarins?	Hung (1996)
	CYP6B5v2 gene	Furanocoumarins?	Hung (1996)
	CYP6B5v1 gene	Furanocoumarins?	Hung (1996)

1 Gene nomenclature: CYP number refers to a gene family of cytochrome P450s, following capital letter refers to subfamily, following number refers to a locus, and v number refers to an allele (see Nelson *et al.* 1995; Waterman *et al.* 1996).

isolated from the pyrethroid-resistant lepidopteran *Heliothis virescens*. However, the paucity of information on substrate specificities and the very small number of known sequences (particularly from closely related taxa) preclude drawing any robust conclusions about the nature of the evolutionary relationships between allelochemical- and insecticide-metabolizing P450s.

Differences in the nature of selection exerted by allelochemicals and insecticides and impacts of that selection on the evolution of resistance

That the course of evolution of resistance to allelochemicals and insecticides may differ is not at all surprising, given the differences in the selection pressures they exert. These striking differences may well lead to qualitative differences in the genetic structure of biochemically based resistance to these two types of toxins. Insecticides are generally applied as pure compounds, whereas secondary compounds typically occur in complex mixtures, often of biosynthetically related structural analogues (Berenbaum & Zangerl 1996). The use of mixtures of toxins, particularly if components of the mixture differ in mode of action, may slow the acquisition of a single biochemically based resistance mechanism (MacDonald *et al.* 1983; Roush & Daly 1990; but see Tabashnik 1989); thus, the evolution of resistance to novel allelochemicals may in general proceed more slowly than does the evolution of resistance to a novel pesticide.

Insecticides also tend to act on a single highly specific target site (Roush & Daly 1990), whereas allelochemicals often act on multiple target sites. Target site (and taxon) specificity are generally highly desirable attributes for an insecticide. In fact, broad-spectrum toxicity could potentially interfere with pesticide approval and licensing. The neurotoxic organophosphates and carbamates, for example, inhibit acetylcholinesterase, a target enzyme that has no analogue in organisms lacking a nervous system (Matsumura 1975). In contrast, the furanocoumarins, allelochemicals found principally in plants in the Apiaceae and Rutaceae, act principally by binding covalently to pyrimidine bases in DNA and interfering with transcription, but these compounds are also capable of causing toxicity by binding to unsaturated lipids in membranes, inactivating enzymes, and generating toxic oxyradicals in the presence of ultraviolet light (Berenbaum 1991). Furanocoumarins are demonstrably toxic not only to a wide range of insects (Berenbaum 1995a) but also mammals, birds, fish, snails, nematodes, fungi, bacteria, plants and viruses (Berenbaum & Zangerl 1996). Representatives of many of these taxa are prospective consumers of plants producing furanocoumarins, so broad-spectrum activity is potentially of greater benefit to an umbellifer in an evolutionary time frame than it would be to a farmer growing umbellifers in an ecological time frame.

Insecticides are generally more acutely toxic than are allelochemicals at concentrations present in the field (McKenzie & Batterham 1994). Applications of insecticide are often made at levels that far exceed population LD50 levels and so tend to select for novel mutations or extremely rare alleles rather than modifications of existing genetic variation; thus, insecticide resistance tends to be due to one or a few major genes rather than polygenic traits (Roush & Daly 1990). Plant allelochemicals at ecological concentrations may simply cause slower growth or reduced fecundity rather than outright mortality and may in turn select for polygenic resistance, with many modifier genes rather than one or two major genes.

The probability of encountering a toxin is a function not only of the distribution of the compound but also of the feeding habits of the consumer. For oligophagous insects with a narrow range of host plants that invariably produce a particular class of allelochemicals, circumstances in which complete susceptibility (e.g. resulting from the presence of a null allele for resistance to that class of chemicals) is beneficial are unlikely ever to arise. As a result, the range of variation for resistance to an allelochemical in its host plants is likely to be narrower than the range of variation for resistance to an insecticide, if only because individuals that are truly susceptible to host plant allelochemicals probably do not exist (Fig. 3.1). In polyphages, such susceptible individuals may experience no fitness reduction at all if encounters with the toxin can be avoided. To cite a mammalian example, Liang *et al.* (1996) created a strain of 'knockout' mice (*Mus musculus*) by homologous recombination in embryonic stem cells deficient in CYP1A2; these mice suffered no detectable adverse effects in the absence of this enzyme as long as known substrates were not present. The human homologue of this enzyme, CYP1A2, is present in a range of allelic forms and is known to metabolize caffeine, among other substrates, with activities ranging from exceedingly low to very high. For individuals who are capable of excluding caffeine from their diet, the absence of a functional enzyme to metabolize that particular compound is unlikely to have serious fitness consequences.

Overall, the differences between allelochemical production and insecticide application suggest that, theoretically, resistance to insecticides should evolve more rapidly, involve changes at fewer loci, and have a greater range of variability within a species than allelochemical resistance.

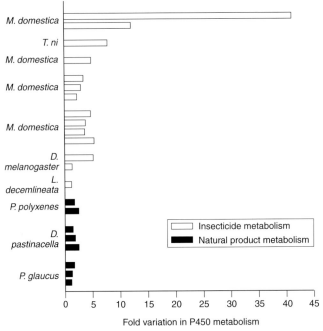

Figure 3.1 Fold variation between families or strains in cytochrome P450-mediated metabolism of xenobiotics. Sources of data are *Papilio polyxenes* (Berenbaum & Zangerl 1993), *Depressaria pastinacella* (Berenbaum & Zangerl 1992), *Papilio glaucus* (Hung & Berenbaum, unpublished), *Musca domestica* (Schonbrod *et al.* 1968; Terriere 1968; Plapp & Casida 1969; Hammock *et al.* 1977), *Trichoplusia ni* (Kuhr 1971), *Drosophila melanogaster* (Hallstrom & Grafstrom 1981) and *Leptinotarsa decemlineata* (Argentine *et al.* 1992).

Theory is far easier to generate than fact; population-level comparisons of allelochemical resistance and insecticide resistance within any one species simply do not exist. Operational challenges have made the measurement of within-species variation difficult. For instance, it is difficult to assay P450-mediated metabolism accurately from individual insects if they are very small (as many insects are); also, because age, diet and experimental conditions all appear to affect metabolism assays, isolating genetic variation from the noise of experimental and environmental variation can be difficult. Ideally, data from a wide range of studies could be pieced together to provide some insight into whether these predictions are supported by reality.

Perhaps best supported by evidence is the idea that genetic variability is greater for insecticide resistance than for allelochemical resistance. While copious data are available on interspecific differences in metabolic rates (aldrin epoxidase activities have been measured in over 100 species of Lepidoptera alone, see Neal 1987a), relatively few data are available on the range of variation within and between populations in any single species. Available studies suggest that insecticide-metabolizing P450s often do display a greater range of variation than allelochemical-metabolizing P450s (Fig. 3.1).

Factors maintaining genetic variation in biochemically based resistance to xenobiotics

Genetically based variation in detoxification is well documented, but there is much speculation as to the forces acting to maintain such extreme variation. Costs or trade-offs are widely assumed to underlie variation in resistance of many forms (Rausher 1996). Among the earliest indications of such costs of biochemically based resistance to xenobiotics was the widely recognized phenomenon of resistance instability — the tendency of resistant populations to revert to susceptible status in the absence of insecticide selection pressure (MacDonald *et al.* 1983). Such instability has been documented for several different classes of insecticides and there are some suggestions that metabolic resistance to allelochemicals can break down in populations out of contact with these compounds for several generations (e.g. laboratory colonies — but see Wheeler *et al.* 1992).

Several types of costs are postulated to exist. Allocation costs result from investments of energy, materials or resources into resistance mechanisms rather than growth or reproduction. Such diversion of resources may be manifested as a decrease in fitness. Genetic costs arise from pleiotropic effects that alter life history attributes, physiological function or reproductive success. Ecological costs come about through conflicting selection pressures from different mortality factors; activation of a resistance mechanism against one type of selective agent may render an individual more susceptible to a different agent of mortality (Kasamatsu 1992). Evidence exists in support of all of these forms of resistance costs, although rarely are they documented from a single system.

Estimating allocation costs

Allocation costs attracted attention very early in discussions of allelochemical resistance, particularly with respect to cytochrome P450-mediated mechanisms. The virtually universal inducibility of these enzymes was assumed to be a mechanism for conserving energy and material costs. These enzymes require reduced nicotinamide adenine dinucleotide phosphate (NADPH) and adenosine triphosphate (ATP) as cofactors; high levels of activity in the absence of relevant substrates would presumably shunt these factors away from other energy-consumptive reactions. Induction can increase cytochrome P450 content many-fold; it seemed plausible that protein synthesis for other functions could be severely compromised were these systems uninducible and maintained at high levels constantly.

Efforts to document or quantify allocation costs of metabolism or of induction have not met with overwhelming success. Brattsten's (1979) theoretical calculations suggest that both material and energy investments in P450 synthesis are minimal relative to the metabolic needs of the whole organism. Early attempts to examine costs of resistance to plant toxins in herbivorous insects involved gravimetric estimates of performance in the presence and absence of toxins in artificial diets (Schoonhoven & Meerman 1978). These early studies failed to take into account toxicological impacts of the compounds, which could be confounded with metabolic costs of disposition. Subsequent studies focused on quantifying investments in specific detoxification costs, such as induction, as a way of separating toxicological impact from metabolic costs. Neal (1987b) demonstrated that a ninefold increase in total P450 content in *Helicoverpa zea*, brought about by indole-3-carbinol, a non-toxic inducer of P450s, was not associated with any measurable reduction in growth or performance. Similarly, α-pinene, a non-toxic inducer of P450s (PSMOs) in *Spodoptera eridania*, did not have a detectable impact on growth and utilization efficiency on artificial diets (Cresswell *et al.* 1992). Cresswell *et al.* (1992), however, did show that P450-mediated detoxification of nicotine, a solanaceous alkaloid not routinely encountered by the generalist noctuid *Spodoptera eridania*, was associated with reduced relative growth rates and efficiencies of conversion of digested food. Appel and Martin (1992) failed to find such effects of the metabolic disposition of nicotine, even in concentrations in excess of 8% of the artificial diet, in *Manduca sexta*, the tobacco hornworm, a specialist on nicotine-containing plants. Differences in effects of nicotine metabolism may be attributable to the degree of specialization of the two herbivores examined. As a specialist on Solanaceae, *M. sexta* has a long evolutionary association with nicotine;

absence of apparent costs may result from investment in energy-efficient detoxification systems. Target site insensitivity in the central nervous system (Morris 1984) as well as rapid excretion (Self *et al.* 1964) may represent less energy-consumptive mechanisms for disposing of a toxicant than extensive P450-mediated metabolism.

In general, studies done with artificial diets are difficult to interpret because nutrients may be provided to insects in excess; when certain nutrients are limiting, as they may well be under natural circumstances, costs of detoxification may be more readily detected. Diet studies in which essential nutrients are provided in limiting supply do demonstrate under some circumstances a reduction in detoxification capacity. Lindroth *et al.* (1993) showed that a reduction in available dietary protein, vitamins and minerals did not affect P450 (= PSMO) activity but limiting vitamin and mineral intake reduced esterase and carbonyl reductase activity in *Papilio glaucus*. In contrast, restricting *D. melanogaster* to a nutritionally inadequate diet of 1% sucrose agar reduced 7-ethoxycoumarin-*O*-de-ethylase activity by as much as 40% relative to levels in larvae consuming nutritionally complete diets (Patil *et al.* 1990). In the parsnip webworm, *Depressaria pastinacella*, reduction of dietary protein by 75% had no effect on constitutive or inducible levels of furanocoumarin metabolism; only when dietary protein was completely eliminated from the diet did P450 activity drop (although inducibility was unaffected). In this insect, growth was reduced 40% on diets with only 25% of the normal protein content; *D. pastinacella* foregoes growth rather than furanocoumarin metabolism in the presence of limiting nutrients (Berenbaum & Zangerl 1994).

Estimating genetic costs

Genetic costs of resistance are in some ways more easily documented than allocation costs. Among the clearest manifestations of genetic costs is the instability of insecticide resistance in the absence of insecticide selection pressure (Minkoff & Wilson 1992; Cochran 1993), although not all forms of resistance show this instability (Raymond *et al.* 1993). Resistance due to gene amplification (e.g. carboxylesterase resistance to organophosphates) is inherently unstable and reverts readily in the absence of insecticide. In these instances, allocation costs may play a role; amplification means that enormous amounts of detoxification enzyme are produced. In organophosphate-resistant *Myzus persicae*, carboxylesterases may make up to 3% of total body protein; in resistant *Culex*, carboxylesterases may constitute up to 6% of total body protein (Roush & Daly 1990). Target site insensitivity may be similarly unstable if the target molecule in its altered state is physiologically less competent to carry out normal functions. Alterations in the structure of juvenile hormone binding protein are thought to be responsible for methoprene resistance in *Drosophila melanogaster*. Resistant strains are less competitive in laboratory experiments due to impaired development, conceivably due to changes in juvenile hormone function (Minkoff & Wilson 1992). Similarly, resistance in diamondback moth *Plutella xylostella* to *Bacillus thuringiensis* endotoxin is attributed to altered toxin-binding sites; altered physiological performance, including low mating success, is thought to be a pleiotropic effect of these alterations in toxin binding sites (Groeters *et al.* 1994). For resistance that is due to a change in a regulatory gene, as has been postulated for some types of P450-mediated insecticide resistance, the potential exists for extensive pleiotropy and subsequent instability due to alterations in other physiological functions, although little definitive experimental evidence exists of such effects. Indeed, it has proved difficult to attribute any form of insecticide resistance exclusively to altered P450 metabolism (but see Ferrari & Georghiou 1981; Roush & Plapp 1982). Generally, changes in rates of P450 metabolism in resistant genotypes are manifested along with other forms of resistance mechanisms (e.g. Campanhola *et al.* 1991).

For P450-mediated resistance, Plapp (1984) has proposed a mechanism that might provide an explanation for some of these phenomena. He suggested that enhanced detoxification rates are the result of changes in regulatory rather than structural genes (see also Brattsten 1988). An alteration of promoter elements that enhances P450 responsiveness broadly could account for numerous instances of cross or multiple resistance. Evidence has generally supported this hypothesis as molecular techniques have become more refined. A *trans*-acting factor encoded on chromosome II, for example, is associated with lpr pyrethroid resistance in *Musca domestica*, while the structural gene encoding the P450 is located on chromosome V (Cohen *et al.* 1994). Diazinon resistance in Rutgers, another resistant strain, also maps to chromosome II (Cluck *et al.* 1990), as does DDT resistance (Tsukamoto & Suzuki 1964), organophosphate resistance (Plapp *et al.* 1976) and naphthalene metabolism (Schaefer & Terriere 1970).

Additional, albeit indirect, support of this notion is the fact that insecticide-resistant strains within a species often appear to have greatly expanded host ranges, or cause more damage on a wide range of crops. If a regulatory gene change is responsible for the acquisition of resistance, then resistant strains might acquire, along with insecticide resistance, the ability to metabolize a greater range of host plant allelochemicals, and thus be capable of colonizing novel host plants. Conversely, if a regulatory gene change allows colonization of a novel host, then the modified strains might be predisposed to acquire insecticide resistance. The B biotype of *Bemisia tabaci*, for example, has both an expanded host range and expanded insecticide resistance capabilities relative to the A biotype (Costa *et al.* 1993). Also, utilization of certain chemically distinct host plants of the greenhouse whitefly *Trialeurodes vaporariorum* appears to enhance insecticide resistance in this species over time; indeed, management of host plant diversity has been suggested as a means for reducing insecticide resistance in the species (Omer *et al.* 1992).

Because P450s are also responsible for both pheromone and hormone metabolism, a change in a master regulatory gene may pleiotropically disrupt such physiological processes as moulting and egg maturation. P450s, for example, catalyse the hydroxylation of ecdysone to 20-hydroxyecdysone, a moulting hormone (Weirich *et al.* 1996). Yu (1995) demonstrated that ecdysone 20-hydroxylase can be induced up to twofold in autumn armyworm by an array of plant allelochemicals (indole-3-carbinol, flavone, β-naphthoflavone, xanthotoxin, menthone and menthol among them), raising the possibility that changes in the regulation of enzymes responsible for xenobiotic metabolism may interfere with primary metabolism.

Estimating ecological costs

Ecological costs come into play when a resistance mechanism renders an individual more susceptible to a totally different mortality agent than that selecting for resistance. The potential for such costs of P450-mediated detoxification is exemplified by the phenomenon of bioactivation. Although the conversion by P450s of lipophilic substrates to hydrophilic metabolites generally results in a reduction in toxicity and in enhanced export from the body, in some cases, due to structural attributes of the substrate, this metabolic conversion actually enhances toxicity. This phenomenon has been documented for synthetic and natural substrates. The conversion of aldrin to dieldrin and parathion to paraoxon produces a metabolite more toxic than the parent compound, and the oxidation of pyrrolizidine alkaloids by P450s enhances their toxicity significantly (Terriere 1968; Brattsten 1979).

Analogue synergism (enhancement of toxicity in the presence of a structural analogue) represents another possible ecological cost of P450-based resistance. Such synergism can be due to mechanism-based inhibition of P450 activity. For example, xanthotoxin and other furanocoumarins inactivate P450s from vertebrates (Woo *et al.* 1983; Fouin-Fortunet *et al.* 1986; Labbe *et al.* 1987; Tinel *et al.* 1987) and invertebrates (Zumwalt & Neal 1993; Neal & Wu 1994). The proposed mechanism of inhibition in vertebrates is that xanthotoxin and structurally related inhibitors are metabolically activated by epoxidation of the double bond in the furan ring to form an unstable radicaloid, which subsequently binds to the P450 active site (Letteron *et al.* 1986). Once the active site is blocked, the enzyme can no longer interact with other substrates and the animal is thus rendered susceptible to compounds that might in the absence of the analogue be efficiently metabolized. Evidence for such synergistic interactions among allelochemicals comes from several systems. Both linear and angular furanocoumarins, bioassayed individually, are toxic as well as deterrent to a wide range of insects (Berenbaum 1995b). A mixture of six furanocoumarins occurring naturally in fruits of *Pastinaca sativa* (wild parsnip), however, was more toxic to *Heliothis* (*Helicoverpa*) *zea* than an equimolar amount of the single most phototoxic constituent, xanthotoxin (Berenbaum *et al.* 1991). Moreover, Berenbaum and Feeny (1981) demonstrated that angelicin, an angular furanocoumarin, when applied topically to foliage of umbellifer species containing linear furanocoumarins, profoundly reduced fecundity of *Papilio polyxenes*, a specialist on plant species containing linear furanocoumarins.

Although *P. polyxenes* is capable of metabolizing xanthotoxin up to 10 times faster than non-adapted polyphagous species such as *Trichoplusia ni* (Berenbaum 1995b), it does not metabolize all furanocoumarins with equal efficiency. Ivie *et al.* (1986) demonstrated that metabolism of angelicin proceeds at a much slower rate than metabolism of its linear analogue psoralen. When three furanocoumarins (bergapten, xanthotoxin and angelicin) are assayed in combination with one another, overall rates of total furanocoumarin metabolism decline, as do rates of metabolism of each individual furanocoumarin. This response is suggestive of metabolic interference among these compounds. Growth rates are also reduced on diets containing combinations of furanocoumarins, compared with diets containing equimolar concentrations of individual furanocoumarins, (Berenbaum & Zangerl 1993). Moreover, recent studies have shown that xanthotoxin induction of its own metabolism is inhibited by sphondin, an angular furanocoumarin found in some host plants of *P. polyxenes*. Such inhibition is consistent with analogue synergism, whereby sphondin is bound to but not turned over by the xanthotoxin-metabolizing P450.

Like *P. polyxenes*, the parsnip webworm *Depressaria pastinacella* feeds exclusively on furanocoumarin-containing plants. However, its hosts, species in the umbelliferous genera *Pastinaca* and *Heracleum*, are unique in producing large quantities of angular furanocoumarins. Moreover, webworms feed exclusively on reproductive tissues (buds, flowers and fruits), which contain much higher concentrations of both linear and angular furanocoumarins than foliage. Nitao (1989) demonstrated that xanthotoxin metabolism in webworms is NADPH-dependent, inhibited by piperonyl butoxide, and inducible by prior exposure to xanthotoxin, indicating the participation of P450s in detoxification. Despite their metabolic capabilities, webworms are not found on *Pastinaca sativa* plants high in xanthotoxin and sphondin; when they are forced to consume flowers from such plants, their rates of furanocoumarin metabolism are depressed relative to webworms who are fed flowers from plants infested by webworms (Zangerl & Berenbaum 1993). This reduction in activity is consistent with analogue synergism, whereby one furanocoumarin interferes with metabolism of another. *In vitro* assays confirm that overall metabolism of a mixture of furanocoumarins is reduced compared with metabolism of individual furanocoumarins (Berenbaum & Zangerl 1992).

Bioactivation costs are not entirely unavoidable. The generalist *Spodoptera eridania* has P450s that are highly induced by foliage of carrot, which contains numerous compounds that are detoxified by P450s, but are hardly induced at all by foliage of *Crotalaria*, a legume with high concentrations of pyrrolizidine alkaloids (Brattsten 1979). This pattern of response maximizes the efficiency of the detoxification system. Similarly, in organophosphate-resistant *Heliothis virescens*, activity levels of P450s in this species that convert the thioate to the oxon are lower than in susceptible strains (Konno *et al.* 1989). Also, irreversible binding of toxins to P450s may serve as a form of detoxification via sequestration.

Specialization for particular substrates may reduce the ability of particular P450s to accommodate multiple functions, thus effecting yet another kind of ecological cost. Such appears to be the case for mammalian P450s. Site-directed mutagenesis studies and sequence comparisons have identified a small number of amino acids in CYP2 enzymes that profoundly influence catalytic properties and substrate specificities. Lindberg and Negishi (1989), for example, demonstrated that a single amino-acid substitution, F209L, confers steroid 15α-hydroxylase activity on CYP2A5 in mice, normally a coumarin-hydroxylase with no detectable steroid-hydroxylating capabilities. Two allelic variants of CYP2C3 are known; both forms catalyse 16α-hydroxylation of progesterone, but only one can catalyse 6β-hydroxylation. One amino-acid substitution, S364T, accounts for the difference (Johnson 1992). Allelic variants with widely disparate activities are known to occur in mammalian P450s (Berenbaum & Zangerl 1998). In insects, allelic variation has been found in CYP6B proteins, but its functional significance has not yet been determined. However, it can be pointed out that CYP6B3, which differs by less than 12% in sequence from CYP6B1 in black swallowtail, is not capable of metabolizing furanocoumarins to the same degree that CYP6B1 can in a heterologous system (Hung 1996); the amino acids responsible for the lower catalytic activity toward these substrates have not yet been identified.

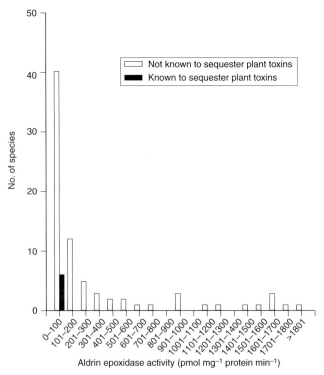

Figure 3.2 Aldrin epoxidase activity (pmol mg^{-1} protein min^{-1}) for species known to sequester plant toxins and those not known to sequester plant toxins. Based on homogenates of insect midguts as determined by Krieger *et al.* (1971) and Rose (1985). From Neal (1987a).

In some circumstances, high levels of metabolic activity against allelo-chemicals may be associated with an ecological cost in the form of reduced capacity for sequestering unmetabolized plant toxins for defence against predators. Metabolic transformation of a plant toxin to a non-toxic form could seriously compromise the defence capabilities of an insect. Neal (1987a) compared published values of aldrin epoxidase activity in species known to sequester toxins for defence with levels of activity in insects not known to sequester toxins and found significantly lower levels of this enzyme activity in sequestering species, a pattern consistent with the existence of such an ecological cost (Fig. 3.2).

One final kind of ecological cost can be considered: this is an oppor-tunity cost, in the economic sense (Clark 1985). An opportunity cost can be defined as 'the value of the best available alternative' (Clark 1985). Maintaining a P450-based detoxification system involves certain fixed costs, whether or not substrates are turned over. In the absence of substrates,

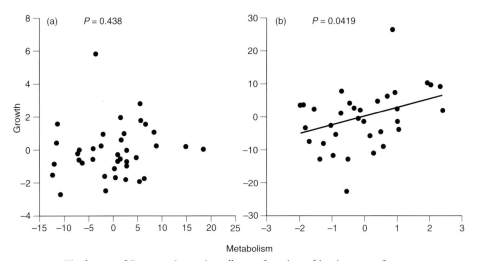

Figure 3.3 Final mass of *Depressaria pastinacella* as a function of *in vitro* rate of metabolism of furanocoumarins. These regressions are based on residuals from multiple regression of final caterpillar mass against initial caterpillar mass, consumption rate and *in vitro* rate of metabolism. Separate regressions were performed for caterpillars fed reproductive parts from wild parsnip plants with (a) below-average furanocoumarin content and (b) above-average furanocoumarin content. Data from Zangerl and Berenbaum (1993).

variable costs (energy and raw materials) may well be diverted from growth (or production, in economic terms) to maintain this system, with no net gain. However, when substrates are present, the net effect of these investments of materials and energy is positive; growth is maintained because P450-mediated metabolism prevents the insects from succumbing to the toxic effects of food plant constituents. This relationship is illustrated in *Depressaria pastinacella*, which feeds exclusively on plants containing furanocoumarins. Variation in metabolism rates of this species accounts for a significant amount of variation in growth only when furanocoumarins are present at high levels (Fig. 3.3).

Conclusions

It is important to emphasize that this discussion has focused on only a single type of biochemical resistance—reactions mediated by P450s. Resistance to both insecticides and allelochemicals undoubtedly involves multiple mechanisms in most insects. Multiple mechanisms of resistance to insecticides may be most likely to occur in those species subjected to intense and widespread insecticide selection pressure; no fewer than eight mechanisms are involved, for example, in pyrethroid metabolism in *Blattella germanica* (Cochran 1993). Multiple mechanisms of resistance to allelochemicals are perhaps most likely to occur in oligophages, for whom allelochemical selection pressure is encountered with every meal. Parsnip webworms, for example, feed exclusively on the fruits and flowers of *Pastinaca* and *Heracleum*, plant parts invariably high in furanocoumarins (Berenbaum 1981; Murray *et al.* 1982). Webworms have multiple mechanisms for dealing with a furanocoumarin-rich diet:
1 selective feeding on less toxic plants (Zangerl & Berenbaum 1993);
2 avoiding ultraviolet light to reduce photoactivation (Carroll *et al.* 1997);
3 sequestration of carotenoids such as lutein to reduce effects of photoactivation (Carroll *et al.* 1997);
4 highly active P450 system to metabolize principal toxicants (Berenbaum & Zangerl 1992);
5 high levels of antioxidant enzymes, inducible upon inhibition of P450s, to provide backup for unmetabolized toxins (Lee & Berenbaum 1990);
6 export of unmetabolized furanocoumarins in silk (Nitao 1990).

One reason for apparent differences in the nature of biochemically based resistance to toxic and synthetic chemicals is that these phenomena are being examined at different points along the evolutionary time scale. Parsnip webworms have had thousands, if not millions, of years to develop mechanisms of resistance for coping with the toxins in its narrow range of host plants. When wild parsnip was a novel host plant for this species, the initial genetic change allowing the insect or its immediate ancestors to colonize the plant was most likely in a major gene; in the intervening millennia, modifier genes, and other genes acting entirely independently of the initial major genes, have come into play, to fine-tune the process. Even so, parsnip webworm genotypes with lower detoxification capabilities persist (Zangerl & Berenbaum 1993), presumably because allocation, genetic and ecological costs serve as constraints.

All allelochemicals must have been biochemical novelties at some point in the evolutionary history of the plant producing them. By the same token, synthetic organic insecticides are biochemical novelties when they are first released for commercial use. This is not to say that the process of encounter by an insect to these two different types of toxins is identical. A novel allelochemical does not suddenly appear in massive quantities in all host plants across extensive landscapes. Genes for new defensive chemicals spread within and between plant populations as a result of selection and must be integrated into the genome with modifier genes because they bring about their own ecological, genetic and allocation costs. Plants (as users of defence chemicals) have an entirely different way of evaluating their economic injury levels than users of insecticide. Cosmetic damage that does not affect seed production may not elicit a substantial investment in defence chemicals by a plant in a natural population, whereas a grower who must contend with consumers fixated on aesthetic concerns may reach for insecticide when confronted by the same level of damage.

A major focus of integrated pest management programmes today is resistance management, by which the acquisition of resistance is delayed so as to prolong the useful lifetime of a control chemical. Examples abound of novel insecticides that fail within only a year of their introduction due to resistance problems (Georghiou & Taylor 1976). Avoiding such situations must involve more than modification of chemical structures—it must involve changes in the ways that insecticides are applied (Berenbaum 1995a). Plants have by and large been extremely successful in deploying their defence chemicals effectively against their insect enemies for thousands of generations. Understanding the differences between the evolution of resistance to synthetic organic insecticides and the evolution of resistance to allelochemicals can lead to insights into improving methods of managing the acquisition of resistance to synthetic organic insecticides in economically important pest species.

Acknowledgements

We thank Chien-Fu Hung for valuable comments on the manuscript. The preparation of this manuscript, as well as much of the work described in it, was supported by NSF DEB 96-28977 and NSF IBN 96-30442.

References

Appel, H.M. & Martin, M.M. (1992). Significance of metabolic load in the evolution of host specificity of *Manduca sexta*. *Ecology*, **73**, 216–228.

Argentine, J.A., Clark, J.M. & Lin, H. (1992). Genetics and biochemical mechanisms of abamectin resistance in two isogenic strains of Colorado potato beetle. *Pesticide Biochemistry and Physiology*, **44**, 191–207.

Berenbaum, M. (1981). Patterns of furanocoumarin production and insect herbivory in a population of wild parsnip (*Pastinaca sativa* L.). *Oecologia*, **49**, 236–244.

Berenbaum, M. (1986). Target site insensitivity in plant insect interactions. In *Molecular Mechanisms in Insect–Plant Associations* (Ed. by V.L. Brattsten & S. Ahmad), pp. 257–272. Plenum Press, New York.

Berenbaum, M.R. (1991). Coumarins. In *Herbivores: Their Interactions with Secondary Plant Metabolites*, Vol. 1 (Ed. by G. Rosenthal & M. Berenbaum), pp. 221–249. Academic Press, New York.

Berenbaum, M.R. (1995a). Chemical defense: theory and practice. *Proceedings of the National Academy of Sciences, USA*, **92**, 2–8.

Berenbaum, M.R. (1995b). Phototoxicity of plant secondary metabolites: insect and mammalian perspectives. *Archives of Insect Biochemistry and Physiology*, **29**, 119–134.

Berenbaum, M.R. & Feeny, P. (1981). Toxicity of angular furanocoumarins to swallowtail butterflies: escalation in a coevolutionary arms race. *Science*, **212**, 927–929.

Berenbaum, M.R. & Zangerl, A.R. (1992). Genetics of physiological and behavioral resistance to host furanocoumarins in the parsnip webworm. *Evolution*, **46**, 1371–1384.

Berenbaum, M.R. & Zangerl, A.R. (1993). Furanocoumarin metabolism in *Papilio polyxenes*: biochemistry, genetic variability, and ecological significance. *Oecologia*, **95**, 370–375.

Berenbaum, M.R. & Zangerl, A.R. (1994). Costs of inducible defense in insects: effects of protein limitation on growth, silk production, and detoxification in parsnip webworms. *Ecology*, 75, 2311–2317.

Berenbaum, M.R. & Zangerl, A.R. (1996). Phytochemical diversity: Adaptation or random variation? In *Phytochemical Diversity and Redundancy in Ecological Interactions* (Ed. by J.T. Romero, J.A. Saunders & P. Barbosa), pp. 1–24. Plenum Press, New York.

Berenbaum, M.R. & Zangerl, A.R. (1998). Population-level adaptation to host-plant chemicals: The role of cytochrome P450 monooxygenases. In *Genetic Structure and Local Adaptation in Natural Insect Populations* (Ed. by S. Mopper & S.Y. Strauss), pp. 91–112. Chapman & Hall, New York.

Berenbaum, M.R., Nitao, J.K. & Zangerl, A.R. (1991). Adaptive significance of furanocoumarin variation in *Pastinaca sativa* (Apiaceae). *Journal of Chemical Ecology*, 17, 207–215.

Berenbaum, M.R., Favret, C. & Schuler, M.A. (1996). On defining "key innovations" in an adaptive radiation: cytochrome P450 and Papilionidae. *American Naturalist*, 148, S139–S155.

Bertilsson, L. (1995). Geographical/interracial differences in polymorphic drug oxidation. *Clinical Pharmacokinetics*, 29, 192–209.

Blum, M. (1981). *Chemical Defenses of Arthropods*. Academic Press, New York.

Bradfield, J.Y., Lee, Y.-H. & Keeley, L.L. (1991). Cytochrome P450 family 4 in a cockroach; molecular cloning and regulation by hypertrehalosemic hormone. *Proceedings of the National Academy of Sciences, USA*, 88, 4558–4562.

Brattsten, L.B. (1979). Ecological significance of mixed function oxidation. *Drug Metabolism Reviews*, 10, 35–58.

Brattsten, L.B. (1987a). Metabolic insecticide defences in the boll weevil compared to those in a resistance-prone species. *Pesticide Biochemistry and Physiology*, 27, 1–12.

Brattsten, L.B. (1987b). Inducibility of metabolic insecticide defences in boll weevils and tobacco budworm caterpillars. *Pesticide Biochemistry and Physiology*, 27, 13–23.

Brattsten, L.B. (1988). Potential role of plant allelochemicals in the development of insecticide resistance. In *Novel Aspects of Insect–Plant Interactions* (Ed. by P. Barbosa & D.K. Letourneau), pp. 313–348. John Wiley and Sons, New York.

Brattsten, L.B. (1992). Metabolic defenses against plant allelochemicals. In *Herbivores: Their Interactions with Secondary Plant Metabolites*, Vol. 2 (Ed. by G. Rosenthal & M. Berenbaum), pp. 175–242. Academic Press, San Diego.

Brattsten, L.B., Wilkinson, C.F. & Eisner, T. (1977). Herbivore–plant interactions; mixed function oxidases and secondary plant substances. *Science*, **196**, 1349–1352.

Campanhola, C., McCutchen, B.F., Baehrecke, E.H. & Plapp, F.W., Jr. (1991). Biological constraints associated with resistance to pyrethroids in the tobacco budworm (Lepidoptera: Noctuidae). *Journal of Economic Entomology*, **84**, 1404–1411.

Carroll, M., Hanlon, A., Hanlon, T., Zangerl, A.R. & Berenbaum, M.R. (1997). Behavioral effects of carotenoid sequestration by the parsnip webworm, *Depressaria pastinacella*. *Journal of Chemical Ecology*, **23**, 2707–2719.

Clark, J.M. (1985). Analysis of costs. In *Economics* (Ed. by P.A. Samuelson & W.D. Nordhaus), pp. 461–473. McGraw Hill, St Louis.

Cluck, T.W., Plapp, F.W. & Johnston, J.S. (1990). Genetics of organophosphate resistance in field populations of the house fly (Diptera: Muscidae). *Journal of Economic Entomology*, **83**, 48–54.

Cochran, D.G. (1993). Decline of pyrethroid resistance in the absence of selection pressure in a population of German cockroaches (Dictyoptera: Blattellidae). *Journal of Economic Entomology*, **86**, 1639–1644.

Cohen, M.B., Schuler, M.A. & Berenbaum, M.R. (1992). A host-inducible cytochrome P-450 from a host-specific caterpillar: molecular cloning and evolution. *Proceedings of the National Academy of Sciences, USA*, **89**, 10920–10924.

Cohen, M.B., Koener, J.F. & Feyereisen, R. (1994). Structure and chromosomal localization of CYP6A1, a cytochrome P450-encoding gene from the house fly. *Gene*, **146**, 267–272.

Costa, H.S., Brown, J.K., Sivasupramaniam, S. & Bird, J. (1993). Regional distribution, insecticide resistance, and reciprocal crosses between the A and B biotypes of *Bemisia tabaci*. *Insect Science Applications*, **14**, 255–266.

Cresswell, J.E., Merritt, S.Z. & Martin, M.M. (1992). The effect of dietary nicotine on the allocation of assimilated food to energy metabolism and growth in fourth-instar larvae of the southern armyworm, *Spodoptera eridania* (Lepidoptera: Noctuidae). *Oecologia*, **89**, 449–453.

Ehrlich, P.R. & Raven, P.H. (1964). Butterflies and plants: a study in coevolution. *Evolution*, **18**, 586–608.

Ferrari, J.A. & Georghiou, G.P. (1981). Effects on insecticidal selection and treatment on reproductive potential of resistant, susceptible, and heterozygous strains of the southern house mosquito. *Journal of Economic Entomology*, **74**, 323–327.

Feyereisen, R., Koener, J.F., Farnsworth, D.E. & Nebert, D.W. (1989). Isolation and sequence of cDNA encoding a cytochrome P450 from an insecticide-resistant strain of the house fly *Musca domestica*. *Proceedings of the National Academy of Sciences, USA*, **86**, 1465–1469.

Feyereisen, R., Andersen, J.F. & Carino, F.A. (1995). Cytochrome P450 in the house fly: structure, catalytic activity and regulation of expression of CYP6A1 in an insecticide-resistant strain. *Pesticide Science*, **43**, 233–239.

Fouin-Fortunet, H., Tinel, M., Descatoire, V., Letteron, P., Larrey, J., Geneve, J. & Pessayre, D. (1986). Inactivation of cytochrome P-450 by the drug methoxsalen. *Journal of Pharmacology and Experimental Therapeutics*, **236**, 237–247.

Gaedigk, A., Blum, M. & Gaedigk, R. (1991). Deletion of the entire cytochrome P450 CYP2D6 gene as a cause of impaired drug metabolism in poor metabolizers of debrisoquine/spartein polymorphism. *American Journal of Human Genetics*, **48**, 943–950.

Georghiou, G.P. & Taylor, C.E. (1976). Pesticide resistance as an evolutionary phenomenon. In *Proceedings of the XVth International Congress of Entomology, 1976*, pp. 759–785. US Department of Agriculture.

Gonzalez, F.J. & Nebert, D.W. (1990). Evolution of the P450 gene superfamily. *Trends in Genetics*, **6**, 182–186.

Gordon, H.T. (1961). Nutritional factors in insect resistance to chemicals. *Annual Review of Entomology*, **6**, 37–54.

Gould, F., Carroll, C.R. & Futuyma, D. (1982). Cross-resistance to pesticides and plant defenses: a study of the two-spotted spider mite. *Entomologia Experimentalis et Applicata*, **31**, 175–180.

Groeters, F.R., Tabashnik, B.E., Finson, N. & Johnson, M.W. (1994). Fitness costs of resistance to *Bacillus thuringiensis* in the diamondback moth (*Plutella xylostella*). *Evolution*, **48**, 197–201.

Hallstrom, I. & Grafstrom, R. (1981). The metabolism of drugs and carcinogens in isolated subcellular fractions of *Drosophila melanogaster*. II. Enzyme induction and metabolism of benzo[a]pyrene. *Chemico Biological Interactions*, **34**, 145–159.

Hammock, B.D., Mumby, S.M. & Lee, P.W. (1977). Mechanisms of resistance to the juvenoid methoprene in the housefly *Musca domestica* L. *Pesticide Biochemistry and Physiology*, **7**, 261–272.

Hodgson, E. (1985). Microsomal monooxygenases. In *Comprehensive Insect Physiology, Biochemistry and Pharmacology*, Vol. 2 (Ed. by G.A. Kerkut & L.I. Gilbert), pp. 225–321. Pergamon Press, New York.

Hung, C.-F. (1996). *Isolation and characterization of cytochrome P450s from Papilio polyxenes and Papilio glaucus*. Doctoral dissertation, University of Illinois at Urbana-Champaign.

Hung, C.-F., Kao, C.H., Liu, C.C., Lin, J.G. & Sun, C.N. (1995). Detoxifying enzymes of selected insect species with chewing and sucking habits. *Journal of Economic Entomology*, **83**, 361–365.

Ivie, G.W., Bull, D.L., Beier, R.C. & Pryor, N.W. (1986). Comparative metabolism of [³H] psoralen and [³H] isopsoralen by black swallowtail

(*Papilio polyxenes* Fabr.) caterpillars. *Journal of Chemical Ecology*, **12**, 871–884.

Janzen, D.H. (1977). Why fruits rot, seeds mold, and meats spoil. *American Naturalist*, **111**, 691–713.

Johnson, E.F. (1992). Mapping determinants of the substrate selectivities of P450 enzymes by site-directed mutagenesis. *Trends in Pharmaceutical Science*, **13**, 122–126.

Kasamatsu, K. (1992). Negative correlation to tetradifon sensitivity in fenpropathrin selected strain of the two-spotted spider mite, *Tetranychus urticae* (Acari: Tetranychidae). *Applied Entomology and Zoology*, **27**, 458–460.

Kimura, S., Umeno, M., Skoda, R.C., Meyer, U.A. & Gonzalez, F.J. (1989). The human debrisoquine 4-hydroxylase (CYP2D) locus: sequence and identification of the polymorphic CYP2D6 gene, a related gene, and a pseudogene. *American Journal of Human Genetics*, **45**, 889–904.

Konno, T., Hodgson, E. & Dauterman, W.C. (1989). Study on methyl parathion resistance in *Heliothis virescens. Pesticide Biochemistry and Physiology*, **33**, 189–199.

Krieger, R., Feeny, P.P. & Wilkinson, C.F. (1971). Detoxification enzymes in the guts of caterpillars: an evolutionary answer to plant defense? *Science*, **172**, 579–581.

Kuhr, R.J. (1971). Comparative metabolism of carbaryl by resistant and susceptible strains of the cabbage looper. *Journal of Economic Entomology*, **64**, 1373–1378.

Labbe, G., Descatoire, V., Letteron, P., Degott, C., Tinel, M., Larrey, D., Carrion-Pavlov, Y., Geneve, J., Amouyal, G. & Pessayre, D. (1987). The drug methoxsalen, a suicide substrate for cytochrome P-450, decreases the metabolic activation, and prevents the hepatotoxicity, of carbon tetrachloride in mice. *Biochemical Pharmacology*, **36**, 907–914.

Lee, K. & Berenbaum, M.R. (1990). Defense of parsnip webworm against phototoxic furanocoumarins: the role of antioxidant enzymes. *Journal of Chemical Ecology*, **16**, 2451–2460.

Letteron, P., Descatoire, V., Larrey, D., Tinel, M., Geneve, J. & Pessayre, D. (1986). Inactivation and induction of cytochrome P450 by various psoralen derivatives in rats. *Journal of Pharmacology and Experimental Therapy*, **238**, 685–692.

Liang, H.C.L., Li, H., McKinnon, R.A., Duffy, J.J., Potter, S.S., Puga, A. & Nebert, D.W. (1996). CYP1A2(–/–) null mutant mice develop normally but show deficient drug metabolism. *Proceedings of the National Academy of Sciences, USA*, **93**, 1671–1676.

Lindberg, R.L.P. & Negishi, M. (1989). Alteration of mouse cytochrome P450 substrate specificity by mutation of a single amino-acid residue. *Nature*, **339**, 632–634.

Lindroth, R.L., Jung, S.M. & Feuker, A.M. (1993). Detoxication activity in the gypsy moth: effects of host CO_2 and NO_3^- availability. *Journal of Chemical Ecology*, **19**, 357–367.

Ma, R., Cohen, M.B., Berenbaum, M.R., & Schuler, M.A. (1994). Black swallowtail (*Papilio polyxenes*) alleles encode cytochrome P450s that selectively metabolize linear furanocoumarins. *Archives of Biochemistry and Biophysics*, **310**, 332–340.

MacDonald, R.S., Surgeoner, G.A., Solomon, K.R. & Harris, C.R. (1983). Effect of four spray regimes on the development of permethrin and dichlorvos resistance, in the laboratory, by the house fly (Diptera: Muscidae). *Journal of Economic Entomology*, **76**, 417–422.

Matsumura, F. (1975). *Toxicology of Insecticides*. Plenum Press, New York.

McKenzie, J. & Batterham, P. (1994). The genetic, molecular, and phenotypic consequences of selection for insecticide resistance. *Trends in Ecology and Evolution*, **9**, 166–169.

Minkoff, C., III & Wilson, T.G. (1992). The competitive ability and fitness components of the Methoprene-tolerant (Met) *Drosophila* mutant resistant to juvenile hormone analog insecticides. *Genetics*, **131**, 91–97.

Morris, C.E. (1984). Electrophysiological effects of cholinergic agents on the CNS of a nicotine-resistant insect, the tobacco hornworm (*Manduca sexta*). *Journal of Experimental Zoology*, **229**, 361–374.

Mullin, C.A. (1985). Detoxification enzyme relationships in arthropods of differing feeding strategies. In *Bioregulators for Pest Control* (Ed. by P.A. Hedin), pp. 267–278. American Chemical Society Symposium Series 276, Washington.

Mullin, C.A. (1986). Adaptive divergence of chewing and sucking arthropods for plant allelochemicals. In *Molecular Aspects of Insect–Plant Associations* (Ed. by L.B. Brattsten & S. Ahmad), pp. 175–209. Plenum Press, New York.

Murray, R.D.H., Mendez, J. & Brown, S.A. (1982). *The Natural Coumarins*. Wiley, New York.

Neal, J.J. (1987a). *Ecological aspects of insect detoxication enzymes and their interaction with plant allelochemicals*. Doctoral dissertation, University of Illinois at Urbana-Champaign.

Neal, J.J. (1987b). Metabolic costs of mixed-function oxidase induction in *Heliothis zea*. *Entomologia Experimentalis et Applicata*, **43**, 175–179.

Neal, J.J. & Wu, D. (1994). Inhibition of insect cytochromes P450 by furanocoumarins. *Pesticide Biochemistry and Physiology*, **50**, 43–50.

Nelson, D.R., Koymans, L., Kamataki, T., Stegeman, J.J., Feyereisen, R., Waxman, D.J., Waterman, M.R., Gotoh, O., Coon, M.J., Estabrook, R.W., Gunsalus, I.C. & Neibert, D.W. (1996) P450 superfamily: update on new sequences, gene mapping, accession numbers, and nomenclature. *Pharmacogenetics*, **6**, 1–41.

Nitao, J.K. (1989). Enzymatic adaptation in a specialist herbivore for feeding on furanocoumarin-containing plants. *Ecology*, **70**, 629–635.

Nitao, J.K. (1990). Metabolism and excretion of the furanocoumarin xanthotoxin by parsnip webworm, *Depressaria pastinacella*. *Journal of Chemical Ecology*, **16**, 417–428.

Omer, A.D., Leigh, T.F. & Granett, J. (1992). Insecticide resistance of greenhouse whitefly (Hom., Aleyrodidae) and fitness on plant hosts relative to the San Joaquin Valley (California) cotton agroecosystem. *Journal of Applied Entomology*, **113**, 244–254.

Patil, T.N., Morton, R.A. & Singh, R.S. (1990). Characterization of 7-ethoxycoumarin-o-deethylase from malathion resistant and susceptible strains of *Drosophila melanogaster*. *Insect Biochemistry*, **20**, 91–98.

Plapp, F.W., Jr. (1984). The genetic basis of insecticide resistance in the house fly: evidence that a single locus plays a major role in metabolic resistance to insecticides. *Pesticide Biochemistry and Physiology*, **22**, 194–201.

Plapp, F.W., Jr. & Casida, J.E. (1969). Genetic control of house fly NADPH-dependent oxidases: relation to insecticide chemical metabolism and resistance. *Journal of Economic Entomology*, **62**, 1174–1179.

Plapp, F.W., Jr., Tate, L.G. & Hodgson, E. (1976). Biochemical genetics of oxidative resistance to diazinon in the house fly. *Pesticide Biochemistry and Physiology*, **6**, 175–182.

Rausher, M.D. (1996). Genetic analysis of coevolution between plants and their natural enemies. *Trends in Genetics*, **12**, 212–217.

Raymond, M., Poulin, E., Boiroux, V., DuPont, E. & Pasteur, N. (1993). Stability of insecticide resistance due to amplification of esterase genes in *Culex pipiens*. *Heredity*, **70**, 301–307.

Riskallah, M.R., Dauterman, W.C. & Hodgson, E. (1986). Host plant induction of microsomal monooxygenase activity in relation to diazinon metabolism and toxicity in larvae of the tobacco budworm *Heliothis virescens* (F.). *Pesticide Biochemistry and Physiology*, **25**, 233–247.

Rose, H.A. (1985). The relationship between feeding specialization and host plants to aldrin epoxidase activities of midgut homogenates in larval Lepidoptera. *Ecological Entomology*, **10**, 455–467.

Rosenheim, J.A., Johnson, M.W., Mau, R.F.L., Welter, S.C. & Tabashnik, B.E. (1996). Biochemical preadaptations, founder events, and the evolution of resistance in arthropods. *Journal of Economic Entomology*, **89**, 263–273.

Rosenthal, G. & Berenbaum, M.R. (1991). *Herbivores: Their Interactions with Secondary Plant Metabolites*, Vol. 1. Academic Press, New York.

Roush, R.T. & Daly, J.C. (1990). The role of population genetics in resistance research and management. In *Pesticide Resistance in Arthropods*

(Ed. by R.T. Roush & B.E. Tabashnik), pp. 97–152. Chapman & Hall, New York.

Roush, R.T. & Plapp, F.W., Jr. (1982). Effects of insecticide resistance on biotic potential of the house fly (Diptera: Muscidae). *Journal of Economic Entomology*, **75**, 708–713.

Saner, C., Weibel, B., Würgler, F.E. & Sengstag, C. (1996). Metabolism of promutagens catalyzed by *Drosophila melanogaster* CYP6A2 enzyme in *Saccharomyces cerevisiae*. *Environmental and Molecular Mutagenesis*, **27**, 46–58.

Schaefer, J. & Terriere, L.C. (1970). Enzymatic and physical factors in house fly resistance to naphthalene. *Journal of Economic Entomology*, **63**, 787–792.

Schonbrod, R.D., Khan, M.A.Q., Terriere, L.C. & Plapp, F.W., Jr. (1968). Microsomal oxidases in the house fly: a survey of fourteen strains. *Life Sciences*, **7**, 681–688.

Schoonhoven, L.M. & Meerman, J. (1978). Metabolic costs of changes in diet and neutralization of allelochemicals. *Entomologia Experimentalis et Applicata*, **24**, 489–493.

Self, L.S., Guthrie, F.E. & Hodgson, E. (1964). Adaptation of tobacco hornworm to ingestion of nicotine. *Journal of Insect Physiology*, **10**, 907–914.

Smith, A.E. & Secoy, D.M. (1975). Forerunners of pesticides in classical Greece and Rome. *Journal of Agricultural and Food Chemistry*, **23**, 1050–1056.

Snyder, M.J., Stevens, J.L., Andersen, J.F. & Feyereisen, R. (1995). Expression of cytochrome P450 genes of the CYP4 family in midgut and fat body of the tobacco hornworm, *Manduca sexta*. *Archives of Biochemistry and Biophysics*, **321**, 13–20.

Tabashnik, B.E. (1989). Managing resistance with multiple pesticide tactics: theory, evidence, and recommendation. *Journal of Economic Entomology*, **82**, 1263–1269.

Tallamy, D.W. (1986). Behavioral adaptations in insects to plant allelochemicals. In *Molecular Aspects of Insect–Plant Associations* (Ed. by L.B. Brattsten & S. Ahmad), pp. 273–300. Plenum Press, New York.

Terriere, L.C. (1968). The oxidation of pesticide: the comparative approach. In *The Enzymatic Oxidation of Toxicants* (Ed. by E. Hodgson), pp. 175–196. North Carolina State University Press, Raleigh, North Carolina.

Tinel, M., Belghiti, J., Descatoire, V., Amouyal, G., Letteron, P., Geneve, J. *et al.* (1987). Inactivation of human liver cytochrome P-450 by the drug methoxsalen and other psoralen derivatives. *Biochemical Pharmacology*, **36**, 951–955.

Tomita, T. & Scott, J.G. (1995). cDNA and deduced protein sequence of

CYP6D1: the putative gene for a cytochrome P450 responsible for pyrethroid resistance in a house fly. *Insect Biochemistry and Molecular Biology*, **25**, 275–283.

Tsukamoto, M. & Suzuki, R. (1964). Genetic analysis of DDT-resistance in two strains of the house fly, *Musca domestica. Botyu-Kagaku*, **31**, 1.

Wang, X-P. & Hobbs, A.A. (1995). Isolation and sequence analysis of a cDNA clone for a pyrethroid inducible cytochrome P450 from *Helicoverpa armigera. Insect Biochemistry and Molecular Biology*, **25**, 1001–1009.

Waterman, M.R., Gotoh, O., Coon, M.J. & Nelson, D.R. (1996). P450 superfamily: update on new sequences, gene mapping, accession numbers, and nomenclature. *Pharmacogenetics*, **6**, 1–41.

Weirich, G.R., Williams, V.P. & Feldlaufer, M.F. (1996). Ecdysone 20-hydroxylation in *Manduca sexta* midgut: kinetic parameters of mitochondrial and microsomal ecdysone 20-monooxygenases. *Archives of Insect Biochemistry and Physiology*, **31**, 305–312.

Wheeler, G.S., Slansky, F., Jr. & Yu, S.J. (1992). Laboratory colonization has not reduced constitutive or induced polysubstrate monoxygenase activity in velvetbean caterpillars. *Journal of Chemical Ecology*, **18**, 1313–1325.

Wheelock, G.D. & Scott, J.G. (1992). The role of cytochrome P450lpr in deltamethrin metabolism by pyrethroid-resistant and susceptible strains of house flies. *Pesticide Biochemistry and Physiology*, **43**, 67–77.

Wilkinson, C.F. & Brattsten, L.B. (1972). Microsomal drug metabolizing enzymes in insects. *Drug Metabolism Reviews*, **1**, 153–228.

Woo, W.S., Shin, K.H. & Lee, C.K. (1983). Effect of naturally occurring coumarins on the activity of drug metabolizing enzymes. *Biochemical Pharmacology*, **32**, 1800–1803.

Yu, S.J. (1995). Allelochemical stimulation of ecdysone 20-monooxygenase in fall armyworm larvae. *Archives of Insect Biochemistry and Physiology*, **28**, 365–375.

Yu, S.J., Berry, R.E. & Terriere, L.C. (1979). Host plant stimulation of detoxifying enzymes in a phytophagous insect. *Pesticide Biochemistry and Physiology*, **12**, 280–284.

Zangerl, A.R. & Berenbaum, M.R. (1993). Plant chemistry, insect adaptations to plant chemistry, and host plant utilization patterns. *Ecology*, **74**, 47–54.

Zumwalt, J.G. & Neal, J.J. (1993). Cytochrome P450 from *Papilio polyxenes*: adaptation to host plant allelochemicals. *Comparative Biochemistry and Physiology*, **106C**, 111–118.

Chapter 4

To leave or to stay, that is the question: predictions from models of patch-leaving strategies

P. Haccou,[1] M. Sjerps[2] and E. van der Meijden[1]

Summary

Many herbivores have patchily distributed resources, especially specialist feeders. They also often face strong competition from conspecifics. We consider optimal patch-leaving strategies for such situations. Traditionally, patch-leaving strategies are studied by means of rate maximization models, using the marginal value theorem. There are several reasons why this theorem cannot always be used for plant–herbivore interactions under competition. Firstly, the theorem is derived for individual patch depletion. Secondly, it applies to food intake, and cannot be used directly for egg laying in patches, since competing females affect the survival chances of each other's offspring. This implies that the payoff of a female is still affected by other's behaviour after she has left. Thirdly, there are many situations where individuals only encounter a few patches during their lifetime, e.g. in larval dispersal.

We present a general model that can be used for studying patch-leaving strategies in the presence of competition. When several competitors deplete a patch simultaneously, a dilemma can arise that is similar to a 'war of attrition': the first one(s) to leave obtain less than the highest possible gain; but when one or more individuals leave, those remaining obtain more than the gain in the environment. The model is based on a generalization of the

1 Institute of Evolutionary and Ecological Sciences, University of Leiden, PO Box 9516, 2300 RA Leiden, The Netherlands
2 Forensic Laboratory, Volmerlaan 17, 2288 GD Rijswijk, The Netherlands

war of attrition and is applied to (i) larval dispersal strategies, (ii) energy rate maximization and (iii) oviposition under competition.

Introduction

Plant size and population structure set a limit to available biomass for herbivores. As a consequence, many herbivore species face strong competition sooner or later. Aphids, for instance, often locally compete for good quality food (Dixon 1973; Dixon *et al.* 1993; Sequira & Dixon 1996). Specialist insect herbivores may lay eggs in large batches, which forces their larvae to compete for food when their numbers exceed the carrying capacity (Morris 1963; Baltensweiler *et al.* 1977; Chaplin 1980). Even during egg laying itself there is often competition between insects (Straw 1989a; Schoonhoven 1990; Dempster 1992). Resources are usually patchily distributed, especially for specialists, and in most cases one patch does not provide enough resource for all competitors (Straw 1989b, 1991; Dempster *et al.* 1995a,b; Briese 1996; Matter 1996; Dempster 1997). Thus, plants shape the behaviour of herbivorous insects: the question arises, when should competitors leave a patch and start searching for a new one?

Patch-leaving strategies are usually studied by means of optimal foraging models, where it is assumed that each individual maximizes its energy gain per time unit. In such models, it is furthermore assumed that individuals encounter many patches sequentially. In each patch, the individual gains food, but as the patch depletes, the energy gain per time unit in the patch decreases progressively. It is furthermore assumed that the expected energy gain per time unit in the environment is fixed. This expected gain rate equals the expected gain per patch divided by the average time spent in patches plus the expected travel time between patches. The well-known marginal value theorem (MVT) (Charnov 1976; Parker & Stuart 1976), states that when there is only one individual in a patch, the optimal time to leave is the moment that the energy gain rate in the patch drops below the expected energy gain rate in the environment. This can easily be understood, since individuals that stay in the patch after this time gain less energy per time unit than they can get in the environment. The MVT cannot, however, be used straightforwardly to predict patch-leaving strategies under competition. In fact, it has been shown to give wrong predictions in a number of such cases (see Sjerps & Haccou 1994b; and below).

The MVT was derived primarily for studying foraging for food. There are many instances in plant–herbivore interactions where the issue is not to find food, but rather to find suitable oviposition sites, e.g. in moths searching for host plants, or fruit flies searching for rotting fruit. In such cases, the gain from a patch (the 'payoff') is the expected number of offspring that will survive until adulthood. This changes the situation drastically, especially when there is competition, since the number of surviving offspring from a patch depends on the total number of eggs that are laid in a patch. This implies that the payoff for a female that has left a patch decreases when incoming females lay additional eggs, which affects optimal leaving strategies.   .

Finally, herbivorous insects may encounter only a few patches during their lifetime. For instance, when host plants are widely distributed insect larvae may feed on just two or three different host plants before pupation. An additional complication here is that there may be a time horizon (e.g. due to the end of a season, or the limited availability of (suitable) leaves, flowers, fruits or seeds), before which larvae have to pupate. Furthermore, dispersal may entail a high mortality risk, due to predation, desiccation and the risk of not finding a suitable new food patch. These factors are not considered in standard rate maximization models. Below we describe a model for larval dispersal strategies proposed by Sjerps and Haccou (1994a) that incorporates these factors.

Observations indicate that, when several individuals are depleting a patch, often not all of them leave at the same time. This is especially surprising in cases where early dispersers leave remaining individuals plenty of food, although risks of dispersal are high (e.g. mealybug crawlers, van Alphen *et al.* 1989; cinnabar moth larvae, Myers & Campbell 1976a,b; van der Meijden 1979; Crawley & Gillman 1989; aphids, several examples in Crawley 1983). Here we shall provide a general model (based on the results of Sjerps & Haccou 1994a,b) that can explain these phenomena and that can be used to study dispersal from patches by competing individuals in a variety of situations, including the ones described above. One basic factor in all these cases is that when several individuals are simultaneously depleting a patch, a dilemma can arise: the first one(s) that leave obtain less than the ones that stay, but once one or more individuals leave the remaining ones obtain more than the gain in the environment. This situation is similar to a 'war of attrition' (WOA) (Maynard Smith 1974). In the original WOA, two contestants compete for one indivisible resource. The one who stays the longest obtains the resource. In this situation, a fixed leaving time is never optimal, since an opponent who stays just a little bit longer always gets the resource. Parker and Stuart (1976) were the first to suggest that an extension of this model could be applied to the case of simultaneous patch depletion by several individuals. However, in order to do that, several generalizations of the model had to be made.

An important generalization, made by Bishop and Cannings (1978), was to consider the payoff difference between the stayer and leaver in a contest as a function of the investment, instead of as a constant. Haigh and Cannings (1989) extended this to games between more than two individuals. Finally, Sjerps and Haccou (1994a,b) extended these models in such a way that patch leaving under competition can be studied. We here give an outline of their main results and apply those to the problems outlined above, ordered on the basis of their mathematical complexity, i.e. larval dispersal when there is a time horizon, energy gain rate maximization and oviposition.

Mathematics: model and general results

We here give a model that can be used for a wide variety of situations where dispersal induced by competition may occur. For specific cases, certain adjustments in formulations have to be made (see later sections for examples), but the basic results that are summarized in this section can be used generally.

Consider a group of individuals depleting a resource together. All choose a hypothetical leaving time. The one who chooses the smallest leaving time will be called the 'leaver'. Those remaining (the 'stayers') can choose whether to stay or to leave too, depending on which option gives the highest expected payoff. Note that we do not assume that all individuals have different leaving times. Several individuals may leave at the same time (although formally we assume that they do not leave at exactly the same moment). Note also that we do not require that any individual leaves before the resource is depleted. Let s be the time at which the resource is depleted if all individuals stay. If all chosen leaving times are larger than s, everyone stays until the resource is depleted.

The evolutionarily stable strategy (ESS) can be derived by considering two payoff functions: $L(t)$ is the payoff of the leaver and $W(t)$ is the payoff of the stayers, if the leaver leaves at time t. To illustrate the interpretation of these two functions, consider the situation with two competitors, say A and B. If A chooses hypothetical leaving time x and B chooses leaving time *y*, then there are three possible situations: (i) if *x* is smaller than *y*, A is the leaver and gets payoff $L(x)$ and B gets $W(x)$; (ii) if *x* is larger than *y*, A is the stayer and gets $W(y)$, whereas B gets $L(y)$; (iii) if *x* equals *y*, A and B each have an expected payoff of $(W(x)+L(x))/2$.

The exact form of the payoff functions L and W depends on the specific situation that is studied. However, some general considerations can be made about the form of the payoff functions. $W(t)$ must be at least as large as $L(t)$, since the stayers have more options. Furthermore, $W(t)$ decreases with *t*, due to resource depletion and, in some cases, the presence of a time horizon. $L(t)$ can have different shapes, depending on the problem under consideration. Here we will only consider cases where $L(t)$ initially increases and then decreases, or where $L(t)$ decreases from the start. More general shapes are considered by Sjerps and Haccou (1994a).

Sjerps and Haccou (1994a) calculated ESS leaving strategies. The general rules for such strategies are summarized below, together with intuitive reasons for the results.

1 Individuals should never leave at times *t* when $W(s)$ is larger than $L(t)$, since they can get more by staying until the end.

2 Individuals should not leave as long as $L(t)$ increases, since they can get more by staying longer.

3 At the earliest moment where $W(t)$ equals $L(t)$ at least one individual should leave, because after this time the payoff is always lower. The stayers can choose to leave or stay, depending on the biological problem under consideration.

4 Once $L(t)$ decreases and is larger than $W(s)$ but smaller than $W(t)$, the leaver gets less than the stayers, but the payoffs of the leaver as well as the stayers decrease with *t*. Here a WOA occurs, because the individuals that stay the longest have a higher payoff. In such periods a stochastic leaving strategy is optimal.

To illustrate these rules we sketch four qualitatively different situations in Fig. 4.1. In situation (a) $W(s)$ is always larger than $L(t)$, so all should stay until the end (rule 1). In situation (b) the maximum of $L(t)$ is larger than $W(s)$, but $L(t)$ becomes smaller than $W(s)$ at some later time. In that case, none should leave before $L(t)$ reaches its maximum (rule 2), or after the time that $L(t)$ becomes less than $W(s)$ (rule 1). In the remaining period there is a WOA (rule 4). There is a positive chance that all individuals are still present at the end of this period, in which case they all stay until time *s*. In situation (c), there is a WOA from the point at which $L(t)$ reaches a maximum until the time that $L(t)$ equals $W(t)$ (rules 2 and 4). If all are still

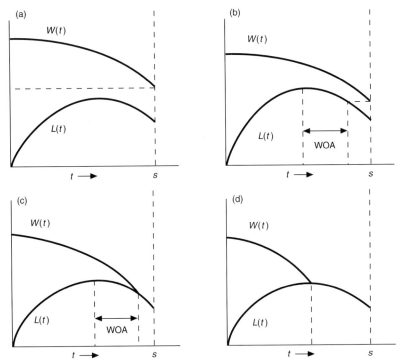

Figure 4.1 Different possible combinations of payoff functions of leavers ($L(t)$) and stayers ($W(t)$), where t is the leaving time of the leaver. At time s the resource is depleted if all individuals stay. In periods indicated by WOA, a war of attrition occurs.

present at the end of that period, at least one should leave (rule 3). In situation (d) at least one should leave when $L(t)$ equals $W(t)$ (rule 3).

In an evolutionary context, the relevant issue is not the payoff of individuals, but the payoff of a strategy, i.e. the expected payoff per individual in a population that uses the strategy. These can be calculated straightforwardly from the functions L and W. If the population chooses leaving times according to a probability distribution $Q(y)$, the expected payoff to an individual who chooses leaving time x in that population is:

$$E(x,Q) = \int_{0}^{x} W(y)\,dQ(y) + L(x)\int_{x}^{\infty} dQ(y) + \frac{1}{2}(L(x)+W(x))P_Q(x=y) \qquad (1)$$

90

where $P_Q(x = y)$ denotes the probability that the chosen leaving times are equal. This expected payoff is constant if $Q(y)$ corresponds to the ESS leaving strategy described above: it equals $W(s)$ in case (a) and the maximum of $L(t)$ in the other cases (proof: see Bishop & Cannings 1978; Haigh & Cannings 1989; Sjerps & Haccou 1994a,b).

Note that the value of t at which $L(t)$ reaches a maximum is the time at which the expected payoff in the environment equals the payoff in the patch. As we will show, situations like that in Fig. 4.1c can occur in rate maximization models. This implies that the mean leaving time can be larger than the value predicted by the MVT.

During periods in which a stochastic leaving strategy is optimal, the leaving tendency (chance per time unit of leaving) equals:

$$\lambda(t) = \frac{-L'(t)}{W(t) - L(t)} \qquad (2)$$

Note that when a leaver leaves, the potential leaving times of the stayers can no longer be observed, i.e. they are censored. For instance, an observation of one individual leaving a group of 10 corresponds to observing one leaving time plus nine censored observations. To analyse such data, for instance to test whether the observed leaving strategy corresponds to a predicted strategy, survival analysis methods have to be used, since these take censors into account (see e.g. Kalbfleisch & Prentice 1980).

Larval dispersal

Although many insects lay their eggs singly (e.g. the majority of British butterflies (Crawley 1983)) some species produce batches of eggs. Some of these latter species adjust the number of eggs in a batch to available biomass, or select plants or plant species that provide enough food for all eggs to develop. In other species no such relationship between egg number and plant size can be observed. Crawley (1983) mentions that in Lepidoptera the habit of laying eggs in batches correlates with unpalatability and aposomatic colouration of larvae (Fisher 1930), with reduced mortality by parasitoids (Dowden 1961) and with associations between caterpillars and ants (Kitching 1981). Laying eggs in batches in these species seems to be positively related to larval survival at early stages, since it can offer protection against predators and parasitoids.

At later stages, however, such behaviour may have disadvantages since local food sources may become exhausted. Examples can be found in the butterfly *Melitaea harrisii* that feeds on *Aster umbellatus* (Dethier 1959), the cinnabar moth *Tyria jacobaeae* on ragwort *Senecio jacobaea* (van der Meijden & van der Veen-van Wijk 1997) and several other lepidopteran species (Rausher *et al.* 1981; Damman 1991). In these species defoliation of the food plant that was selected for oviposition is almost a rule. *Melitaea* lays up to 400 eggs per batch and *Tyria* up to 150. These numbers are far too high to allow the caterpillars to develop on the selected plant.

Local resource depletion is usually accompanied by dispersal of individuals to find a new resource. Crawley (1983) finds dispersal in larvae to be widespread among Lepidoptera and aphids following resource depletion or a change in quality of the food source. An evolutionarily relevant question is: when should an individual leave its current resource?

In some species, dispersal only seems to occur when the food plant is completely defoliated (e.g. *Pieris rapae*, Jones 1977; *Omphalocera munroei*, Damman 1991). In others, however, some individuals leave before this occurs. In many such cases, they leave remaining individuals enough food to survive until adulthood or pupation. The problem can be nicely illustrated with the behaviour of crawlers of the cassave mealybug. These parthenogenic offspring can choose to settle on the plant where they are born, or to leave the plant and settle elsewhere. Dispersal in this species is inactive by wind (Nwanze *et al.* 1979) and the majority of dispersing crawlers will die before they have reached a new food plant. Although one would expect these crawlers to go on feeding on their first food plant until it is depleted as a resource, many crawlers appear to leave a plant long before it seems to be exhausted (van Alphen *et al.* 1989). Also in aphids we find similar examples of individuals leaving host plants before they (appear to) become unsuited for further exploitation (Crawley 1983; P. Harrewijn, personal communication, 1997). Other examples are found in Lepidoptera that produce large batches of eggs (Dempster 1971; Myers & Campbell 1976a,b; van der Meijden 1976; Crawley 1983; Crawley & Gillman 1989). Here, too, dispersal usually involves huge risks, due to increased risk of attack by ground predators, such as ants (van der Meijden 1973; Haccou & Hemerik 1985), and desiccation and starvation risks (van der Meijden 1973; Crawley & Gillman 1989; Damman 1991).

Since the behaviour of early dispersers is counterintuitive, we have sought a functional explanation for this phenomenon. The question to be answered is: why is there variation in patch-leaving time and why do some individuals leave before their resource is depleted, incurring a high risk of mortality either by natural enemies or by not finding another resource?

We now apply the model for competitive patch-leaving games discussed above to the analysis of dispersal strategies when two larvae are playing the game. First, we consider the shape of the payoff functions for winner and leaver in this particular case. Apart from dispersal risks, an important factor which can affect the shape of the payoff functions is that larvae may be constrained by a time horizon T, which corresponds to 'the end of the season', through cessation of the availability of suitable food, temperature conditions, etc.

The shape of $L(t)$ is determined by the predation risk and weight loss during dispersal, the expected weight gain from a new plant and the time left to search for a new plant before the season ends. $L(t)$ will increase initially, since as larvae get larger, their searching efficiency improves (Rausher 1979; Damman & Feeny 1988; Damman 1991) and predation risk during dispersal decreases. Furthermore, since the endurance of larvae increases with weight, their effective search time will be larger. After a while, however, $L(t)$ will decrease, due to the time horizon, T, which limits the search time as well as the time that is left to eat from a new plant. Furthermore, the density of surrounding host plants will decrease in the course of the season, due to herbivory by conspecifics (or other species). Based on these considerations, we can assume that $L(t)$ will have a shape as in Fig. 4.1, starting at zero and initially increasing, while decreasing later.

The shape of $W(t)$ is largely determined by plant nutrient content. If the leaver leaves early, the other larva may have enough food until time T. Thus, $W(t)$ may be constant at first. After a while, however, $W(t)$ will decrease, since the later the leaver leaves, the less time and food there will be left for the stayer. Thus, the shape of $W(t)$ will be as in Fig. 4.1. Note that, once the leaver has left, the stayer can choose between the options (i) leave too, (ii) leave at a later time or (iii) eat all the remaining food and then pupate or disperse. Since the stayer can choose the option with the highest gain, its payoff is never less than that of the leaver. Furthermore, note that $W(s)$ does not refer to obtaining extra food, because the plant is defoliated at that time. Larvae either pupate or disperse between s and T, and $W(s)$ corresponds to the maximum payoff of these two options.

We can now make some general predictions about dispersal. First of all, it is now clear why larvae sometimes disperse while leaving enough food for remaining larvae. This can occur whenever the ultimate payoff if both stay until complete defoliation is less than the maximum payoff to the leaver, while the payoff of stayers is larger at the time when the leaver's payoff is maximal (i.e. in situations as given in Fig. 4.1b, c). Furthermore, we can conclude that dispersal will only occur when $L(t)$ decreases. Thus, we should expect dispersal in periods where predation risk increases, surrounding host plant density decreases or when weather conditions deteriorate. This seems counterintuitive but it can be understood from the results presented above, since, as long as conditions for leavers improve with time it is better to wait a bit longer before dispersing. As soon as conditions deteriorate, however, a critical situation occurs since the longer they wait, the worse it becomes.

Changes in plant quality (i.e. biomass or nutrient content) will shift $W(t)$ up or down. When host plant quality is high, we will be in a situation as in Fig. 4.1a, and both larvae should stay until the plant is defoliated. As plant quality decreases, we will approach the situation of Fig. 4.1b, where there is a stochastic leaving strategy during a certain period, and a positive chance that both larvae stay until time s. A further decrease in plant quality will produce the situation of Fig. 4.1c, with a stochastic leaving strategy during which at least one larva should leave. Note that the dispersal tendency (see Eq. 2) also increases, since the difference between $L(t)$ and $W(t)$ decreases. Finally, we reach the situation of Fig. 4.1d, where at least one larva leaves at a fixed time. If plant quality also affects the rate at which $W(t)$ decreases, all these effects are enhanced. To summarize, variations in plant quality can determine whether leaving time is fixed or stochastic. Furthermore, it can effect the dispersal tendency and the ends of dispersal periods. Plant quality does not affect the beginnings of migration periods, since these are determined by the shape of $L(t)$.

Note that vertical shifts in $L(t)$ and $W(t)$ are equivalent: shifting $L(t)$ upwards has the same effect as shifting $W(t)$ downwards. Thus, for instance, an overall decrease in predation risk or surrounding vegetation will have the same effects as a decrease in plant quality. Time varying factors may change the starts of dispersal periods, e.g. if predation risk increases sooner in the season, the maximum of $L(t)$ will shift to the left and thus dispersal should start sooner. Furthermore, such factors can change the rate at which $L(t)$ decreases and therefore may affect the dispersal tendency as well as the end of dispersal periods.

Since larvae on a plant are usually closely related, stayers and leavers may profit to some extent from each other's payoffs, through their inclusive fitness. This alters the payoff functions. Sjerps and Haccou (1994a) examined this by replacing the payoff functions $L(t)$ and $W(t)$ by respectively:

$$L_r(t) = L(t) + rW(t)$$
$$W_r(t) = W(t) + rL(t)$$

(3)

where r is the coefficient of relatedness ($0 < r < 1$). They show that relatedness between the larvae increases the duration of dispersal periods as well as the dispersal tendency. Furthermore, since $W(t)$ decreases, it may sometimes occur that $L_r(t)$ decreases even though $L(t)$ increases. This implies that when individuals are closely related, dispersal can also occur when the payoff of leavers (without taking relatedness into account) increases. In the extreme case where individuals are genetically identical, $r = 1$ and the payoffs $L_r(t)$ and $W_r(t)$ are equal. We are then in a totally different situation, since there is no competition. Then, all individuals should leave at the same time, namely when $L(t)$ plus $W(t)$ is maximal. Such extreme situations may sometimes occur in aphids when a group consists of one clone. Note that effects of relatedness can only be modelled in this way under specific assumptions, such as fair meiosis, additive gene action, weak selection, diploid genetics and non-overlapping generations (see Taylor 1988).

When predation risks and weather conditions fluctuate, the shapes of $L(t)$ and $W(t)$ may be less simple. In such cases, these functions may have local maxima. Sjerps and Haccou (1994a) derived leaving strategies for such, more general, shapes of $L(t)$ and $W(t)$. Their main conclusions are similar to those given above. The main difference is that for less simple payoff functions there can be several dispersal periods, separated by gaps during which no dispersal occurs. Such gaps correspond to periods in which $L(t)$ increases.

When there are more than two larvae on a host plant, the process proceeds in stages; each time one individual leaves, a new stage begins and the payoff functions of stayers and leavers depend on the stage. Thus, the leaving strategy becomes conditional on the number of larvae remaining and the amount food that is left the moment that leavers leave. The general considerations concerning effects of plant quality, predation risks, etc., that are summarized in this section will remain valid. However, to get more detailed predictions about leaving strategies we will need to specify effects of those factors on the payoff functions. In the following section we will give an example of an *n*-player game in a slightly different context, namely competitive patch depletion with sequential patch visits.

n-player games: rate maximization under competition

Above, we considered situations in which mobility was relatively low, and lifespan short compared with the travel time between patches, so that individuals encounter only a few patches during their lifetime. When herbivores are relatively mobile and/or have a long lifespan, they can encounter many food patches. Examples are, for instance, migratory locusts and herbivorous beetle species, such as the flea beetle *Longitarsus jacobaeae* (E. van der Meijden, personal observations). In such cases, it can be assumed that foraging behaviour is aimed at maximizing expected energy gain per time unit. The MVT states that the optimal time to leave is the moment the gain rate in the patch drops below the average gain rate in the surrounding habitat. This theorem is derived under several assumptions. One of those assumptions is that there is only one individual in the patch. It has often been presumed that the MVT can also be used when several individuals are depleting a patch simultaneously, for example by Bernstein *et al.* (1988, 1991), Lucas and Schmid-Hempel (1988), and Visser (1991). However, this is not necessarily true, since each individual's optimal time to leave may depend on the behaviour of the others. Sjerps and Haccou (1994b) derived optimal leaving strategies when there is competition for resources. They found that the MVT does not always predict (average) optimal leaving times accurately. We will briefly summarize their results here.

Following the approach of McNamara and Houston (1987a), we consider a particular patch, and assume that after the animals have left this patch, they return to a background foraging process, which yields a fixed average energy gain per time unit, denoted by γ. Let n denote the number of individuals remaining in the patch: at stage n, there are n individuals in the patch, once one has left we proceed to stage $n-1$, etc. The leaving strategy depends on the number of remaining individuals, n, and the amount of resource at the beginning of the stage. Let t be the time since the beginning of stage n. The payoff function of the leaver corresponds to the difference between the energy gain in the patch up to time t and the expected energy gain in the environment during a period of time t (which equals γt). This payoff function can have two possible shapes, depending on the initial gain rate in the patch.

1 At the start of stage n, the gain rate in the patch is smaller than γ. In this case, individuals that leave the patch immediately have a payoff of zero, so $L(0)=0$. Furthermore, the payoff of the leaver decreases with t, since the gain rate in the patch decreases as the patch gets depleted.

2 At the start of stage n, the gain rate in the patch is larger than γ. In this

case the payoff of the leaver increases until the time at which the gain rate in the patch equals γ, denoted by t^*. After this time the payoff of the leaver decreases.

Once the leaver has left, the stayers choose the optimal leaving strategy corresponding to stage $n-1$. Sjerps and Haccou (1994b) proved that the stayers of stage n get the payoff of the leaver of stage n plus the maximum payoff of the leavers of stage $n-1$. Therefore, the payoff of stayers depends on the situation in stage $n-1$. Here, again, there are two possibilities:

I At the start of stage $n-1$, the gain rate in the patch is smaller than γ. If this occurs, the stayers of stage n do not have any advantage when the leaver leaves. Thus, the payoff of the stayers in stage n equals the payoff of the leaver.

II At the start of stage $n-1$, the gain rate in the patch is larger than γ. In that case, the stayers of stage n can get an additional payoff in stage $n-1$. Thus, the payoff of the stayers in stage n is larger than the payoff of the leaver.

Whether we are in situation I or II depends on the resource that is left at the beginning of stage $n-1$. Thus, it depends on the time at which the leaver leaves in stage n.

There are four possible ways in which situations 1 or 2 can be combined with I or II, which are illustrated in Fig. 4.2. Note that these are all similar to the situations in Fig. 4.1c and d. Situations like those in Fig. 4.1a or b cannot occur here, since the payoffs of leavers and stayers are always equal once the patch is depleted. In situation (a), the gain rate in the patch at the beginning of stage n is larger than γ. At t^* the gain rate in the patch equals γ, but if one of the individuals leaves at this time the gain rate at the start of stage $n-1$ is larger than γ. At t' the gain rate at the start of stage $n-1$ equals γ, so from that point on the payoff of the stayers of stage n becomes equal to that of the leaver. From the results discussed above it follows that in this case none should leave before t^*. In the period from t^* until t' a WOA occurs, so there is a stochastic leaving strategy. In situation (b), the gain rate in the patch at the beginning of stage n is smaller than γ, but if one of the individuals leaves right away the gain at the start of stage $n-1$ is larger than γ. Here, there is a WOA in the period from $t=0$ until t' (the point at which the gain rate at the start of stage $n-1$ equals γ). In situation (c), the gain rate in the patch at the beginning of stage n is larger than γ until t^*, where the gain rate in the patch equals γ. However, at this point the gain rate at the beginning of stage $n-1$ is also less than or equal to γ. In situation (d), the gain rate at the beginning of stage n as

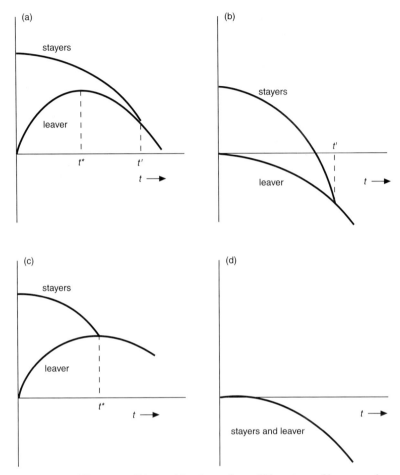

Figure 4.2 Different possible combinations of payoff functions of leavers and stayers in the case of rate maximization under competition.

well as the gain rate at the beginning of stage $n-1$ is smaller than or equal to γ.

In situations (a) and (b), there is a positive chance that all individuals stay until t'. In that case, several individuals leave at t', until the situation is such that the gain rate of remaining individuals is larger than γ if one of them leaves (so that we are back in situations (a) or (b)). Similarly, in situations (c) and (d), several individuals should leave at respectively t^* and $t=0$, until we are back in a stage where situations (a) or (b) occur.

Whether or not all these situations can actually occur depends on the way the gain rate in the patch changes with the number of individuals, n. When there is no interference, the number of individuals in a patch only determines the rate at which food is depleted. In that case, the departure of one (or more) individuals makes the gain rate in the patch decrease more slowly. It does not increase the gain rate of the remaining ones, as in situations (a) or (b). Therefore, once the gain rate in the patch has dropped below γ, it stays below γ, irrespective of whether or not individuals leave. Therefore we are initially in situation (c) and at following stages always in situation (d). This implies that all individuals should leave at the time when the gain rate has dropped to γ, which is equivalent to the MVT.

When there is interference, all situations in Fig. 4.2 can occur, so there are periods with stochastic leaving strategies. Note that when this happens the expected leaving time is larger than the value predicted by the MVT (which equals t^*).

The leaving process proceeds according to a semi-Markov process: as soon as one individual leaves the situation is adjusted. The leaving strategy depends on how many individuals are currently in the patch and how much resource is left, not on past events. Survival analysis methods (see e.g. Kalbfleisch & Prentice 1980) can be used to test such predictions.

The main conclusions of this section are that as long as there is no interference, the MVT accurately predicts the optimal leaving time. As soon as there is interference, however, this is only true in situations where the optimal strategy is a fixed leaving time (as in Fig. 4.2c,d). When stochastic leaving strategies are optimal, the MVT leaving time corresponds to the time t^* at which the period of a WOA starts. In such cases, average leaving times can be expected to be much longer than the MVT leaving time. In general, leaving tendencies increase with the amount of interference, since that will increase the difference between the payoff functions of leaver and stayers (cf. Eq. 2).

Oviposition under competition: stealing each other's resources

At first sight, the situation where several insects are laying eggs on patches of host plants or fruits may seem identical to the n-player game discussed above. However, this is not true, as the payoffs are now related to the expected numbers of offspring from the patch. The survival chance of offspring depends on how many eggs have been laid in the patch. Therefore, stayers can 'steal' resource from leavers after they have left. This complicates the case considerably. Here we will provide an outline of how such situations can be studied with our approach. As above, we will assume that after the animals have left this patch they return to a background foraging process, which yields a fixed average rate of gain γ.

Consider two females laying eggs in the same patch. Let t be the time since they arrived in the patch. (For the sake of simplicity we will assume that they arrived simultaneously.) The payoff of the leaver as well as the stayer now depends on the additional number of eggs that the stayer lays after the leaver has left. Thus, it depends on the additional time that the stayer spends in the patch. If the stayer stays for y additional time units, the payoffs are:

$$L(t,y) = x_1(t,y) - \gamma t$$
$$W(t,y) = x_2(t,y) - \gamma(t+y)$$

(4)

Here $x_1(t,y)$ will decrease in y, due to the additional eggs that the stayer lays, and $x_2(t,y)$ may or may not increase initially in y, depending on how much resource is left at time t. If y equals zero, both leave at the same time and the payoffs will be equal, so $x_1(t,0)$ equals $x_2(t,0)$.

Once the leaver has left, the stayer will choose y in such a way that she gets a maximal payoff, so the payoff functions are $L(t,y^*)$ and $W(t,y^*)$, where y^* is the value of y that maximizes $W(t,y)$. Note that this value will depend on t. At large values of t, there will not be much resource left, and then the cost of spending additional time in the patch (i.e. γy) will be larger than the gain. Thus, there will be a time t' after which y^* is zero. From that time on the payoffs of leaver and stayer are equal. When the patch offers an initial gain rate that is larger than γ, however, the stayer gets more than the leaver for low values of t.

The results of the basic model can be applied directly to this case, if $L(t)$ and $W(t)$ are replaced by $L(t,y^*)$ and $W(t,y^*)$. In principle, all situations given in Fig. 4.2 can occur here, depending on the shape of the functions x_1 and x_2. The situation in Fig. 4.2d corresponds to the case where the initial gain rate in the patch is less than γ. In that case both should leave immediately.

Note that the MVT leaving time is the value of t at which $L(t,0)$ is maximal, since this is the time at which for both individuals the gain rate in the patch is equal to the gain rate in the environment. This is not necessarily the same as the value of t^*, where $L(t,y^*)$ is maximal. Therefore, there is an essential difference with the results of the n-player game above. When situations like that in Fig. 4.2a occur, the MVT leaving time does not always correspond to the time at which the WOA begins. Similarly, whereas in *n*-player games optimal fixed leaving times (Fig. 4.2c) correspond to the MVT leaving time, this is not necessarily true when competitors can steal each other's resources. Effects of the functions x_1 and x_2 on leaving strategies are currently under study (P. Haccou *et al.*, unpublished data).

Discussion

We have presented several examples of patch leaving under competition. The main conclusion is that competition may induce stochastic leaving strategies. This has several counterintuitive implications. Firstly, even though dispersal is risky, some individuals may leave a patch while others remain with plenty of food. Secondly, dispersal only occurs in periods where the expected payoff of dispersing decreases. As a consequence, individuals may remain in patches even though the environment offers a higher expected gain.

The basic model can be applied to a variety of situations by making various adjustments. In this paper, we give three examples of such applications. There are, however, more possibilities. For instance, the 'larval dispersal' model can be generalized to consider situations with more than two individuals and sequential patch visits. The '*n*-player game' and 'oviposition under competition' models can be generalized to the situation where the expected gain rate in the environment is not fixed, but, for instance, varies in time, or is affected by the leaving strategy itself (Sjerps & Haccou 1994b). Also, a discounting factor may be included in these models, to account for mortality risks during travel between patches (Schoener 1971). Note that the 'oviposition under competition' model can also be used for studying patch-leaving strategies in superparasitizing larval parasitoids.

Although *t* is the time during which the competitors are in the patch together, this does not necessarily imply that they should arrive simultaneously. It implies that *t* is the time since the latest arrival. However, non-simultaneous arrival does have consequences for our model and results that we have not yet discussed. Firstly, there may be asymmetries between competitors. This can occur in the 'larval dispersal' case, where such asymmetries may involve differences in age and weight of the larvae, and in the 'oviposition competition' case, since previously arrived females have already laid eggs in the patch whereas later arrivals have not. In the 'rate maximization' case, however, non-simultaneous arrival does not cause asymmetries, since there the payoff only depends on the status quo. Here, a new arrival only implies that the situation shifts from stage n to stage $n+1$ and the optimal leaving strategy is adjusted accordingly.

Asymmetries may change the ESS strategy. Selten (1980) showed that when competitors perceive the asymmetries correctly only pure optimal strategies can exist. Hammerstein and Parker (1982) showed that there may be mixed optimal strategies in the WOA if there are errors in perception of asymmetries. Generalizations like these still need to be developed for our model.

A second consequence of non-simultaneous arrivals is that additional competitors may arrive before the leaver has left. Leaving strategies may be adjusted to this possibility. Until now, we have not considered this possibility.

In principle, the models can be extended to account for these consequences of non-simultaneous arrival of competitors. The leaving strategy then becomes conditional on the resource in the patch at the moment that later competitors arrive and, in the case of asymmetries, the states of the competitors. This can be modelled in a similar way to the n-player game, where the strategy is conditional on the resource and the stage of the game. Furthermore, the possibility of additional arrivals before the leaver has left can be taken into account in the calculation of the payoff functions. However, calculation of the optimal strategies for such cases may be tedious and involves additional theoretical work.

Finally, in oviposition competition, the payoff of a female may be affected by competitors that arrive after she has left. Females may anticipate this, and adjust the number of eggs that they lay accordingly. This is a considerable complication, which remains to be studied. Sjerps and Haccou (1993) studied this situation for the simplest case, where only two females compete.

In our model several assumptions are made concerning the information available to competitors. Specifically, it is assumed that the payoff functions are known to competitors or that they can get an accurate impression of those functions.

Impressions about the payoffs of stayers can be acquired through encounters with competitors. For instance, larvae growing up on the same host plant might have information about the number of competitors. There are indications that this is indeed the case for cinnabar moth larvae (van der Meijden 1976; Crawley & Gillman 1989; Sjerps 1994). Sometimes, rough estimates of the numbers of competitors might be available, e.g. moths or fruit flies laying eggs in the same patch may react to each other's oviposition-deterring pheromones (Straw 1989a; Schoonhoven 1990; Dempster 1992). On the other hand, it is not always necessary for individuals to have information on the numbers of competitors. For instance, in the *n*-player game, individuals only have to react to their current gain rate in a patch and the amount of interference. It is reasonable to assume that they get a fairly good impression of those.

We also assumed that the payoff of leavers is known. For larval dispersal this implies knowledge about environmental factors such as dispersal risks. This assumption can only be expected to be valid if environments have been stable over a large number of generations (so that there has been some selection), or if there is information about their current values. Such information might be available. For instance, weather conditions may give indications about the expected chances of finding new food plants or predator activity. Furthermore, in areas with ants, predator activity might be observable to larvae since ants often crawl over larvae when they are on plants, without attacking. On the other hand, the shapes of the larval dispersal payoff functions we considered are quite general: payoffs will always decrease towards the end of the season, and the end of the season is a fairly constant feature over the generations or may be predictable on the basis of weather conditions or changing photoperiod. The '*n*-player game' and 'oviposition under competition' models assume the environmental gain rate is known. Individuals may be able to 'estimate' such gain rates after many patches have been visited (see e.g. McNamara & Houston 1987b).

Whereas the assumptions may not always be satisfied, the models do give an indication of the type of phenomena to expect. Especially, they provide an explanation for seemingly counterintuitive observations on dispersal under competition. The results indicate that leaving tendencies should be affected by experiences in a patch, such as current individual gain rate and encounters with others. The effect of such factors on leaving tendencies can be examined through survival analysis methods as done by, for example, Sjerps (1994) and, in a different context, by Haccou *et al.* (1991) and Hemerik *et al.* (1993).

Acknowledgements

We thank Chris Cannings, Jacques van Alphen and Olivier Glaizot for stimulating discussions and suggestions, especially with respect to modelling oviposition competition.

References

Baltensweiler, W., Benz, G., Bovey, P. & Delucchi, V. (1977). Dynamics of the larch budmoth populations. *Annual Review of Ecology and Systematics*, **22**, 79–100.

Bernstein, C., Kacelnik, A. & Krebs, J.R. (1988). Individual decisions and the distribution of predators in a patchy environment. *Journal of Animal Ecology*, **57**, 1007–1026.

Bernstein, C., Kacelnik, A. & Krebs, J.R. (1991). Individual decisions and the distribution of predators in a patchy environment II. The influence of travel costs and structure of the environment. *Journal of Animal Ecology*, **60**, 205–225.

Bishop, D.T. & Cannings, C. (1978). A generalized war of attrition. *Journal of Theoretical Biology*, **70**, 85–124.

Briese, D.T. (1996). Oviposition choice by the *Onopordum* capitulum weevil *Larinus latus* (Coleoptera: Curculionidae) and its effect on the survival of immature stages. *Oecologia*, **105**, 464–474.

Chaplin, S.C. (1980). An energetic analysis of hostplant selection by the large milkweed bug, *Oncopeltus fasciatus*. *Oecologia* **46**, 254–261.

Charnov, E.L. (1976). Optimal foraging: the marginal value theorem. *Theoretical Population Biology*, **9**, 129–136.

Crawley, M.J. (1983). *Herbivory, the Dynamics of Plant–Animal Interactions*. Blackwell Science Ltd, Oxford.

Crawley, M.J. & Gillman, M.P. (1989). Population dynamics of cinnabar moth and ragwort in grassland. *Journal of Animal Ecology*, **58**, 1035–1050.

Damman, H. (1991). Oviposition behaviour and clutch size in a group-feeding pyralid moth *Omphalocera munroei*. *Journal of Animal Ecology*, **60**, 193–204.

Damman, H. & Feeny, P. (1988). Mechanisms and consequences of selective oviposition by the zebra swallowtail butterfly. *Animal Behaviour*, **36**, 563–573.

Dempster, J.P. (1971). The population ecology of the cinnabar moth, *Tyria jacobaeae* L. (Lepidoptera, Arctiidae). *Oecologia*, **7**, 27–67.

Dempster, J.P. (1992). Evidence of an oviposition-deterring pheromone in the orange-tip butterfly, *Anthocharis cardamines* (L.). *Ecological Entomology*, **14**, 443–454.

Dempster, J.P. (1997). The role of larval food resources and adult movement in the population dynamics of the orange-tip butterfly (*Anthocharis cardamines*). *Oecologia*, **111**, 549–556.

Dempster, J.P., Atkinson, D.A. & Cheesman, O.D. (1995a). The spatial population dynamics of insects exploiting a patchy food resource. I. Population extinctions and regulation. *Oecologia*, **104**, 340–353.

Dempster, J.P., Atkinson, D.A. & French, M.C. (1995b). The spatial population dynamics of insects exploiting a patchy food resource. II. Movements between patches. *Oecologia*, **104**, 354–362.

Dethier, V.G. (1959). Egg-laying habits of Lepidoptera in relation to available food. *Canadian Entomologist*, **91**, 554–561.

Dixon, A.F.G. (1973). *Biology of Aphids*. Edward Arnold, London.

Dixon, A.F.G., Wellings, P.W., Carter, C. & Nichols, J.F.A. (1993). The role of food quality and competition in shaping the seasonal cycle in the reproductive activity of the sycamore aphid. *Oecologia*, **95**, 89–92.

Dowden, P.B. (1961). The gypsy moth egg parasite *Ocencyrus kuanai*, in southern Connecticut in 1960. *Journal of Economic Entomology*, **54**, 876–878.

Fisher, R.A. (1930). *The Genetic Theory of Natural Selection*. Clarendon Press, Oxford.

Haccou, P. & Hemerik, L. (1985). The influence of larval dispersal in the cinnabar moth (*Tyria jacobaeae*) on predation by the red wood ant (*Formica polyctena*): an analysis based on the proportional hazards model. *Journal of Animal Ecology*, **54**, 755–769.

Haccou, P., De Vlas, S.J., van Alphen, J.J.M. & Visser, M. (1991). Information processing by foragers: effects of intra-patch experience on the leaving tendency of *Leptopilina heterotoma*. *Journal of Animal Ecology*, **60**, 93–106.

Haigh, J. & Cannings, C. (1989). The n-person war of attrition. *Acta Applicata Mathematica*, **14**, 59–74.

Hammerstein, P. & Parker, G.A. (1982). The asymmetric war of attrition. *Journal of Theoretical Biology*, **96**, 647–682.

Hemerik, L., Driessen, G. & Haccou, P. (1993). Effects of intra-patch experiences on patch time, search time and searching efficiency of the parasitoid *Leptopilina clavipes*. *Journal of Animal Ecology*, **62**, 33–44.

Jones, R.E. (1977). Search behaviour: a study of three caterpillar species. *Behaviour*, **60**, 237–259.

Kalbfleisch, J.D. & Prentice, R.L. (1980). *The Statistical Analysis of Failure Time Data*. John Wiley & Sons, New York.

Kitching, R.L. (1981). Egg clustering and the southern hemisphere lycaenids: comments on a paper by N.E. Stamp. *American Naturalist*, **118**, 423–425.

Lucas, J.R. & Schmid-Hempel, P. (1988). Diet choice in patches: time-constraint and state-space solutions. *Journal of Theoretical Biology*, **131**, 307–332.

Matter, S.F. (1996). Interpatch movement of the red milkweed beetle, *Tetraopes tetraophthalmus*: individual responses to patch size and isolation. *Oecologia*, **105**: 447–453.

Maynard Smith, J. (1974). The theory of games and the evolution of animal conflicts. *Journal of Theoretical Biology*, **47**, 209–221.

McNamara, J.M. & Houston, A.I. (1987a). A general framework for understanding the effects of variability and interruptions of foraging behaviour. *Acta Biotheoretica*, **36**, 3–22.

McNamara, J.M. & Houston, A.I. (1987b). Memory and the efficient use of information. *Journal of Theoretical Biology*, **125**, 385–395.

Morris, R.F. (1963). The dynamics of endemic spruce budworm populations. *Memoirs of the Entomological Society of Canada*, **31**, 1–332.

Myers, J.H & Campbell, B.J. (1976a). Indirect measures of larval dispersal in the cinnabar moth, *Tyria jacobaeae* (Lepidoptera: Arctiidae). *Canadian Entomologist*, **108**, 967–972.

Myers, J.H. & Campbell, B.J. (1976b). Distribution and dispersal in populations capable of resource depletion—a field study on cinnabar moth. *Oecologia*, **24**, 7–20.

Nwanze, K.F., Leuschner, K. & Ezrumah, H.C. (1979). The cassava mealy bug, *Phenococcus manihoti*. *Tropical Pest Management*, **28**, 27–32.

Parker, G.A. & Stuart, R.A. (1976). Animal behaviour as a strategy optimizer: evolution of resource assessment strategies and optimal dispersal thresholds. *American Naturalist*, **110**, 1055–1076.

Rausher, M.D. (1979). Egg recognition: its advantage to a butterfly. *Animal Behaviour*, **27**, 1034–1040.

Rausher, M.D., Mackay, D.A. & Singer, M.C. (1981). Pre- and post-alighting host discrimination by *Euphydryas editha* butterflies: the behavioural mechanisms causing clumped distributions of egg clusters. *Animal Behaviour*, **29**, 1220–1228.

Schoener, T.W. (1971). Theory of feeding strategies. *Annual Review of Ecology and Systematics*, **2**, 360–404.

Schoonhoven, L.M. (1990). Host-marking pheromones in Lepidoptera with special reference to 2 *Pieris* spp. *Journal of Chemical Ecology*, **16**, 3043–3052.

Selten, R. (1980). A note on evolutionarily stable strategies in asymmetric animal conflicts. *Journal of Theoretical Biology*, **84**, 93–101.

Sequira, R. & Dixon, A.F.G. (1996). Life history responses to host quality changes and competition in the Turkey-oak aphid, *Myzocallis boerneri* (Hemiptera: Sternnorryncha: Callaphidae). *European Journal of Entomology*, **93**, 153–158.

Sjerps, M. (1994). *ESS's at different life stages, applied to the cinnabar moth.* PhD thesis, Leiden University.

Sjerps, M. & Haccou, P. (1993). Information determines the optimal clutch sizes of competing insects: Stackelberg versus Nash equilibrium. *Journal of Theoretical Biology*, **163**, 473–483.

Sjerps, M. & Haccou, P. (1994a). A war of attrition between larvae on the same host plant: stay and starve or leave and be eaten? *Evolutionary Ecology*, **8**, 269–287.

Sjerps, M. & Haccou, P (1994b). Effects of competition on optimal patch leaving: a war of attrition. *Theoretical Population Biology*, **3**, 300–318.

Straw, N. (1989a). Evidence for an oviposition-deterring pheromone in *Tephrites bardanae* (Schrank) (Diptera: Tephritidae). *Oecologia*, **78**, 121–130.

Straw, N. (1989b). The timing of oviposition and larval growth by two tephritid fly species in relation to host-plant development. *Ecological Entomology*, **14**, 443–454.

Straw, N. (1991). Resource limitation of tephritid flies on lesser burdock, *Arctium minus* (Hill) Bemh. *Oecologia*, **86**, 492–502.

Taylor, P.D. (1988). An inclusive fitness model for dispersal of offspring. *Journal of Theoretical Biology*, **130**, 363–378.

van Alphen, J.J.M., Neuenschander, P., van Dijken, M.J., Hammond, W.N.O. & Herren, H.R. (1989). Insect invasions: the case of the cassava mealy bug and its natural enemies evaluated. *The Entomologist*, **108**, 38–55.

van der Meijden, E. (1973). Experiments on dispersal, late-larval predation, and pupation in the cinnabar moth (*Tyria jacobaeae* L.) with a radioactive label (192Ir). *Netherlands Journal of Zoology*, **23**, 430–445.

van der Meijden, E. (1976). Changes in the distribution pattern of *Tyria jacobaeae* during the larval period. *Netherlands Journal of Zoology*, **26**, 136–161.

van der Meijden, E. (1979). Herbivore exploitation of a fugitive plant species: local survival and extinction of the cinnabar moth and ragwort in a heterogeneous environment. *Oecologia*, **42**, 307–323.

van der Meijden, E. & van der Veen-van Wijk, C.A.M. (1997). Tritrophic metapopulation dynamics. A case study of ragwort, the cinnabar moth and the parasitoid *Cotesia popularis*. In Metapopulation Biology: *Ecology, Genetics and Evolution* (Ed. by I. Hanski & M.E. Gilpin), pp. 307–405. Academic Press, San Diego.

Visser, M.E. (1991). *Foraging decisions under patch depletion: an ESS approach to superparasitism in solitary parasitoids*. PhD thesis, Leiden University.

Chapter 5

The evolution of direct and indirect plant defence against herbivorous arthropods

M.W. Sabelis, M. Van Baalen, F.M. Bakker, J. Bruin,
B. Drukker, M. Egas, A.R.M. Janssen, I.K. Lesna, B. Pels,
P.C.J. Van Rijn and P. Scutareanu[1]

Summary

Plant communities generally retain a green appearance despite the presence of herbivores. This may well be explained by the role of predators in suppressing the number of herbivores, but a critical evaluation of the role of plant defences is still lacking. This is not only because the relative importance of direct plant defences against herbivores is poorly known, but also because the impact of indirect plant defences has been largely underrated. In the last 10–15 years, it has become increasingly clear that plants can promote the effectiveness of the herbivore's enemies in various ways and that these so-called indirect defences are in fact widespread in the plant kingdom.

In this chapter we classify and review the various modes of indirect plant defence. In particular, we will put emphasis on experimental evidence for their impact on herbivore suppression at the population level. In addition we review possibilities for cheaters to make good use of the plant's facilities without returns or even to the detriment of the plant. Finally, we analyse the interaction between a plant, its neighbours, its herbivores and the herbivore's enemies as a game of defence (between neighbouring plants), escape (between herbivores) and resource exploitation (between herbivores and between predators). We show that the mobility of a herbivore and that

1 All authors at: Institute of Systematics and Population Biology, University of Amsterdam, Kruislaan 320, 1098 SM Amsterdam, The Netherlands

of its predator plays a crucial role in determining the extent to which a plant can reap the benefits from investing in direct or indirect defence.

In addition, we discuss how prevailing resource exploitation strategies may change due to metapopulation structure, dynamics and dispersal via its impact on whether local populations consist of single strains or are composed of various strains differing in the way they exploit their resource. We conclude that the interplay between plant defence, plant competition, exploitation of host plant and prey in tritrophic systems results in evolution towards low investment in plant defence. Natural selection may favour predators to overexploit local populations of herbivores, as long as overall herbivore densities do not become too low. This game–theoretic analysis needs to be extended to models of evolutionary and population dynamics in order to specify the conditions under which 'the world is green' hypothesis holds.

Introduction

Ecologists generally agree that under natural conditions plant communities retain a green appearance despite attacks by herbivores (Hairston *et al.* 1960; Strong *et al.* 1984). This has inspired two lines of thought: (i) plants defend themselves effectively against many herbivores and together exhibit such great diversity in defence mechanisms that 'super' herbivores mastering all plant defences did not evolve and those breaking defences of some plants are limited by the availability of these plants; (ii) predators suppress herbivore populations to very low levels. Hairston *et al.* (1960) tacitly ignored the first and emphasized the latter of the two in formulating their hypothesis on why 'the world is green'. Strong *et al.* (1984) more carefully considered both hypotheses in their review on the impact of herbivorous arthropods on plants, but also favoured the view that pre- dators have an overriding impact on herbivorous arthropods, thereby releasing plants from herbivore attack. However, the two explanatory mechanisms (plant defence vs. predator impact) may well act in concert. Ever since the seminal review paper by Price *et al.* (1980) there is a growing awareness among ecologists that plant defences include more than just trickery to reduce the herbivore's capacity for (population) growth; the plant may provide facilities to promote the foraging success of the herbivore's enemies. In this chapter we will focus on this mutualistic conspiracy between plants and predatory arthropods against herbivorous arthropods.

This form of defence is termed *indirect* as opposed to direct defence against herbivores. Examples of direct defences are plant structures that hinder (feeding by) the herbivore (cuticle thickness, 'smooth' cuticle surfaces that do not provide a holdfast, impenetrable masses of leaf trichomes, glandular trichomes acting as sticky traps) and secondary plant compounds that modify the quality of ingested plant food (digestion inhibitors), intoxicate the herbivore or signal the plant's well-defended state to 'discourage' the herbivore (feeding deterrents). Indirect plant defences bypass the direct defence route against the second trophic level by promoting the effectiveness of the third. These involve the provisioning of protection, food and alarm signals that appear to be utilized by the enemies of the herbivore. When Price *et al.* (1980) published their review paper, certain ant–plant interactions provided the best-known examples of indirect defences (Janzen 1966; Bentley 1977; see reviews by Beattie 1985 and Jolivet 1996). Now, almost 20 years later, it has become increasingly clear that mutualistic interactions between plants and predatory arthropods are widespread (Beattie 1985; Dicke & Sabelis 1988, 1989, 1992; Drukker *et al.* 1995; Turlings *et al.* 1995; Jolivet 1996; Walter 1996).

The aim of this chapter is to review the evidence for mutualism interactions between plants and predatory arthropods, and to provide a rationale for their evolution based on the theory of natural selection. We will argue that, to understand the evolution of mutualistic interactions between plants and the natural enemies of herbivorous arthropods, one should identify the advantages to the individual plant and the individual predator, predict consequences for the population dynamics of herbivorous and predatory arthropods, and elucidate how dynamics in turn may affect the evolution of the mutualistic interaction. This may provide an answer to the fundamental question whether plants optimize investments in direct and indirect defence, and to whether these optimal defensive allocations provide a rationale for 'the world is green' hypothesis.

Firstly, we briefly explore the complexities arising from the fact that the players in the tritrophic game operate on very different spatial and temporal scales. Secondly, we will introduce a simple model of *local* predator–prey dynamics on an individual plant based on specific scale assumptions and use this model to identify the main categories of defensive strategies of a plant. Thirdly, we review the empirical evidence for net benefits to individual plants from investing in different modes of indirect defence. Finally, we provide a game–theoretical analysis to identify evolutionarily stable strategies of exploitation (predator–herbivore, herbivore–plant), dispersal (predator, herbivore) and defence (plant) in tritrophic systems, with particular emphasis on the (direct and indirect) defence games neighbouring plants can play against each other.

Spatial and temporal scales of interaction

Plants and herbivorous and predatory arthropods are engaged in interactions with widely different temporal scales. Plants have usually much longer generation times than arthropods. Hence, plant population change tends to be slow relative to that of the arthropods. This is why models of arthropod predator–prey dynamics are usually decoupled from plant population dynamics by assuming a pseudo-steady state. The generation times of herbivorous and predatory arthropods may also differ, but are usually close enough to justify modelling as a ditrophic system (Hassell 1978; Sabelis 1992).

The spatial scale of predator–prey interaction is set by the distribution of herbivorous arthropods. Many herbivorous arthropods have strongly clumped distributions over their host plants. This may be the result of:

1 aggregation towards weakened host plants or hosts whose defences are overwhelmed by pioneer attacks, as in bark beetles (Berryman *et al.* 1985; De Jong & Sabelis, 1988, 1989);

2 large egg clutches deposited by a female, as in various species of moths (larch bud moths, gypsy moths, ermine moths, tent caterpillars, browntail moths); or

3 multigeneration congregations resulting from one or a few founders with a high intrinsic capacity of population increase (relative to the rate of dispersal) and low *per capita* food demands, as in many small herbivorous arthropods with short generation times (scale insects, mealybugs, aphids, leafhoppers, whiteflies, thrips, spider mites and rust mites).

These groups of herbivorous arthropods may occupy a leaf area less than a plant, or cover several individual plants. Moreover, group size and the leaf area occupied will increase with the number of generations spent in the group and with the extent to which the herbivores move to neighbouring host plants instead of dispersing far away. These traits are of crucial importance to understanding plant defences because individual plants (or kin groups of plants) are the units of selection and the selective advantage of defence depends on the extent to which a plant can influence local dynamics of herbivores and predators by direct and/or indirect defences.

Indeed, this influence is limited because predators and herbivores are independent players in the game: they may decide to stay, to move to neighbouring plants or to disperse far away. Hence, one may ask whether the defence of an individual plant initially affects the herbivore population it harbours itself (and much later the herbivore population as a whole) or whether its impact on the herbivores permeates population-wide and without delay (as a consequence of high herbivore mobility). Much the same questions can be formulated with respect to the degree in which individual plants can monopolize the advantages from attracting natural enemies of the herbivores. To analyse the complexities arising from spatial and temporal scales we will start by assuming that plants investing in direct and indirect defence will acquire all benefits and then later consider the case where neighbouring plants may profit too.

Predator–herbivore dynamics on individual plants

To understand the range of possible plant defence strategies it is instructive to model the dynamics of small arthropod herbivores at the scale of an individual plant (or a coherent group of clonal plants). For reasons of simplicity we assume that the spatial scale covered by a local predator and prey population is smaller than that of an individual plant (or group of ramets) and that the individual plant does not impose a carrying capacity to the arthropod population inhabiting the plant. Moreover, we will limit ourselves to the case of local populations with multiple and overlapping generations, as is realistic for small arthropods. Our aim is to find the simplest possible model for the dynamics of herbivorous and predatory arthropods and then ask the question how an individual plant can reduce damage by influencing the herbivores and their predators.

We assume that prey form patchy aggregations, consisting of clusters of herbivore-colonized leaves, and that predators entering such patches can freely move around, and spend little time in moving between herbivore-colonized leaves relative to the time spent within the herbivore colonies. Thus, the cluster of herbivore-colonized leaves can be considered as a coherent and homogeneous arena for strongly coupled predator–prey interactions (Fig. 5.1).

We further assume that the predators search for the highest prey densities and minimize interference with (intraspecific) competitors. In addition, we assume that within newly (and expanding) infested leaf areas herbivore density is typically constant, a characteristic determined by the herbivore or

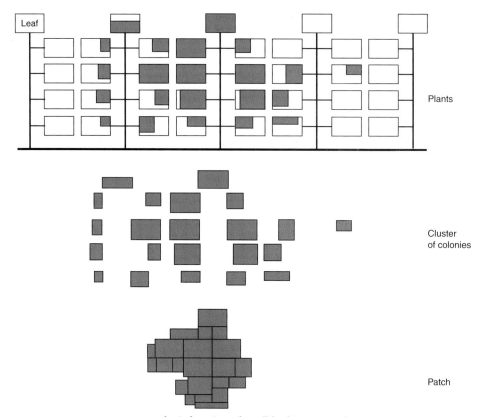

Figure 5.1 A patchy infestation of small herbivorous arthropods in a row of plants (top side-view). It is inspired by observations on spider mites, but in essence applies to many other herbivorous arthropods. The infested leaf parts are excised (middle) and then put together as a jigsaw puzzle (bottom). The latter forms the patch or arena within which the interaction between predator and prey takes place (assuming negligible time spent in moving between infested leaves).

the combination of herbivore and plant (Sabelis 1990; Sabelis & Janssen 1994). Thus, per unit of plant area herbivores raise a fixed amount of offspring, and predators reaching a freshly colonized leaf site will continue to eat prey until they do better by moving to another site nearby on the same plant. These assumptions lead to a constant rate of predation, which is much like 'eating a pancake': a constant amount of food at each bite until there is nothing left. As long as the pancake is not completely eaten, predators can maximize the rate of predation, development and reproduction. Hence, under conditions of a stable age distribution they will reach their intrinsic rate of population increase. Similar assumptions are made with respect to herbivore growth capacity in absence of the predators. For the case where predators stay until all prey are eaten, the dynamics of predator and herbivore numbers can be described by the following two linear differential equations, which differ from the classic Lotka–Volterra models in that the predation term now only depends on the number of predators (Sabelis 1992; Janssen & Sabelis 1992):

$$\frac{dN}{dt} = \alpha N - \beta P$$

$$\frac{dP}{dt} = \gamma P$$

(1)

This is the so-called 'pancake predation' model where t is the time since start of the predator–prey interaction, $N(t)$ is the number of prey at time t, $P(t)$ is the number of predators at time t, α is the rate of prey population growth, β is the maximum predation rate and γ is the rate of predator population growth. Analytical solutions for the number of predators and prey since the start of the interaction are readily obtained:

$$N(t) = N(0)e^{\alpha t} - P(0)\frac{\beta}{\gamma - \alpha}(e^{\gamma t} - e^{\alpha t})$$

$$P(t) = P(0)e^{\gamma t}$$

(2)

This model exhibits exponential growth of the predators until a time τ when all prey are eaten (and the predators move on to another food source). The time from predator invasion to prey extinction can be expressed as function of α, β and γ and the initial numbers of predator and prey, $N(0)$ and $P(0)$:

$$\tau = \frac{1}{\gamma - \alpha}\ln\left(1 + \frac{\gamma - \alpha}{\beta}\frac{N(0)}{P(0)}\right)$$

(3)

Three types of dynamical behaviour of the prey population may occur.
1 continuous decay until extinction (Fig. 5.2c, *immediate decline*);
2 initial increase, followed by decrease until extinction (Fig. 5.2b, *delayed decline*); and
3 continuous increase (but at a pace slower than the intrinsic rate of prey

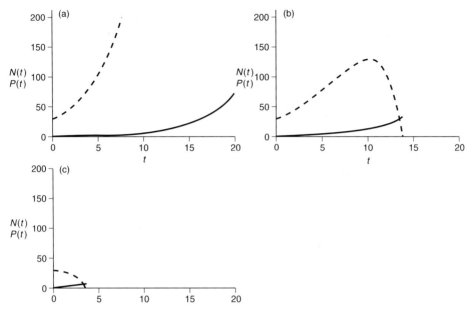

Figure 5.2 Three types of local predator–prey dynamics according to the pancake predation model (with parameters $\alpha = 0.3$, $\beta = 1$, $\gamma = 0.25$, $\mu = 0$, and $N(0) = 25$): (a) prey increase ($P(0) = 1$), (b) delayed prey decline ($P(0) = 3$) and (c) immediate prey decline ($P(0) = 8$). The general conditions for each of these types of dynamics are discussed in the text.

population growth) (Fig. 5.2a, *continued increase; predator surfing on a prey wave*).

In the latter case, predatory arthropods cannot suppress local prey population outbreaks.

Immediate decline of the herbivores occurs when the net growth rate of the herbivore population is negative:

$$\alpha N(0) < \beta P(0) \tag{4}$$

or:

$$\frac{P(0)}{N(0)} > \frac{\alpha}{\beta} \tag{5}$$

The conditions for immediate or delayed herbivore decline are found by calculating the condition for which the time to prey extinction has a finite value. Provided that the plant is not overexploited during predator–prey interaction, this condition is:

$$\frac{P(0)}{N(0)} > \frac{\alpha - \gamma}{\beta} \tag{6}$$

Thus, for the herbivore population to decline, the ratio of predators to herbivores should exceed the ratio of the *per capita* population growth rate of the herbivore and the maximum *per capita* predation rate. If the latter condition is not met, herbivores continue to increase and predators 'surf' on the 'population growth wave' of the herbivore. Inevitably, this increase will stop when the plant becomes overexploited.

The total damage incurred by the plant over the whole interaction period can be expressed in the number of herbivore-days $D(\tau)$ or the area under the curve expressing the temporal changes in the size of the herbivore population:

$$D(\alpha, \beta, \gamma, N(0), P(0)) = \frac{1}{\alpha}\left[P(0)\frac{\beta}{\gamma}\left(e^{\gamma t} - 1\right) - N(0)\right] \tag{7}$$

Note that this measure of the damage strongly depends on the exponential term and thus on the time to prey extinction (τ) and the *per capita* growth rate of the predator population (γ).

Thus, given the initial population sizes ($N(0)$ and $P(0)$) and estimates of the parameters (α, β, γ) it is possible to assess the overall damage by the herbivore and the predator's potential to suppress the prey population immediately, with a delay or not at all. Now, we may ask how a plant can influence the local dynamics so as to minimize herbivore damage. Under the assumption that the parameters can be modified independently, the answer is straightforward: (i) it should keep the predator/herbivore ratio as high as possible; and/or (ii) it should decrease the growth rate of the herbivore as much as possible. However, trade-off relations between parameters may complicate matters. For example, decreasing the growth rate of the herbivore (α) by toxins may also decrease the predation rate (β) and the growth rate of the predator (γ). In the extreme, the herbivore may even use the plant-provided toxins to defend itself against predators. Thus, the plant will not always profit from decreasing the growth rate of the herbivore. It will only profit if it decreases the herbivore's growth rate proportionally stronger than the predation rate and the growth rate of the predator.

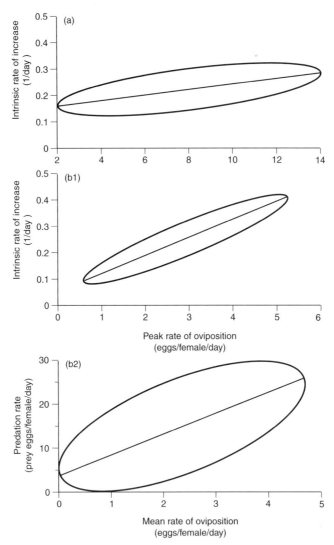

Figure 5.3 Examples of ranges of the *per capita* population growth (prey α; predator γ) and predation (β) parameters in the pancake predation model for different groups of plant-inhabiting arthropods (herbivores, predators). The ranges are presented as ellipses around regression lines of the model parameters against ovipositional rate (as in a, b) or against adult size (as in c): (a) intrinsic rate of population increase (females/day) of spider mites (Acari: Tetranychidae) plotted against the peak rate of oviposition (Sabelis 1990); (b1) intrinsic rate of

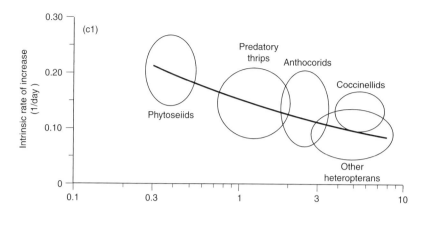

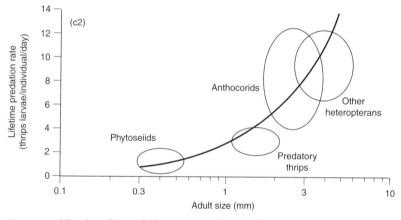

Figure 5.3 (*Continued*) population increase (females/day) of predatory mites (Acari: Phytoseiidae) plotted against the peak rate of oviposition (Sabelis and Janssen 1994); (b2) *per capita* predation rate of adult female, predatory mites on eggs/larvae of spider mites, plotted against the mean rate of oviposition (Janssen & Sabelis 1992); (c1) intrinsic rate of population increase (females/day) of predators of thrips (Acari: Phytoseiidae; Insecta: Aeolothripidae, Anthocoridae, Miridae, Nabidae, Coccinellidae), plotted against adult size (Sabelis & Van Rijn 1997); (c2) *per capita* predation rate of arthropod predators on young thrips larvae, plotted against adult size (Sabelis & Van Rijn 1997). Note that b2 shows the predation rate of the adult females only, whereas c2 gives a weighted estimate, calculated as the sum of the products of predator-stage-specific predation rates and the share in the predator population assuming a stable age distribution.

Another message gleaned from the equations is that plants may benefit from promoting the presence of predators with high predation rates and high growth rates. Often, these demands are in conflict with each other, because the predation rate tends to increase with body size, whereas the intrinsic rate of population increase tends to decrease with body size (Sabelis 1992). Such relations with body size are clear from analysing published data on predators of phytophagous thrips, such as mirids, anthocorids, predatory thrips and predatory mites (Fig. 5.3c) (Sabelis & Van Rijn 1997). At lower taxonomic levels (within-family, within-genus) the picture may be different. For example, within the Phytoseiidae — a family of plant-inhabiting predatory mites — positive high correlations between β and γ exist (Fig. 5.3a,b) (Janssen & Sabelis 1992; Sabelis & Janssen 1994). One may wonder whether the plant could selectively attract one species of predator over the other and thereby profit from selecting the more effective predators. However, how a plant could do so given that predators will go for the most profitable prey remains to be elucidated by experiment.

Predators are independent players in the tritrophic game and they are the ones that decide whether it is profitable to stay on a plant or not. The pancake predation model is based on the assumption that predators are strongly arrested and stay until all prey are eaten. This scenario is not implausible, because it may be risky to disperse and search for new herbivore patches. In fact, it is frequently observed, e.g. in interactions among predatory mites and spider mites (Fig. 5.4a) (Sabelis & Van der Meer 1986). One may, of course, expect predators to leave somewhat earlier than the exact moment of prey extinction. This would relieve the herbivores from predation pressure and predator/prey ratios may become so low that the herbivore population increases again, thereby giving rise to cyclic dynamics. The plant would then accumulate damage over the predator–prey cycles whereas it would be better off when predators exterminate the herbivores or maintain them at a very low level. Examples of these various outcomes are shown in Fig. 5.4 for interactions between phytoseiid predators and spider mites, phytoseiid mites and thrips, and anthocorid predators and thrips (Sabelis & Van Rijn 1997). All these examples convincingly show that local herbivore populations are strongly suppressed by predators. Thus, the dynamics of the 'pancake predator' model seems to be a good caricature of the initial predator–prey population cycle, and in some special cases also for the end phase (Fig. 5.4a–c), while in others the end phase differs (e.g. a stable, non-zero, but low level) (Fig. 5.4d). Thus, for all cases where colonization of predator and prey go through one cycle before ending in extermination or reaching a very low equilibrium, this model is a useful approximation of reality to understand the role of indirect and direct plant defences.

Predator arrestment during the interaction with the herbivores plays a decisive role in the success of indirect defence strategies of the plant. For example, if we extend the pancake predator model with a constant predator dispersal rate (i.e. independent of prey availability), then dispersal acts so as to decrease the effective population growth rate of the predators. As shown in Fig. 5.5a, small decreases of the growth rate have dramatic consequences for the duration of the interaction and even more for the number of herbivores attacking the plant. The overall damage to the plant increases non-linearly with a reduction of the predator's population growth rate. Hence, it is important to observe that several species of predator are strongly arrested in herbivore-colonized patches and tend not to leave until the prey population is driven to near extinction (Sabelis & Van der Meer 1986; Sabelis & Van Rijn 1997).

The herbivore can always make the last move in the tritrophic game. They may not only develop resistance against the predators and overcome barriers raised by the plant, but ultimately, they may decide to leave the plant. If the plant gets rid of herbivores by attracting predators, and also by stimulating herbivore dispersal, then it benefits disproportionately, as illustrated in Fig. 5.5b.

In conclusion, there is a variety of ways in which a plant can benefit from influencing behaviour and dynamics of predators and herbivores. Direct plant defences do not merely slow down the herbivore's growth rate, they may also affect predator impact, either positively (higher predator/herbivore ratio) or negatively (plant toxins protecting herbivores against predators). Indirect defences also do not merely affect predator performance, they may also affect selection on herbivores, be it positive (enemy avoidance) or negative (resistance to predators) for the plant. Hence, to understand the plant's allocation to direct and indirect defences we should not only assess the costs, but also elucidate how these two types of defences interact in their impact on overall herbivore damage: are they synergistic or antagonistic?

Plant–predator mutualism: costs and benefits
Why did plants evolve traits to promote survival, reproduction and colonization of predatory arthropods? It is not at all self-evident that an

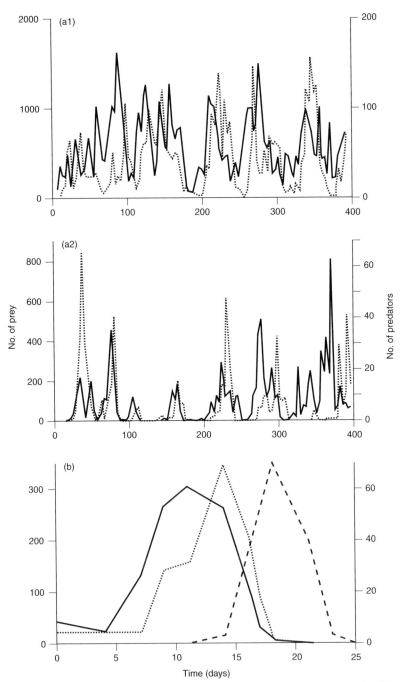

Figure 5.4 Predator–prey dynamics at various spatial scales. The examples (a–d) show overall persistence of predator–prey populations (a1), but frequent extinctions at a local scale (a2, b) irrespective of whether the local populations reside in open (b, c1) or closed (c2) space.

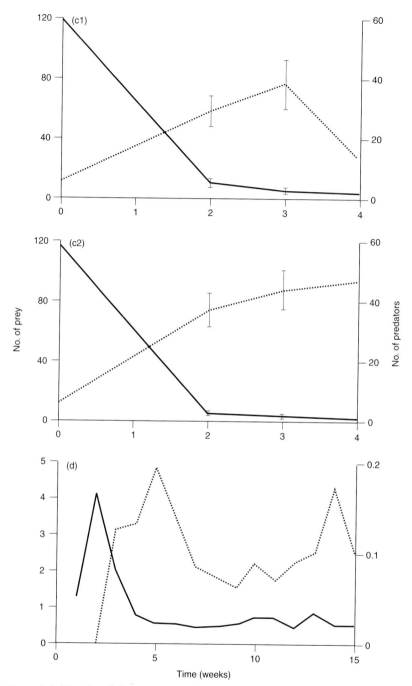

Figure 5.4 (*Continued*) In some cases, where the prey have invulnerable stages or may hide in refuges, local dynamics does not end with extinction and converge to stable equilibria, but these stable levels can be very low due to the presence of alternative food for the predators (d). (*Continued p. 124*)

individual plant gains selective advantage through indirect defence. A rare mutant plant investing to promote the effectiveness of predators may also benefit its neighbours, since they do not pay the costs. Selection can only act on relative fitness differences between mutant and resident plants. Hence, the rare mutant will only spread in the plant population if it (or its progeny) profits more from predator promotion than other plant genotypes. The crucial question is, therefore, how a plant manages to reap the benefits at least to some degree.

Figure 5.4 (*Continued*) (a1) Overall population fluctuations of the predatory mite *Phytoseiulus persimilis* Athias-Henriot (dotted line) and the two-spotted spider mite *Tetranychus urticae* Koch (solid line) in a circular system of six interconnected trays, each with 20 Lima-bean plants maintained in the two-leaf stage (by frequent removal of the apex and replacement of plants exhausted as a food source) (Janssen *et al.* 1997b). (a2) Predator–prey dynamics showing several events of prey (and next also predator) extinction on one of the six interconnected trays from the experiment shown in Fig. a1 (Janssen *et al.* 1997b; see also Van der Klashorst *et al.* 1992). (b) Local dynamics of the two-spotted spider mite *T. urticae* (solid line) and the predatory mite *P. persimilis* (dotted line) showing that first the prey population and then the predator population become extinct and that aerial dispersal of the predatory mites (dashed line) takes place near the moment of prey extermination (B. Pels and M.W. Sabelis, unpublished observations; see also Sabelis & Van der Meer 1986). (c) Dynamics of the bulb mite *Rhizoglyphus robini* Claparède (solid line) and the soil predatory mite *Hypoaspis aculeifer* Canestrini (dotted line) in either open (c1) or closed (c2) jars with scales of three lily bulbs (Lesna *et al.* 1996), showing that the populations of bulb mites are driven close to extinction in closed as well as open jars, and that the population of predatory mites in the open jars starts to decline (due to dispersal), only after the bulb mite population is decimated, whereas the predator population in the closed jars does not decline. (d) Local dynamics of populations of the phytophagous thrips *Thrips palmi* Karny (solid line) and the predatory bug *Orius* sp. (dotted line) on eggplants, showing that the thrips population first goes through a cycle and then persists at a very low level, whereas the predator population after an initial cycle persists for several months at a higher level than its prey (Kawai 1995). This may well be an example where the predator population is promoted due to alternative food (i.e. pollen), thereby strongly suppressing the prey. The prey is not driven to extinction, but persists possibly due to the presence of invulnerable stages (i.e. eggs inserted in leaves and pupae in the soil) (Sabelis & Van Rijn 1996).

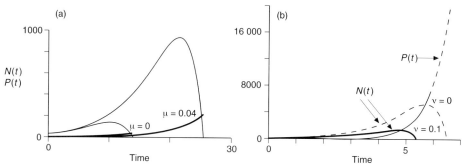

Figure 5.5 Sensitivity of local predator–prey dynamics according to the pancake model for changes in (a) the *per capita* population growth rate of the predator (γ) (or similarly, when predators disperse during the interaction period with rate μ, the effective predator growth rate $\gamma - \mu$) ($\mu = 0$ or 0.04, $\alpha = 0.3$, $\beta = 3$, $\gamma = 0.25$, $N_0 = 30$, $P_0 = 1$) and (b) the *per capita* population growth of the herbivore (α) (or similarly, when prey disperse during the interaction period with rate v, the effective prey growth rate $\alpha - v$) ($v = 0$ (dashed lines) or 0.1 (drawn lines), $\alpha = 1$, $\beta = 1$, $\gamma = 1.5$, $N_0 = 50$, $P_0 = 1$).

In principle, a plant could delay acquisition of natural enemies until the actual attack occurs, because small herbivorous arthropods tend to colonize the plant in low numbers and their *per capita* food demands are low. However, this tactic would only be successful if a sufficient number of predators can be lured at any moment. Such a response is unlikely for many types of predatory arthropods. For instance, predatory mites are very small (<1mm), have a low capacity for ambulatory movement and for long-distance orientation to prey patches. Hence, delaying predator acquisition until attack is risky because a plant can only attract predatory mites from nearby, say within a radius of several tens of metres (probably not hundreds and certainly not thousands). Moreover, the initial ratio of predators to herbivores is an important determinant of the time needed to suppress the local herbivore population and the amount of damage inflicted upon the plant, as shown for the 'pancake' model (Janssen & Sabelis 1992a; Sabelis 1992). This is probably why plants also invest in acquisition before being attacked by phytophagous mites. An additional reason is that, precisely because predatory mites are small, have a low capacity for active displacement and incur high mortality during dispersal, plants need only a small investment to encourage these predators to stay.

125

Below we will review the many and varied ways in which plants may promote the effectiveness of predatory arthropods and the extent to which these plant traits are exploited by other 'players in the tritrophic game' that do not in turn benefit the plant. Finally, we will evaluate the balance of costs and benefits and discuss possibilities for the evolution of plant–predator mutualisms.

Plants provide protection

It is well known that predatory arthropods visit arthropod-borne cavities or galls. These structures may provide prey and possibly also protection. However, plants also create specialized structures, called domatia, that are inhabited by predatory arthropods, fungivores, but rarely herbivores (Beattie 1985; O'Dowd & Willson 1989; Pemberton & Turner 1989; Jolivet 1996; Walter 1996). Examples are plant structures exclusively inhabited by some ant species, such as leaf pouches (e.g. on *Cola* plants), swollen petioles (e.g. of *Piper*), hollow stems (e.g. of *Cecropia*), hollow thorns (e.g. of *Acacia*), hollow roots and tubers (e.g. of *Pachycentria* and *Myrmecodia*). Other examples are the so-called 'mite houses' (acarodomatia) on leaves. These are found where leaf veins bifurcate and consist of small, sometimes very specialized cuticle modifications: hair tufts, pockets, pits and pouches with entrance holes only slightly larger than a mite (O'Dowd & Willson 1989). In general, domatia are thought to provide protection against adverse weather conditions or relatively larger hyperpredators (Beattie 1985; Walter 1996).

There are strikingly simple reasons why domatia selectively promote the survival of predatory arthropods and not that of herbivores, which in principle could enter as well. The critical feature is that domatia cover only a very small proportion of the plant and by their limited availability impose selection on any arthropod seeking protection. However, herbivorous arthropods would not gain, as they would increase their rate of encounter with predatory arthropods seeking protection in the same domatia.

There is some evidence for a positive impact of domatia on the survival of predatory arthropods and a negative impact on the population growth of herbivorous arthropods (Fig. 5.6a) (Grostal & O'Dowd 1994; Agrawal & Karban 1997). An interesting example of plant structures with a domatium-like function is the apex of cassava plants. These clusters of (often hairy) leaf primordia are exclusively inhabited by *Typhlodromalus aripo*, a predatory mite that suppresses cassava green mite populations on young leaves near to the apex (Fig. 5.6b) (Bakker & Klein 1998). This shows that

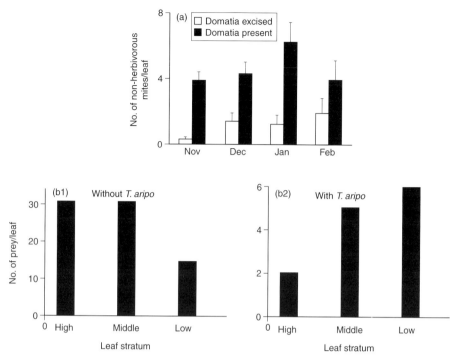

Figure 5.6 Population level effects of plant traits involving protective structures, food and alarm signals to the predators of herbivorous arthropods. (a) The effect of excision of hair tufts (acarodomatia) from leaves of the laureltinus (*Viburnum tinus* L.) on the abundance (mean and standard error) of non-herbivorous mites (10% predatory mites, 90% microbivores) per leaf (Grostal & O'Dowd 1994). Herbivores were only rarely found inside domatia. (b) The impact of the apex-inhabiting predatory mite *Typhlodromalus aripo* (b2 vs. b1) on the frequency of per plant presence of the cassava green mite *Mononychellus tanajoa* Bondar in three leaf strata (high, middle, low) of a cassava plant (Bakker & Klein 1998). The two histograms show the distribution of absolute frequencies (77 plants in b1; 13 plants in b2) over the strata. For comparison of the two histograms, however, one should only consider the relative frequency distributions over the strata! (*Continued p. 128*)

the crucial question is why plants create specific domatia-like structures on leaves. Possibly, domatia provide the proximity needed for predatory arthropods to act immediately when the plant leaves are being attacked by herbivorous arthropods.

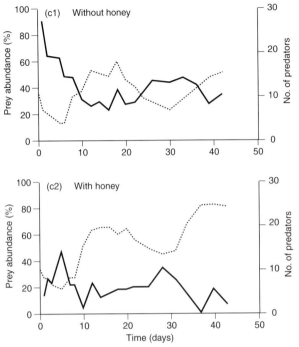

Figure 5.6 (*Continued*) (c) Effect of adding honey droplets (to mimick cassava exudates) on the dynamics of the leaf-inhabiting predatory mite *Typhlodromalus manihoti* (= *limonicus*) and the cassava green mite (*Mononychellus tanajoa* Bondar) on cassava plants in a screenhouse (Bakker & Klein 1992). The abundance of the predatory mites is expressed as numbers per plant, whereas that of the cassava green mites is expressed as a percentage relative to the maximum abundance class, which was taken to be 50 or more mobile mites per leaf.

It should be stressed that domatia are not the only structures influencing the effectiveness of the third trophic level, as plant architecture in general can play a significant role. A particularly elegant demonstration is provided by Kareiva and Sahakian (1990) who observed increased rates of aphid predation by coccinellids on pea plants that were leafless and hence lacked the smooth and slippery leaves of the control plants.

Plants provide food

Some plants produce food bodies that are utilized by ants (Beattie 1985; Jolivet 1996). Examples are lipid- and protein-rich multicellular tissues known as Beltian bodies (*Acacia*), Müllerian bodies (*Cecropia*), Beccarian

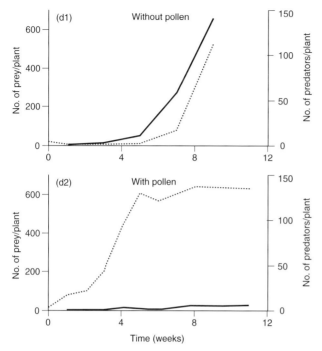

Figure 5.6 (*Continued*) (d) The dynamics of western flower thrips *Frankliniella occidentalis* Pergande (solid line) and the predatory mite *Amblyseius degenerans* (Berlese) (dotted line) for the case where pollen was virtually absent (male-sterile cucumber variety) (d1) and where cat-tail pollen was provided on the youngest full grown leaves every other week (d2). Shown are the means of two replicate experiments, carried out by P.C.J. Van Rijn, Y.M. Van Houten and M. Raats (unpublished data, 1997). (*Continued p. 130*)

bodies (*Macaranga*) and Pearl bodies (*Ochroma*). Another striking example is the induction of single-celled food bodies along petiole margins of *Piper cenocladum* by the presence of the ant *Pheidole bicornis* (Risch & Rickson 1981). The food bodies complement dietary requirements for lipids and proteins, and are sometimes essential for reproduction of ants.

Many plants produce nectaries that are not involved in attracting pollinators. Although they may occur in flowers outside the perianth (stalks and bracts of inflorescences), they are most frequently found on leaves, stems, petioles, stipules and sometimes on sepals and petals (Elias 1983). Parasitoids, ants and various other predatory arthropods commonly feed on the nectar, and as a byproduct of tending the nectaries they attack

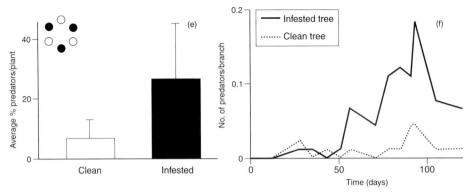

Figure 5.6 (*Continued*) (e) Ability of females of the predatory mite, *Phytoseiulus persimilis* Athias-Henriot, to locate cucumber plants infested with two-spotted spider mites (*Tetranychus urticae* Koch) (A.R.M. Janssen, unpublished data, 1996). Predatory mites were released in the centre of a circle of six cucumber plants with three infested and three uninfested, arranged alternately (as in the top left of the histogram). Shown are the mean and standard deviation of the percentage of centre-released predators reaching the clean plants (open columns) and the infested plants (solid columns) over five independent trials. (f) Number of adult anthocorid predators per branch of pear trees, standing next to an encaged pear tree that had either a heavy infestation of pear psyllids (solid line), or that had very little damage of pear psyllids (dotted line) (Drukker *et al.* 1995). Before day 50 (early June) the anthocorids are found feeding on aphids in other trees (e.g. *Alnus*), but, thereafter, they immigrate into pear orchards, thereby evidently attracted by volatiles produced by trees infested by *Psylla* (Scutareanu *et al.* 1997).

herbivorous arthropods on the plant (Bentley 1977; Beattie 1985; Rogers 1985; Koptur 1992; Pemberton & Lee 1996). Some tiny leaf nectaria are not visited by ants but possibly by smaller arthropods, such as predatory mites (Bakker & Klein 1992). Predatory mites have been observed to feed on extrafloral nectaria of various plants (e.g. nectaria on leaf petiole of *Prunus* spp.; M.W. Sabelis, personal observation) and this can greatly prolong their survival when other food sources are absent (Van Rijn & Tanigoshi, personal communication).

Plants may also provide nutrient-rich foods in other ways than via nectaries. Examples are plant exudates along the leaf ribs and petioles of cassava plants that are consumed by the phytoseiid mite, *Typhlodromalus manihoti* De Moraes (= *limonicus*), and promote its survival (Bakker & Klein 1992, 1998). Addition of honey droplets (to simulate exudates) to cassava plants had a clear positive effect on the population growth of predatory mites and a negative impact on cassava green mites (Fig. 5.6c) (Bakker & Klein 1992). This predatory mite is not arrested at a particular site on the plant, but moves around on the plant or group of cassava plants. The exudates may provide fuel for active search, thereby leading to a higher rate of contact with prey. It is possible that the plants profit more from an actively patrolling population of predatory mites than from arresting them near the site of exudate release.

A final example of how the plant feeds predators is the production of pollen that can be utilized by various arthropod predators, such as hover flies, flower and pirate bugs, and predatory mites. Several plant species provide pollen with an exine layer thin enough to enable feeding. Some pollen provides nutrition of such high quality that predatory mites can develop and reproduce on this diet just as well as on their most preferred prey (Overmeer 1985). Moreover, there is evidence for arrestment of predatory arthropods in areas with pollen, as shown for flower bugs and some predatory mites (Sabelis & Van Rijn 1997). Van Rijn, Van Houten and coworkers (unpublished data) found a clear, positive impact of the addition of *Typha* pollen on the abundance of the predatory mite *Amblyseius degenerans* and a negative impact on the abundance of the western flower thrips (Fig. 5.6d).

For the plant to profit from 'making edible pollen' some conditions should be met. The plant-provided food should not be of too high quality, since otherwise predators may preferentially feed on these foods and leave the plant pests unharmed. When fed exclusively on *Vicia* pollen, dietary deficiencies have been reported for the phytoseiid mite, *Amblyseius andersoni* Chant, as these pollen lack a sufficient amount of carotenoids to enter diapause (Van Zon *et al.* 1981). These and other deficiencies may help explain why many phytoseiid mites tend to prefer herbivorous mites to pollen when offered a choice and also why they respond to odours emanating from a broader range of herbivorous arthropods (Dicke *et al.* 1986).

Another unsolved problem is that pollen can be utilized by non-predatory arthropods, including herbivores. The western flower thrips, *Frankliniella occidentalis* Pergande, is an example. Both larvae and adults of this species feed on pollen which promotes their development and oviposition. Van Rijn and Sabelis (1998) discuss the conditions under which the plant may benefit from producing pollen despite the fact that these can be utilized by predatory mites and their prey, the western flower thrips. They show that when predator and herbivore population are in equilibrium, alternative food always increases the population size of the

predators and decreases that of the herbivore. When, however, the interacting populations are not in equilibrium but fluctuate, the profit to the
plant depends on the impact on the population growth of predators relative
to herbivores. The important point to note is that plants may gain by
providing food under some circumstances, but not in others. The overall
benefit to the plant critically depends on population dynamics.

Plants call for help

With some notable exceptions (induction of extrafloral nectar production
by predatory ants (Beattie 1985) or by herbivorous arthropods (F. Wäckers,
personal communication, 1997) the provisioning of protection and food by
plants mainly represents examples of indirect, constitutive defence, i.e.
defence operating before herbivore attack. However, plants also generally
possess inducible indirect defences. They release volatile chemicals in
response to attack by herbivorous arthropods and these signals appear
to act as alarm calls that attract and arrest predatory arthropods. This
mechanism is now demonstrated for several plant–herbivore–predator
systems. As an example, we will focus on spider mites and predatory mites
(Sabelis & Van de Baan 1983; Sabelis & Dicke 1985; Dicke & Sabelis 1988,
1989, 1992; Dicke *et al.* 1990a,b). Two lines of evidence for the active
involvement of plants have been presented. Firstly, the chemicals released
upon attack by phytophagous mites (e.g. methyl salicylate and several
monoterpenes) are known to occur in plants. Secondly, not only infested
leaves release the volatiles, but also uninfested leaves attached to the same
plant. Thus, there is local, as well as systemic induction of volatiles upon
spider mite attack, and the quest for the elicitor involved in local transduction from the herbivore's oral secretions to the plant (Hopke *et al.*
1994; Mattiacci *et al.* 1995; Alborn *et al.* 1997; Paré & Tumlinson 1997) and
systemic signal transduction from leaf to leaf has begun (Boland *et al.* 1992;
Dicke *et al.* 1993; Dicke 1994).

Also, there is strong evidence that olfaction is implicated in responses
of predatory mites to odours from plants infested by spider mites and
other phytophagous mites. Many chemical compounds are identified (Dicke
et al. 1990a), the predator's chemosensors have been described (e.g. for
predatory mites; see Jagers Op Akkerhuis *et al.* 1985), olfactory responses to
various single blend components have been assessed (Dicke *et al.* 1990a),
but we know little of how the information contained in the blend is
integrated. As quantitative and qualitative blend composition varies with
the species of spider mite, the species or cultivar of host plant, the age

and condition of the host plant (Takabayashi *et al.* 1991, 1994; Scutareanu *et al.* 1997), one may wonder how predatory mites can cope with this bewildering variety of signals. Do they learn the association between odour and prey or are they genetically predisposed? Margolies *et al.* (1997) have shown that it is possible to select for a response to odour. Thus, there is at least a partial genetic determinant. Also, there are some indications for learning behaviour (Dicke *et al.* 1990c), but the degree to which they can learn to associate particular odours with the presence of prey is still unknown.

The orientation mechanisms that may help a predatory mite to find and stay near the odour source have been studied in predatory mite, *Phytoseiulus persimilis* Athias-Henriot. This predator has been shown to orientate in odour gradients (Sabelis *et al.* 1984), to respond differentially to wind direction with and without herbivore-induced plant odour (Sabelis & Van der Weel 1993), and to suppress its tendency to disperse aerially in response to herbivore-induced plant odour even when severely starved (Sabelis & Afman 1994). Efficient location of spider-mite-infested plants has been demonstrated in wind tunnels (M.W. Sabelis & P. Schippers, unpublished data), in the laboratory (Sabelis & Van der Weel 1993) and under greenhouse conditions (Fig. 5.6e) (A.R.M. Janssen, unpublished data).

To what extent infested plants continue to attract predators, is not clear. Janssen *et al.* (1997a) showed that there may well be limitations to the number of predatory mites that a plant can attract. When predatory mites are offered a choice between odours from spider mite colonies with and without conspecific predators, they prefer odours from the latter. This probably means that predatory mites are initially attracted to a plant infested by spider mites, but once the infested leaves are discovered and colonized others will move on to neighbouring plants with spider mite colonies free of predators. Thus, predatory mites make their own foraging decisions. They not only use information to locate their prey, but also to avoid competition. An open question is where this information on competitors is coming from. Janssen *et al.* (1997a) show that the production of the odour is associated with the presence of adult spider mites, and they suggest that it is an alarm pheromone of the spider mites.

The ability to produce herbivore-induced alarm signals is probably widespread in the plant kingdom, as there is supporting evidence that spider mites elicit such responses across plant families (Dicke 1994). This ability is not limited to the plant's interaction with herbivorous mites and predatory mites. Apart from the extensive work on the response of parasitoids to plant synomones induced by feeding of caterpillars (Turlings *et al.* 1995), there is also evidence for responses of anthocorids to odours from bean plants infested by spider mites (Dwumfour 1992) and pear leaves infested by psyllids (Drukker *et al.* 1995) and to components identified in the odour blends of infested pear leaves, such as (E,E)-α-farnesene and methyl-salicylate (Scutareanu *et al.* 1996, 1997).

The olfactory response of anthocorid predators to odours from pear leaves infested by psyllids is not only demonstrated in the laboratory, but also in the field. Drukker *et al.* (1995) found a strong aggregative response of anthocorid predators to pear trees that were infested by psyllids, but provided with a cage made of fine mesh gauze to prevent the predators making direct contact with their prey (Fig. 5.6f) (Drukker *et al.* 1995). This response is implicated in the immigration of anthocorid predators starting in July by the time that populations of psyllids in pear orchards increase. Experiments with traps for monitoring immigration of anthocorids using dispensers with plant synomones may well demonstrate a novel way of manipulating predator/prey ratios (Molleman *et al.* 1997).

Cheaters of plant–predator mutualisms

Mutualistic interactions are never foolproof (Bronstein 1994). This invariably applies to plant–predator mutualisms, because plants may provide domatia, nectar, exudates, edible pollen and signals, but they cannot directly control which organisms will utilize them. To determine whether or not they gain a net benefit, all potential utilizers should be taken into account and their separate and joint impact on plant fitness should be assessed.

(Hyper-)predators

Although rare, there are reports of predators that utilize plant-provided facilities, but do not act against herbivores attacking the plant. Janzen (1975) reported that the ant *Pseudomyrmex nigripilosa* collects food on *Acacia* trees, but does not provide protection against herbivorous arthropods and may therefore be a parasite of the mutualistic interaction between plants and predators. Gaume and McKey (1997) found that two species of ants (*Petalomyrmex phylax* and *Cataulacus mckeyi*) on a nectar- and nest-providing understorey tree (*Leonardoxa africana*) of coastal rainforests in Cameroon mutually excluded each other. *Petalomyrmex* was found on two out of three plants, patrolled the young leaves day and night, and chased away or killed any phytophagous insects encountered. *Cataulacus* occurred on the remainder of the trees, but was active only during the day (when nectar production was most intense) and did not attack phytophagous insects in many cases. *Petalomyrmex* therefore is a mutualist whereas *Cataulacus* seems to be a parasite of the mutualistic interaction.

Sometimes, the interests of the plant and its bodyguards conflict, as is the case when its bodyguards would also kill beneficial visitors such as pollinators. However, one recently discovered example illustrates that this conflict can be resolved by special adaptations. *Crematogaster* ants inhabiting the modified thorns of *Acacia* plants are deterred by a volatile chemical signal from young flowers at the crucial stage of dehiscence, allowing bees and other pollinators to visit and transfer pollen. The ants patrol the young (undehisced) buds, and also return to the flowers after dehiscence, protecting the fertilized ovules and developing seeds of the plants they are associated with (Willmer & Stone 1997).

One may also wonder if plants attract and arrest predators of herbivores that also feed on other predators present on the plant (intraguild predation, cannibalism), thereby decreasing the impact on herbivorous arthropods (Polis *et al.* 1989; Polis & Holt 1992; Holt & Polis 1997). This brings us to the most crucial assumption underlying the hypothesis of plant–predator mutualisms (Price *et al.* 1980): the negligible importance of a fourth trophic level. When hyperpredators were abundant, intermediate level predators would decrease and therefore herbivore numbers would increase (Hairston *et al.* 1960) and harm the plant. Perhaps plants can decouple the interaction between hyperpredators and intermediate-level predators by providing small domatia as refuges from larger predators. However, one may object that these hyperpredators may open domatia or lie in wait next to them, being easy-to-detect foraging sites. There is every reason to critically test the hypothesis on domatia as refuges from hyperpredators.

Herbivores

Herbivores are unlikely to remain passive subjects of the defensive measures taken by plants. Selection may favour mutant herbivores capable of going unnoticed by the plant's alarm-recognition system. The increased predation pressure exerted by the plant's bodyguards may also favour mutant herbivores that are better able to resist attack by natural enemies, and once this has been achieved, the bodyguards create competitor-free space for the well-defended herbivore. For example, *Polyhymno* caterpillars feed on ant-defended *Acacia* in the tropical lowlands of Veracruz (Mexico) and can cause serious damage. They construct sealed shelters by silking together the pinnae of *Acacia* leaves. These shelters serve as partial refuges from ants, thereby providing access to ant-guarded but otherwise poorly defended plants (Eubanks *et al.* 1997). There are many more ways in which herbivores can defend themselves against ants or other plant bodyguards; examples are hard exocuticles, toxins, escape behaviour, ant mimicry and symphily.

This list of herbivore defences would not be complete without mentioning the often-cited and perhaps most debatable of all: sugar-rich excretions that encourage ant attendance (Way 1963). This phenomenon occurs in many homopterans. However, nothing can stop the ants from eating not only the honeydew, but also its producer. Indeed, some studies have shown that ants may eat the homopterans they tend, and these homopterans seem to be rather scarce under natural conditions (Beattie 1985; Buckley 1986, 1987). It may be that the homopterans decrease the risk of being eaten by ants by providing excess food in the form of honeydew. Alternatively, the honeydew may attract the enemies of their competitors (Evans & England 1996). Aphids might sacrifice sugar and amino acids, excrete them into the honeydew and gain by reducing predation risk and competition with other herbivores. As neither honeydew composition nor population growth of the homopterans can be controlled by the plant, it is quite unlikely that plants manipulate these herbivores as prey to feed ant colonies. Moreover, via extrafloral nectaries and food bodies the plants can achieve the same goal, while keeping control over the negative and positive effects. Homopterans are therefore more likely to be intruders in the plant–predator system thanks to an old trick: they offer nutrient-rich honeydew, thereby lowering the risk of being eaten themselves. Possibly they gain more in terms of protection by ants from other natural enemies (ladybeetles, aphid parasitoids), than they lose via predation by ants.

One may speculate whether plants can gain advantage by tolerating the presence of herbivorous arthropods that inflict light damage to the plant and serve as alternative prey to predators of other more harmful herbivores. Vagrant eriophyoid mites may be an example. Due to their short stylets they cause superficial damage to the leaves. As shown by Collyer (1964) the addition of apple rust mites reduces the final population size of European red mites as well as the total damage to the plant when these two phytophagous mite species have the same predatory mites as enemies. These results are reminiscent of theoretical predictions from models of two prey shared by one predator (Holt 1977; Sabelis & Van Rijn 1997). Karban *et al.* (1994) studied effects of adding an alternative prey to an acarine predator–prey system in vineyards and reported results with a high between-year and between-site variation. Hence, if any benefit from 'mild' herbivores accrues to the plant, it may only become manifest at sufficiently long time scales. It is more parsimonious to hypothesize that eriophyoid mites have more chance to escape from predation because they are very small (*c.* 0.1 mm) and represent a much smaller amount of food.

The examples of shelter-building caterpillars, honeydew-producing homopterans and tiny eriophyoids illustrate that herbivores may find ways to settle on a plant despite the mutualistic interaction between plant and predator. These herbivores are not necessarily cheaters of the mutualism *per se*, even though their presence has negative effects due to feeding damage and perhaps also positive side-effects on the maintenance of bodyguards. Positive and negative effects of these intruders may or may not cancel out for the plant. Thus, indirect effects may span the range from mutualistic to parasitic.

There are other cases where herbivores are genuine 'parasites' of the mutualistic plant–predator interaction in that they use the plant's defences to their own advantage. For example, *Domatiathrips cunninghamii* Mound, is a species of thrips so minute (*c.* 1 mm) that it is close to the size of a mite. It can invade acarodomatia on the leaves of an understory tree in Costa Rica, *Psychotria gracilifora* (Mound 1993). Another interesting example is that of herbivorous thrips utilizing pollen, thereby promoting their own rate of population increase. Depending on the relative impact of pollen on predator vs. thrips population growth and the rate of predation on thrips, time to prey suppression may change and thrips may or may not escape control by predators (Sabelis & Van Rijn 1997; Van Rijn & Sabelis 1998). Other examples of herbivores taking advantage of the plant-provided facilities are herbivores sequestering toxic secondary plant compounds to decrease their profitability as a food source for their natural enemies, and herbivores attracted by the plant's alarm signals induced by the same or another herbivore. Herbivore-induced plant volatiles have been shown to attract (Loughrin *et al.* 1995) and deter other herbivores (Dicke 1986). Pallini *et al.* (1997) found that, when uninfested, competitor-free plants were the alternative, two-spotted spider mites showed a weak, but significant, positive response to odours from cucumber plants infested by conspecific mites. However, they avoided odours from cucumber plants under attack of western flower thrips, a herbivore that can also act as a predator of two-spotted spider mites. No doubt, studies on yet other herbivores may reveal similar positive or negative responses to plants betraying their weakened state or signalling the presence of superior competitors by releasing volatile chemicals as alarm signals. Such studies may prove essential in elucidating constraints to communication between the first and the third trophic level.

Neighbouring plants

Neighbouring plants may benefit from plants investing in attraction and arrestment of predators of the herbivores. This is because predators will decrease the chance that the attractive plants will become a source of infection in the future and because the predators attracted may also wander around, thereby patrolling the neighbouring plants as well. For example, Drukker *et al.* (1995) observed increased concentrations of anthocorid predators on uninfested pear trees next to *Psylla*-infested pear trees that are known to release volatile chemicals in response to feeding by psyllids (Scutareanu *et al.* 1997).

If the neighbouring plants are close kin (as in clonal plants), investment in attracting and arresting predators increases the inclusive fitness of the investor. Plant–predator mutualisms will then evolve, even when an individual plant can be killed in one or a few 'bites' by a herbivore. However, when plants are not identical by descent, neighbours may be competitors. Hence, the investment of one plant may benefit its competing neighbour as well, thereby decreasing its own fitness (Sabelis & De Jong 1988). In addition, neighbouring plants may profit by initiating a defensive action in response to alarm signals of herbivore-infested or otherwise diseased plants nearby (Farmer & Ryan 1990; Bruin *et al.* 1992, 1995; Shonle & Bergelson 1995; Shulaev *et al.* 1997; Adviushko *et al.* 1997).

Do direct and indirect defences act synergistically?

Whether increased direct plant defences interfere with the effectiveness of natural enemies has rarely been investigated (Boethel & Eikenbary 1986). Increased direct defence may have positive and negative effects on the impact of natural enemies. On the positive side is that reduced population growth rates—all other things being equal—enable the natural enemies to suppress the pest most quickly. In addition, the vulnerable phases to enemy attack may last longer (Isenhour *et al.* 1989; Loader & Damman 1991) or the individual herbivores become weaker and more vulnerable, thereby leading to stronger pest suppression (Price *et al.* 1980). This is what we referred to earlier, as the indirect gain of direct plant defence. On the negative side are direct plant defences that decrease the rate of mortality due to natural enemies more than they reduce the growth and feeding rate of the herbivores. Such a negative balance arises, for example, when herbivores sequester toxic plant compounds and, as a byproduct, gain by the negative impact on natural enemies (intoxication, deterrence, etc.), sticky and/or toxic glandular secretions of the plant cause direct mortality of the natural enemies (Van Haren *et al.* 1987), and when plant/leaf morphology (e.g. pubescent or glabrous and very smooth leaves) reduce the effectiveness of the predator's search by impeding movement or reduced adhesion (Eigenbrode & Espelie 1995). The critical point, however, is not so much whether the indirect effects on natural enemies are negative *per se*, but rather whether the overall balance of defensive efforts turns out to be positive for the plant. If direct plant defences appear to alleviate predation pressure on the herbivores, one may still wonder how they compare with the direct and indirect gains of reducing the population growth of the herbivore.

This can be illustrated using the 'pancake' predation model introduced in Eq. 1. Consider joint proportional changes (p) in population growth rate ($p\alpha$) and plant feeding rate of the herbivores ($p\delta$), as well as joint proportional changes (q) in predation rate ($q\beta$) and predator population growth ($q\gamma$). Taking the product of the overall number of herbivore days (D) and the herbivore's consumption rate as a measure for the damage to the plant it can be shown that:

$$\frac{\delta D(\alpha,\beta,\gamma)}{p\delta D(p\alpha,q\beta,q\gamma)} = \frac{D(\alpha,\beta,\gamma)}{\dfrac{p}{q}D\left(\dfrac{p}{q}\alpha,\beta,\gamma\right)} \tag{8}$$

Thus, when the proportional changes in all parameters ($p\delta$, $p\alpha$, $q\beta$, $q\gamma$) are equal ($p=q$), then the amount of plant damage remains unaltered. This defines the critical borderline where positive and negative effects of proportional changes cancel out. When $p<q$, the plant suffers less damage, whereas for $p>q$ it suffers more. In words, plants do not always profit from increased direct defence against herbivores. They can only profit when the reduction in feeding and growth rates of the herbivore is stronger than the concomitant reduction in predation and growth rate of the predator.

When direct and indirect defences interact in an antagonistic way, then a decrease in p causes q to drop below unity. For synergism to occur a decrease in p should cause q to become larger than unity. This may occur when the plant's direct defences make the herbivore more vulnerable to predation. For example, when herbivorous arthropods get temporarily caught in the secretions of glandular trichomes, they are an easy prey for predatory arthropods, which are not much hindered by the secretions; Braman and Beshear (1994) report this type of positive interaction for mirid bugs on Florida azaleas. Plants well defended by glandular hairs also provide enemy-free space. For a predator with a juvenile phase vulnerable to intraguild predators it may be profitable to deposit its eggs on such plants despite the negative effects of glandular secretions. This may explain why some anthocorid predators readily deposit eggs on tomato (Coll 1996; Ferguson & Schmidt 1996), despite negative effects on foraging success due to dense mass of glandular hairs (Coll & Ridgway 1995). There may be many more manifestations of synergism between direct and indirect defences. In a review of published studies on interactions between plant resistance and biological control, Hare (1992) found that out of 16 cases there were six antagonistic, eight neutral and two synergistic. The underlying mechanisms are poorly understood, however.

Do benefits of plant defence outweigh costs?

To date, there have been hundreds of attempts to measure a reduction in fitness of resistant plants in absence of selection by herbivores. These studies have generally been inconclusive, suggesting that plants generally invest most of their energy in growth and differentiation and spend only a small portion of their nutrient and energy budget in direct defences (Simms & Rausher 1987, 1989; Herms & Mattson 1992; Simms 1992; Bergelson and Purrington 1996). This holds even more strongly for indirect defences. For example, Beattie (1985) discusses work of O'Dowd showing that extra-floral nectar production in *Ochroma* took only 1% of the energy invested in leaves. The nectaries themselves are usually simple in structure and closely connected to the vascular system, which contains primary metabolites under pressure, making the secretion process very cheap. Phloem exudation does not even require a specialized morphological structure, such as a nectarium. Also, production of edible pollen does not seem very expensive. It has to be produced anyway, its road to success in fertilization is extremely narrow, making the production of larger amounts a viable option. As a final example, the metabolic costs of producing alarm signals is probably a negligible fraction of the costs involved in producing a leaf. Dicke and Sabelis (1989) estimated that the instantaneous costs of producing alarm chemicals (monoterpenes and methyl-salicylate) are equivalent to *c.* 0.001% of leaf-production costs per day.

Although these calculations are consistent in showing low metabolic expenditure in direct and indirect defences, they may underestimate the real costs: they underrate the large effect of small initial costs on exponential leaf growth of leaves (Dicke & Sabelis 1989), and they completely ignore the costs of making the infrastructure for producing domatia, nectaries, food bodies, pollen, alarm chemicals, etc., as these are hard to estimate. In some special cases, e.g. production of food bodies, one may argue that costs are not very low. Another reason for underestimating the costs of defences is that experimental analyses invariably suffered from linkage problems. The only way to measure unambiguously the fitness cost of a resistance gene is to use genetic engineering to introduce the resistance gene into a controlled genetic background without an associated linkage block. This method has been successfully explored with respect to herbicide resistance by Bergelson *et al.* (1996).

One may ask why resistance genes are often polymorphic in natural plant populations, when the costs of defence are so low (Parker 1990; Bergelson & Purrington 1996). It seems more likely that investment in defence depends on the risks of herbivory and on underlying trade-off relations with, for example, seed set (Bergelson *et al.* 1996). However, it is also possible that plants play defence games against each other and that the best defence depends not only on what others are doing, but — as we will see later — also on the mobility of the herbivores between plants. Similarly, one may ask why several plant defence mechanisms are inducible, if the direct costs of the defensive plant products are so low (Beattie 1985; Dicke & Sabelis 1989). The answer to this question may be found in a more careful consideration of the role of cheaters (hyperpredators, other herbivores, neighbouring plants). If it is not because of construction costs or metabolic costs that some indirect defences are inducible, then it is possibly the indirect costs due to the misuse by organisms that harm the plant.

Recognizing that the exact costs of indirect defence are hard to measure, it is important to stress that the benefits can be huge. For example, ants routinely reduce herbivores and seed predators by 50–100% (Schemske 1980; Stephenson 1982; Koptur 1984; Putz & Holbrook 1988) and the consequences of ant defence are increased plant growth, survival and reproduction (Inouye & Taylor 1979; Schemske 1980; Messina 1981; Horvitz & Schemske 1984). Other predators than ants also play a role in suppressing mealybugs, scale insects, leafhoppers, whiteflies, spider mites and rust mites

(Helle & Sabelis 1985; Minks & Harrewijn 1989; Rosen 1990; Lindquist *et al.* 1996; Ben-Dov & Hodgson 1997). In absence of these predators, each of these plant parasites is capable of completely defoliating the plant. Whenever the benefits of indirect defences are hard to detect or even absent under field circumstances, this low benefit may well be due to a low availability of predators in the environment, competition between plants for acquiring bodyguards and defence/escape strategies of the herbivores (Rashbrook *et al.* 1992).

Did plant–predator mutualisms arise from (co)evolution?

As biologists, we observe traits of organisms and ask why these traits have the exact form observed, and not another one. Answers may be derived from direct causation, developmental processes (ontogeny), phylogenetic history, or from considerations of function or adaptive value. A useful approach to assessing functions is to demonstrate that for certain traits there is a resident population that cannot be invaded by any other mutant trait. Such a trait is called an evolutionarily stable strategy (ESS) (Maynard Smith 1982). In multitrophic systems one should demonstrate that there exists an unbeatable combination of coevolutionarily stable strategies at each trophic level (CoESSs). However, such an analysis shows the resistance to invasion of a population with a certain trait, but not how the trait evolved in the first place. The exact evolutionary route may well start from traits evolved for totally different reasons. For example, once pollen evolved to serve its function in sexual reproduction, the thickness of the exine layer may have varied from plant to plant, which in turn may have influenced the possibility for predators to feed on the pollen content.

No one could draw the conclusion that *right from the start* plants provided protection, food and signals *with the sole aim* to arrest and attract predators as bodyguards (see, for example, the citations from a letter written by Jermy in Harrewijn *et al.* 1995). In some cases such a conclusion can even be rejected *a priori* on theoretical grounds. For example, the first mutant plant to release volatile chemicals in response to herbivore attack will not profit at all as there were no predators that could possibly associate this signal with the presence of prey. For such a mutant to spread in the plant population, a number of very special conditions should be met (such as associative learning of predators in response to chemical signals), and even if these were met, it can never have been the 'aim' of the mutant plant to attract predators, as it could not possibly 'know' the effect. Hence, it seems likely that herbivore-induced release of volatile chemicals has evolved as a byproduct of a trait that evolved for totally different reasons than to attract predators. For example, several of the herbivore-induced plant volatiles are known to have an antibiotic effect, suggesting that they may have evolved as a direct defence against phytopathogens (Harrewijn *et al.* 1995; Turlings *et al.* 1995). In fact, some of the plant-emitted terpenoids have been shown to slow down development of phytophagous insects or even kill them (Turlings *et al.* 1995). This may have paved the way for subsequent coevolution of signals betraying the presence of herbivores to their enemies. The traits under selection are then the quantity and quality of the chemicals released from the herbivore-damaged leaves and—via leaf-to-leaf signal transduction—from elsewhere in the plant. It is not self-evident that a plant should release volatile chemicals that function as antibiotics or are present in plants for other reasons. If their release would have the effect of attracting even more herbivores, there would be selection on the plants to avoid release, but if plants benefit all-in-all from releasing chemicals as volatiles, then selection will favour plants that fine-tune the signal (qualitative and quantitative composition of the blend), and will favour predators that discriminate between signals and respond. Plant–predator communication has a great potential for conflict of interest between signaller and receiver and, for the signalling system to be evolutionary stable, the signals are expected to be costly. Godfray (1995) argues that to gain some future protection from attracting natural enemies, evolutionary stability of the signalling system demands that the signal is costly, but if the signals are true 'calls for help', the minimum costs—for the signalling system to be evolutionarily stable—decrease. The cost of the signals may be direct (e.g. biosynthetic costs) as well as indirect, e.g. because they attract other herbivores. These indirect costs could be so high that they make the alarm system counterproductive, but if not, they should be considered part of the overall costs of signal emission and then 'paradoxically' they contribute to the evolutionary stability of the signalling system (Godfray 1995).

The coevolutionary route of plant–predator mutualisms is for a large part concealed in the chemistry of alarm chemicals and the behavioural responses of predatory arthropods. As these are difficult to retrieve, inferences on reciprocal selection underlying coevolution are therefore doomed to remain conjectural. Domatia, food bodies, extrafloral nectaries and perhaps pollen may offer better prospects in unravelling the coevolutionary history of plant–predator interactions. On the one hand, it seems feasible to show selective benefits to the plant both theoretically and experimentally. On the other hand, these morphological structures stand a good chance of being preserved in the fossil record (O'Dowd *et al.* 1991), there is interspecific and intraspecific, genetic and phenotypic variation among plants in quantity, quality and structure of domatia, nectaries and food bodies (Beattie 1985; Jolivet 1996), and it seems possible to artificially select for (isofemale) lines of predators that either do or do not use domatia, nectaries and food bodies. Thus, perhaps it is possible to reconstruct 'this coevolutionary game'.

Tritrophic game theory and metapopulation dynamics

The evolution of direct and indirect plant defences against arthropod herbivores is by no means a simple process. For one thing we have to take the defence and exploitation strategies of all three trophic levels into account. For another, time scale and spatial scale arguments force us to consider metapopulation models with the full tritrophic structure, and incorporating plant dynamics will add new dynamical behaviour to the repertoire of the otherwise ditrophic models.

Indeed, evolution in structured populations can only be properly understood by taking metapopulation structure into account. Although the strategies play against each other at the patch level, their relative success determines metapopulation dynamics, and metapopulation processes in turn determine which strategies will meet and compete again at the patch level. This chain of processes will be further referred to as the population-dynamical feedback.

To gain insight into this complex problem, carefully planned simplifications are required. Our strategy is first to consider the evolution of relevant traits at one trophic level in pairwise interactions with one other level (predator–herbivore, herbivore–predator, plant–herbivore, herbivore–plant, plant–predator, predator–plant), thereby assuming a steady state at the metapopulation level. We will end with a tentative, verbal discussion of what may happen evolutionarily in the full tritrophic system, again assuming steady-state metapopulation conditions for reasons of simplicity.

The predator's dilemma: to milk or to kill?

It is one thing for a plant to attract and arrest predators, but it is another to lure predators that are also effective in suppressing the herbivore population on the plant. Clearly, it is the predator who reacts to the lure, and decides how fast it will consume herbivores, how fast it will multiply and how long it will stay. In other words, the effectiveness of indirect plant defences depends on the herbivore-exploitation strategies present in the predator population. In principle, these exploitation strategies can be many and varied. To illustrate this point it is instructive to make use of the 'pancake' model for local predator–herbivore dynamics. The strategy set is determined by combinations of the *per capita* growth rate of the predator (γ) and the *per capita* predation rate (β). However, in a spatially structured environment there is one more parameter: the *per capita* dispersal rate of the predator. Of course, all predators will disperse when herbivores are exterminated and there is no other food, but they can also decide to move away during the interaction. Both decreased predation and increased dispersal of the predator relieve the prey of predation pressure, thereby causing the prey population to represent a larger future food source for the predator (Gilpin 1975; Van Baalen & Sabelis 1995). Increased dispersal during the interaction, however, seems a more sensible strategy, as it increases the founding of new populations by the dispersers and in what follows we will focus on this dispersal trait alone.

Let us assume for simplicity that the *per capita* dispersal rate is a constant μ. The effective *per capita* population growth rate of the predators is reduced by μ so that the predator growth rate now equals:

$$\frac{dP}{dt} = (\gamma - \mu)P \tag{9}$$

As a decreased value of γ has been shown to drastically alter local predator–prey dynamics, so will increased values of μ. As illustrated in Fig. 5.5a, a small increase of μ from 0 to 0.04 day^{-1}, greatly alters the area under the prey curve and thereby also the damage to the plant. Hence, a plant would benefit from stimulating predators to stay until all prey are eliminated, but whether it succeeds depends on what is best for the predators (as well as on the population-dynamical feedback). The local success of a predator's exploitation strategy may be expressed as the number of dispersers produced during the interaction with prey, plus those dispersing after prey extermination. As shown in Fig. 5.7, the production of dispersers increases disproportionally with μ and reaches an asymptote when the *per capita* dispersal rate is so high that the predators cannot suppress the growth of the prey population any more, i.e. when $\mu = \gamma - \alpha + \beta\, P(0)/N(0)$. Thus, predators suppressing dispersal during the interaction reach their full capacity to suppress the local prey population, but they produce the lowest number of dispersers per prey patch. In terms of production of dispersers

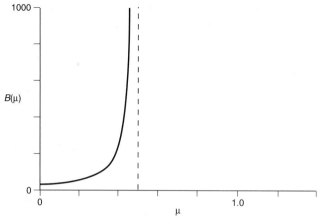

Figure 5.7 The relation between the overall production of predator dispersers (*B*) and the *per capita* rate of dispersal (μ) during the predator–prey interaction period (from predator invasion to prey elimination). From Van Baalen and Sabelis (1995).

this so-called *killer* strategy does less well than the strategy of a *milker*, which typically has a non-zero dispersal rate. However, if killers enter a prey patch with milkers, then they would steal much of the prey the milkers had set aside for future use. Therefore, if there is a risk of invasions by killers, it pays to anticipate such events and selection will favour exploitation strategies that are less milker type and more killer type.

The outcome of the milker–killer dilemma is determined by a complex interplay of local competition between the exploitation strategies and global or metapopulation dynamics. It depends on the probability of coinvasions in the same prey patches (or alternatively on the probability of exploiting a prey patch alone), on the resulting production of dispersers per prey patch and on metapopulation dynamics, as this in turn determines the probability of (co)invasion. The complexity of this population-dynamical feedback is staggering; to continuously keep track of the numbers of each strategy type when competing in local populations, dispersing into the global population and invading into local populations requires a massive book-keeping procedure. Hence, one is bound to simplify to get some insight. For example, Van Baalen and Sabelis (1995) assumed that all patches start with exactly the same number of predators and prey (the assumption of a metapopulation-wide equilibrium) and that the predators had enough time to reach their full production potential per prey patch (the assumption of sequential interaction rounds). In this setting they considered the reproductive success of one mutant predator clone with a *per capita* dispersal rate, μ_m, relative to the mean success of predators in the population of the resident clone which possesses another *per capita* rate of dispersal (μ_{res}). The question is then whether there exists a value of μ_{res} for which it does not pay any mutant to deviate (Maynard Smith 1982). In particular, Van Baalen and Sabelis (1995) calculated for which combination of parameters ($P(0)$, $N(0)$, α, β, γ) it does not pay to increase μ away from zero, i.e. the conditions favouring selection for killers. As illustrated in Fig. 5.8, the general outcome is that killers are usually favoured by selection, except when the number of predator foundresses is low and the number of prey foundresses is high. In other words, the milkers are favoured as long as they have a sufficiently large share in their local populations to maintain control over the time to prey elimination (τ).

This analysis whets the appetite for more elaborate work taking into account:

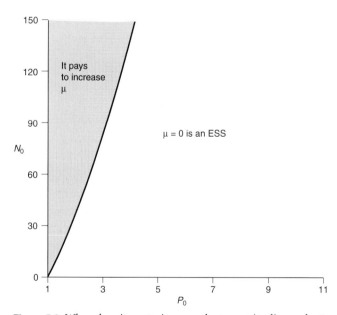

Figure 5.8 When does it pay to increase the *per capita* dispersal rate of the predator (μ) away from zero? A diagram showing that milker strategies are only possible when $P(0)$ is low and $N(0)$ sufficiently high. ESS, evolutionarily stable strategy. Modified after Van Baalen and Sabelis (1995).

1 asynchrony in local dynamics;

2 stochastic variation in predator and prey colonization rates (since these are probably low!);

3 an upper boundary to prey population size set by the local amount of food; and

4 metapopulation dynamics.

Such extensions are likely to show that milkers achieving a longer inter-action period are also exposed longer to subsequent predator invasions (and thus, face competition with killers sooner or later), that stochastic rather than uniform invasions will help to isolate milkers, thereby gaining full advantage of their exploitation strategy, and that limits to the amount of food available for the prey will decrease opportunities for a milker, as it loses full control over the interaction period (τ). As these factors have opposite effects it is not immediately clear whether killers or milkers will win the battle or may even coexist.

Apart from the need for more theoretical work, experimental analysis of variation in exploitation strategies of predators seems a promising avenue for future research. Such an analysis carried out for the case of the predatory mite *Phytoseiulus persimilis* Athias-Henriot, revealed that laboratory cul-tures harbour exclusively predators of the killer type (Sabelis & Van der Meer 1986), whereas field-collected populations in the Mediterranean (Sicily) exhibit some variation in the onset of dispersal before or after prey elimination (B. Pels & M.W. Sabelis, unpublished observation). Most interestingly, early dispersal was found in a strain collected inland where local populations are scarce and hence more isolated! Possibly, ants that tend colonies of honeydew-producing homopterans represent an example of prudent predation. The ants present on an individual plant may well be kin, as they are likely to originate all from the same nest. Hence, the conditions for the evolution of decreased predation on aphids are met.

Obviously, the prevalence of killers is of great importance for the evolution of indirect plant defences. By providing protection and food to predatory arthropods and by signalling herbivore attack to predators, plants increase the predator invasion rate into young colonies of the herbivorous arthropods and this promotes the probability of coinvasions of milkers and killers, which—other things being equal—is ultimately in favour of the latter. Yet there may be a pitfall in that, so far, neither theoretical nor experimental analyses addressed the possibility of more flexible strategies, such as 'milk when exploiting the prey patch alone, and kill when other (e.g. non-kin) predators have entered the same patch'.

148

The herbivore's dilemma: to stay or to leave?

Just like the predators, the herbivores are independent players in the tritrophic game. When their local populations are discovered and invaded by predators of the milker type, there are still possibilities for achieving reproductive success, especially when the milker has such a high dispersal rate that it cannot suppress the herbivore population. However, when killers enter the herbivore population, it may pay the herbivores to invest in defence against the killer-like predators or to leave the prey patch in search for enemy-free space. For simplicity we will consider the last type of response only.

Consider the pancake predation model again (Eq. 1), but now extended with a *per capita* dispersal rate (v) of the herbivore:

$$\frac{dN}{dt} = (\alpha - v)N - \beta P$$

$$\text{(10)}$$

$$\frac{dP}{dt} = \gamma P$$

As shown in Fig. 5.5b an increase in v causes the time to prey elimination to become shorter, and so will the area under the herbivore curve and the number of predators that will disperse. We may now ask whether there is an evolutionarily stable dispersal rate. To get an answer we should first define reproductive success as the *per capita* dispersal rate (v) multiplied by the area (A) under the herbivore curve (which itself depends on v), divided by the initial number of herbivores ($N(0)$). This fitness measure always shows a maximum for intermediate values of v, because A decreases rapidly with v. Suppose all patches start synchronously with the same initial number of predators and herbivores (assumption of metapopulation-wide equilibrium) and that each patch is invaded by $N(0)$ herbivore clones with v_{res} and just one herbivore mutant clone with v_m. Further assume that the two types of herbivore clones are attacked in proportion to their relative abundance, but that the mutant is so rare that $N_{res} + N_m \approx N_{res}$. This makes herbivore dynamics in the patch and time to prey elimination (τ) entirely dependent on the resident population. The mutant's presence does not influence the growth of the resident herbivore population. Now, we may ask whether there is a resident herbivore population with v_{res} that cannot be invaded by a mutant with another value of v_m. The results presented in Fig. 5.9 show that the ESS dispersal rate v^* increases with:

1 decreasing *per capita* population growth rate of the herbivores (α);

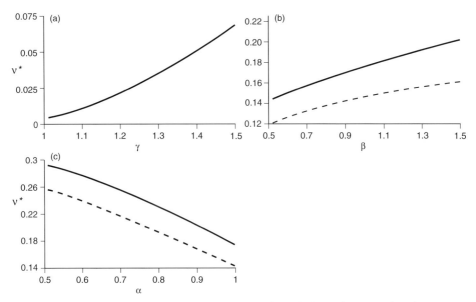

Figure 5.9 (a) The optimal prey dispersal rate (v^*) as a function of predator growth rate (γ). A milker-like predator strategy (low γ) favours a low prey dispersal rate (predation tolerant prey). A killer-like predator strategy (high γ) favours a high prey dispersal rate (predator avoidance behaviour). Other parameter values: $N1(0) = 49$, $N2(0) = 1$, $P(0) = 1$, $\alpha = 1$, $\beta = 1$. (b) The relation between evolutionary stable strategy (ESS) dispersal rate and α. Drawn lines: $N1(0) = 99$, dashed lines: $N1(0) = 199$. In both cases the mutant $N2(0) = 1$. Other parameter values: $P(0) = 1$, $\beta = 1$, $\gamma = 1.5$. (c) The relation between ESS dispersal rate and β. Drawn lines: $N1(0) = 99$, dashed lines: $N1(0) = 199$. In both cases the mutant $N2(0) = 1$. Other parameter values: $P(0) = 1$, $\alpha = 1$, $\gamma = 1.5$.

2 increasing *per capita* predation rate (β); and
3 increasing *per capita* population growth rate of the predators (γ), and, thus, with a decrease in time to prey elimination (τ).

Moreover, the larger the initial number of herbivores, the longer the time to prey elimination and the smaller the ESS dispersal rate v^*.

These results provide some important clues as to how the ESS dispersal rate v^* will change with the exploitation strategy of the predators. This is because milkers are predators with an effectively lower *per capita* rate of population growth due to non-zero dispersal ($\gamma - \mu$) and the lower the predator's population growth rate, the lower the ESS dispersal rate v^* of the herbivore will be. Thus, a prevalence of milkers in the predator population will cause selection for lower dispersal rates of the herbivores and thus in-creased tendency to stay in the herbivore aggregation, whereas a prevalence of killers will cause selection for higher dispersal rates of the herbivores out of the herbivore aggregation. These higher ESS dispersal rates of the herbivores can, of course, not increase beyond their *per capita* population growth (α). Thus, the existence of herbivore aggregations in the presence of killer-like predators is not in conflict with our theory.

There is much to be learned as to how the evolution of plant defence strategies interfere with the evolution of herbivore dispersal. Increased efforts in direct plant defence will decrease the *per capita* rate of herbivore population growth (α), and as a byproduct this will trigger selection for a higher ESS dispersal rate of the herbivores v^*. Thus, the effective *per capita* rate of herbivore population growth ($\alpha - v^*$) is decreased even more. This will pave the way for the evolution of feeding deterrents. The same applies to increased investments in indirect plant defences. When plants promote the *per capita* predation rate (β) or the *per capita* rate of predator population growth (γ), the byproduct will be that selection favours increased ESS dispersal rates of the herbivores, thereby lowering the effective rate of herbivore population growth ($\alpha - v^*$). Thus, plants may also invest in releasing herbivore deterrents signalling a high risk of being eaten by predators, and the herbivores will be selected for vigilance in detecting the actual presence of predators. There are several examples of herbivorous arthropods that prefer plants with lower risk of falling victim to natural enemies despite a lower food quality (Fox & Eisenbach 1992; Ohsaki & Sato 1994). However, it has never been really clarified why the low-quality plants are visited less frequently by the natural enemies. A particularly nice illustration of herbivore vigilance is the observation that, when exposed to odour from plants, predators and conspecific herbivores, larvae of the western flower thrips seek refuge in webs covering leaf sections infested by spider mites, whereas they prefer undamaged leaves when exposed to odour from plants and conspecific larvae alone (A. Pallini *et al.*, unpublished observations).

The plant's dilemma: direct/indirect defence or no defence at all?
By investing in direct and indirect defences, a plant gains protection against herbivory, but in doing so, it also benefits its neighbours. If these are close kin, an individual plant also increases its inclusive fitness by investment in defence, but if not, it may well promote the fitness of its competitors for the same space and nutrient sources. As it were, the neighbour gains

associational protection (Atsatt & O'Dowd 1976; Hay 1986; Pfister & Hay 1988; Fritz & Nobel 1990; Fritz 1995; Hjältén & Price 1997). This leads to the plant's dilemma: should it defend itself, thereby benefiting its neighbours as well, or decrease its defensive efforts? The solution is simple: the defences should protect the plant without benefiting palatable neighbours too much.

When plants are either undefended or well defended and herbivores do not discriminate between them, there are three possibilities:

1 all plants are palatable;
2 all plants are well-defended; or
3 there is a stable mixture of palatable and well-defended plants.

Coexistence of the two types is possible when either of them increases when rare; in a population of well-defended plants a rare palatable plant can easily gain cheap protection by associating close to a well-defended plant, whereas in a population of palatable plants, a rare, well-defended plant does better as it benefits only few of the palatable individuals and, hence, increases the average fitness of the palatables only very little. When either of the two types gradually increases its share in the total population, the benefits wane and ultimately balance the costs thereby giving rise to a polymorphic plant population (Sabelis & De Jong 1988; Augner *et al.* 1991; Tuomi & Augner 1993; Augner 1994).

The conditions for polymorphisms are quite broad, but some of the critical assumptions are not generally valid. For example, herbivores are likely to distinguish between palatable and well-defended plants. In that case, an individual plant is likely to benefit from direct defences and may even drive the selective herbivores towards the palatable plants. It should be noted that this mechanism does not apply to indirect defences. Predatory arthropods are usually more mobile than the prey (stages) they attack and they will readily move from the plant that employs them as bodyguards to a neighbour plant when the latter is under herbivore attack. Thus, even when the herbivore is a selective feeder, palatable plants will profit more easily from settling close to a plant defended by predatory arthropods as bodyguards. Clearly, this mechanism promotes polymorphisms (Sabelis & De Jong 1988). However, when defensive plant strategies are not discrete, but continuous (i.e. they cover the full range of possible investment levels), then there may be no polymorphism because ultimately all plants will exhibit the best average defensive response.

The latter case was analysed by Tuomi *et al.* (1994). They assumed that the cost of defence increases linearly with the probability of killing the herbivore, the slope being referred to as the marginal cost of defence (i.e. how fast costs increase with the impact of defence on the herbivore). The ESS lethality level depends on the risk of herbivory, the marginal cost of defence, and the mobility of the herbivores between neighbouring plants, as shown in Fig. 5.10. When herbivory risks are low and marginal costs of defence high, then it does not pay to kill the herbivore. But when the risk of herbivore damage is high and marginal defence costs are sufficiently low, then it pays to kill the herbivore. For intermediate ratios of marginal defence costs and risk of herbivory, ESS lethality depends on the mobility of the herbivore between neighbouring plants (Tuomi *et al.* 1994). Obviously, high mobility will cause neighbouring plants to share the same herbivores and this will select for lower lethality levels, whereas low mobility will select for increased lethality. Now that it is becoming increasingly clear that neighbouring plants may communicate via damage-related signals (Farmer & Ryan 1990; Bruin *et al.* 1992, 1995; Shonle & Bergelson 1995; Adviushko *et al.* 1997; Shulaev *et al.* 1997), it may well be

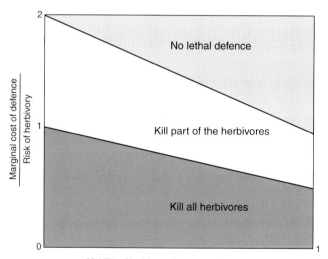

Figure 5.10 The effect of herbivore mobility (*x*-variable) on the parameter areas where the evolutionarily stable strategy is to kill the herbivore (lower shaded region), to kill part of the herbivores (intermediate, unshaded region) and not to invest in killing herbivores at all (upper shaded region). The areas depend on the *y*-variable expressed in terms of the ratio of marginal costs of plant defence and the initial risk of herbivory. The cost of herbivory equals unity if the herbivore survives, and 0.5 if it dies. From Tuomi *et al.* (1994).

that plant defences include strategies conditional upon the neighbour's state, as defined by (i) whether it is actually under attack, and by (ii) its defensive response. This is a largely open problem in need of further theoretical and experimental work.

Whether the effects of defensive efforts go in discrete jumps or are more gradual is an important determinant for the existence of polymorphisms, but the most relevant message is that in both cases associational protection may lead to a lower average investment in defences. This applies to direct defences (Tuomi *et al.* 1994), as well as indirect defences (Sabelis & De Jong 1988).

But coevolution may act as a boomerang …

The most elusive unknown of all is the interplay between metapopulation dynamics and evolution at all three trophic levels. To illustrate this, it is worthwhile to carry out an (admittedly speculative) thought experiment by considering what will happen when plants evolve to invest more in direct and indirect defences, and predators are initially of the killer type. First and foremost, increased defensive efforts by the plant will decrease the size of the herbivore's metapopulation. Subsequently, the size of the predator's metapopulation will decrease which, in turn, will cause a drop in the rate of the invasion of predators into herbivore patches. As a consequence, the probability of coinvasion of predators with different prey-exploitation strategies into the same herbivore patch will decrease, thereby providing a selective advantage to predators that are more milker-like. This will render indirect plant defences less effective. Hence, plants will be selected to lower their investment in indirect defences. The cycle of cause and effect will continue, but it seems reasonable to expect that ultimately coevolution will favour plants with a lower investment in indirect defences and predators with a more milker-like exploitation strategy.

A similar thought experiment can be carried out for the case that a plant does not exclusively benefit from its own investment in direct and indirect defences, but also its neighbours do, and they may well be competitors (Sabelis & De Jong 1988; Augner *et al.* 1991; Tuomi *et al.* 1994). Again, increased investment in plant defence will cause a boomerang effect because neighbouring plants profit and allocate the energy saved directly to increase their seed output or indirectly by increasing their competitive ability.

Boomerang coevolution arises through the impact of plant defences on alternative allocation strategies of neighbouring plants and via the positive effect on the milker-like prey exploitation strategies. This may well be the evolutionary explanation for why plants generally channel so little of their energy resources into defence against herbivores, whether it concerns direct defence (Simms & Rausher 1987, 1989; Herms & Mattson 1992; Simms 1992) or indirect defence (p. 52 in Beattie 1985; Dicke & Sabelis 1989). Hence, we hypothesize that boomerang coevolution may constrain the plant's investment in direct and indirect defences. Low investment, however, does not necessarily imply that plant defences have a low impact. This will entirely depend on the quantitative details of where the offensive and defensive traits of the interacting organisms at all three trophic levels will settle evolutionarily. In other words, the impact of the plant's defences will increase if herbivores become more milker-like and predators more killer-like, and the impact will decrease when herbivores become more killer-like and predators more milker-like.

Epilogue

This chapter could have been entitled 'Why are plants edible?'. Indeed, one may wonder why plants are not armoured such that herbivory in arthropods would have been an unprofitable lifestyle. One easy answer is that such a harness would be very costly. We have tried to provide other answers in an attempt to explain why many plants spend only very little of their energy budget in defence. Firstly, by investing more in direct defence, plants do not only protect themselves but also their neighbours, especially when herbivores can easily move from one plant to the other and cannot discriminate between well-defended and ill-defended plants. Hence, the ill-defended plants can gain associational protection when settling close to well-defended plants. As these plants can spend their energy for other purposes (seed set and/or competitiveness), on average, it will not pay plants to invest much in defence. Secondly, by investing in indirect defence, neighbouring plants will profit too, depending on the degree of mobility of the predator and the location of profitable prey patches. Hence, plants densely packed with bodyguards are also an unlikely outcome of evolution in tritrophic interations. Finally, increased investment in indirect plant defence may trigger a population-dynamical feedback, where the number of local herbivore populations decreases and milker-like predators will gain selective advantage, as they face less competition with killers within the local populations. Where the tritrophic system will settle evolutionarily is hard to predict, because tritrophic systems are known to exhibit complex population dynamics (Sabelis *et al.* 1991; Jansen and Sabelis 1992, 1995; Klebanoff & Hastings 1994; Jansen 1995; Kuznetsov & Rinaldi 1996). A definitive explanation for why 'the world is green' (Hairston *et al.* 1960) requires understanding of why plants invest 'little' in direct defence against herbivores and gain protection 'cheaply' from killer-like predators.

References

Adviushko, S.A., Brown, G.C., Dahlman, D.L. & Hildebrand, D.F. (1997). Methyl jasmonate exposure induces insect resistance in cabbage and tobacco. *Environmental Entomology*, **26**, 642–854.

Agrawal, A.A. & Karban, R. (1997). Domatia mediate plant-arthropod mutualism. *Nature* **387**, 562–563.

Alborn, H.T., Turlings, T.C.J., Jones, T.H., Stenhagen, G., Loughrin, J.H. & Tumlinson, J.H. (1997). An elicitor of plant volatiles from beet armyworm oral secretion. *Science*, **276**, 945–949.

Atsatt, P.R. & O'Dowd, D.J. (1976). Plant defense guilds. *Science*, **193**, 24–29.

Augner, M. (1994). Should a plant always signal its defence against herbivores? *Oikos*, **70**, 322–332.

Augner, M., Fagerström, T. & Tuomi, J. (1991). Competition, defence and games between plants. *Behavioural Ecology and Sociobiology*, **29**, 231–234.

Bakker, F.M. & Klein, M.E. (1992). Transtrophic interactions in Cassava. *Experimental and Applied Acarology*, **14**, 299–311.

Bakker, F.M. & Klein, M.E. (in press). Extrafloral nectaries and extrafoliar domatia. In *Acarology IX*, Vol. 2 (Ed. by R. Mitchell, D.J. Horn, G.R. Needham & W.C. Welbourn). Ohio Biological Survey, Columbus.

Beattie, A.J. (1985). *The Evolutionary Ecology of Ant–Plant Mutualisms.* Cambridge University Press, Cambridge.

Ben-Dov, Y. & Hodgson, C.J. (1997). *Soft Scale Insects. Their Biology, Natural Enemies and Control.* World Crop Pests, Vol. 7, Elsevier Science Publishers, Amsterdam.

Bentley, B.L. (1977). Extrafloral nectaries and protection by pugnacious bodyguards. *Annual Review of Ecology and Systematics*, **8**, 407–427.

Bergelson, J. & Purrington, C.B. (1996). Surveying the costs of resistance in plants. *American Naturalist*, **148**, 536–558.

Bergelson, J., Purrington, C.B., Palm, C.J. & Lopez-Gutierrez, A. (1996). Costs of resistance: a test using transgenic *Arabidopsis thaliana*. *Proceedings of the Royal Society B*, **263**, 1659–1663.

Berryman, A.A., Dennis, B., Raffa, K.F. & Stenseth, N.Ch. (1985). Evolution of optimal group attack, with particular reference to bark beetles (Coleoptera: Scolytidae). *Ecology*, **66**, 898–903.

Boethel, D.J. & Eikenbary, R.D. (Eds) (1986). *Interactions of Plant Resistance and Parasitoids and Predators of Insects*. Ellis Horwood, Chichester.

Boland, W., Feng, Z., Donath, J. & Gäbler, A. (1992). Are acyclic C11 and C16 homoterpenes plant volatiles indicating herbivory? *Naturwissenschaften*, **79**, 368–371.

Braman, S.K. & Beshear, R.J. (1994). Seasonality of predaceous plant bugs (Heteroptera: Miridae) and phytophagous thrips (Thysanoptera: Thripidae) as influenced by host plant phenology of native azaleas. *Environmental Entomology*, **23**, 712–718.

Bronstein, J.L. (1994). Our current understanding of mutualism. *The Quarterly Review of Biology*, **69**, 31–51.

Bruin, J., Dicke, M. & Sabelis, M.W. (1992). Plants are better protected against spider mites after exposure to volatiles from infested conspecifics. *Experientia*, **48**, 525–529.

Bruin, J., Sabelis, M.W. & Dicke, M. (1995). Do plants tap SOS-signals from their infested neighbours? *Trends in Ecology and Evolution*, **10**, 167–170.

Buckley, R.C. (1986). Ant–plant–Homoptera interactions. *Advances in Ecological Research*, **16**, 53–85.

Buckley, R.C. (1987). Interactions involving plants, Homoptera and ants. *Annual Review of Ecology and Systematics*, **18**, 111–135.

Coll, M. (1996). Feeding and oviposition on plants by an omnivorous insect predator. *Oecologia*, **105**, 234–240.

Coll, M. & Ridgway, R.L. (1995). Functional and numerical responses of *Orius insidiosus* (Heteroptera: Anthocoridae) to its prey in different vegetable crops. *Annals of the Entomological Society of America*, **88**, 732–738.

Collyer, E. (1964). The effect of an alternative food supply on the relationship between two *Typhlodromus* species and *Panonychus ulmi* (Koch) (Acarina). *Entomologia Experimentalis et Applicata*, **7**, 120–124.

De Jong, M.C.M. & Sabelis, M.W. (1988). How bark beetles avoid interference with squatters: An ESS for colonization by *Ips typographus*. *Oikos*, **51**, 88–96.

De Jong, M.C.M. & Sabelis, M.W. (1989). How bark beetles avoid interference with squatters: a correction. *Oikos*, **54**, 128.

Dicke, M. (1986). Volatile spider-mite pheromone and host-plant kairomone, involved in spaced-out gregariousness in the spider mite *Tetranychus urticae*. *Physiological Entomology*, **11**, 251–262.

Dicke, M. (1994). Local and systematic production of volatile herbivore-induced terpenoids. *Journal of Plant Physiology* **143**, 465–472.

Dicke, M. & Sabelis, M.W. (1988). How plants obtain predatory mites as bodyguards. *Netherlands Journal of Zoology*, **38**, 148–165.

Dicke, M. & Sabelis, M.W. (1989). Does it pay plants to advertise for bodyguards? Towards a cost–benefit analysis of induced synomone production. In *Variation in Growth Rate and Productivity of Higher Plants* (Ed. by H. Lambers, M.L. Cambridge, H. Konings & T.L. Pons), pp. 341–358. SPB Academic Publishing BV, The Hague.

Dicke, M. & Sabelis, M.W. (1992). Costs and benefits of chemical information conveyance: proximate and ultimate factors. In *Insect Chemical Ecology, an Evolutionary Approach* (Ed. by B. Roitberg & M. Isman), pp. 122–155. Chapman & Hall, London.

Dicke, M., Sabelis, M.W. & Groeneveld, A. (1986). Vitamin A deficiency modifies response of predatory mite *Amblyseius potentillae* to volatile kairomone of two-spotted spider mite, *Tetranychus urticae*. *Journal of Chemical Ecology*, 12, 1389–1396.

Dicke, M., Van Beek, T.A, Posthumus, M.A., Ben Dom, N., van Bokhoven, H. & De Groot., AE. (1990a). Isolation and identification of volatile kairomone that affects acarine predator–prey interactions: involvement of host plant in its production. *Journal of Chemical Ecology*, 16, 381–396.

Dicke, M., Sabelis, M.W., Takabayashi, J., Bruin, J. & Posthumus, M.A. (1990b). Plant strategies of manipulating predator–prey interactions through allelochemicals: prospects for application in pest control. *Journal of Chemical Ecology*, 16, 3091–3118.

Dicke, M., Van der Maas, K.J., Takabayashi, J. & Vet, L.E.M. (1990c). Learning affects response to volatile chemicals by predatory mites. *Proceedings of the Experimental and Applied Entomology Meeting, Nederlandse Entomologische Vereniging*, Amsterdam, 1, 31–36.

Dicke, M., Van Baarlen, P., Wessels, R. & Dijkman, H. (1993). Herbivory induces systemic production of volatiles that attract herbivore predators: extraction of an endogenous elicitor. *Journal of Chemical Ecology*, 19, 581–599.

Drukker, B., Scutareanu, P. & Sabelis, M.W. (1995). Do anthocorid predators respond to synomones from *Psylla*-infested pear trees under field conditions? *Entomologia Experimentalis et Applicata*, 77, 193–203.

Dwumfour, E.F. (1992). Volatile substances evoking orientation in the predatory flower bug *Anthocoris nemorum* (Heteroptera: Anthocoridae). *Bulletin of Entomological Research*, 82, 465–469.

Eigenbrode, S.D. & Espelie, K.E. (1995). Effects of plant epicuticular lipids on insect herbivores. *Annual Review of Entomology*, 40, 171–194.

Elias, T.S. (1983). Extrafloral nectaries: their structure and distribution. In *The Biology of Nectaries* (Ed. by B.L. Bentley & T.S. Elias), pp. 174–203. Columbia University Press, New York.

Eubanks, M.D., Nesci, K.A., Petersen, M.K., Liu, Z. & Sanchez, H.B. (1997). The exploitation of an ant-defended host plant by a shelter-building herbivore. *Oecologia*, 109, 454–460.

Evans, E.W. & England, S. (1996). Indirect interactions in biological control of insects: Pests and natural enemies in alfalfa. *Ecological Applications*, **6**, 920–930.

Farmer, E.E. & Ryan, C.A. (1990). Interplant communication: Airborne methyl jasmonate induces synthesis of proteinase inhibitors in plant leaves. *Proceedings of the National Academy of Sciences, USA*, **87**, 7713–7716.

Ferguson, G.M. & Schmidt, J.M. (1996). Effect of selected cultivars on *Orius insidiosus*. *IOBC/WPRS Bulletin*, **19**, 39–42.

Fox, L.R. & Eisenbach, R. (1992). Contrary choices: possible exploitation of enemy-free space by herbivorous insects in cultivated *vs* wild crucifers. *Oecologia*, **89**, 574–579.

Fritz, R.S. (1995). Direct and indirect effect of plant genetic variation on enemy impact. *Ecological Entomology*, **20**, 18–26.

Fritz, R.S. & Nobel, J. (1990). Host plant variation in mortality of the leaf-folding sawfly on the arryo willow. *Ecological Entomology*, **15**, 25–35.

Gaume, L. & McKey, D. (1997). To be or not to be a mutualist? Comparison of two ant species inhabiting the same plant species. *ESEB-VI Abstracts*.

Gilpin, M.E. (1975). *Group Selection in Predator–Prey Communities*. Princeton University Press, Princeton, New Jersey.

Godfray, H.C.J. (1995). Communication between the first and third trophic levels: an analysis using biological signalling theory. *Oikos*, **72**, 367–374.

Grostal, P. & O'Dowd, D.J. (1994). Plants, mites and mutualism: leaf domatia and the abundance and reproduction of mites on *Viburnum tinus* (Caprifoliaceae). *Oecologia*, **97**, 308–315.

Hairston, N.G., Smith, F.E. & Slobodkin, L.B. (1960). Community structure, population control and competition. *American Naturalist*, **94**, 421–425.

Hare, J.D. (1992). Effects of plant variation on herbivore–natural enemy interactions. In *Plant Resistance to Herbivores and Pathogens. Ecology, Evolution and Genetics* (Ed. by R.S. Fritz & E.L. Simms), pp. 278–298. University of Chicago Press, Chicago.

Harrewijn, P., Minks, A.K. & Mollema, C. (1995). Evolution of plant volatile production in insect–plant relationships. *Chemoecology*, **5/6**, 55–73.

Hassell, M.P. (1978). *Dynamics of Arthropod Predator–Prey Systems*. Monographs in Population Biology. Princeton University Press, Princeton, NJ.

Hay, M.E. (1986). Associational defenses and the maintenance of species diversity: Turning competitors into accomplices. *American Naturalist*, **128**, 671–641.

Helle, W. & Sabelis, M.W. (1985). *Spider Mites: Their Biology, Natural Enemies and Control*. World Crop Pests Series, Vol. 1A, B. Elsevier Science Publishers, Amsterdam.

Herms, D.A. & Mattson, W.J. (1992). The dilemma of plants: to grow or to defend. *Quarterly Review of Biology*, **67**, 283–335.

Hjältén, J. & Price, P.W. (1997). Can plants gain protection from herbivory by association with unpalatable neighbours? A field experiment in a willow-sawfly system. *Oikos*, **78**, 317–322.

Holt, R.D. (1977). Predation, apparent competition and the structure of prey communities. *Theoretical Population Biology*, **12**, 197–229.

Holt, R.D. & Polis, G.A. (1997). A theoretical framework for intraguild predation. *American Naturalist*, **14**, 745–764.

Hopke, J., Donath, J., Blechert, S. & Boland, W. (1994). Herbivore-induced volatiles: the emission of acyclic homoterpenes from leaves of *Phaseolus lunatus* and *Zea mays* can be triggered by a β-glucosidase and jasmonic acid. *FEBS Letters*, **352**, 146–150.

Horvitz, C.C. & Schemske, D.W. (1984). Effects of ants and ant-tended herbivores on seed production of a neotropical herb. *Ecology*, **65**, 1369–1378.

Inouye, D.W. & Taylor, O.R., Jr. (1979). A temperate region plant–ant–seed predator system: consequences of extrafloral nectar secretion by *Helianthella quinquenervis*. *Ecology*, **60**, 1–7.

Isenhour, D.J., Wisman, B.R. & Layton, R.C. (1989). Enhanced predation by *Orius insidiosus* (Hemiptera: Anthocoridae) on larvae of *Heliothis zea* and *Spodoptera frugiperda* (Lepidoptera: Noctuidae) caused by prey feeding on resistant corn genotypes. *Environmental Entomology*, **18**, 418–422.

Jagers Op Akkerhuis, G., Sabelis, M.W. & Tjallingii, W.F. (1985). Ultrastructure of chemoreceptors of the pedipalps and first tarsi of *Phytoseiulus persimilis*. *Experimental and Applied Acarology*, **1**, 235–251.

Jansen, V.A.A. (1995). Effects of dispersal in a tri-trophic metapopulation model. *Journal of Mathematical Biology*, **34**, 195–224.

Jansen, V.A.A. & Sabelis M.W. (1992). Prey dispersal and predator persistence. *Experimental and Applied Acarology*, **14**, 215–231.

Jansen, V.A.A. & Sabelis, M.W. (1995). Outbreaks of colony-forming pests in tri-trophic systems: consequences for pest control and the evolution of pesticide resistance. *Oikos*, **74**, 172–176.

Janssen, A. & Sabelis, M.W. (1992). Phytoseiid life-histories, local predator–prey dynamics and strategies for control of tetranychid mites. *Experimental and Applied Acarology*, **14**, 233–250.

Janssen, A., Bruin, J., Jacobs, G., Schraag, R. & Sabelis, M.W. (1997a). Predators use volatiles to avoid prey patches with conspecifics. *Journal of Animal Ecology*, **66**, 223–232.

Janssen, A., Van Gool, E., Lingeman, R., Jacas, J. & Van de Klashorst, G. (1997b). Metapopulation dynamics of a persisting predator–prey system in the laboratory: time series analysis. *Experimental and Applied Acarology*, **21**, 415–430.

Janzen, D.H. (1966). Coevolution of mutualism between ants and acacias in Central America. *Evolution*, **20**, 249–275.

Janzen, D.H. (1975). *Pseudomyrmex nigripilosa*: a parasite of mutualism. *Science*, **188**, 936–937.

Jolivet, P. (1996). *Ants and Plants: An Example of Coevolution*. Backhuys Publishers, Leiden.

Karban, R., Hougen-Eitzmann, D. & English-Loeb, G. (1994). Predator-mediated apparent competition between two herbivores that feed on grapevines. *Oecologia*, **97**, 508–511.

Kareiva, P. & Sahakian, R. (1990). Tritrophic effects of a simple architectural mutation in pea plants. *Nature*, **345**, 433–434.

Kawai, A. (1995). Control of *Thrips palmi* Karny (Thysanoptera: Thripidae) by *Orius* spp. (Heteroptera: Anthocoridae) on greenhouse eggplant. *Applied Entomology and Zoology*, **30**, 1–7.

Koptur, S. (1984). Experimental evidence for defense of *Inga* (Mimosoideae) saplings by ants. *Ecology*, **65**, 1787–1793.

Koptur, S. (1992). Extrafloral nectary-mediated interactions between insects and plants. In *Insect–Plant Interactions IV* (Ed. by E.A. Bernays), pp. 81–129. CRC Press, Boca Raton, USA.

Klebanoff, A. & Hastings, A. (1994). Chaos in three-species food chains. *Journal of Mathematical Biology*, **32**, 427–451.

Kuznetsov, Yu. A. & Rinaldi, S. (1996). Remarks on food chain dynamics. *Mathematical Biosciences*, **134**, 1–33.

Lesna, I., Sabelis, M.W. & Conijn, C.G.M. (1996). Biological control of the bulb mite, *Rhizoglyphus robini*, by the predatory mite, *Hypoaspis aculeifer*, on lilies: predator–prey interactions at various spatial scales. *Journal of Applied Ecology*, **33**, 369–376.

Lindquist, E.E., Sabelis, M.W. & Bruin, J. (1996). *Eriophyoid Mites: Their Biology, Natural Enemies and Control*. World Crop Pests Series, Vol. 6, Elsevier Science Publishers, Amsterdam.

Loader, C. & Damman, H. (1991). Nitrogen content of food plants and vulnerability of *Pieris rapae* to natural enemies. *Ecology*, **72**, 1586–1590.

Loughrin, J.H., Potter, D.A. & Hamilton-Kemp, T.R. (1995). Volatile compounds induced by herbivory act as aggregation kairomones for the Japanese beetle (*Popilia japonica* Newman). *Journal of Chemical Ecology*, **21**, 1457–1467.

Margolies, D.C., Sabelis, M.W. & Boyer, J.E., Jr. (1997). Response of a phytoseiid predator to herbivore-induced plant volatiles: Selection on attraction and effect on prey exploitation. *Journal of Insect Behaviour*, **10**, 695–709.

Mattiacci, L., Dicke, M. & Posthumus, M.A. (1995). β-Glucosidase: an elicitor of herbivore-induced plant odor that attracts host-searching parasitic wasps. *Proceedings of the National Academy of Sciences, USA*, **92**, 2036–2040.

Maynard Smith, J. (1982). *Evolution and the Theory of Games*. Cambridge University Press, Cambridge.

Messina, F.J. (1981). Plant protection as a consequence of an ant-membracid mutualism: interactions in goldenrod (*Solidago* sp.). *Ecology*, **62**, 1433–1440.

Minks, A.K. & Harrewijn, P. (1989). *Aphids: Their Biology, Natural Enemies and Control*. World Crop Pests Series, Vol. 2B. Elsevier Science Publishers, Amsterdam.

Molleman, F., Drukker, B. & Blommers, L. (1997). A trap for monitoring pear psylla predators using dispensers with the synomone methylsalicylate. In *Proceedings in Experimental and Applied Entomology*, Vol. 8, pp. 177–182. NEV, Amsterdam.

Mound, L.A. (1993). The first species (Insecta) inhabiting leaf domatia: *Domatiathrips cunninghamii* gen. et sp. nov. (Thysanoptera: Phlaeothripidae). *Journal of the New York Entomological Society*, **101**, 424–430.

O'Dowd, D.J. & Willson, M.F. (1989). Leaf domatia and mites on Australasian plants: ecological and evolutionary implications. *Biological Journal of the Linnean Society*, **37**, 191–236.

O'Dowd, D.J., Brew, C.F.R., Christophel, D.C. & Norton, R.A. (1991). Mite–plant associations from the eocene of southern Australia. *Science*, **252**, 99–101.

Ohsaki, N. & Sato, Y. (1994). Food plant choice of *Pieris* butterflies as a trade-off between parasitoid avoidance and quality of plants. *Ecology*, **75**, 59–68.

Overmeer, W.P.J. (1985). Alternative prey and other food sources. In *Spider Mites. Their Biology, Natural Enemies and Control*, Vol. 1B (Ed. by W. Helle & M.W. Sabelis), pp. 131–140. Elsevier, Amsterdam.

Pallini, A., Janssen, A. & Sabelis, M.W. (1997). Odour-mediated responses of phytophagous mites to conspecific and heterospecific competitors. *Oecologia*, **110**, 179–185.

Paré, P.W. & Tumlinson, J.H. (1997). *De novo* biosynthesis of volatiles induced by insect herbivory in cotton plants. *Plant Physiology*, **114**, 1161–1167.

Parker, M.A. (1990). The pleiotropy theory for polymorphisms of disease resistance genes in plants. *Evolution*, **44**, 1209–1217.

Pemberton, R.W. & Turner, C.E. (1989). Occurrence of predatory and fungivorous mites in leaf domatia. *American Journal of Botany*, **76**, 105–112.

Pemberton, R.W. & Lee, J.-H. (1996). The influence of extrafloral nectaries on parasitism of an insect herbivore. *American Journal of Botany*, **83**, 1187–1194.

Pfister, C.A. & Hay, M.E. (1988). Associational plant refuges: convergent patterns in marine and terrestrial communities result from different mechanisms. *Oecologia*, **77**, 118–129.

Polis, G.A. & Holt, R.D. (1992). Intraguild predation: the dynamics of complex trophic interactions. *Trends in Ecology and Evolution*, **7**, 151–154.

Polis, G.A., Myers, C.A. & Holt, R.D. (1989). The ecology and evolution of intraguild predation: Potential competitors that eat each other. *Annual Review of Ecology and Systematics*, **20**, 297–330.

Price, P.W., Bouton, C.E., Gross, P., McPheron, B.A., Thompson, J.N. & Weiss, A.E. (1980). Interactions among three trophic levels: influence of plants on interactions between insect herbivores and natural enemies. *Annual Review of Ecology and Systematics*, **11**, 41–65.

Putz, F.E. & Holbrook, N.M. (1988). Further observations on the dissolution of mutualism between *Cecropia* and its ants: the Malaysian case. *Oikos*, **53**, 121–125.

Rashbrook, V.K., Compton, S.G. & Lawton, J.H. (1992). Ant–herbivore interactions: reasons for the absence of benefits to a fern with foliar nectaries. *Ecology*, **73**, 2167–2174.

Risch, S.J. & Rickson, F.R. (1981). Mutualism in which ants must be present before plants produce food bodies. *Nature*, **291**, 149–150.

Rogers, C.E. (1985). Extrafloral nectar: entomological implications. *Bulletin of the Entomological Society of America*, **31**, 15–20.

Rosen, D. (1990). *Armoured Scale Insects: Their Biology, Natural Enemies and Control*. World Crop Pests Series, Vol. 4B. Elsevier Science Publishers, Amsterdam.

Sabelis, M.W. (1990). Life history evolution in spider mites. In *The Acari: Reproduction, Development and Life-History Strategies* (Ed. by R. Schuster & P.W. Murphy), pp. 23–50. Chapman & Hall, New York.

Sabelis, M.W. (1992). Arthropod predators. In *Natural Enemies, The Population Biology of Predators, Parasites and Diseases* (Ed. by M.J. Crawley), pp. 225–264. Blackwell Scientific Publications, Oxford.

Sabelis, M.W. & Afman, B.P. (1994). Synomone-induced suppression of take-off in the phytoseiid mite *Phytoseiulus persimilis* Athias-Henriot. *Experimental and Applied Acarology*, **18**, 711–721.

Sabelis, M.W. & De Jong, M.C.M. (1988). Should all plants recruit bodyguards? Conditions for a polymorphic ESS of synomone production in plants. *Oikos*, **53**, 247–252.

Sabelis, M.W. & Dicke, M. (1985). Long-range dispersal and searching behaviour. In *Spider Mites. Their Biology, Natural Enemies and Control*, Vol. 1B (Ed. by W. Helle & M.W. Sabelis), pp. 141–160. Elsevier, Amsterdam.

Sabelis, M.W. & Janssen, A. (1994). Evolution of life-history patterns in the Phytoseiidae. In *Mites. Ecological and Evolutionary Analyses of Life-History Patterns* (Ed. by M.A. Houck), pp. 70–98. Chapman & Hall, New York.

Sabelis, M.W. & Van de Baan, H.E. (1983). Location of distant spider-mite colonies by phytoseiid predators: Demonstration of specific kairomones emitted by *Tetranychus urticae* and *Panonychus ulmi* (Acari: Phytoseiidae, Tetranychidae). *Entomologia Experimentalis et Applicata*, **33**, 303–314.

Sabelis, M.W. & Van der Meer, J. (1986). Local dynamics of the interaction between predatory mites and two-spotted spider mites. In *Dynamics of Physiologically Structured Populations* (Ed. by J.A.J. Metz and O. Diekmann), pp. 322–344. Lecture Notes in Biomathematics, **68**. Springer-Verlag, Berlin.

Sabelis, M.W. & Van der Weel, J.J. (1993). Anemotactic responses of the predatory mite, *Phytoseiulus persimilis* Athias-Henriot, and their role in prey finding. *Experimental and Applied Acarology*, **17**, 1–9.

Sabelis, M.W. & Van Rijn, P.C.J. (1996). Eriophyoid mites as alternative prey. In *Eriophyoid Mites: Their Biology, Natural Enemies and Control* (Ed. by E.E. Lindquist, M.W. Sabelis & J. Bruin), pp. 757–764. World Crop Pests, Vol. 6. Elsevier Science Publishers, Amsterdam.

Sabelis, M.W. & Van Rijn, P.C.J. (1997). Predation by insects and mites. In *Thrips as Crop Pests* (Ed. by T. Lewis), pp. 259–354. CAB International, Wallingford.

Sabelis, M.W., Vermaat, J.E. & Groeneveld, A. (1984). Arrestment responses of the predatory mite, *Phytoseiulus persimilis*, to steep odour gradients of a kairomone. *Physiological Entomology*, **9**, 437–446.

Sabelis, M.W., Diekmann, O. & Jansen, V.A.A. (1991). Metapopulation persistence despite local extinction: predator–prey patch models of the Lotka–Volterra type. *Biological Journal of the Linnean Society*, **42**, 267–283.

Schemske, D.W. (1980). The evolutionary significance of extrafloral nectar production by *Costus woodsonii* (Zingiberaceae): an experimental analysis of ant protection. *Journal of Ecology*, **68**, 959–967.

Scutareanu, P., Drukker, B., Bruin, J., Posthumus, M.A. & Sabelis, M.W. (1996). Leaf volatiles and polyphenols in pear trees infested by *Psylla pyricola*. Evidence for simultaneously induced responses. *Chemoecology*, **7**, 34–38.

Scutareanu, P., Drukker, B., Bruin, J., Posthumus, M.A. & Sabelis, M.W. (1997). Isolation and identification of volatile synomones involved in the interaction between *Psylla*-infested pear trees and two anthocorid predators. *Journal of Chemical Ecology*, **23**, 2241–2260.

Shonle, I. & Bergelson, J. (1995). Interplant communication revisited. *Ecology*, 76, 2660–2663.

Shulaev, V., Silverman, P. & Raskin, I. (1997). Airborne signalling by methyl salicylate in plant pathogen resistance. *Nature*, 385, 718–721.

Simms, E.L. (1992). Costs of plant resistance to herbivory. In *Plant Resistance to Herbivores and Pathogens. Ecology, Evolution and Genetics* (Ed. by R.S. Fritz and E.L. Simms), pp. 392–425. University of Chicago Press, Chicago.

Simms, E.L. & Rausher, M.D. (1987). Costs and benefits of plant resistance to herbivory. *American Naturalist*, 130, 570–581.

Simms, E.L. & Rausher, M.D. (1989). Natural selection by insects and costs of plant resistance to herbivory. *Evolution*, 43, 573–585.

Stephenson, A.G. (1982). The role of the extrafloral nectaries of *Catalpa speciosa* in limiting herbivory and increasing fruit production. *Ecology*, 63, 663–679.

Strong, D.R., Lawton, J.H. & Southwood, R. (1984). *Insects on Plants*. Harvard, Cambridge, MA.

Takabayashi, J., Dicke, M. & Posthumus, M.A. (1991). Variation in composition of predator-attracting allelochemicals emitted by herbivore-infested plants: relative influence of plant and herbivore. *Chemoecology*, 2, 1–6.

Takabayashi, J., Dicke, M. & Posthumus, M.A. (1994). Volatile herbivore-induced terpenoids in plant–mite interactions: variation caused by biotic and abiotic factors. *Journal of Chemical Ecology*, 20, 1329–1354.

Tuomi, J. & Augner, M. (1993). Synergistic selection of unpalatability in plants. *Evolution*, 47, 668–672.

Tuomi, J., Augner, M. & Nilsson, J. (1994). A dilemma of plant defences: is it really worth killing the herbivore? *Journal of Theoretical Biology*, 170, 427–430.

Turlings, T.C.J., Loughrin, J.H., McCall, P.J., Röse, U.S.R., Lewis, W.J. & Tumlinson, J.H. (1995). How caterpillar-damaged plants protect themselves by attracting parasitic wasps. *Proceedings of the National Academy of Sciences, USA*, 92, 4169–4174.

Van Baalen, M. & Sabelis, M.W. (1995). The milker–killer dilemma in spatially structured predator–prey interactions. *Oikos*, 74, 391–413.

Van der Klashorst, G., Readshaw, J.L., Sabelis, M.W. & Lingeman, R. (1992). A demonstration of asynchronous local cycles in an acarine predator–prey system. *Experimental and Applied Acarology*, 14, 179–185.

Van Haren, R.J.F., Steenhuis, M.M., Sabelis, M.W. & de Ponti, O.B.M. (1987). Tomato stem trichomes and dispersal success of *Phytoseiulus persimilis* relative to its prey *Tetranychus urticae*. *Experimental and Applied Acarology*, 3, 115–121.

Van Rijn, P.C.J. & Sabelis, M.W. (1998). Should plants provide food for predators when it also benefits the herbivores? Effects of pollen on a thrips–predatory mite system. In *Acarology IX*, Vol. 2 (Ed. by R. Mitchell, D.J. Horn, G.R. Needham & W.C. Welbourn). Ohio Biological Survey, Columbus.

Van Zon, A.Q., Overmeer, W.P.J. & Veerman, A. (1981). Carotenoids function in photoperiodic induction of diapause in a predacious mite. *Science*, **213**, 1131–1133.

Walter, D.E. (1996). Living on leaves: Mites, tomenta, and leaf domatia. *Annual Review of Entomology*, **41**, 101–114.

Way, M.J. (1963). Mutualism between ants and honeydew-producing Homoptera. *Annual Review of Entomology*, **8**, 307–344.

Willmer, P.G. & Stone, G.N. (1997). How aggressive ant-guards assist seed set in *Acacia* flowers. *Nature*, **388**, 165–167.

Part 2

Herbivores and vegetation succession

Introductory remarks

J. van Andel[1]

This section focuses on the impact of herbivores on the structure, composition and dynamics of vegetation. Herbivores remove and/or damage plant biomass, thus having a direct effect on vegetation structure. They interfere also with plants in an indirect way, by speeding up the nutrient supply rate due to their excrements. These interactions affect the competitive hierarchies among plants, thus having an impact on succession.

Plants and herbivores can be considered functional components of their ecosystem, the dynamics of which can be measured in terms of primary and secondary production, and of their role in the flow of energy and matter. This represents the systems approach to plant–herbivore interactions. Plants and herbivores can also be described in terms of a population–community approach, emphasizing the different role of species and population dynamics, affecting several aspects of biodiversity. An important trend in current research is the progress in integration between these two approaches. They are, indeed, complementary. Evolution takes place in a 'battlefield' called the ecosystem, and nutrient cycling at the system level depends on the efficiency of nutrient processing by the organisms. Feedback mechanisms at the ecosystem level may result from dramatic changes in populations of so-called keystone species. The meeting point between the population–community approach and the ecosystems approach was to be expected in the research on plant–animal interactions in particular, and fortunately the linkage happened to occur.

Long-term field observations, initiated some decades ago, revealed a sound basis for current experiments to test hypotheses about mechanisms of change. Without such permanent monitoring we would not, for example, have been able to distinguish whether actual spatial patterns in vegetation really represent a chronosequence. This knowledge is justification for comparative experiments in different successional stages. Such experiments

1 Department of Plant Biology, University of Groningen, PO Box 14, 9750 AA Haren, The Netherlands

include exclosures, which are indispensable in field studies on the effect of herbivores on plant succession.

Above-ground herbivores, ranging from insects to mammals can feed on shoots and roots. As far as their above-ground effects are concerned, they are known to retard succession, while optimizing their food supply at a particular successional stage. Below-ground herbivores feed on roots, which process is known to accelerate succession, at least in early stages. Plant pathogens, both above and below ground, may accelerate succession still further, if they kill dominant plants. Effects on the rate and direction of succession apparently differ between above-ground and below-ground herbivores. This is an interesting phenomenon, because it reflects the differences between plant competition for light and for nutrients. Without entering the debate as to whether or not there is a trade-off between the two, we could state that competition for nutrients is less asymmetric than competition for light. One-sided competition (for light) can be affected by herbivores in a number of different ways, depending on abundance and edibility of the component species, and on the food preferences of the herbivore. If we assume non-limiting nutrient supply, then the competitive abilities for light are decisive. However, in the case of below-ground herbivory, which in itself may affect the competitive abilities of the root systems in a way similar to above-ground herbivores affecting the canopy, light cannot remain non-limiting because of its non-homogeneous supply. So, in the case of root herbivory, the consequences are determined by above-ground competition for light.

In the present section, we do not consider top-down control of herbivores by predators, as this is the subject of other sections. We are aware of focusing on only part of the biotic interactions in ecosystems, and notice that a study of more complex trophic relationships (food chains, food webs) necessarily implies further simplifications in the plant–herbivore interactions.

Ritchie and Olff review existing theory and recent field studies to explore effects of above-ground herbivore diversity on plant species composition, succession, diversity and spatial heterogeneity. Diverse herbivore species may have compensatory or additive effects. Shifts in limiting resources for plants along a successional sere imply that herbivores may have different types of effects on succession at different stages. In cases where herbivores consume competitive dominants, i.e. retard succession, they should increase plant diversity and spatial heterogeneity in plant species' abundance. Otherwise, herbivores may accelerate succession to a particular set of dominant plant species, homogenize plant species composition and reduce diversity. Multiple interacting plant and herbivore species may produce unexpected positive feedbacks and alternative successional outcomes that depend on initial densities. Such condition-dependent outcomes may contribute to increased species diversity and strong spatial heterogeneity in species composition.

Mortimer *et al.* review the effects of subterranean insect and nematode herbivory on plant community dynamics and point at future directions in this field. Below-ground herbivores have the same ability to alter the composition and structure of plant communities as has been widely documented for vertebrate and invertebrate herbivores feeding on above-ground plant parts. Generalist herbivores and plant parasites may have important impacts on plant community structure through differences in feeding preferences and plant susceptibilities. However, even generalist herbivores show some preference for feeding on particular species of plant and thus alter the competitive balance between species. Regarding the role of plant parasitic nematodes in vegetation succession, research in ecosystems other than sand dunes is just beginning.

Huisman *et al.* investigate mechanisms of plant species replacement and herbivores along productivity gradients, using a model that combines standard plant–herbivore theory with a mechanistic model of competition for light. Recently published plant–herbivore theory that incorporates nutrient competition between the plants predicts a general increase in herbivore abundance with increasing productivity, as well as a change in plant species composition from good nutrient competitors at low productivity to good grazing tolerators at high productivity. The authors show that competition for light may lead to different patterns. For example, increased productivity may lead to the dominance of tall plants that shade small plant species preferred by the herbivore. In this case, the herbivore may decrease in abundance or even disappear with a further increase in productivity. Such patterns of decreased herbivore abundance with increasing productivity have actually been observed in salt-marsh ecosystems grazed by intermediate-sized vertebrate herbivores like rabbits, hares and geese.

Drent and van der Wal review current progress regarding ideas on cyclic grazing of herbivores and grazing lawns. They illustrate the importance of this phenomenon on the Schiermonnikoog salt marsh, which is grazed by natural populations of hares and geese. They show that the timing of visits of the geese to certain spots on the salt marsh is close to the maximization of harvest of nutritious regrowth. On longer time spans, they show that the successional dynamics of the salt-marsh vegetation is also affected by the repeated returns of the herbivores, with different herbivores having different effects. Where hares seem to delay succession because they graze the relatively poor-quality, late-successional dominants during the winter, the geese affect succession less, because they prefer the high-quality, early-successional species. Therefore, the hares facilitate for the geese by delaying the competitive displacement of their food species.

Jefferies reviews interactions between herbivores and microorganisms (microbial gut flora and soil microflora) affecting nutrient fluxes in the ecosystem, and discusses the special conditions where increased dominance from herbivores overrides the regulatory controls imposed by other organisms, which leads to trophic cascades and discontinuous vegetation states. There are few studies of trophic cascades in terrestrial ecosystems. Where trophic cascades have been reported, they share a number of common characteristics. Top-down dominance is often manipulated by human activities. As in aquatic systems, the terrestrial trophic cascades occur when one or two species represent different trophic levels of a short trophic ladder. Where herbivore numbers are increasing as a result of external energy and nutrient subsidies, strong top-down dominance has led to the destruction of vegetation, soil degradation and alternative stable states. Effects on vegetation occur primarily as a result of changes in soil properties. Although most of the case studies describe strong top-down control, the processes that result in alternative stable states are driven by changes in the soil–microbial–plant system which involve a large abiotic component.

Predictive ecology is the most difficult part of our discipline. Progress may only be made by exploring the interface between theoretical and experimental approaches. Prediction depends on knowledge of mechanisms, which can be obtained from experiments or on theoretical assumptions representing such knowledge for the time being (assumptions are preferably testable to give progress in understanding a chance). Predictions should be evaluated in the context of long-term field processes. Simulation models are a useful tool to develop scenarios of ecosystem development as resulting from knowledge or assumptions on plant–animal interactions. Here again, long-term field studies are an indispensable reference for evaluating the ecological significance of the results from such simulations. As a concluding remark, it seems useful to emphasize that the debate of whether or not there is a trade-off, within a plant, between competitive ability for light and competitive ability for nutrients, needs to be solved. In the present section, effects of above-ground and below-ground herbivory on plant communities have been reviewed in view of their impact on competitive hierarchies among plants. Their combined effects have hardly been investigated. Only if the mechanisms of plant competition are well known can further understanding of the impact of single and multiple herbivores make any progress. At least it is obvious that competition for light demands more complex models than competition for nutrients. Furthermore, it is clear that, while modelling approaches in general assume equilibrium conditions being once achieved, empirical evidence reminds us of heterogeneity in space and time affecting the system dynamics indefinitely.

Chapter 6

Herbivore diversity and plant dynamics: compensatory and additive effects

M.E. Ritchie[1] and H. Olff[2]

Summary

Ecosystems typically contain a variety of herbivore species but the effects of diverse herbivore assemblages on plant communities and succession are not well understood. Theory predicts that herbivores might have opposing or compensatory effects by selectively consuming different competing plant species. Alternatively, different herbivore species may consume the same plant species, thereby having additive effects on particular plant species. We review several recent field studies that manipulate the separate and combined effects of different mammalian and insect herbivore species within grasslands. We found evidence for both compensatory and additive effects of multiple herbivore species on plant species composition, diversity and spatial heterogeneity. Compensatory effects typically occurred when dominant plants competed for soil nutrients in the absence of herbivory, while additive effects occurred when dominant plants competed for water or light. When compensatory effects occurred, large herbivores typically consumed different plant species than small herbivores. These patterns suggest that the impact of herbivore diversity on plant communities will depend on the resource for which plants compete and the body size range of available herbivores. The resource for which plants compete may depend on soil fertility and the supply ratios of different limiting resources. These conclusions represent hypotheses that require future testing in field studies.

1 Ecology Center and Department of Fisheries and Wildlife, Utah State University, Logan UT 84322-5210, USA
2 Department of Terrestrial Ecology and Nature Conservation, Wageningen Agricultural University, Bornsesteeg 69, 6708 PD Wageningen, The Netherlands

Introduction

Herbivores often influence plant communities, sometimes dramatically (Harper 1977; Crawley 1983, 1989; Huntly 1991; Davidson 1993). Herbivores can directly consume plant species, indirectly release other plant species from competition, alter nutrient cycling, and disturb soils or other substrates. Through these direct and indirect effects, herbivores can alter succession, plant species diversity, structural heterogeneity and productivity. Herbivore effects can be important even when herbivores consume a small proportion of total plant biomass, because relatively rare plant species may become dominant in the absence of herbivores (Inouye *et al.* 1980; Edwards & Gillman 1987; Ritchie & Tilman 1995). Thus, different herbivore species within a community, whether abundant or rare, have the potential to influence plant species composition and succession.

Ecosystems can contain a variety of herbivore species. For example, the number of vertebrate herbivores can range from one, e.g. lesser snow geese, *Chen caerulescens*, on Hudson Bay (Bazely & Jefferies 1989), to more than 30 grazers larger than 4 kg on the Serengeti (Sinclair & Arcese 1995; Prins & Olff 1998). Invertebrate herbivores can also influence plant biomass and species composition and their effects will combine with those of vertebrates. An important question is therefore: how do the effects of multiple herbivores combine to influence plant communities and succession? The answer depends on how interactions among herbivore species influence plant consumption by each species, and how direct and indirect effects of different herbivore species combine. Competition or facilitation among herbivore species (Bell 1970; Belovsky 1986; Prins & Douglas-Hamilton 1990; Laca & Demment 1996; Prins & Olff 1998) may cause some species to specialize on fewer plant species or to generalize and include more plant species in their diets (Morris 1996). Thus, the effect a single herbivore species has on a plant community may depend on whether (and which) other herbivore species are present. Different herbivore species may consume different plant species, such that their effects on the plant community balance each other. In this case, multiple herbivore species may consume more total biomass but have compensating effects on plant composition. Thus, a single herbivore species may shift a plant community dramatically (Bakker 1985; Huntly & Inouye 1988; Crawley 1990; Hik *et al.* 1992; Hill 1992), while multiple herbivore species may have little net effect on the community (Belsky 1983; McNaughton 1986). Alternatively, different herbivore species may tend to consume the same plant species, such that effects of different herbivore species are additive. In this case, multiple herbivores may alter a plant community even more strongly than a single herbivore.

These ideas can be viewed in a successional context (Davidson 1993; Olff *et al.* 1997; Chapter 8). Single herbivore species may retard succession by consuming primarily late successional plant species, or accelerate succession by consuming primarily early successional species. Multiple herbivore species can have the same effects, depending on whether their effects are compensatory or additive. Herbivore assemblages that consume species characteristic of different successional stages may have compensatory effects, and arrest succession at an intermediate stage. Alternatively, additive effects of multiple herbivores that consume the same plant species may strongly retard or accelerate succession.

The alternative effects of herbivores on plant communities may depend on the type of resources limiting plant growth, e.g. water, nitrogen, phosphorus, light, etc. (Breman & de Wit 1983; Milchunas *et al.* 1988; Jefferies *et al.* 1994; Ritchie *et al.* 1998). These effects also may depend on adaptations of herbivore species, such as size, digestive morphology, enzyme biochemistry, etc. (van Soest 1982; Davidson 1993; Chapter 11). Body-size variation among herbivore species may be particularly important (Bell 1970; Jarman 1974; Belovsky 1986; Prins & Olff 1998) because herbivores of different body size potentially consume different plant species. The long-term history of herbivory in an ecosystem must also be considered (Stebbins 1981; Milchunas *et al.* 1988; Milchunas & Lauenroth 1993). Thus, effects of multiple herbivores may differ among ecosystems, because plant adaptations to the limiting resources in each ecosystem may constrain the ability of a herbivore to consume plants (Mattson 1980; Coley *et al.* 1985; Chapin *et al.* 1986; Gulmon & Mooney 1986).

These ideas suggest a rich array of hypotheses about the effects of herbivore diversity on plant species composition, succession, diversity and spatial heterogeneity. However, these hypotheses have not been tested, because few studies have compared effects of single vs. multiple herbivore species. Here, we explore existing theory and empirical studies that address five major questions.

1 Are effects of multiple herbivores on plant species composition and succession additive or compensatory?

2 Do herbivore species shift their diets in response to interactions with other herbivores?

3 How do multiple herbivores alter plant species diversity?

4 How do multiple herbivores influence spatial heterogeneity of plants?

5 Do effects of multiple herbivores depend on the nature of resource limitation?

To explore these questions and further refine our hypotheses, we review several recent field studies (Table 6.1). These studies feature manipulation of different herbivore species within a variety of grassland ecosystems. In a Minnesota old field (Ritchie & Tilman 1993), experimental populations of different combinations of three species of grasshoppers (Orthoptera: Acrididae) were established in replicate field cages placed over existing vegetation. In the other studies (Cid *et al.* 1991; Bowers 1993; Ritchie & Wolfe 1994; Ritchie & Tilman 1995; Ritchie *et al.* 1998; H. Olff unpublished observations), replicate size-selective fences excluded progressively more herbivore species. While many other studies have explored effects of one or more herbivore species on plants (Bakker 1985; Huntly 1991; Hill 1992; Davidson 1993; Jefferies *et al.* 1994; Hulme 1996), Table 6.1 presents the only studies of which we are aware that directly manipulate herbivore diversity.

The experimental treatments in each of these studies produce a gradient of herbivore diversity. For example, in a floodplain meadow in The Netherlands (Junner Koeland, H. Olff, unpublished observations), replicate 20×20 m fences were erected to exclude the two major grazers. One type excluded both cattle (*Bos tauros*) and rabbits (*Oryctolagus cuniculus*), while the other excluded only rabbits. Thus, the diversity of vertebrate herbivore species grazing plots ranged from one (the vole, *Microtus agrestis*) in

Table 6.1 Summary of studies that assess effects of different combinations of herbivore species.

Site	Vegetation type	Herbivores manipulated	Body size range (kg)	Reference
South Dakota	Mixed-grass prairie	Bison, prairie dogs	0.5–600	Cid *et al.* (1991)
Minnesota	Old field	Grasshoppers	0.0005–0.001	Ritchie & Tilman (1993)
Minnesota	Oak savanna	Insects, white-tailed deer	0.0001–95	Ritchie *et al.* (1998)
Utah	Sagebrush steppe	Elk, mule deer, jack-rabbits, cattle	1.5–500	Ritchie & Wolfe (1994)
Virginia	Old field	White-tailed deer, rabbits, voles	0.035–95	Bowers (1993)
Junner Koeland, The Netherlands	Floodplain meadow	Cattle, rabbits	1–500	H. Olff, unpublished

plots excluding both cattle and rabbits to three in unfenced control plots. The other studies established similar designs relevant to the herbivores dominant at their sites. Note that, with the exception of the Minnesota grasshopper study, the designs do not permit the effects of each individual species to be evaluated. Rather, the number of herbivore species increases in a biased manner across treatments, i.e. progressively adding larger species. However, this is the best and most practical design in most cases, because it is rarely possible to exclude small herbivores without also excluding large ones. In all these studies, plant species composition was sampled one or more times per year within each plot. All studies were of relatively short duration (<4 years), and different effects might have been observed in longer-term studies. Thus, these field studies provide important insights and suggest new hypotheses, but much work remains to understand fully the effect of herbivore diversity on plant dynamics.

Food web interactions

Simple food web models provide valuable insights into the consequences of multiple vs. single herbivores. Although they do not account for the rich diversity of species and interactions in real communities (Strong 1992; Polis & Strong 1996), simple webs form a template for exploring the complexities that arise as more herbivore species are added to an ecosystem (Fig. 6.1).

In the absence of herbivores and other disturbances, different plant species (e.g. early vs. late successional) are likely to compete for one or more limiting resources (e.g. de Wit 1960; Tilman 1982; Berendse 1994; Huisman 1994). The species that reduces the availability of a given limiting resource to the lowest level is expected to win in competition (Fig. 6.1a) (MacArthur 1969; Tilman 1976, 1982; Grover 1994). By feeding preferentially on weaker competitors, a single herbivore can accelerate species replacement in favour of stronger competitors (Armstrong 1979). Alternatively, a herbivore may feed preferentially on stronger competitors, thereby producing a positive indirect effect on weaker plant species, allowing them to coexist with or even exclude the stronger competitor (Harper 1969; Armstrong 1979; Pacala & Crawley 1992; Grover 1994). Viewed in a successional context, a single herbivore species can either accelerate or retard succession (Fig. 6.1b).

The addition of a second herbivore species to the food web model (Fig. 6.1c,d) illustrates the potential consequences of herbivore diversity. If the second herbivore consumes different plant species than the first, then

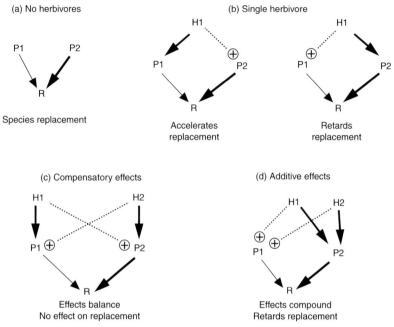

Figure 6.1 Hypothetical effects and depiction of the consequences of adding herbivore species to a community. Plant species (Ps) consume a shared limiting resource (R), with the superior competitor indicated by a thick arrow. Plants are consumed by herbivores (Hs), as indicated by thin arrows. Dashed lines with pluses inside circles indicate indirect positive effects. Note that herbivore effects are compensatory when herbivore species consume different plant species and are additive when they consume the same plants.

its direct and indirect effects on the plant community may counter those of the first (Fig. 6.1c). For example, two plant species may compete for a limiting resource and be consumed by two herbivores. If one herbivore consumes primarily one plant species and the other consumes primarily the second species, then the two herbivores may indirectly increase each other's resource and act as indirect mutualists (Levine 1976; Vandermeer 1980; Holt 1977, 1984; Ritchie & Tilman 1992, 1993). In this case, herbivores may have *compensatory* effects that produce little net effect on the relative abundance of plant species, i.e. the effect of each herbivore species is balanced by the effects of the other. Such effects correspond to those expected in species-rich, highly reticulated food webs (Strong 1992; Polis & Strong 1996; Chapter 19). If two or more herbivore species consume the same plant species, they may compete if their shared resources are limiting. Effects of individual herbivore species on the plant community will be *additive* even in the presence of competition. The net effect of herbivores is the sum of the product of *per capita* consumption and density for each herbivore species, weighted by the effect of consumption by each species (Fig. 6.1d). Additive effects can occur even when multiple herbivore species eat different parts (e.g. roots vs. leaves) or developmental stages of the same plant species, because the accumulated negative effects of multiple species on plant fitness should be greater than that of a single herbivore species.

We found evidence for both compensatory and additive effects of herbivore diversity. For example, voles and voles plus rabbits decreased cover of perennial plants and increased cover of annuals in a Virginia old field (Fig. 6.2a), based on aggregating significant effects of treatments on cover of individual species. These two species likely consumed perennials and thereby relieved annuals from competition. However, deer browsing diminished these effects by grazing annuals, such that grazing by all three herbivores resulted in a lower net effect of herbivory on the relative abundances of plant species. Similar effects occurred in a Utah sagebrush steppe (Fig. 6.2b), where ungulates (elk and mule deer) and jackrabbits significantly increased biomass of grasses by grazing forbs and shrubs (Ritchie & Wolfe 1994). Adding cattle negated these effects due to their intense consumption of grasses. These examples support the idea that when herbivore species differ in their effects on plant species due to different diet choices, their combined effects may be compensatory.

Examples of additive effects of multiple herbivores occurred in a South Dakota mixed-grass prairie (Cid *et al.* 1991), in a Minnesota oak savanna (Ritchie & Tilman 1995; Ritchie *et al.* 1998), and in Junner Koeland, a floodplain meadow in The Netherlands (H. Olff, unpublished observations). In South Dakota, prairie dogs and bison both grazed and reduced dominant grasses, and neither had significant impacts on forbs (Fig. 6.2c). When combined, these two herbivores reduced grasses the same as each herbivore species by itself, and again had little impact on forbs. In a Minnesota oak savanna (Ritchie & Tilman 1995; Ritchie *et al.* 1998), insects damaged and reduced woody plants (Fig. 6.2d). When white-tailed deer were added, they strongly reduced legumes. Both species favoured increased grass abundance, and although the magnitudes of their effects on different plant types varied, their combined effects reflected the sum of their individual effects. In the Dutch floodplain meadow (Junner

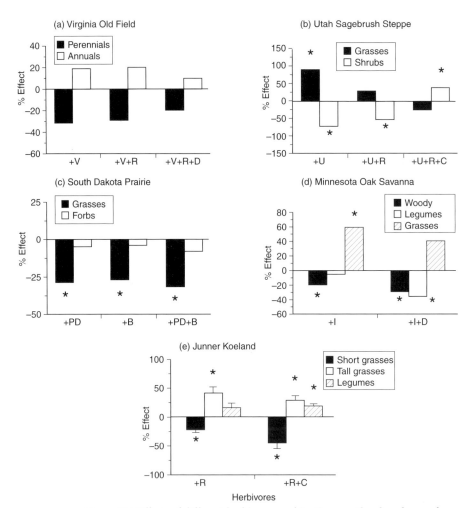

Figure 6.2 Effects of different herbivore combinations on the abundance of groups of plant species, expressed as percentage effect, the proportional change in the biomass or change in percentage cover, from adding herbivores relative to exclosures of all herbivores. (a) Effects of voles (V) *Microtus pennsylvanicus*, voles plus cottontail rabbits *Sylvilagus floridanus* (V+R), and voles, rabbits and white-tailed deer *Odocoileus virginianus* (V + R + D) on annuals and perennial herbaceous plants in a recently abandoned old field in Virginia (Bowers 1993). Statistical tests were not possible because we aggregated significant ($P < 0.05$) differences across replicate plots. (b) Effects of ungulates (U), mainly Rocky Mountain elk *Cervus elaphus* and mule deer *Odocoileus hemionus*, ungulates plus jackrabbits (U+R) *Lepus townsendi*, and ungulates, jackrabbits and cattle (U+R+C) *Bos taurus* in a sagebrush steppe in Utah (Ritchie & Wolfe 1994).

Koeland), rabbits significantly decreased percentage cover of short grasses and indirectly increased tall grasses, and forbs plus legumes after 2 years (Fig. 6.2e). Cattle grazing in addition to rabbit grazing did not change these effects, suggesting that combined effects of rabbits and cattle were additive. Because these results are from early in the experiment, they may represent transient effects.

If herbivore species compete for the same plant species, their densities will be reduced. If so, the effect of multiple species may have the same strong effect of a single species, but individual species' effects will still be additive. In this case, herbivores are likely to have strong effects on plant community composition, depending on which plant species the combined herbivore species prefer. If, however, herbivore species diverge facultatively in diet to avoid competition (MacArthur & Levins 1967; Abrams 1990; Dawson & Ellis 1996), the potential for additive effects may be reduced. Thus, shifts in diet choice that emerge from herbivore interactions may induce compensatory effects of herbivore diversity on plant communities.

A short-term (1-year) study of grasshoppers in field cages in a Minnesota old field provided a system to test these ideas (Ritchie & Tilman 1993). Dietary shifts occurred in response to interactions among the three grasshopper species that exhibited an order of magnitude of variation in body size. The survivorship of one species, the forb-feeding *Melanoplus femur-rubrum*, increased in the presence of two grass-feeding species, *Spharagemon collare* and *Phoetaliotes nebrascensis* (Fig. 6.3a). However, each of these grass-feeding species experienced competition, as their survivorship declined in the presence of the other two species. In response to these interactions, *M. femur-rubrum* increased the proportion of forbs

---

Figure 6.2 (*Continued*) (c) Effects of prairie dogs (PD) *Cynomys leucurus*, bison (B) *Bison bison*, and prairie dogs plus bison (PD+B) on grasses and forbs at Wind Cave National Park in South Dakota (Cid *et al.* 1991). (d) Effects of insects (I) and insects plus white-tailed deer (I+D) on legumes and woody plants in a Minnesota oak savanna (Ritchie *et al.* 1998). (e) Effects of rabbits (R) and rabbits plus cattle (R+C) on short grasses, tall grasses and legumes in a floodplain meadow, Junner Koeland, in The Netherlands (H. Olff, unpublished observations). Positive effects indicate that the herbivore combination increased the abundance of a plant group, while negative effects imply that the plant group decreased in abundance. Asterisks indicate significant effects ($P < 0.05$). Note that herbivore species differ in their effects in (a) and (b) but were similar in effects in (c–e).

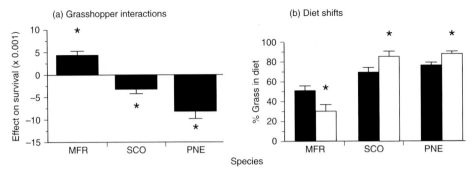

Figure 6.3 Interactions among three grasshopper species (Orthoptera: Acrididae) in cages placed over existing vegetation in an old field in Minnesota: *Melanoplus femur-rubrum* (MFR), *Spharagemon collare* (SCO), and *Phoetaliotes nebrascensis* (PNE). (a) The difference in average daily survivorship of each species between cages where the other two species were present vs. cages where it was by itself. (b) The proportion of grass in the diet of each species when by itself (alone, ■) and in the presence of the other two species (+all, □). Asterisks indicate significant effects ($P < 0.05$, ANOVA). Data are from Ritchie and Tilman (1993).

in its diet, while the other two species consumed proportionately more grasses (Fig. 6.3b). Thus, interactions among herbivores led to greater differentiation in plant consumption among herbivores. These diet shifts produced compensatory effects of herbivore diversity (see p. 180). Body size variation and adaptive diet shifts may therefore help produce compensatory effects of herbivore diversity.

Plant diversity

Herbivores can sometimes have dramatic effects on plant diversity. Single herbivores that reduce the abundance of a competitively dominant plant species can act as 'keystone' species and increase diversity by relieving other species from competition (Paine 1966; Harper 1969; Crawley 1983; Pacala & Crawley 1992). However, herbivores that select competitively inferior species can reduce diversity. In communities with multiple herbivores, the effect of herbivores on diversity is likely to depend on whether herbivore effects are compensatory or additive. The small net impact on plant composition by herbivores with compensatory effects may allow competitively dominant plant species to maintain their advantage, such that plant diversity should decline or remain unchanged in response to multiple herbivores. Thus, herbivore diversity may reduce plant diversity under some conditions. On the other hand, multiple herbivore species with additive effects may have strong effects on plant diversity. In this case, multiple herbivores act similarly to a single, abundant herbivore species that reduces either competitively dominant or inferior plant species. Thus, entire guilds of herbivores may have a simple keystone effect on plant diversity, even in species-rich, highly reticulated food webs (Strong 1992; Polis & Strong 1996).

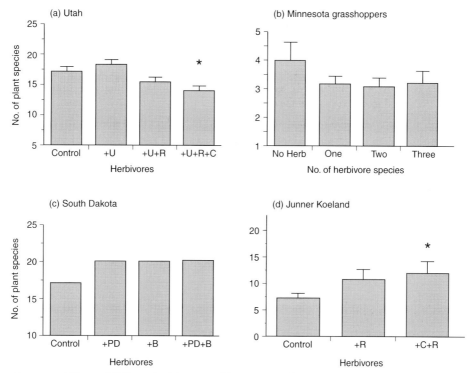

Figure 6.4 Effects of different herbivore combinations on plant species richness. Combinations for (a) and (c) are the same as in Fig. 6.2 (Cid *et al.* 1991; Ritchie & Wolfe 1994). In (b), effects of different numbers of grasshopper species in a Minnesota old field, as discussed in Fig. 6.2 (Ritchie & Tilman 1993), are shown. Effects are averaged over all replicates with 1, 2 or 3 grasshopper species. In (d), mean plant species richness is compared among control (all herbivores excluded), plots grazed by rabbits (R), and plots grazed by both cattle and rabbits (C+R) in a pasture in Junner Koeland in The Netherlands (H. Olff, unpublished observations). Significant effects are indicated by asterisks.

185

At sites with compensatory effects of multiple herbivores on plant abundances, increasing herbivore diversity did not increase plant diversity (Fig. 6.4a,b). Greater numbers of mammalian herbivores in Utah (Ritchie & Wolfe 1994) decreased plant diversity, while, in Minnesota, a greater diversity of grasshopper species (Ritchie & Tilman 1993) and the combined effects of insects and deer did not significantly alter plant diversity. In contrast, herbivores increased diversity at sites with additive effects of herbivore diversity on plant abundances (Fig. 6.4c,d). However, adding more than one herbivore had no additional effect on plant diversity in South Dakota. In the Junner Koeland, plant diversity increased in plots grazed by rabbits, but adding cattle had no significant additional effect. These results suggest that greater herbivore diversity may either increase or decrease plant diversity, depending on whether herbivores are compensatory or additive in their effects on plants.

Even if herbivores have additive effects, however, they are unlikely to increase diversity if they do not relieve plant species from competition. Thus, herbivores may reduce or not change plant diversity in ecosystems in which tolerance of abiotic conditions (e.g. low pH, high salinity) are more important than competition in structuring plant communities (Grime 1979). For example, cattle, geese, rabbits and hares decreased plant diversity in a Netherlands salt marsh, despite removing over 80% of primary productivity (Bakker 1989). Plants in frequently inundated salt marshes may be limited more by disturbance (flooding) and soil salinity than competition for limiting resources such as nitrogen or light (Snow & Vince 1984; Bakker 1985; Olff et al. 1997). Herbivore species with additive effects may also have little effect on diversity if intense past herbivory has reduced the plant species pool to primarily grazing-tolerant species (Milchunas et al. 1988).

Spatial heterogeneity in plant structure and species composition

In addition to overall effects on plant species composition and diversity, herbivores can affect the spatial distribution of species' abundances and their physical structure (Bakker et al. 1984; van den Bos & Bakker 1990; Coughenour 1991; Bullock 1996). At a given locality, a single, selective herbivore may have strong positive indirect effects on some species and strong negative direct effects on others. The preference of an individual for a given species may change across space, depending on which species of plants coexist at a given locality. Consequently, selective herbivory may generate considerable spatial variability in the abundance of plant species.

A diverse herbivore assemblage with compensating effects may reduce such heterogeneity because all plant species at a locality are likely to be negatively affected by some herbivore species. Consequently, for a given biomass of herbivore species, greater herbivore diversity may reduce spatial structure in plant abundance (Milchunas & Lauenroth 1989). In many cases, herbivore biomass, and therefore total plant biomass consumption, may be positively correlated with diversity, and further enhance these homogenizing effects. Multiple herbivore species with additive effects on plant abundances, however, may enhance spatial heterogeneity if their preferences remain similar across space (Bakker *et al.* 1984).

These hypotheses can be tested in two studies. Firstly, the biomass of a dominant prairie grass, *Schizachyrium scoparium*, in the Minnesota old field varied across replicate cages with different numbers of grasshopper species present (Fig. 6.5). Without grasshoppers, *S. scoparium* biomass

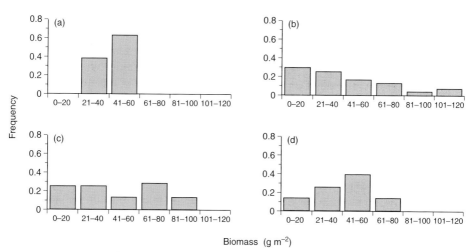

Figure 6.5 Frequency distributions of the biomass of the perennial prairie grass *Schizachyrium scoparium* in field cages with (a) zero, (b) one, (c) two or (d) three grasshopper species in a Minnesota old field. Distributions for one grasshopper species present were from one single-species treatment, selected at random (cages with *Melanoplus femur-rubrum*) and distributions for two grasshopper species present were from one two-species treatment, again selected at random (cages with *M. femur-rubrum* and *Spharagemon collare*). The variance in *Schizachyrium scoparium* biomass differs significantly among herbivore diversity levels ($F = 4.69$, df = 1,31, $P = 0.02$).

exhibited a tight distribution about the mean, but with one grasshopper species present, its biomass varied by an order of magnitude across replicates. This large variance declined as the number of grasshopper species increased. Secondly, the coefficient of variation in plant height in the Junner Koeland (Fig. 6.6a) also increased in the presence of a single herbivore (rabbits), but not in the presence of both rabbits and cattle. Rabbits created patchy short swards intermingled with taller, litter-

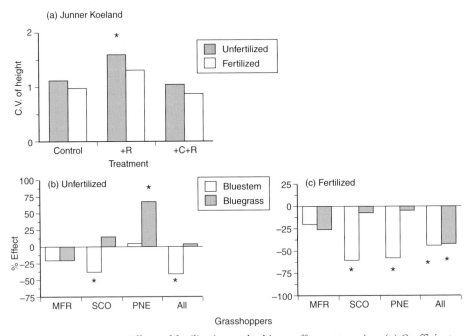

Figure 6.6 Effects of fertilization on herbivore effects at two sites. (a) Coefficient of variation in plant height (CV of height) among 10×10 cm cells in 2×2 m plots exposed to no grazing, grazing by rabbits (R), and grazing by rabbits plus cattle (C+R) in Junner Koeland. Note that CV of height was greater in plots grazed only by rabbits, and in fertilized plots. (b, c) Effects of different grasshopper combinations on the biomass of *Schizachyrium scoparium* (bluestem) and *Poa pratensis* (bluegrass) in field cages in a Minnesota old field. In (b), the responses of unfertilized cages are shown. In (c), plants were fertilized with 13 g m^{-2} of ammonium nitrate 1 week before grasshoppers were added to cages. Asterisks indicate significant effects. Note that in fertilized cages, grasshoppers had different effects, but in fertilized cages, each grasshopper species' effects were similar. For definition of abbreviations, see Fig. 6.3 legend.

dominated ungrazed clumps. The combination of cattle and rabbit grazing, however, removed these clumps. These results suggest that herbivore diversity homogenizes plant community structure, i.e. it reduces the variance in species' abundances among different localities.

Resource limitation and herbivory

The varying results of increasing herbivore diversity in the few studies we reviewed raise the following question: why do the consequences of herbivore diversity differ across sites? More specifically, how and why do compensating vs. additive effects of herbivore diversity vary across sites? Major developments in the theory of plant–herbivore interactions over the past 15 years suggest two major hypotheses. Firstly, ecosystems with a greater range of herbivore body sizes may feature large body-size differences among species and therefore be more likely to have herbivore species that differ in diet. Larger herbivores have the potential for slower digestive passage rates of food, but larger total digestive capacity (van Soest 1982; Demment and van Soest 1985; Hofmann 1989), while having greater total metabolic energy requirements to satisfy. A number of studies suggest that larger herbivores can persist on low-quality but abundant plants, while smaller herbivores can persist on rare but high-quality plants (Bell 1970; Jarman 1974; Breman & de Wit 1983; Belovsky 1986; Laca & Demment 1996; Prins & Olff 1998). Because of trade-offs in nutrient allocation to achieve different growth, competitive and stress-tolerant strategies (Grime 1979; Tilman 1990), plant species are likely to differ in these features. When plant species with such different strategies coexist in an area, herbivore species of different size may consume different plant species and have compensating effects.

A second hypothesis is that the resource for which plants compete, e.g. water, light, nitrogen, may affect palatability of dominant plant resource competitors (Coley *et al.* 1985; Gulmon & Mooney 1986) and thus the abundance, diversity and grazing intensity of herbivores (Milchunas *et al.* 1988; Davidson 1993). However, the full implications of this logic have not yet been explored. Specifically, the environmental characteristics of ecosystems and successional stages (e.g. water availability, temperature, soil fertility) may determine the abundance of herbivores and their effects on plant communities. For example, in systems where plants compete for soil nutrients, competitive dominants are likely to have high nutrient-use efficiencies and therefore low tissue nutrient levels. Furthermore, they often reduce their nutrient losses through long-lived plant parts supported by lignified tissues and through strong nutrient resorption from senescing tissues (Berendse & Jonasson 1992). Low nutrient and high lignin levels in plant tissue are often correlated with low palatability and poor herbivore performance (Mattson 1980; Robbins 1992; Chapter 11). Thus, ecosystems dominated by strong nutrient competitors are also likely to be avoided by small herbivores but utilized by large ones. Because high-quality plants are rare, larger herbivores may be forced to utilize abundant low-quality plants, while smaller herbivores with food requirements may be able to persist only on high-quality plants. Thus, different-sized herbivore species in nutrient-limited systems may consume different plant species, which will induce compensatory effects.

In systems where plants compete for water, plant traits may depend on soil fertility. On rich soils, plants may accumulate 'excess' nutrients in leaf tissues because nutrient: water ratios in the soil solution are likely to be high (Chapin et al. 1986; Koerselman & Meuleman 1996). Competitively dominant plants therefore may be highly palatable and utilized by herbivores of all sizes (Breman & de Wit 1983; McNaughton 1985; Milchunas et al. 1988). Under these conditions, multiple herbivores may therefore produce additive effects because all herbivores will consume the same group of palatable species, whether rare or not. On poor soils, soil nutrient: water ratios may remain low, so that competitive dominants are those that store water through development of woody tissues, reduce thermal stress with thorns, and/or maintain high leaf water potential with secondary compounds (Grubb 1992). In this case, dominant plants are likely to be unpalatable to many herbivores.

When both water and nutrients are abundant, such as on fertile, wet soils, plants may compete for light. Superior light competitors may feature lignified tissues to stiffen stems and leaves. These features reduce these plants' palatability to herbivores (Olff et al. 1997; Chapter 8). However, grazing can prevent the accumulation of stem biomass and litter that con-tribute to light limitation (Knapp & Seastedt 1986). A recent model suggests that the interaction of herbivory and light limitation can lead to alternative stable states (van de Koppel et al. 1996). If herbivore densities are initially high while plant biomass is low, the system is likely to feature high herbivore abundances, intense herbivory, and grazing 'lawns' in which all herbivore species consume virtually the same plant species (McNaughton 1985) and thus have additive effects. If herbivore densities are initially low, plant biomass may accumulate sufficiently to favour dominant, unpalatable light competitors. If so, the system may operate similarly to one with nutrient limitation, i.e. large herbivores consuming abundant, unpalatable dominant plants, and small herbivores consuming rare, palatable plants (Ritchie & Tilman 1995; Ritchie *et al.* 1998).

The consequences of variation in herbivore body size and resource competition by plants can be combined to better understand the consequences of herbivore diversity for plant dynamics. The range of herbivore body size combines with the palatability of competitively dominant plants to determine whether herbivore species differ sufficiently in diet to produce compensatory effects. For example, in nutrient-limited systems with a narrow range in body sizes, i.e. only small herbivores, a diversity of herbivores may yield additive effects because all species focus more or less on high-quality plants. A larger range in body sizes may lead to compensatory effects because large herbivores may reduce competitive dominants, thus balancing the effects of smaller herbivores. Comparison of herbivore effects in oak savanna in Minnesota with tall grass prairie provides one example. Browsing by white-tailed deer and insect herbivory additively reduce highly palatable legumes and woody plants and increase prairie grasses (Fig. 6.2d). However, the addition of cattle or bison, both large grazers, leads to a reduction in competitively dominant grasses and the formation of grazing lawns (Fahnestock & Knapp 1993, 1994). Similar interactions can be seen in systems with potential light competition, where large grazers may be necessary to produce compensatory effects. For example, geese, rabbits and hares cannot prevent the replacement of short, palatable species by a superior light competitor, the grass *Elymus athericus*, during salt-marsh succession (Olff *et al.* 1997). However, the addition of cattle reduces *Elymus* sufficiently to prevent competitive exclusion of these palatable species and facilitates grazing by the smaller herbivores (Olff *et al.* 1997). In systems in which plants compete for water on rich soils, palatable plants are abundant, so herbivores of all sizes can be relatively unselective and exert uniform pressure on all species. This may produce additive effects by multiple herbivores. Ecosystems with different size ranges of herbivores may therefore have different effects on communities, depending on the resource for which plants compete.

Comparing effects of herbivore diversity under different nutrient supply rates can test these ideas. In the Junner Koeland, fertilized, rabbit-grazed plots were less heterogeneous in vegetation height than unfertilized, rabbit-grazed plots (Fig. 6.6a). In the Minnesota old field, which is known to be nitrogen limited (Tilman 1987), a subset of replicate plots were fertilized with $13\,\mathrm{g\,m^{-2}}$ of ammonium nitrate 1 week before grasshoppers were added. In unfertilized plots (Fig. 6.6b), grasshoppers exhibited compensating effects, as *Phoetaliotes nebrascensis* increased the abundance of the C3 grass *Poa pratensis*, while *Spharagemon collare* decreased the abundance of *Schizachyrium scoparium*. The combined effects of all three grasshoppers represented an average of the various grasshoppers' effects on plant species. In fertilized plots, however, grasshopper species had similar effects on plants, and the total effect of all grasshoppers appeared to be additive (Fig. 6.6c). This shift in herbivore effects probably resulted because all plant species became similarly palatable with fertilization. Thus, nutrient addition altered the consequences of herbivore diversity even though the same range of body sizes was available. These experimental results appear to be most consistent with the hypothesis that limiting resources interact with variation in body size among herbivores to determine the consequences of herbivore diversity. However, they must be interpreted with caution because they involve adding nutrients to species adapted to low-nutrient conditions. The same results might not have been obtained if species responded to plants typical of high-nutrient conditions.

Discussion

Consequences of herbivore diversity: a synthesis

The previous discussion of plant–herbivore interactions across environmental gradients, coupled with the early results from field experiments, suggest a possible synthesis of ideas about herbivore effects on plant communities and the consequences of herbivore diversity. We propose a conceptual model that identifies three major types of liming resources for plants (water, nutrients, light) and their constraints on the effects of large vs. small herbivores (Table 6.2). Within this framework, we consider the influence of the evolutionary history of herbivory through its effect on the regional pool of plant species available to respond to changes created by herbivores (see Milchunas *et al.* 1988; Pärtel & Zobel 1996; Zobel 1997). These ideas are admittedly speculative, but offer a new set of hypotheses for addressing *why* and *how* the consequences of herbivore diversity differ among sites, rather than addressing *whether* effects exist.

As we explored in detail in the previous section, herbivore effects on plant community composition and diversity differentiate across four different simplified types of environments:

Table 6.2 Conceptual model of predicted effects of the consequences of multiple herbivore species for plant diversity in different environments.

Environment		Major limiting resource[1] without herbivory	Herbivore characteristics[2]	Joint effects on vegetation composition	Effect on vegetation diversity
Precipitation	Soil				
Dry	Infertile	Water/nutrients	Rare, small	Compensatory	0/–
Dry	Fertile	Water	Abundant, diverse[3]	Compensatory	–
Wet	Infertile	Nutrients, light	Intermediately abundant, large	Additive	++
Wet	Fertile	Light	Abundant, diverse[3]	Additive or compensatory	+/–

1 Resource competed for by ungrazed plants
2 Whether herbivores are rare, intermediately abundant, or abundant and whether the herbivore community is dominated by small or large herbivores or both (diverse)
3 A diverse herbivore assemblage depends on the presence of large herbivores to facilitate smaller herbivores — otherwise herbivores will be rare

1 dry, poor soils in which plants compete jointly for water and nutrients;
2 dry, rich soils in which plants compete primarily for water;
3 moist, poor soils in which plants compete for soil nutrients, and
4 moist, rich soils in which plants compete for light.

In environments with nutrient-poor soils (both moist and dry), the plant communities may exist as a mosaic of abundant low-quality plants and rare, high-quality plants. Large herbivores may consume low-quality plants while small herbivores consume high-quality plants, thus yielding compensatory effects of multiple herbivores and weak or negative effects on plant species composition and diversity. Because dominant plant competitors for soil nutrients may be of poor quality to herbivores (Gulmon & Mooney 1986; Grubb 1992; Wedin 1994), herbivore communities would be expected to feature abundant large herbivores at moist sites where productivity is sufficiently high to support their high metabolic requirements.

At dry sites, low productivity and low quality are likely to have prevented abundant large herbivores over evolutionary time. Plant species in these environments may therefore be poorly adapted to grazing by large herbivores, and thus have a short evolutionary history with introduced large grazers such as livestock (Milchunas *et al.* 1988). Poor grazing tolerance may also be a feature of perennial plant communities in environments where precipitation occurs outside periods thermally favourable for growth (Mediterranean climates). Therefore, in these environments, a diverse herbivore assemblage may have dramatic negative effects on plant diversity and strong effects on species composition (Milchunas *et al.* 1988; Milchunas & Lauenroth 1993), because the few plant species in the regional plant species pool have tolerant regrowth strategies.

On dry rich soils, dominant plants are likely to have high tissue nutrient concentrations (e.g. Breman & de Wit 1983; McNaughton 1985; Koerselman & Meuleman 1996). When subject to intense herbivory, dominant competitors for these resources can tolerate herbivory through regrowth. Consequently, multiple abundant herbivores may consume the dominant, palatable, tolerant species in an additive fashion. Such environments appear to have a long evolutionary history of intense grazing (Milchunas *et al.* 1988, Milchunas & Lauenroth 1993), which will influence the traits of plant species available to respond to herbivore effects. Because of this past history, most plant species may be tolerant of herbivory, such that herbivores have relatively little effect on plant species composition.

On moist, rich soils where plants potentially compete for light, two alternative stable states may exist: 'top-down' control of plants by herbivores vs. 'bottom-up' control of herbivore abundance by poor plant quality (van der Koppel *et al.* 1996; Chapter 8). These two outcomes depend on initial herbivore densities and plant biomass. In situations of top-down control, strong reduction of dominant light competitors by herbivores yields a plant community with abundant high-quality plants in which multiple herbivores have additive effects and increased plant diversity. In situations of bottom-up control, abundant low-quality plants are likely to be consumed only by large herbivores, thus yielding compensatory effects of multiple herbivore species. Because physical disturbance by herbivores affects the availability of light, multiple herbivores should increase plant diversity in light-limited systems, regardless of the plant species they select. These outcomes may also be affected by evolutionary history, because young plants subject to herbivory are likely to be growing rapidly to reach the canopy and exhibit a tolerant-regrowth defence strategy.

These alternative consequences of limiting resources for plants imply that herbivore diversity will differ in its effects across environments. Herbivore diversity plays a crucial role in some systems, e.g. when plants compete for light, because it may determine whether large vs. small herbivores are available to consume dominant plants. Evolutionary history of herbivory may be crucial in predicting ecosystem responses to multiple herbivore species when production is limited by water, particularly on nutrient-poor soils. This synthesis therefore promotes a pluralistic view of herbivore effects on plant communities, in which ecosystem-level constraints on plant traits may be of paramount importance.

Succession

The hypotheses and experimental results we have reviewed thus far can be synthesized to suggest a hypothesis of the effects of herbivore diversity on plant succession. The conceptual model we develop here refines the ideas of Davidson (1993). Specifically, we consider how limiting resources, palatability of competitive dominants and effects of herbivore diversity change along a successional sere.

Early in succession on nutrient-poor soils, superior nutrient competitors with low tissue nutrient concentrations, carbon-based defences, and generally low palatability replace early colonizing plant species with nutrient-rich tissues. Herbivores are therefore likely to accelerate this phase of succession. However, herbivore body size comes into play, as smaller herbivores may consume rarer high-quality plants, while larger herbivores may consume the abundant low-quality plants. Thus, different herbivore species may have opposing, compensatory effects, so that diverse herbivore assemblages may have little net influence on succession, as in the case of the Virginia old field and the Utah sagebrush steppe (Fig. 6.2).

On fertile soils, where plants compete for water, herbivore diversity may have additive effects on grazing-tolerant competitive dominants. Because nutrients are not limiting, these plants can respond to herbivory through regrowth. Their ability to tolerate herbivory and strong competitive ability for water mean that herbivores will accelerate early succession to grazing–tolerant species and slow later succession to strong light competitors. Such patterns may explain the dominance of short grasses in the western Great Plains of North America and the maintenance of open savannas in East Africa.

On moist, rich soils, succession in the absence of herbivory may be determined by competition for light as plant and litter biomass reach high levels. As mentioned earlier, strong light competitors may have low palatability in their mature stage, particularly to small herbivores. However, young plants of this type may be highly palatable, because, in their effort to reach the plant canopy, they maintain high nutrient concentrations in tissues and thin leaves to sustain high relative growth rates (Grime & Hunt 1975; Lambers & Poorter 1992). Thus, multiple herbivore species occurring at high initial densities may have additive effects on young or early successional plants and retard succession (Chapter 8). If herbivores occur at low initial density, their diversity may strongly influence succession. Diverse herbivore assemblages that include large species may retard invasion of strong light competitors (tall grasses or woody plants), because large herbivores are able to consume these plant species. If herbivore diversity is low and large species are absent, selective, additive herbivory by small herbivore species may accelerate succession. Thus, the effects of herbivores on succession involving plants competing for light are somewhat unpredictable and may depend heavily on herbivore diversity and initial conditions.

On some soils, such as in the boreal forest, the resource limiting succession may shift from light back to nitrogen as succession proceeds (Pastor & Naiman 1992; Davidson 1993). Dominant woody nitrogen competitors, e.g. conifers, are well-defended through high carbon:nitrogen ratios and secondary compounds. These species may induce additive effects of herbivore diversity on early successional shrubs and trees with high tissue nutrient and thus accelerate succession (Naiman 1988; Chapter 12).

Overall, shifts in the resource that limits plant growth across a successional sere can modify the effects of herbivores on succession and the consequences of herbivore diversity. In cases where herbivores consume competitive dominants, i.e. retard succession, they should increase plant diversity and spatial heterogeneity in plant species' abundances. Otherwise, herbivores may accelerate succession to a particular set of dominant plant species, homogenize plant species composition and reduce diversity. The results from Minnesota grasshoppers as well as rabbits and cattle in Junner Koeland suggest (Fig. 6.5) that greater herbivore diversity should reduce spatial heterogeneity, thereby increasing the predictability of succession across space.

With compensatory effects, herbivores have the potential to induce a highly predictable succession from a variety of initial conditions (Elton 1958; Whittaker 1975). However, multiple interacting plant and herbivore species may produce unexpected positive feedbacks and alternative successional outcomes that depend on initial densities (May 1973; DeAngelis 1992; Wilson & Agnew 1992). Such condition-dependent outcomes may contribute to increased species diversity across larger spatial scales and strong spatial heterogeneity in species composition (Wilson & Agnew 1992). Multiple herbivores exhibiting additive effects on plants should not have complex, interacting indirect effects, so, under these conditions, succession may be relatively rapid and predictable.

Conclusions

Single herbivore species may have dramatic effects on plant communities through both direct and indirect effects, as demonstrated in numerous field studies. Diverse herbivore communities sometimes have *compensatory* effects, where herbivore species consume different plant species and have balancing direct and indirect effects. Alternatively, multiple herbivore species may consume the same plant species, leading to *additive* or compounding effects of herbivore diversity. These different patterns of consumption appear to be linked to the herbivores' effects on plant diversity and spatial heterogeneity. They also appear to be linked to the regime of resource limitation for plants. In particular, when plants compete for soil nutrients, and in some cases for light, large quality differences among plant species favour compensatory effects among herbivores of different size. Conditions during early or inhibited succession on rich soils with uniformly high-quality plants may favour additive effects. Shifts in limiting resources for plants along a successional sere imply that herbivores may have different types of effects on succession at different stages. The consequences of herbivore diversity are therefore likely to change during succession, such that a diverse herbivore assemblage may retard succession during early stages (e.g. from ruderals to perennials), but accelerate it during later stages (e.g. herbaceous to woody species). These hypothesized patterns imply variation in the consequences of herbivore diversity across space that probably correlate with variation in soil fertility and evolutionary grazing history across a landscape. Alternative effects of herbivore diversity suggest that herbivore and plant diversity may trade-off in nutrient-limited systems, but diverse herbivore assemblages may help reduce plant species losses by preventing intense light competition following nutrient enrichment. The effect of herbivore diversity on overall biodiversity may therefore depend heavily on the degree of differentiation in plant species consumption among herbivore species. Large grazers, for example, may be beneficial to biodiversity in some environments (Bakker 1989; Fahnestock & Knapp 1993, 1994; Olff *et al.* 1997), but not in others (Ritchie & Wolfe 1994; Ritchie & Tilman 1993, Ritchie *et al.* 1998; Olff &

Ritchie 1998). Nature conservation authorities should be aware of these alternatives.

Acknowledgements

We thank James K. Detling, and two other reviewers for helpful comments. H.O. thanks Maurits Gleichman for collecting several of the Junner Koeland data. Field data collection from Minnesota and Utah was supported by the US National Science Foundation (BSR-8811884, DEB-9007125, DEB-9411972 and DEB-9527250) and the Utah Agricultural Experiment Station (Project 831).

References

Abrams, P.A. (1990). Adaptive responses of generalist herbivores to competition: convergence or divergence. *Evolutionary Ecology*, **4**, 103–114.

Armstrong, R.A. (1979). Prey species replacement along a gradient of nutrient enrichment: a graphical approach. *Ecology*, **60**, 76–84.

Bakker, J.P. (1985). The impact of grazing on plant communities, plant populations and soil conditions on salt marshes. *Vegetatio*, **62**, 391–398.

Bakker, J.P. (1989). *Nature Management by Cutting and Grazing*. Kluwer Academic Publishers, Dordrecht.

Bakker, J.P., De Leeuw, J. & Van Wieren, S.E. (1984). Micropatterns in grassland vegetation created and sustained by sheep-grazing. *Vegetatio*, **55**, 153–161.

Bazely, D.R. & Jefferies, R.L. (1989). Lesser snow geese and the nitrogen economy of a grazed salt marsh. *Journal of Ecology*, **77**, 24–34.

Bell, R.H.V. (1970). The use of the herb layer by grazing ungulates in the Serengeti. In *Animal Populations in Relation to Their Food Resources* (Ed. by A. Watson), pp. 111–123. Blackwell Scientific Publications, Oxford.

Belovsky, G.E. (1986). Generalist herbivore foraging and its role in competitive interactions. *American Zoologist*, **26**, 51–69.

Belsky, A.J. (1983). Small-scale pattern in grassland communities in the Serengeti National Park, Tanzania. *Vegetatio*, **55**, 141–151.

Berendse, F. (1985). The effect of grazing on the outcome of competition between plant species with different nutrient requirements. *Oikos*, **44**, 35–39.

Berendse, F. (1994). Competition between plant populations at low and high nutrient supply. *Oikos*, **71**, 253–260.

Berendse, F. & Jonasson, S. (1992). Nutrient use and nutrient cycling in

Northern ecosystems. In *Arctic Ecosystems in a Changing Climate—and Ecological Perspective* (Ed. by F.S. Chapin, R.L. Jefferies, J. Reynolds, J.H. Shaver & G.R. Svoboda), pp. 337–358. Academic Press, New York.

Bowers, M.A. (1993). Influence of herbivorous mammals on an old field community 1–4 years after disturbance. *Oikos*, **67**, 129–141.

Breman, H. & de Wit, C.T. (1983). Rangeland productivity and exploitation in the Sahel. *Science*, **221**, 1341–1347.

Bullock, J.B. (1996). Plant competition and population dynamics. In *The Ecology and Management of Grazing Systems* (Ed. by J. Hodgson & A.W. Illius), pp. 69–100. CAB International, Wallingford.

Chapin, F.S., III, Vitousek, P.M. & Van Cleve, K. (1986). The nature of nutrient limitation in plant communities. *American Naturalist*, **127**, 48–58.

Cid, M.S., Detling, J.K., Whicker, A.D. & Brizuela, M.A. (1991). Vegetational responses of a mixed-grass prairie site following exclusion of prairie dogs and bison. *Journal of Range Management*, **44**, 100–105.

Coley, P.D., Bryant, J.P. & Chapin, F.S., III (1985). Resource availability and plant antiherbivore defense. *Science*, **230**, 895–899.

Crawley, M.J. (1983). *Herbivory, the Dynamics of Plant–Animal Interactions*. Blackwell Scientific Publications, Oxford.

Crawley, M.J. (1989). Insect herbivores and plant population dynamics. *Annual Review of Entomology*, **34**, 531–564.

Crawley, M.J. (1990). Rabbit grazing, plant competition and seedling recruitment in acid grassland. *Journal of Applied Ecology*, **27**, 803–820.

Coughenour, M.B. (1991). Spatial components of plant–herbivore interactions in pastoral, ranching, and native ungulate ecosystems. *Journal of Range Management*, **44**, 530–542.

Davidson, D.W. (1993). The effect of herbivory and granivory on terrestrial plant succession. *Oikos*, **68**, 23–35.

Dawson, T.J. & Ellis, B.A. (1996). Diets of mammalian herbivores in Australian arid, hilly shrublands: Seasonal effects on overlap between euros (hill kangaroos), sheep and feral goats, and on dietary niche breadths and electivities. *Journal of Arid Environments*, **34**, 491–506.

de Wit, C.T. (1960). *On Competition*. Verslag Landbouwkundig Onderzoek no. 66.8, Wageningen.

DeAngelis, D.L. (1992). *Dynamics of Nutrient Cycling and Food Webs*. Chapman & Hall, London.

Demment, M.W. & Van Soest, P.J. (1985). A nutritional explanation for the body-size patterns of ruminant and non-ruminant herbivores. *American Naturalist*, **125**, 641–672.

Edwards, P.J. & Gillman, M.P. (1987). Herbivores and plant succession. In *Colonization, Succession, and Stability* (Ed. by A.J. Gray, M.J. Crawley &

P.J. Edwards), pp. 295–314. Blackwell Scientific Publications, Oxford.

Elton, C.S. (1958). *The Ecology of Invasion by Animals and Plants*. Methuen, London.

Fahnestock, J.T. & Knapp, A.K. (1993). Water relations and growth of tallgrass prairie forbs in response to selective grass herbivory by bison: interactions between herbivory and water stress. *International Journal of Plant Science*, **154**, 432–440.

Fahnestock, J.T. & Knapp, A.K. (1994). Responses of grasses to selective herbivory by bison: interactions between herbivory and water stress. *Vegetatio*, **115**, 123–131.

Grime, J.P. (1979). *Plant Strategies and Vegetation Processes*. John Wiley & Sons, Chichester.

Grime, J.P. & Hunt, R. (1975). Relative growth rate, its range and adaptive significance in a local flora. *Journal of Ecology*, **63**, 393–422.

Grover, J.P. (1994). Assembly rules for communities of nutrient-limited plants and specialist herbivores. *American Naturalist*, **136**, 771–789.

Grubb, P.J. (1992). A positive distrust in simplicity — lessons from plant defences and from competition among plants and among animals. *Journal of Ecology*, **80**, 585–610.

Gulmon, S.L. & Mooney, H.A. (1986). Costs of defence on plant productivity. In *On the Economy of Plant Form and Function* (Ed. by T.J. Givnish), pp. 681–698. Cambridge University Press, Cambridge.

Harper, J.L. (1969). The role of predation in vegetational diversity. *Brookhaven Symposia in Biology*, **22**, 48–62.

Harper, J.L. (1977). *Population Biology of Plants*. Academic Press, London.

Hik, D.S., Jefferies, R.L. & Sinclair, A.R.E. (1992). Foraging by geese, isostatic uplift and symmetry in the development of salt-marsh plant communities. *Journal of Ecology*, **80**, 395–406.

Hill, M.O. (1992). Long-term effect of excluding sheep from hill pastures in North Wales. *Journal of Ecology*, **80**, 1–13.

Hofmann, R.R. (1989). Evolutionary steps of ecophysiological adaptation and diversification of ruminants: a comparative review of their digestive system. *Oecologia*, **78**, 443–457.

Holt, R.D. (1977). Predation, apparent competition and the structure of prey communities. *Theoretical Population Biology*, **12**, 197–229.

Holt, R.D. (1984). Spatial heterogeneity, indirect interactions and the coexistence of prey species. *American Naturalist*, **124**, 377–406.

Huisman, J. (1994). The models of Berendse and Tilman: two different perspectives on plant competition? *Functional Ecology*, **8**, 282–288.

Hulme, P.E. (1996). Herbivores and the performance of grassland plants: a comparison of arthropod, mollusc, and rodent herbivory. *Journal of Ecology*, **84**, 43–51.

Huntly, N.J. (1991). Herbivores and the dynamics of communities and ecosystems. *Annual Review of Ecology and Systematics*, **22**, 477–503.

Huntly, N.J. & Inouye, R. (1988). Pocket gophers in ecosystems: patterns and mechanisms. *BioScience*, **38**, 786–793.

Inouye, R.S., Byers, G.S. & Brown, J.H. (1980). Effects of predation and competition on survivorship, fecundity, and community structure of desert annuals. *Ecology*, **61**, 1344–1351.

Jarman, P.J. (1974). The social organization of antelope in relation to their ecology. *Behaviour*, **48**, 215–267.

Jefferies, R.L., Klein, D.R. & Shaver, G.R. (1994). Vertebrate herbivores and northern plant communities — reciprocal influences and responses. *Oikos*, **71**, 193–206.

Knapp, A.K. & Seastedt, T.R. (1986). Detritus accumulation limits productivity of tallgrass prairie. *BioScience*, **36**, 662–668.

Koerselman, W. & Meuleman, A.F.M. (1996). The vegetation N:P ratio: a new tool to detect the nature of nutrient limitation. *Journal of Applied Ecology*, **33**, 1441–1450.

Laca, E.A. & Demment, M.W. (1996). Foraging strategies of grazing animals. In *The Ecology and Management of Grazing Systems* (Ed. by J. Hodgson & A.W. Illius), pp. 137–158. CAB International, Wallingford.

Lambers, H. & Poorter, H. (1992). Inherent variation in growth rate between higher plants: a search for physiological causes and ecological consequences. *Advances in Ecological Research*, **23**, 187–261.

Levine, S.H. (1976). Competitive interactions in ecosystems. *American Naturalist*, **110**, 903–910.

MacArthur, R.H. (1969). Species packing, and what interspecific competition minimizes. *Proceedings of the National Academy of Sciences, USA*, **64**, 1369–1371.

MacArthur, R.H. & Levins, R. (1967). The limiting similarity, convergence and divergence of coexisting species. *American Naturalist*, **101**, 377–385.

Mattson, W.J. (1980). Herbivory in relation to plant nitrogen content. *Annual Review of Ecology and Systematics*, **11**, 119–161.

May, R.M. (1973). *Stability and Complexity in Model Ecosystems*. Princeton University Press, Princeton.

McNaughton, S.J. (1985). Ecology of a grazing ecosystem: the Serengeti. *Ecological Monographs*, **55**, 259–294.

McNaughton, S.J. (1986). On plants and herbivores. *American Naturalist*, **128**, 765–770.

Milchunas, D.G. & Lauenroth, W.K. (1989). Three dimensional distribution of plant biomass in relation to grazing and topography in the short grass steppe. *Oikos*, **55**, 82–86.

Milchunas, D.G. & Lauenroth, W.K. (1993). Quantitative effects of grazing on vegetation and soils over a global range of environments. *Ecological Monographs*, **63**, 327–366.

Milchunas, D.G., Sala, O.E. & Lauenroth, W.K. (1988). A generalized model of the effects of grazing by large herbivores on grassland community structure. *American Naturalist*, **132**, 87–106.

Morris, D.W. (1996). Coexistence of specialist and generalist rodents via habitat selection. *Ecology*, **77**, 2352–2364.

Naiman, R.J. (1988). Animal influences on ecosystem dynamics. *BioScience*, **38**, 750–752.

Olff, H. & Ritchie, M.E. (1998). Effects of herbivores on grassland plant diversity. *Trends in Ecology and Evolution*, **13**, 261–265.

Olff, H., De Leeuw, J., Bakker, J.P., Platerink, R.J., van Wijnen, H.J. & De Munck, W. (1997). Vegetation succession and herbivory on a salt marsh: changes induced by sea level rise and silt deposition along an elevational gradient. *Journal of Ecology*, **85**, 799–814.

Pacala, S.W. & Crawley, M.J. (1992). Herbivores and plant diversity. *American Naturalist*, **140**, 243–260.

Paine, R.T. (1966). Food web complexity and species diversity. *American Naturalist*, **100**, 65–75.

Pärtel, M. & Zobel, M. (1996). The species pool and its relation to species richness—evidence from Estonian plant communities. *Oikos*, **75**, 111–117.

Pastor, J. & Naiman, R.J. (1992). Selective foraging and ecosystem processes in boreal forests. *American Naturalist*, **140**, 243–260.

Polis, G.A. & Strong, D.R. (1996). Food web complexity and community dynamics. *American Naturalist*, **147**, 813–846.

Prins, H.H.T. & Douglas-Hamilton, I. (1990). Stability in a multi-species assemblage of large herbivores in East Africa. *Oecologia*, **83**, 392–400.

Prins, H.H.T. & Olff, H. (1998). Species-richness of African grazer assemblages: towards a functional explanation. In *Dynamics of Tropical Communities* (Ed. by D. M. Newbery, H.H.T. Prins & N. Brown), pp. 449–490. Blackwell Science Ltd, Oxford.

Ritchie, M.E. & Tilman, D. (1992). Interspecific competition among grasshoppers and their effect on plant abundance in experimental old-field environments. *Oecologia*, **89**, 524–532.

Ritchie, M.E. & Tilman, D. (1993). Predictions of species interactions from consumer-resource theory: experimental tests with grasshoppers and plants. *Oecologia*, **94**, 516–527.

Ritchie, M.E. & Tilman, D. (1995). Responses of legumes to herbivores and nutrients during succession on a nitrogen-poor soil. *Ecology*, **76**, 2648–2655.

Ritchie, M.E. & Wolfe, M.L. (1994). Sustaining rangelands: application of ecological models to determine the risks of alternative grazing systems. *USDA Forest Service General Technical Report* RM247, pp. 328–336. Fort Collins, USA.

Ritchie, M.E., Tilman, D. & Knops, J.M.H. (1998). Herbivore effects on plant and nitrogen dynamics in oak savanna. *Ecology*, **79**, 165–177.

Robbins, C.T. (1992). *Wildlife Feeding and Nutrition*, 2nd edn. Academic Press, New York.

Sinclair, A.R.E. & Arcese, P. (Eds) (1995). *Serengeti II: Dynamics, Management, and Conservation of an Ecosystem*. Chicago University Press, Chicago.

Snow, A.A. & Vince, S.W. (1984). Plant zonation in an Alaskan salt marsh. II. An experimental study of the role of edaphic conditions. *Journal of Ecology*, **72**, 669–684.

Stebbins, G.L. (1981). Coevolution of grasses and herbivores. *Annals of the Missouri Botanical Garden*, **68**, 75–86.

Strong, D.R. (1992). Are trophic cascades all wet? The redundant differentiation in trophic architecture of high diversity ecosystems. *Ecology*, **73**, 747–754.

Tilman, D. (1976). Ecological competition between algae: experimental confirmation of resource-based competition theory. *Science*, **192**, 463–465.

Tilman, D. (1982). *Resource Competition and Community Structure*. Princeton University Press, Princeton.

Tilman, D. (1987). Secondary succession and the pattern of dominance along experimental nutrient gradients. *Ecological Monographs*, **57**, 189–214.

Tilman, D. (1990). Constraints and tradeoffs: toward a predictive theory of competition and succession. *Oikos*, **58**, 3–15.

van de Koppel, J., Huisman, J., van de Wal, R. & Olff, H. (1996). Patterns of herbivory along a gradient of primary productivity: an empirical and theoretical investigation. *Ecology*, **77**, 736–745.

van den Bos, J. & Bakker, J.P. (1990). The development of vegetation patterns by cattle grazing at low stocking density in the Netherlands. *Biological Conservation*, **51**, 263–272.

van Soest, P.J. (1982). *Nutritional Ecology of the Ruminant: Ruminant Metabolism, Nutritional Strategies, the Cellulolytic Fermentation and the Chemistry of Forages and Plant Fibres*. O & B Books, Corvallis.

Vandermeer, J.H. (1980). Indirect mutualism: variations on a theme by Stephen Levine. *American Naturalist*, **116**, 441–448.

Wedin, D. (1994). Species, nitrogen, and grassland dynamics: the constraints of stuff. In *Linking Species and Ecosystems* (Ed. by C. Jones & J.H. Lawton), pp. 253–262. Chapman & Hall, New York.

Whittaker, R.H. (1975). *Communities and Ecosystems*. Macmillan, London.

Wilson, J.B. & Agnew, A.D.Q. (1992). Positive-feedback switches in plant communities. *Advances in Ecological Research*, **23**, 263–336.

Zobel, M. (1997). The relative role of species pools in determining plant species richness: an alternative explanation of species coexistence? *Trends in Ecology and Evolution*, **12**, 266–269.

Chapter 7

Insect and nematode herbivory below ground: interactions and role in vegetation succession

S.R. Mortimer,[1] W.H. Van der Putten[2] and V.K. Brown[1]

Summary

The role of below-ground herbivory in structuring natural and seminatural communities is becoming increasingly apparent. Many groups of invertebrates feed on subterranean plant parts, but insects and nematodes are the most important below-ground herbivores. Studies of interactions between below-ground herbivores have shown both competition and facilitation. Competition may be minimized through niche differentiation or by temporal or spatial avoidance. Studies have shown low levels of niche occupancy and no clear evidence for exclusive guilds or species packing. Facilitation may occur through substrate modification or the increased attractiveness of damaged roots to other herbivores. Specialist herbivores cause changes in plant community composition through their effects on specific host plants primarily as a result of the alteration of competitive interactions between coexisting plant species. Generalist herbivores can similarly affect plant community composition as a result of differences in their preference for plant species or differences in host plant susceptibility. Experimental studies have demonstrated the impact of below-ground herbivory on the rate and direction of vegetation succession, with below-ground herbivores acting as keystone species in certain ecosystems. As a result of the focus of research on pest species or potential biological control agents, research has tended to focus on particular herbivore species and the effects on their host plant. This review shows that the role of below-ground herbivory in structuring plant communities and the interactions between

1 CABI Bioscience UK Centre (Ascot), Silwood Park, Ascot, Berks, SL5 7TA, UK
2 Netherlands Institute of Ecology (NIOO-CTO), PO Box 40, 6666 ZG Heteren, The Netherlands

such herbivores and other components of the biotic and physical environment of the soil remain important and fruitful areas for ecological research.

Introduction

The paucity of work on the role of below-ground herbivory in natural and seminatural communities, compared with the abundance of studies on foliar feeders, has been highlighted by reviews of the subject (Andersen 1987; Stanton 1988; Brown & Gange 1990). In the 10 years since these reviews were published, there has been little work carried out to address this gap in our knowledge. Research has continued on below-ground herbivores of economic importance in agricultural systems, and there have recently also been a number of studies on the potential role of soil-dwelling herbivores as biological control agents of invasive weed species. In spite of the limited extent of the research, it is apparent that feeding on below-ground plant parts by either insects or nematodes can have important effects on community structure. The aims of this chapter are to assess level of knowledge, and provide post-1990 update on the effects of subterranean insect and nematode herbivory in natural communities; to compare the characteristics of below-ground insect and nematode herbivores; to assess the effects of subterranean insect and nematode herbivory on plant community dynamics and particularly succession; and thus to identify gaps in our current knowledge and areas for future studies.

Below-ground herbivores

Animals feeding on below-ground plant parts include both vertebrates and invertebrates. Rodents are the main vertebrates that consume below-ground plant parts; other groups of vertebrates also feed on plant roots, but do not burrow through the soil in order to gain access to food, rather they browse on roots after disturbing the soil surface. As the energy costs of burrowing through the soil are high, fossorial vertebrates tend to be restricted to vegetation types with considerable root systems or storage organs in the soil, such as arid or alpine vegetation (Andersen 1987).

Below-ground plant parts are an important food resource for many groups of soil-dwelling invertebrates. Amongst the soil-dwelling meso- and macro-invertebrates, earthworms, potworms (Enchytraeidae), molluscs, millipedes, isopods and mites consume plant roots. However, in spite of the wide range of taxa that feed on below-ground plants parts, roots comprise a significant proportion of diet for only a few groups. Indeed, feeding on roots may even be considered to be 'accidental' in some groups; one study showing that the main consumers of transparent rootlets were earthworms, consumption occurring as the worms ingested soil (Gunn & Cherrett 1993). Herbivorous insects and nematodes are the most important root feeders and it is these groups that have received most attention as a result of their agricultural and economic importance.

Characteristics of below-ground insect herbivores

Apterygotes are the most abundant soil-dwelling insects in many communities and contain herbivorous taxa, such as springtails (Collembola) and bristletails (Diplura). Amongst the higher orders of insects, root feeders can be found in the Orthoptera, Isoptera, Hemiptera, Thysanoptera, Neuroptera, Coleoptera, Diptera and Lepidoptera. Certain large insects, such as mole crickets, also damage roots indirectly by their burrowing activity (Potter & Braman 1991). In spite of the variety of insects that consume below-ground plant parts as part of their diet, it is the larvae of three orders of holometabolous insects (Coleoptera, Diptera and Lepidoptera) that are the most important herbivores.

The feeding strategies of soil-dwelling insects are difficult to study *in situ*, so knowledge is limited. Studies have involved the use of glass-fronted viewing chambers or 'rhizotrons' (Muller 1989a; Gunn & Cherrett 1993). The feeding sites of different groups are likely to be determined by characteristics of the herbivore (size, mouthpart structure) and characteristics of plant tissues (food quality, root architecture, defences). The feeding strategy of a particular species may even vary with the developmental stage of the insect. Quinn and Hower (1986) found that early instars of clover root curculio fed on root nodules of alfalfa, whilst later instars fed on the epidermis of the tap root.

Collembola and Diplura feed on young roots and root hairs; young roots may be severed or their active growing tips completely eaten away (Gunn & Cherrett 1993), such damage commonly leads to the death of seedlings (Edwards 1962). Collembola also have important effects on the growth of established plants as a result of their feeding on mycorrhizal hyphae. The feeding strategies of the larger insect taxa can be categorized according to feeding site into internal chewers, external chewers and sap feeders (Table 7.1). Internal chewers burrow into large roots or subterranean storage organs (bulbs, corms, tubers, rhizomes). They can be further grouped according to

207

Table 7.1 Typical characteristics of below-ground insect and nematode herbivores.

	Insect	Nematode
Body size	0.2–5 mm (Collembola, Diplura) 1–20 mm (other groups)	0.1–1 mm
Feeding strategy	Mycorrhizae Young roots Root hairs Internal chewers External chewers Sap feeders	Algae/moss/lichen feeders Root hair feeders Epidermal cell feeders Endoparasites Semi-endoparasites Ectoparasites

Specificity
  Complex
  ↓
  Non-specific
	
	Alternating generations with
	  two different host plants
	Single plant species
	Many plant species
	Plants and other trophic levels
		
		Single plant species
		Many plant species

Dispersal	Active long-distance dispersal in
	  adult stages
	Limited local dispersal through soil
	  (also dispersal on soil surface)
		Passive long-distance dispersal
		  by wind, water, etc.
		Limited local dispersal through
		  water film in soil

Distribution	Heterogenous in space and time	Heterogenous in space and time

the type of tissue on which they feed. For example, in a study of the root-feeding insect fauna of *Centaurea* species, Muller (1989a) found different groups of species feeding on the cortex, the central vascular tissue and the meristematic tissue at the root crown. External chewers may consume whole roots or feed selectively on the epidermis or root cortex of larger roots, they too can be divided into guilds according to feeding location (Muller 1989a). Sap feeders, comprising mainly aphids and cicadas, remove sap from vascular tissue through their stylets.

Host specificity varies widely between below-ground insect herbivores, from those feeding on a single plant species, such as some species of weevil (Coleoptera: Curculionidae) and root aphids (Hemiptera: Aphididae), to generalist chewers, such as chafer (Coleoptera: Scarabaeidae) and crane fly (Diptera: Tipulidae) larvae. Some groups of herbivorous insect, such as root-feeding aphids, have complex life cycles, with generations alternating between herbaceous and woody species (Moran *et al.* 1993). Certain taxa do not even restrict their feeding to a single trophic level; apterygotes (Collembola, Diplura) will also feed on dead root matter or fungi, whilst wireworms (Elateridae larvae) are occasionally predators of other members of the soil fauna (Gunn & Cherrett 1993).

The prevalence of flighted adult stages of most below-ground insect herbivores allows good dispersal or migration. This can be seen in the rapid colonization of insect pests introduced to areas outside their natural range, such as *Popilla japonica* (Coleoptera: Scarabaeidae) in the USA and *Sitona discoideus* (Coleoptera: Curculionidae) in Australia and New Zealand. The occurrence of a highly mobile adult phase allows not only dispersal between sites or patches of host plant (Blossey 1993), but also provides the species with the ability to vary its habitat use to avoid seasonally unfavourable conditions. Studies have shown seasonal movement of root-feeding weevil (Culik & Weaver 1994) and midge (Delettre & Lagerlof 1992) species between woodlands and adjacent agricultural fields. In contrast to their mobile adult stages, larval movement of below-ground insect herbivores within the soil is generally limited, although dispersal rates in excess of $4\,cm\,h^{-1}$ have been recorded for some root-feeding pests (Strnad & Bergman 1987).

The local distribution of below-ground insect herbivores tends to be very heterogeneous, as demonstrated by the large body of literature on sampling methodologies for predicting threshold densities for pest management (Blackshaw 1990; Williams *et al.* 1992). This aggregation is a result of a combination of adult oviposition behaviour and the poor dispersal of larval stages mentioned above. Choice of sites for oviposition is influenced by the soil texture, soil moisture, vegetation structure or the presence of specific host plants (Allsopp *et al.* 1992). Often complete egg batches will be deposited at a single site. Aggregation often declines as the season progresses as larvae move away from the oviposition site in response to intraspecific competition for limited food resources (Quinn & Hower 1986; Seal *et al.* 1992). The distribution and abundance of below-ground insect herbivores also shows great variation between years (Prins *et al.* 1992).

Characteristics of below-ground nematode herbivores

Herbivory, usually referred to as plant parasitism, occurs in three of the 17 orders of Nematoda, namely the Tylenchida, Dorylaimida (class Adenophorea), and Aphelenchida (class Secernentia) (Poinar 1983; Wharton 1986). All plant-feeding nematodes have a stylet or a spear, which is used to collect the fluid contents of plant cells. Most herbivorous nematode taxa feed on below-ground plant parts, mosses, algae or fungi, although some genera (*Ditylenchus*, *Aphelenchoides* and *Anguina*), also parasitise the aerial parts of plants (Bongers & Bongers 1998).

The feeding strategies of herbivorous nematodes are grouped according to whether they enter the plant completely (endoparasites), partly (semi-endoparasites), or only with their stylets (ectoparasites). Yeates *et al.* (1993) grouped plant-feeding nematodes into sedentary and migratory endoparasites, semi-endoparasites, ectoparasites, epidermal and root hair feeders, and algal, lichen and moss feeders (Table 7.1). Endoparasites occur in the Tylenchida and may be migratory (e.g. *Pratylenchus* spp.) with the potential to enter more than one root system during their life cycle, or sedentary (e.g. *Heterodera* spp. and *Meloidogyne* spp.), the females of which get attached to a feeding or giant cell and remain there for the rest of their life cycle. A number of nematode species (Tylenchidae) feed on epidermal cells and root hairs. Some nematodes that are facultative in feeding on higher plants can also feed on fungal hyphae, mosses, lichens or algae (Yeates *et al.* 1993). In contrast to facultative soil pathogens (Jarosz & Davelos 1995), facultative plant-feeding nematodes are thought to be relatively harmless to plants.

The degree of host plant specificity varies between the different nematode feeding types. The interaction between sedentary endoparasitic nematodes and their plant hosts is much more complex than in the case of ectoparasites (Lewis 1987). Specificity may be expressed during a number of steps in the process between sensing a host and the final reproduction of the nematode: sensing, host finding, recognition, infection and postinfection processes. Carbon dioxide, heat, ionic flux and (possibly microorganism-influenced) root exudates play a role in host finding (Lewis 1987). Generally, migratory endoparasites are the most destructive in plant roots; their ability to exploit several host specimens during their life cycle makes them more virulent than sedentary endoparasites (MacDonald 1979; Zacheo 1993).

The active dispersal of nematodes is limited, estimates ranging from less than 10 cm year^{-1} in some agricultural soils (Wallace 1963) to 1 m year^{-1} in, for example, coastal sand dunes (Van der Putten *et al.* 1989). However, their ability to explore plant root zones is remarkable; in New Mexico, nematodes were found in the root zone of desert plants at 11–12 m depth (Freckman & Virginia 1989). Pore size is an important determinant for the active small-scale dispersal of nematodes that migrate through water films (Wallace 1963). In contrast, most long-distance dispersal occurs passively. Cyst nematodes and anhydrobiotic nematodes are dispersed by wind and may thus be easily transported from field to field (White 1953; Orr & Newton 1971). However, transport of living nematodes by wind, such as during sand storms, causes mortality (De Rooij-Van der Goes *et al.* 1997). Other factors leading to dispersal are agricultural machinery, water (including horizontal ground-water flows), transported soil particles and plant parts.

Studies show plant-feeding nematodes to have patchy and clumped distributions (e.g. Fenwick 1961). In addition, vertical stratification of nematodes is common in, for example, forest soils with different soil horizons. Most studies on nematode distribution patterns have been performed in agricultural fields where spatial patterns depend on soil texture, soil moisture and drainage patterns (Norton 1978; Goodell & Ferris 1980; Ferris *et al.* 1990). Distribution patterns change throughout the cropping season (Noe & Campbell 1985; Robertson & Freckman 1995). While the distributions of some taxa are spatially autocorrelated (De Rooij-Van der Goes *et al.* 1995; Rossi *et al.* 1996), reflecting the fact that different taxa may share the same host plants, others are differently affected by the environment (Edwards & Kimpinski 1997). Given their patchiness in arable land, the distribution of plant-feeding nematodes in natural vegetation may be supposed to be clumped as well. In successional gradients spatial heterogeneity can be due to dispersal limitations (Zoon *et al.* 1993).

Impact of below-ground herbivores on plants

The loss of roots to insect and nematode herbivores affects the vegetative growth of the plant through limiting its ability to acquire water or nutrients (Gange & Brown 1989) and the potential for future growth may also be affected through the loss of storage tissue. Some studies have shown the effects of root herbivory on growth to be short-lived, the damaged plants quickly recovering as a result of compensatory regrowth (Rea & Wallace 1981; Quinn & Hall 1992; Steinger & Muller-Scharer 1992). In some conditions, low levels of root herbivory even lead to overcompensatory regrowth (Crutchfield & Potter 1995a). However, other work has shown feeding on below-ground plant parts to have a greater impact than foliar herbivory resulting in a significant reduction in plant growth and fecundity (Reichman & Smith 1991; Prins *et al.* 1992). Differences in response of plants to below-ground herbivory can be related to characteristics of their life histories, such as levels of stored resources. Long-lived clonal species appear to be able to respond to herbivory by compensatory regrowth, whilst annual and monocarpic species are usually more susceptible and respond with reduced growth or fecundity.

Damage to below-ground plant parts leads to altered source–sink relationships within the plant; compensatory root growth occurring at the expense of shoot growth (Schmid *et al.* 1990; Steinger & Muller-Scharer 1992). Feeding on below-ground plant parts can also lead to significant changes in the structure of the above-ground plant parts. Feeding by weevil larvae on the meristematic tissue at the root crown led to release from apical dominance in *Centaurea* species and an increase in shoot number (Saner & Muller-Scharer 1994), whereas total shoot biomass was not significantly altered. Feeding damage to below-ground plant parts such as rhizomes can limit the ability of a plant to reproduce vegetatively. In addition, consumption of below-ground plant parts used for storage (bulbs, corms, tubers) will limit the growth of the plant in subsequent seasons (Blossey 1993). There is considerable evidence that the response of a plant to root herbivory, in particular its ability to show compensatory regrowth, is strongly influenced by the availability of nutrients in the soil (Schmid *et al.* 1990; Steinger & Muller-Scharer 1992). The relative importance of root herbivory is likely to be greatest on infertile soils (D'Arcy-Burt & Blackshaw 1991). There is similar evidence that the relative effects of root herbivory are also greater in drought conditions.

Impact of below-ground insect herbivores

The competitive environment of a plant is important in influencing its response to below-ground herbivory. The impact of below-ground insect herbivory on *Centaurea* species was found to be greater when the host plant was competing with grass (Muller-Scharer 1991). Such effects are likely to be greatest when specialist herbivores attack one species of plant growing in a multispecies community, or when the plant species differ in their susceptibility to herbivory. Even generalist herbivores, such as Scarabaeidae larvae, exhibit feeding preferences on different species and are likely to alter competitive interactions between coexisting grass species (Potter *et al.* 1992; Crutchfield & Potter 1995b). However, the competitive ability of a plant species is not universally reduced by root herbivory. Ramsell *et al.* (1993) found that, even though Tipulidae larvae prefer to graze on *Lolium perenne* than *Rumex obtusifolius*, root grazing of *Lolium* by Tipulids increased the competitive ability of *Lolium* over *Rumex*. The explanation of this finding may be that the young regrowth roots of *Lolium* are more efficient in resource acquisition than the old roots that were consumed.

Whilst the effects of below-ground insect herbivory on plant growth vary between species as outlined earlier, effects on seed production and seedling recruitment are generally negative. Root feeding by insects generally leads to reduced reproductive output, measured as flower production, seed production or seed weight, especially in monocarpic species (Reichman & Smith 1991; Prins *et al.* 1992; Saner & Muller-Scharer 1994). Seedling survival is also reduced by root herbivory both before and after emergence of the young shoot (Gange *et al.* 1991). Similarly, damage to root systems of tree seedlings by termites limits their ability to establish as adults in response to canopy opening (Fensham & Bowman 1992).

Apart from the direct effects of below-ground insect herbivores on plant growth and reproduction, plants may also suffer from indirect effects. For example, Bibionid larvae disrupt the soil surface leading to secondary root damage as a result of desiccation and frost; birds searching for larvae on which to feed further exacerbate this damage (D'Arcy-Burt & Blackshaw 1991). Herbivory or other root damage also affects the susceptibility of a plant to foliar feeding insects. Gange and Brown (1989) found that root herbivory led to increased aphid weight, growth, fecundity and longevity, probably caused by increased soluble nitrogen in the leaves as a result of amino-acid mobilization. Similarly, root damage caused by agricultural machinery was found to lead to greater damage by foliar feeding weevils on *Fraxinus excelsior* (Foggo *et al.* 1994). However, damage by below-ground herbivores may lead to the production of induced defences and thus reduce susceptibility to herbivore attack on other plant parts (Schultz 1988). Such indirect interactions between above- and below-ground herbivores were discussed further in a previous volume (Masters & Brown 1997).

Impact of below-ground nematode herbivores

The effects of nematodes on plants depend on the feeding type of the nematode, its population density, interactions with other nematodes or soil organisms and interactions with abiotic factors. The effects of reductions in the number and/or longevity of root hairs or the destruction of cortical cells are local, plants being able to compensate for the loss of considerable root tissue without any apparent reduction in above-ground biomass. Thus, plants have considerable tolerance to nematode damage (Rea & Wallace 1981). However, disruption of vascular tissue, induction of thickening of cortical cell walls and granulation of the endodermis, as well as the formation of compounds such as phenols by migratory endoparasites, is very destructive, as is giant-cell formation by *Meloidogyne* spp., which causes extensive disruption of the xylem (Wilcox-Lee & Loria 1987).

Nematodes can also affect photosynthesis, but the mechanisms are complex and poorly understood, possibly involving a chain-like sequence and interaction of indirect metabolic processes (Wallace 1987). For example, *Meloidogyne* spp. may affect the photosynthesis of its host plant by inhibition of phytohormone production as a result of root damage or reduced xylem transport into the leaves due to stomatal closure (Bird 1974). The effects of nematode damage on below-ground plant parts often resemble the effects of reduced water availability (Wilcox-Lee & Loria 1987); leaf water potential is usually reduced, stomatal conductance decreased and root conductivity reduced (especially for the root knot *Meloidogyne* spp.). However, reports on the effects of nematode damage on transpiration show little consistency (Wilcox-Lee & Loria 1987).

Differential impacts on plants of contrasting root architecture

Plants of contrasting root architecture are likely to differ in their susceptibility and responses to below-ground herbivory. Differences in the characteristics of root systems, such as meristem positioning and branching pattern, occur between plants of different life history and between monocotyledonous or dicotyledonous species. In annual species, the main functions of the root system are anchorage and the acquisition of water and nutrients in order to support rapid growth and reproduction. In species with a longer life cycle, below-ground plant parts are also used for other functions such as storage and vegetative reproduction. Many monocarpic species store resources in large tap roots in order to support reproduction in a later year. Similarly, some perennial species pass through unfavourable conditions completely below ground, with resources stored in corms, bulbs or other organs. Such storage tissues make attractive food sources for soil-dwelling herbivores and are often protected by chemical defences. Longer-lived woody species tend to have low root/shoot ratios and large root systems play an important role in anchorage.

The feeding of below-ground herbivores shows interesting relationships with the size and architecture of the root systems. Most generalist herbivores feed on the network of fine roots close to the soil surface. Thus, shallow-rooted grass species were found to be more susceptible to herbivory by Bibionid larvae than deep-rooted grass species (D'Arcy-Burt & Blackshaw 1991). Insect larvae move deeper into the soil in response to reductions in soil moisture and therefore their feeding site may be linked to temporal changes in soil conditions. In contrast, specialist herbivores feed not only on a single species or small group of species, but on specific

214

niches within the below-ground plant structures, e.g. the larvae feeding on the central vascular tissue of *Centaurea* tap roots (Muller 1989a), the storage tissues of purple loosestrife (Blossey 1993) or the root nodules of many legume species (Quinn & Hall 1992). The effects of below-ground herbivory are most severe in young plants with small root systems (Muller-Scharer 1991; Prins *et al.* 1992). The mortality of lupin bushes as a result of herbivory by Hepialid larvae was found to be related to plant age, younger plants suffering greater mortality, whilst older plants were better able to withstand damage (Strong *et al.* 1995; Chapter 19). However, adults may prefer to lay eggs close to larger individuals of their host plants in order to maximize larval food supply (Prins *et al.* 1992).

Interactions between below-ground herbivores

Interactions between below-ground insect herbivores

Most work on below-ground insect herbivores has been concerned with their management as pests or potential for use as biocontrol agents and, as such, has focused on their effects on their host plant. There are few examples in the literature of studies investigating the interactions between different herbivores. The highly aggregated distribution of many below-ground insect herbivores, especially pronounced in the early larval instars, leads to intraspecific competition for limited food resources. Competition between larvae for root nodules was found to be the greatest mortality factor for early instars of clover root curculio on alfalfa. Later instars feed on the epidermis of the tap roots, with competition possibly occurring for surface area (Quinn & Hower 1986; Quinn & Hall 1992). The patchy distribution of many below-ground insect herbivores would lead one to expect that interspecific competition is less important than intraspecific competition. Differences in temporal as well as spatial distribution also reduce the encounter rates between species. Crutchfield and Potter (1995a) found seasonal differences in the relative importance of different chafer species feeding on turf grass. Even so-called generalist herbivores, such as chafer larvae, exhibit preferences for the roots of different species (Potter *et al.* 1992; Crutchfield & Potter 1995a,b). However, it is amongst the stenophagous insect fauna of particular host plants that interspecific competition is likely to be greatest.

In a comprehensive survey of the below-ground herbivorous insect fauna of *Centaurea* species, Muller (1989a) identified five structural feeding niches on the root system. Each of these niches could be occupied by a number of insects, for example the central vascular tissue of the tap root was fed on by Bupresstidae, Curculionidae and Lepidoptera larvae. Three types of interaction between herbivores in the same guild were proposed by Muller. Firstly, niche competition with substrate modification (e.g. gall formation) leading to exclusion or displacement of one of the species; the larvae of these species exhibited regular distributions with increasing larval density. In contrast, some species competed for the same niche but lacked the ability to modify the substrate; the larvae of these species tending to exhibit a clumped distribution. Finally, some species showed positive associations, usually as a result of adults exhibiting similar preferences for sites for oviposition. The below-ground entomofauna of *Centaurea* fed on roots in the autumn and spring when the nitrogen and water content of the roots is highest (Muller *et al.* 1989). However, in spite of the absence of temporal niche differentiation, low levels of occupancy of niches led to low encounter rates between herbivores and the conclusion that interspecific competition between herbivores is unimportant (Muller 1989a). Similar conclusions were reached by Gunn and Cherrett (1993), who, in spite of finding preference for feeding on different parts of the root system amongst the soil-dwelling insects they studied, found no clear evidence for exclusive guilds or species packing.

Interactions between below-ground nematode herbivores

A large amount of work has been carried out on interactions between nematodes in agro-ecosystems, which has been extensively reviewed by Eisenback (1993); this review is used as a basis for the following section. Examples of competitive, neutral and facilitative interactions have been reported for all possible combinations of feeding types (Eisenback 1993). The outcome of nematode–nematode interactions are affected by a large number of possible factors. Experimental conditions, such as soil temperature, soil texture, length of the experiment, order of inoculation and time lag between inoculations are of crucial importance for the outcome of the interaction studies (Eisenback 1993). Many studies have been carried out in greenhouses, but these results are often quite different from interactions under field conditions. For example, Miller and Wihrheim (1968) found that *Tylenchorhynchus claytoni* and *Globodera tabaccum* impaired each other's reproduction in the greenhouse, but that *G. tabacum* dominated in the field.

Biotic factors, such as host suitability (Khan *et al.* 1986a,b) and relative population densities, also affect the outcome of interactions. In general, nematode species having a complex interaction with their plant host are more competitive than species with a less complicated host relationship. However, this is not always the case, for example damage to the root system by ectoparasites can result in a negative impact on sedentary endoparasites because of a reduction in the density of suitable feeding sites. The competitive interaction between nematodes is also affected by differences in length of the life cycles (Rao & Prasad 1971), as well as by different host compatibilities (McGawley & Chapman 1983). However, results of competition experiments with nematode species of different degrees of compatibility to the shared plant host have demonstrated facilitation of the incompatible species by the compatible species, rather than competition (W.H. Van der Putten and H. Duyts, unpublished results).

Many studies are based on nematodes coexisting in arable soils, in which the nematode community is strongly influenced by the cropping history (Ferris *et al.* 1990). In perennial pastures with *Trifolium repens* (white clover), *Heterodera* spp., *Meloidogyne* spp. and *Pratylenchus* spp. shared a similar niche in time (Yeates *et al.* 1985). Differences in stylet lengths of coexisting species offer opportunities for niche differentiation by allowing feeding on different cell layers (Bongers & Bongers 1998). The strong spatial heterogeneity of nematode occurrence in the field, both horizontally and vertically, the differentiation in the timing and location of feeding patterns, suggest that spatiotemporal niche segregation is an important factor in the coexistence of nematode species.

Interactions between below-ground insect and nematode herbivores

Herbivorous insects and nematodes which feed on below-ground plant parts share the same food resource and might therefore be expected to interact. However, as outlined earlier, both groups of herbivores exhibit great temporal and spatial heterogeneity in abundance. In addition, there is evidence for both niche differentiation, with taxa varying in their specificity for host plant and feeding sites. Interactions between these two groups of herbivores may, therefore, be the exception rather than the rule. The ability to draw conclusions about interactions between them is limited by the paucity of published studies in the literature, although similar interactions such as that between root-feeding nematodes and stem-boring insects (Russin *et al.* 1993) have already been demonstrated.

Studies of plant-feeding nematode communities have demonstrated the occurrence of competitive, mutualistic and neutral interactions (Eisenback 1993) and it is probable that a similar suite of interactions occur between insects and nematodes feeding on below-ground plant parts (Table 7.2). Negative interactions may be direct, for example by interference competition for food, or indirect, for example through alteration of the food quality of plant tissues as a result of the induction of plant defences. Positive interactions may occur when one herbivore facilitates or enhances the performance of a second herbivore as a result of the effects of damage on tissue structure, changed root structure due to compensatory growth, or food quality.

Table 7.2 Potential interactions between below-ground herbivores.

Interaction	Mechanism	Insect examples	Nematode examples
Avoidance			
Spatial avoidance	Patchy distribution	Blackshaw (1990)	Fenwick (1961)
	Differences in feeding site	Muller (1989a)	Bongers & Bongers (1998)
Temporal avoidance	Seasonality	Crutchfield & Potter (1995a,b)	Yeates et al. 1985
Competition			
Intraspecific	—	Quinn & Hall (1992)	De Rooij-Van der Goes (1995)
Interspecific	—	Muller (1989a)	Miller & Wihrheim (1968)
Inhibition			
Feeding by Herbivore 1 leads to reduced performance of Herbivore 2	Mechanisms may include: • induction of plant defences • structural alteration of plant tissue	Muller (1989a)	Eisenback (1993)
Facilitation			
Feeding by Herbivore 1 leads to increased performance of Herbivore 2	Mechanisms may include: • changes in the food quality of plant tissue • increased attractiveness as a result of root exudates • structural alteration of plant tissue		Eisenback (1993)

Interaction between below-ground herbivores and other soil organisms

Interactions between below-ground herbivores and plant pathogens

The effects of herbivory by insects or nematodes on below-ground plant parts may be affected by interactions with plant pathogens. Damage to below-ground plant parts caused by soil-dwelling herbivores can increase the chances of infection of the plant by bacteria, viruses or fungal pathogens. The plant is therefore exposed to the dual problems of herbivory and disease. The interaction between insect herbivores and pathogens has received little research attention, although some studies of crop species have been reported. Leath and Hower (1993) found that damage to roots of *Trifolium pratense* by weevil larvae predisposed the roots to fungal infection. Infection was found to spread out from the feeding site, transmission of the fungus to other parts of the plant occurring if feeding damaged the xylem vessels. Soil-dwelling insect herbivores also act as dispersal agents for bacteria, viruses or fungal spores. The interaction between herbivores and pathogens also operates in the opposite direction. Adult bark beetles (Coleoptera: Scolytidae) were found to prefer to feed on diseased red clover roots, in keeping with the general feeding preference of this beetle family for dead or decaying matter (Leath & Byers 1973).

In contrast, a vast amount of literature describes interactions between plant-feeding nematodes and pathogens (Khan 1993a). Certain nematode species act as transmitters of viruses, and nematodes can also be vectors for bacteria and fungal propagules. Both direct and indirect changes of the invasibility or susceptibility of plants to fungal or bacterial pathogens as affected by nematodes have been reported (Khan 1993b). The majority of studies, however, concern agricultural or horticultural crops and few studies on these nematode–pathogen interactions, such as that by De Rooij-Van der Goes (1995) exist for natural communities. Such studies are needed in order to determine the importance of these interactions in natural plant communities. Moreover, in natural ecosystems, both the order and timing of the participating organisms in the pathogen complexes are still the result of evolutionary history, rather than of agricultural practices, breeding and resistance breaking. The effects of pathogens on nematodes should also be studied. Suppression of hatching of nematodes by some pathogenic fungi, or changed sex ratios as a result of the changed nutritional status of the host plant by others, affect nematode population development (Hasan 1993).

Interactions between below-ground herbivores and plant mutualists

The growth of plants that depend on mycorrhizal associations for their nutrient acquisition is likely to be affected by members of the soil fauna which feed on ectomycorrhizal hyphae. Soil-dwelling insects in the orders Collembola (springtails) and Diplura (bristletails) and nematodes in the genera *Aphelenchus*, *Aphelenchoides*, *Bursaphelenchus* and *Ditylenchus* (Francl 1993) are the most important ectomycorrhizal grazers. Other insects and nematodes feeding on roots infected with either arbuscular mycorrhizae or endomycorrhizae also affect the efficacy of the infection by root removal. Other members of the soil fauna, such as earthworms, molluscs and isopods, also feed on mycorrhizae but are relatively less important (Gunn & Cherrett 1993), although they play a role in the dispersal of spores. The presence of such fungivores will therefore affect the competitive relationships between mycorrhizal and non-mycorrhizal plants. Studies on the impact of the interaction between mycorrhizae and their herbivores in succession are absent from the literature. However, experimental manipulation of mycorrhizal infection in an early successional community has shown that exclusion of mycorrhizae retarded colonization by perennial forbs (Gange *et al.* 1990, 1993). It might be predicted that the presence of mycorrhizal feeders will slow the rate of succession by altering the balance between early-successional plant species (typical non-mycorrhizal) and mycorrhizal perennial species typical of later stages of succession.

In addition to the well-established effects of mycorrhizae on plant growth, mycorrhizal infection can confer additional benefits. Gange *et al.* (1994) found a significant interaction between the density of weevil larvae and mycorrhizal infection on the biomass of pot-grown plants. Mycorrhizal infection was found to confer a degree of resistance, or tolerance, to herbivory. At low herbivore densities, the effects of larval feeding were insignificant. In addition, the survival and growth of weevil larvae was found to be greatest on uninfected plants. Little and Maun (1996) found a positive effect of arbuscular mycorrhizal fungi in reducing the effects of endoparasitic nematodes on the sand binder *Ammophila breviligulata* under greenhouse conditions. This effect only occurred when the plants were subjected to moderate sand deposition, but not when the plants remained unburied.

When exploring the possible use of endomycorrhizae against soil-dwelling herbivores in crop protection, both positive and neutral effects have been reported (Pinochet *et al.* 1996; Roncadori 1997). Small-scale experimental results on the application of mycorrhizal fungi as biological crop protection agents are often difficult to validate at large, practical scales (Roncadori 1997). Therefore, results of ecological experiments on mycorrhiza-soil herbivore interactions need to be verified in the field in order to obtain fair estimations of the ecological consequences. Whether the findings of root protection by mycorrhizae are the result of direct effects of the fungus on the chemistry of the root tissue, or differences in root morphology between infected and uninfected plants is not clear. Further discussion of the interaction between subterranean insects, nematodes and mycorrhizae can be found in a previous volume (Gange & Bower 1997; Roncadori 1997), as well as in other reviews (e.g. Francl 1993; Pinochet *et al.* 1996).

Plants infected with endophytic fungi may have enhanced resistance to herbivory. Several studies in grasslands have shown that endophyte-infected grasses were less susceptible to feeding by below-ground herbivores, especially chafer larvae, than uninfected neighbours (Potter *et al.* 1992; Clay *et al.* 1993; Crutchfield & Potter 1995a). The mechanisms for the resistance (or tolerance) conferred on plants by endophyte infection are probably linked to changes induced by phytohormones. Endophyte infection leads to increased chitinase production, which can confer greater protection against pathogens. Thickening of the root endodermal layer due to endophyte presence leads to increased resistance against nematode and insect herbivory (Joost 1995).

Impact of below-ground herbivory on vegetation succession

Factors important in determining the rate and direction of vegetation succession include interactions between plants such as competition, facilitation and inhibition, and the interaction between plants and their physical environment and higher trophic levels. Much work has been carried out on these factors in both primary and secondary succession. Animals play an important role in plant succession through their effects on plant performance, survival and establishment. The effects of above-ground vertebrate and invertebrate grazers in successional systems have been widely studied. However, in the same way that there has been limited work on the effects of below-ground herbivory on plant performance, the implications of these effects for changes in plant community structure have been neglected to an even greater extent.

Successional trends in below-ground insect herbivores

Differences in the below-ground insect herbivore assemblages are apparent between communities of different successional age. Different groups of herbivores show different trends, determined by the interaction of the habitat requirements of the herbivore and its life history. Few studies have examined the trends in composition of the subterranean herbivorous insect fauna as succession progresses. The density and diversity of below-ground insect herbivores was found to increase in the early years of secondary succession (Brown & Gange 1992). Increases in diversity are often the result of the gradual colonization of successional communities by specialist herbivores. Studies of newly created grasslands on ex-arable land have shown that specialist herbivores remain absent even after colonization of the site by their host plant (Mortimer *et al.* 1998), such patterns caused by limited dispersal are to be expected in patchily distributed habitats in agricultural landscapes.

Inferences on the effects of succession on the distribution of below-ground insect herbivores can also be made from studies of agricultural grasslands. The below-ground insect herbivores of older pastures studied by Clements and co-workers tended to be those with long-lived larval stages, whilst those of young grass leys were species with widely dispersed adults and short-lived larval stages (Clements *et al.* 1987). Populations of herbivores may increase with age of grasslands and then decline as a result of the build-up of pathogens of the herbivore. The density of Bibionid larvae was found to increase with increasing content of organic matter in the soil, a parameter which often increases with successional age (D'Arcy-Burt & Blackshaw 1991). However, different species exhibited different patterns in relation to grassland age, some preferring young leys and others old permanent pasture. Frouz (1994) found that the density of Chironomid larvae decreased as succession progressed, the main factors responsible being soil moisture and vegetation cover. In contrast, other species, such as root aphids, may be absent from young communities as a result of their dependence on the presence of a particular host plant species or on an association with ants.

Successional trends in below-ground nematode herbivores

Successional trends in plant-feeding nematodes have been described for the Kampinos forest by Wasilewska (1970, 1971) and inland sand dunes by De Goede *et al.* (1993a,b). Obligate plant feeders typify the community of nematodes parasitizing higher plants in the early colonization stages in the succession from drift sand to forest, whilst the dominance of these obligates declined strongly both in numbers and in biomass towards later stages (Wasilewska 1970, 1971). De Goede *et al.* (1993b) found obligate plant feeders to dominate the 0–10 cm layer of mineral soil of both drift sand and *Spergulo–Corynephoretum* communities characteristic of early successional stages, whereas facultative plant feeders reached dominance in the later Scots pine (*Pinus sylvestris* L.) forest stages.

In coastal foredunes, plant-feeding nematodes occur in all stages of dune vegetation succession from early foredune stages through to vegetation dominated by *Hippophaë rhamnoides* shrubs. The abundance of plant-feeding nematodes in the rhizosphere increased towards later stages, whereas populations in the bulk soil remained constant (Zoon *et al.* 1993). Obligatory plant feeders are also well represented in the root zone of the first colonizing dune plant species, even in young stands established from seeds (De Rooij-Van der Goes *et al.* 1995). In the first stages of foredune vegetation succession, the root zones of most dominant grass species were characterized by specific combinations of migratory and sedentary endoparasitic nematodes (Van der Putten & Duyts 1998).

Effects of below-ground herbivory on vegetation and succession

The responses of plant communities to below-ground herbivory are likely to be just as varied as those to herbivores feeding on above-ground plant parts, being determined by the characteristics of the plant species, the feeding preferences of the herbivores concerned and the prevailing environmental conditions both during and following herbivory. Below-ground herbivores will affect plant community structure through their influence on plant growth, reproduction, establishment and competitive interactions. Insect larvae feeding on the roots of particular plant species may severely limit the size and/or age structure of populations of their host. Strong *et al.* (1995) have shown that root herbivory by Hepialid larvae is responsible for patchy die-back of *Lupinus arboreus* in a coastal shrub community leading to large temporal and spatial fluctuations in *Lupinus* populations (see also Chapter 19). Insects and nematodes feeding on the roots of *Trifolium repens* cause patchy die-back in pasture swards 2–3 years after sowing (Pederson *et al.* 1991). Generalist root feeders also have significant effects on plant community structure. Work in pasture communities has shown that below-ground insect herbivores can lead to the death of sown grasses in patches and the subsequent colonization of the sward by unsown grass and weed species (D'Arcy-Burt & Blackshaw

1991). Even generalist herbivores show some preference for feeding on particular species of plant and thus alter the competitive balance between species (Potter *et al.* 1992; Crutchfield & Potter 1995b).

Whilst the effects of below-ground herbivores on local temporal and spatial variations in plant community structure have received some recent attention in the pest management and biological control literature, ecological studies on the long-term effects of below-ground insect herbivory on vegetation succession are still rare. Working in early successional old fields, Brown and Gange (1989a) found that both foliar and soil herbivores have a significant impact on primary productivity. Root-feeding insect herbivores were found to limit the establishment of annual and perennial forb species through seed predation and seedling mortality (Gange *et al.* 1991). Chemical exclusion of below-ground insect herbivores using a soil insecticide resulted in increased species richness (Brown & Gange 1989b). This contrasts with the effect of foliar herbivores which increase species richness by limiting growth of perennial grasses, thereby promoting the establishment of forb species through increasing microsite availability (Brown & Gange 1989a). In conclusion, using information on the differential responses of foliar feeders and below-ground insect herbivores, foliar feeders were considered to delay succession and root feeders to accelerate old-field succession (Brown & Gange 1992).

In spite of unequivocal evidence for the impact of below-ground insect herbivores on early successional plant communities, their importance in more mature grasslands is less clear. In a study of six mature and six mid-successional grasslands, two of the authors found persistent increases in vegetation cover after the exclusion of soil-dwelling insects at only four of the 12 grasslands studied (S.R. Mortimer & V.K. Brown, unpublished data). At most of these sites, the effect was the result of an increase in the cover of perennial grasses. Other life-history groupings and species showed significant responses to the treatment at some sites in some years, but the effect was only persistent in a few cases. No significant effect of the application of the soil-acting insecticide could be found on the species richness of the plant communities studied, even after 3 years of the treatment. However, at those sites which showed increased cover of perennial grasses, it is expected that reductions in species richness will occur with prolonged treatment, as a result of the loss of species with low competitive ability or those with the requirement for regeneration from seed. Similar results have been found in studies of the effects of below-ground insect herbivores in mature tall-grass prairie (Seastedt *et al.* 1987; Gibson *et al.* 1990; Todd *et al.* 1992).

Various approaches have been followed to determine the role of plant-feeding nematodes in ecosystem functioning. In North American prairies, the importance of nematodes as below-ground herbivores (Stanton 1988) and their role in below-ground food web functioning (Moore *et al.* 1991) have been studied. The total nematode biomass in near-climax *Festuca idahoensis* grassland communities is of $0.83–0.88\,\mathrm{g\,m^{-2}}$, of which 35% was herbivorous (Weaver & Smolik 1987); by way of comparison, these grasslands can sustain $4–5\,\mathrm{g\,m^{-2}}$ of cattle biomass (Weaver & Smolik 1987). Ingham and Detling (1990) estimated that the reduction of above-ground net primary production by root-feeding nematodes was 16 times more than they consumed and concluded that nematodes are among the main herbivores in these ecosystems.

In coastal dunes of the northern hemisphere, the role of plant-feeding nematodes has been studied in relation to plant species die-back and succession (Seliskar & Huettel 1993; Van der Putten *et al.* 1993; Little & Maun 1996). The specificity of successive soil pathogen complexes and their effects on the relative competitive abilities of host plants are important in controlling the succession of these dune vegetations (Van der Putten *et al.* 1993). The effect of the soil pathogens was strongly correlated to that of endoparasitic nematodes (Van der Putten & Peters 1997). Van der Putten and Van der Stoel (1998) found that the specificity of successive pathogen complexes, as observed by Van der Putten *et al.* (1993), may be explained by plant-feeding nematodes acting as keystone species or by the successive plant species having different threshold values. Recent surveys (C.D. Van der Stoel, H. Duyts & W.H. Van der Putten, unpublished results) show that specialized endoparasitic nematodes occur in the sequence of dominant foredune plant species, which supports the hypothesis of the role of nematodes as keystone species that may either interact with or transmit other harmful soil organisms, such as soil fungi.

Other examples of the role of plant-feeding nematodes on changes in vegetation composition concern the involvement of endoparasites (root lesion, root knot and cyst nematodes) in the persistence of *Trifolium repens* (white clover) in pastures (Ennik *et al.* 1964, 1965; Yeates *et al.* 1985) or on the die-off of *Boutaloua gracilis* (blue grama) in prairies (Stanton *et al.* 1984). Similarly, unpublished results of H. Olff, B. Hoorens, R.G.M. De Goede, W.H. Van der Putten and M. Gleichman in a species-rich inland sand dune grassland show that oscillation of the codominant clonal mono-cotyledonous species may be correlated with a biotic soil-borne inhibition of one of the codominants. The susceptible species appeared to have a lower threshold value to the plant-feeding nematode community, which was exacerbated by presence of the unsusceptible species.

Estimation of the importance of plant-feeding nematodes in natural and seminatural communities is hindered by methodological problems. For example, in pot trials, excluding nematodes with the selective nematicide Oxamyl resulted in considerable growth enhancement of test plants compared with those grown in untreated natural field soil (Zoon 1995; Van der Putten *et al.* 1990). However, inoculation of two dominant foredune plant species grown in sterilized field soil with ecto- and endoparasitic nematodes did not result in substantial growth reduction, even when the density of plant-feeding nematodes in the sterilized soil was similar to that in unsterilized field soil (Maas *et al.* 1983; De Rooij-Van der Goes 1995). Therefore, the role of nematodes as plant feeders may be over-estimated if based on results of nematicide treatments alone. This may be the result of the direct effect of nematicides on the fungal community at the rhizoplane (Zoon 1995), the role of nematodes as carriers of viruses, bacteria and fungal propagules, or their involvement in complex multi-trophic interactions.

As most of the prairie studies were performed by chemical perturbations of the soil ecosystem, such as by selective nematicides (see e.g. Ingham *et al.* 1986a,b), it might be that the direct effects of nematodes in these ecosystems have been overestimated. On the other hand, prairie systems are relatively stable in terms of vegetation composition, whereas in coastal dunes the rate of vegetation succession is faster, with stages characterized by single dominant plant species, especially the early stages, on which most studies have focused. These differences in characteristics of the plant community both influence, and are influenced by, the nematode community (Yeates 1987). Thus, understanding the role of plant-feeding nematodes in food webs and vegetation succession is a challenge for eco-logists, the current state of the art being obscured by both the methodology used and the contrasting ecosystems on which the studies have focused.

Future considerations

The decade since the reviews of Andersen (1987), Stanton (1988) and Brown and Gange (1990) has seen an increase in the amount of research on below-ground herbivory in non-crop species. However, the implications of such herbivory for multispecies plant communities and the extent of interactions between subterranean herbivores, and between them and plant pathogens and mutualists, are still poorly understood and offer useful opportunities for improving our knowledge of the forces structuring communities. The role of plant-feeding nematodes and their interactions with pathogens in controlling the direction of vegetation succession on dunes is now apparent (Van der Putten *et al.* 1993; Van der Putten & Van der Stoel 1998). In addition to its impact on succession, spatiotemporal variation in below-ground herbivory may account for local fluctuations in vegetation composition. The extent to which subterranean herbivores contribute to the patchy die-back of plant species, and thereby to patch dynamics, allowing coexistence of species in the plant community is not known, although small-scale disturbances are thought to play an important role in maintaining the species richness of many plant communities.

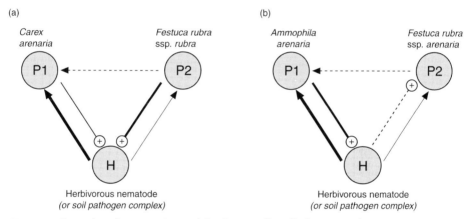

Figure 7.1 Examples of community modules from studies of below-ground herbivory in coastal sand dune systems. Direct negative effect ⟶; direct positive effect —⊕; indirect negative effect - - - - →; indirect positive effect - - - - - ⊕; (the width of the line indicates the strength of the effect). (a) Plant species P1 is susceptible to the generalist herbivore H, whilst plant species P2 is relatively tolerant. However, P2 creates conditions under which the nematode flourishes thereby exerting an indirect negative effect on P1 (H. Olff, B. Hoorens, R. G. M. De Goede, W. H. Van der Putten & M. Gleichman, unpublished results); (b) Plant species P1 is susceptible to the specific nematode (or soil pathogen complex) H and creates conditions under which the nematode flourishes, whilst P2, is relatively tolerant. The herbivore therefore has an indirect positive effect on plant species P2. In the presence of the herbivore, P2 has a competitive advantage and hence replaces P1 in the successional sequence (see Van der Putten & Peters 1997).

227

The tendency for studies of below-ground herbivory to focus on specific interactions between herbivores and their host plants, especially in the biological control literature, may lead to an overemphasis of the importance of specialist herbivores. Generalist herbivores may have important impacts on plant community structure through differences in feeding preferences of the herbivore and the susceptibility of the plant (Table 7.3). Recent work has suggested that in multispecies plant communities, differences in damage thresholds between plant species may be as important as host plant specificity. For example, a plant species that is reasonably tolerant of herbivore damage may provide conditions in which the herbivore flourishes and thus have an effect on the performance of a second plant species (Fig. 7.1). Thus, the tolerant species has an indirect negative effect on the susceptible species. Communities that contain a range of soil-dwelling herbivorous taxa are likely to be characterized by indirect interactions between component plant species as a result of differential effects

Table 7.3 Effects of herbivory by specialist and generalist feeders on below-ground plant parts (after Van der Putten & Van der Stoel 1998).

Specificity of herbivore	Mechanism	Effect on composition of plant community?
Specialist	Alteration of competitive interactions Plant death leading to gaps in the vegetation for colonization or expansion of other species	Yes
Generalist	Frequency-dependent herbivory (plants are consumed in proportion to their abundance in the community)	No
	Degree of herbivory varies between plant species as a result of differences in root architecture (e.g. depth)	Yes
	Degree of herbivory varies between species as a result of differences in feeding preferences of herbivore	Yes
	Degree of herbivory varies between species as a result of differences in thresholds in host plant susceptibility	Yes
	Impact of herbivory varies between plant species as a result of effects of damage on pathogen infection	Yes

of herbivory on competitive interactions (Miller 1994). The likelihood of finding complex direct and indirect effects of below-ground herbivory is increased if interactions with plant pathogens or mutualistic species are considered (Table 7.3). However, the interaction between below-ground herbivory and fungal, bacterial or viral diseases is poorly understood in non-agricultural systems.

Similarly, very little is known about interactions between the soil-dwelling insect and nematode herbivores. There are many examples of interactions between nematodes, although almost exclusively based on studies in agro-ecosystems. Niche segregation, competition and facilitation have all been demonstrated. The field of research on the role of below-ground interactions in natural and seminatural communities is clearly wide open. There is an urgent need for multidisciplinary approaches at multiple sites, since climatic conditions, ecosystem characteristics and species interactions are all involved in determining the importance of soil herbivory.

Acknowledgements

Thanks are due to the helpful comments of an anonymous referee. The experimental work described was supported by the Natural Environment Research Council and the Environment and Climate programme of the European Commission, contract no. ENV4-CT95-0002.

References

Allsopp, P.G., Klein, M.G. & McCoy, E.L. (1992). Effect of soil moisture and soil texture on oviposition by Japanese beetle and rose chafer (Coleoptera: Scarabaeidae). *Journal of Economic Entomology*, **85**, 2194–2200.

Andersen, D.C. (1987). Below-ground herbivory in natural communities: a review emphasizing fossorial animals. *Quarterly Review of Biology*, **62**, 261–285.

Bird, A.F. (1974). Plant response to root knot nematode. *Annual Review of Phytopathology*, **12**, 69–85.

Blackshaw, R.P. (1990). Application of a sequential sampling technique to DIY assessment of leatherjacket (Diptera: Tipulidae) populations in grassland. *Grass and Forage Science*, **45**, 257–262.

Blossey, B. (1993). Herbivory below-ground and biological weed control — life history of a root-boring weevil on purple loosestrife. *Oecologia*, **94**, 380–387.

Bongers, T. & Bongers, M. (1998). Functional diversity of nematodes. *Applied Soil Ecology* (in press).

Brown, V.K. & Gange, A.C. (1989a). Differential effects of above- and below-ground insect herbivory during early plant succession. *Oikos*, **54**, 67–76.

Brown, V.K. & Gange, A.C. (1989b). Herbivory by soil-dwelling insects depresses plant species richness. *Functional Ecology*, **3**, 667–671.

Brown, V.K. & Gange, A.C. (1990). Insect herbivory below-ground. *Advances in Ecological Research*, **20**, 1–58.

Brown, V.K. & Gange, A.C. (1992). Secondary plant succession — how is it modified by insect herbivory? *Vegetatio*, **101**, 3–13.

Clay, K., Marks, S. & Cheplick, G.P. (1993). Effects of insect herbivory and fungal endophyte infection on competitive interactions among grasses. *Ecology*, **74**, 1767–1777.

Clements, R.O., Bentley, B.R. & Nuttall, R.M. (1987). The invertebrate population and response to pesticide treatment of two permanent and two temporary pastures. *Annals of Applied Biology*, **111**, 399–407.

Crutchfield, B.A. & Potter, D.A. (1995a). Damage relationships of Japanese beetle and southern masked chafer (Coleoptera: Scarabaeidae) grubs in cool-season turfgrasses. *Journal of Economic Entomology*, **88**, 1049–1056.

Crutchfield, B.A. & Potter, D.A. (1995b). Tolerance of cool-season turfgrasses to feeding by Japanese beetle and southern masked chafer (Coleoptera: Scarabaeidae) grubs. *Journal of Economic Entomology*, **88**, 1380–1387.

Culik, M.P. & Weaver, J.E. (1994). Seasonal crawling activity of adult root-feeding insect pests (Coleoptera: Curculionidae; Scolytidae) of red clover. *Environmental Entomology*, **23**, 68–75.

D'Arcy-Burt, S. & Blackshaw, R.P. (1991). Bibionids (Diptera: Bibionidae) in agricultural land: a review of damage, benefits, natural enemies and control. *Annals of Applied Biology*, **118**, 695–708.

De Goede, R.G.M., Georgieva, S.S., Verschoor, B.C. & Kamerman, J.W. (1993a). Changes in the nematode community structure in a primary succession of blown-out areas in a sand drift landscape. *Fundamental and Applied Nematology*, **16**, 501–513.

De Goede, R.G.M., Verschoor, B.C. & Georgieva, S.S. (1993b). Nematode distribution, trophic structure and biomass in a drift sand landscape. *Fundamental and Applied Nematology*, **16**, 525–538.

Delettre, Y.R. & Lagerlof, J. (1992). Abundance and life-history of terrestrial Chironomidae (Diptera) in 4 Swedish agricultural cropping systems. *Pedobiologia*, **36**, 69–78.

De Rooij-Van der Goes, P.C.E.M. (1995). The role of plant-parasitic nematodes and soil-borne fungi in the decline of *Ammophila arenaria* L. Link. *New Phytologist*, **129**, 661–669.

De Rooij-Van der Goes, P.C.E.M., Van der Putten, W.H. & Van Dijk, C. (1995). Analysis of nematodes and soil-borne fungi from *Ammophila arenaria* (Marram grass) in Dutch coastal foredunes by multivariate techniques. *European Journal of Plant Pathology*, **101**, 149–162.

De Rooij-Van der Goes, P.C.E.M., Van Dijk, C., Van der Putten, W.H. & Jungerius, P.D. (1997). The effects of sand movement by wind in coastal foredunes on nematodes and soil-borne fungi. *Journal of Coastal Conservation*, **3**, 133–142.

Edwards, C.A. (1962). Springtail damage to bean seedlings. *Plant Pathology*, **11**, 67–68.

Edwards, L.M. & Kimpinski, J. (1997). Relationships between soil penetration resistance and soil nematode burden in barley on Prince Edward Island. *Biology and Fertility of Soils*, **24**, 13–17.

Eisenback, J.D. (1993). Interactions between nematodes in cohabitance. *Nematode Interactions* (Ed. by M.W. Khan), pp. 134–174. Chapman & Hall, London.

Ennik, G.C., Kort, J. & Luesink, B. (1964). The influence of soil disinfection with DD, certain components of DD and some other compounds with nematicidal activity on the growth of white clover. *Netherlands Journal of Plant Pathology*, **70**, 117–135.

Ennik, G.C., Kort, J. & Van de Bund, C.F. (1965). The clover cyst nematode (*Heterodera trifolii* Goffart) as the probable cause of white clover dieback in a sward. *Journal of the British Grassland Society*, **20**, 258–262.

Fensham, R.J. & Bowman, D.M.J.S. (1992). Stand structure and the influence of overwood on regeneration in tropical eucalypt forest on Melville Island. *Australian Journal of Botany*, **40**, 335–352.

Fenwick, D.W. (1961). Estimation of field populations of cyst-forming nematodes of the genus *Heterodera*. *Journal of Helminthology*, **35**, 63–76.

Ferris, H., Mullens, T.A. & Foord, K.E. (1990). Stability and characteristics of spatial pattern description parameters for nematode populations. *Journal of Nematology*, **22**, 427–439.

Foggo, A., Speight, M.R. & Gregoire, J.C. (1994). Root disturbance of common ash, *Fraxinus excelsior* (Oleaceae), leads to reduced foliar toughness and increased feeding by folivorous weevil, *Stereonychus fraxini* (Coleoptera, Curculionidae). *Ecological Entomology*, **19**, 344–348.

Francl, L.J. (1993). Interactions of nematodes with mycorrhizae and mycorrhizal fungi. In *Nematode Interactions* (Ed. by M.W. Khan), pp. 203–216. Chapman & Hall, London.

Freckman, D.W. & Virginia, R.A. (1989). Plant-feeding nematodes in deep-rooting desert ecosystems. *Ecology*, **70**, 1665–1678.

Frouz, J. (1994). Changes in the terrestrial Chironomid community

(Diptera, Chironomidae) during secondary succession in old fields. *Pedobiologia*, **38**, 334–343.

Gange, A.C. & Bower, E. (1997). Interactions between insects and mycorrhizal fungi. In *Multitrophic Interactions in Terrestrial Systems* (Ed. by A.C. Gange & V.K. Brown), pp. 115–132. Blackwell Science Ltd, Oxford.

Gange, A.C. & Brown, V.K. (1989). Effects of root herbivory by an insect on a foliar-feeding species, mediated through changes in the host plant. *Oecologia*, **81**, 38–42.

Gange, A.C. Brown, V.K. & Farmer, L.M. (1990). A test of mycorrhizal benefit in an early successional plant community. *New Phytologist*, **115**, 85–91.

Gange, A.C., Brown, V.K. & Farmer, L.M. (1991). Mechanisms of seedling mortality by subterranean insect herbivores. *Oecologia*, **88**, 228–232.

Gange, A.C., Brown, V.K. & Sinclair, G.S. (1993). Vesicular-arbuscular mycorrhizal fungi: a determinant of plant community structure in early succession. *Functional Ecology*, **7**, 616–622.

Gange, A.C., Brown, V.K. & Sinclair, G.S. (1994). Reduction of black vine weevil larval growth by vesicular-arbuscular mycorrhizal infection. *Entomologia Experimentalis et Applicata*, **70**, 115–119.

Gibson, D.J., Freeman, C.C. & Hulbert, L.C. (1990). Effects of small mammal and invertebrate herbivory on plant species richness and abundance in tallgrass prairie. *Oecologia*, **84**, 169–175.

Goodell, P. & Ferris, H. (1980). Plant-parasitic nematode distributions in an alfalfa field. *Journal of Nematology*, **12**, 137–141.

Gunn, A. & Cherrett, J.M. (1993). The exploitation of food resources by soil meso-invertebrates and macro-invertebrates. *Pedobiologia*, **37**, 303–320.

Hasan, A. (1993). The role of fungi in fungus–nematode interactions. *Nematode Interactions* (Ed. by M.W. Khan), pp. 273–287. Chapman & Hall, London.

Ingham, E.R. & Detling, J.K. (1990). Effects of root-feeding nematodes on above-ground net primary production in a North American grassland. *Plant and Soil*, **121**, 279–281.

Ingham, E.R., Trofymow, J.A., Ames, R.N., Hunt, H.W., Morley, C.R., Moore, J.C. & Coleman, D.C. (1986a). Trophic interactions and nitrogen cycling in a semi-arid grassland soil. I. Seasonal dynamics of the natural populations, their interactions and effects of nitrogen cycling. *Journal of Applied Ecology*, **23**, 597–614.

Ingham, E.R., Trofymow, J.A., Ames, R.N., Hunt, H.W., Morley, C.R., Moore, J.C. & Coleman, D.C. (1986b). Trophic interactions and nitrogen cycling in a semi-arid grassland soil. II. System responses to removal of different groups of soil microbes or fauna. *Journal of Applied Ecology*, **23**, 615–630.

Jarosz, A.M. & Davelos, A.L. (1995). Effects of disease in wild plant populations and the evolution of pathogen aggressiveness. *New Phytologist*, **129**, 371–387.

Joost, R.E. (1995). *Acremonium* in fescue and ryegrass—boon or bane—a review. *Journal of Animal Science*, **73**, 881–888.

Khan, M.W. (1993a). *Nematode Interactions*. Chapman & Hall, London.

Khan, M.W. (1993b). Mechanisms of interactions between nematodes and other plant pathogens. In *Nematode Interactions* (Ed. by M.W. Khan), pp. 55–78. Chapman & Hall, London.

Khan, R.M., Khan, A.M. & Khan, M.W. (1986a). Interaction between *Meloidogyne incognita*, *Rotylenchus reiniformis* and *Tylenchorhynchus brassicae* on tomato. *Revue de Nematologie*, **9**, 245–250.

Khan, R.M., Khan, M.W. & Khan, A.M. (1986b). Interactions between *Meloidogyne incognita*, *Rotylenchus reiniformis* and *Tylenchorhynchus brassicae* as cohabitants on eggplant. *Nematologica Mediterranea*, **14**, 201–206.

Leath, K.T. & Byers, R.A. (1973). Attractiveness of diseased red clover roots to the clover root borer. *Phytopathology*, **63**, 428–431.

Leath, K.T. & Hower, A.A. (1993). Interaction of *Fusarium oxysporum* f. sp. *medicaginis* with feeding activity of clover root curculio larvae in alfalfa. *Plant Disease*, **77**, 799–802.

Lewis, S.A. (1987). Nematode–plant compatibility. In *Vistas on Nematology* (Ed. by J.A. Veech & D.W. Dickson), pp. 246–252. E.O. Painter Printing Co., DeLeon Springs, FL.

Little, L.R. & Maun, M.A. (1996). The "*Ammophila* problem" revisited: a role for mycorrhizal fungi. *Journal of Ecology*, **84**, 1–7.

Maas, P.W.Th., Oremus, P.A.I. & Otten, H. (1983). Nematodes (*Longidorus* sp. and *Tylenchorhynchus microphasmis* Loof) in growth and nodulation of sea buckthorn (*Hippophaë rhamnoides* L.). *Plant and Soil*, **73**, 141–147.

MacDonald, D. (1979). Some interactions of plant parasitic nematodes and higher plants. In *Ecology of Root Pathogens* (Ed. by S.V. Krupa & Y.R. Dommergues), pp. 157–178. Elsevier Scientific Publishing Company, Amsterdam.

Masters, G.J. & Brown, V.K. (1997). Host-plant mediated interactions between spatially separated herbivores: effects on community structure. In *Multitrophic Interactions in Terrestrial Systems* (Ed. by A.C. Gange & V.K. Brown), pp. 217–237. Blackwell Science Ltd, Oxford.

McGawley, E.C. & Chapman, R.A. (1983). Reproduction of *Criconemoides similis*, *Helicotylenchus pseudorobustus*, and *Paratylenchus projectus* on soybean. *Journal of Nematology*, **19**, 542.

Miller, P.M. & Wihrheim, S.E. (1968). Mutual antagonism between

Heterodera tabacum and some other parasitic nematodes. *Plant Disease Reporter*, **52**, 57–58.

Miller, T.E. (1994). Direct and indirect species interactions in an early old field community. *American Naturalist*, **143**, 1007–1025.

Moore, J.C., Hunt, H.W. & Elliott, E.T. (1991). Ecosystem perspectives, soil organisms, and herbivores. In *Microbial Mediation of Plant–Herbivore Interactions* (Ed. by P. Barbosa, V.A. Krischik & C.G. Jones), pp. 105–140. John Wiley & Sons, New York.

Moran, N., Seminoff, J. & Johnstone, L. (1993). Genotypic variation in propensity for host alternation within a population of *Pemphigus betae* (Homoptera: Aphididae). *Journal of Evolutionary Biology*, **6**, 691–705.

Mortimer, S.R., Hollier, J.A. & Brown, V.K. (1998). Interactions between plant and insect diversity in the restoration of lowland calcareous grasslands in southern Britain. *Applied Vegetation Science*, **1**, 101–114.

Muller, H. (1989a). Structural analysis of the phytophagous insect guilds associated with the roots of *Centaurea maculosa* Lam., *C. diffusa* Lam. and *C. vallesiaca* Jordan in Europe. *Oecologia*, **78**, 41–52.

Muller, H. (1989b). Growth pattern of diploid and tetraploid spotted knapweed, *Centaurea maculosa* Lam. (Compositae), and effects of the root-mining moth *Agapeta zoegana* (L.) (Lepidoptera: Cochylidae). *Weed Research*, **29**, 103–111.

Muller, H., Stinson, C.S.A., Marquardt, K. & Schroeder, D. (1989). The entomofaunas of roots of *Centaurea maculosa* Lam., *C. diffusa* Lam. and *C. vallesiaca* Jordan in Europe. Niche separation in space and time. *Journal of Applied Entomology*, **107**, 83–95.

Muller-Scharer, H. (1991). The impact of root herbivory as a function of plant density and competition: survival, growth and fecundity of *Centaurea maculosa* in field plots. *Journal of Applied Ecology*, **28**, 759–776.

Noe, J.P. & Campbell, C.L. (1985). Spatial pattern analysis of plant-parasitic nematodes. *Journal of Nematology*, **17**, 86–93.

Norton, D.C. (1978). *Ecology of Plant Parasitic Nematodes*. John Wiley & Sons, New York.

Orr, C.C. & Newton, O.H. (1971). Distribution of nematodes by wind. *Plant Disease Reporter*, **55**, 61–63.

Pederson, G.A., Windham, G.L., Ellsbury, M.M., McLaughlin, M.R., Pratt, R.G. & Brink, G.E. (1991). White clover yield and persistence as influenced by cypermethrin, benomyl and root knot nematode. *Crop Science*, **31**, 1297–1302.

Pinochet, J., Calvet, C., Camprubi, A. & Fernandez, C. (1996). Interactions between migratory endoparasitic nematodes and arbuscular mycorrhizal fungi in perennial crops: A review. *Plant and Soil*, **185**, 183–190.

Poinar, G.O., Jr. (1983). *The Natural History of Nematodes*. Prentice-Hall,

234

Englewood Cliffs, NJ.

Potter, D.A. & Braman, S.K. (1991). Ecology and management of turfgrass insects. *Annual Review of Entomology*, **36**, 383–406.

Potter, D.A., Patterson, C.G. & Redmond, C.T. (1992). Influence of turfgrass species and tall fescue endophyte on feeding ecology of Japanese beetle and southern masked chafer grubs (Coleoptera: Scarabaeidae). *Journal of Economic Entomology*, **85**, 900–909.

Prins, A.H., Nell, H.W. & Klinkhamer, P.G.L. (1992). Size-dependent root herbivory in *Cynoglossum officinale*. *Oikos*, **65**, 409–413.

Quinn, M.A. & Hall, M.H. (1992). Compensatory response of a legume root-nodule system to nodule herbivory by *Sitona hispidulus*. *Entomologia Experimentalis et Applicata*, **64**, 167–176.

Quinn, M.A. & Hower, A.A. (1986). Effects of root nodules and taproots on survival and abundance of *Sitona hispidulus* (Coleoptera: Curculionidae) on *Medicago sativa*. *Ecological Entomology*, **11**, 391–400.

Ramsell, J., Malloch, A.J.C. & Whittaker, J.B. (1993). When grazed by *Tipula paludosa*, *Lolium perenne* is a stronger competitor of *Rumex obtusifolius*. *Journal of Ecology*, **81**, 777–786.

Rao, B.H.K. & Prasad, S.K. (1971). Population studies on *Meloidogyne javanica* and *Rotylenchus reniformis* occurring together and separately and their effects on the host. *Indian Journal of Entomology*, **32**, 194–200.

Rea, F. & Wallace, H.R. (1981). The response of some Australian native plants to *Meloidogyne javanica*. *Nematologica*, **27**, 449–457.

Reichman, O.J. & Smith, S.C. (1991). Responses to simulated leaf and root herbivory by a biennial, *Tragopogon dubius*. *Ecology*, **72**, 116–124.

Robertson, G.P. & Freckman, D.W. (1995). The spatial distribution of nematode trophic groups across a cultivated ecosystem. *Ecology*, **76**, 1425–1432.

Roncadori, R.W. (1997). Interactions between arbuscular mycorrhizas and plant parasitic nematodes in agro-ecosystems. In *Multitrophic Interactions in Terrestrial Systems* (Ed. by A.C. Gange & V.K. Brown), pp. 101–114. Blackwell Science Ltd, Oxford.

Rossi, J.-P., Delaville, L. & Quénéhervé, P. (1996). Microspatial structure of a plant-parasitic nematode community in a sugarcane field in Martinique. *Applied Soil Ecology*, **3**, 17–26.

Russin, J.S., McGawley, E.C. & Boethel, D.J. (1993). Population development of *Meloidogyne incognita* on soybean defoliated by *Pseudoplusia includens*. *Journal of Nematology*, **25**, 42–49.

Saner, M.A. & Muller-Scharer, H. (1994). Impact of root-mining by *Eteobalea* spp. on clonal growth and sexual reproduction of common toadflax, *Linaria vulgaris* Mill. *Weed Research*, **34**, 199–204.

Schmid, B., Miao, S.L. & Bazzaz, F.A. (1990). Effects of simulated root herbivory and fertilizer application on growth and root biomass allocation in the clonal perennial *Solidago canadensis*. *Oecologia*, **84**, 9–15.

Schultz, J.C. (1988). Plant responses induced by herbivores. *Trends in Ecology and Evolution*, **3**, 45–49.

Seal, D.R., McSorley, R. & Chalfant, R.B. (1992). Seasonal abundance and spatial distribution of wireworms (Coleoptera: Elateridae) in Georgia sweet potato fields. *Journal of Economic Entomology*, **85**, 1802–1808.

Seastedt, T.R., Todd, T.C. & James, S.W. (1987). Experimental manipulations of arthropod, nematode and earthworm communities in a North American tallgrass prairie. *Pedobiologia*, **30**, 9–17.

Seliskar, D.M. & Huettel, R.N. (1993). Nematode involvement in the dieout of *Ammophila breviligulata* (Poaceae) on the mid-Atlantic coastal dunes of the United States. *Journal of Coastal Research*, **9**, 97–103.

Stanton, N.L. (1988). The underground in grasslands. *Annual Review of Ecology and Systematics*, **19**, 573–589.

Stanton, N.L., Morrison, D. & Laycock, W.A. (1984). The effect of phytophagous nematode grazing on Blue grama die-off. *Journal of Range Management*, **37**, 447–450.

Steinger, T. & Muller-Scharer, H. (1992). Physiological and growth-responses of *Centaurea maculosa* (Asteraceae) to root herbivory under varying levels of interspecific plant competition and soil nitrogen availability. *Oecologia*, **91**, 141–149.

Strnad, S.P. & Bergman, M.K. (1987). Movement of first-instar western corn rootworms (Coleoptera: Chrysomelidae) in soil. *Environmental Entomology*, **16**, 975–978.

Strong, D.R., Maron, J.L., Connors, P.G., Whipple, A., Harrison, S. & Jefferies, R.L. (1995). High mortality, fluctuation in numbers and heavy subterranean insect herbivory in bush lupine, *Lupinus arboreus*. *Oecologia*, **104**, 85–92.

Todd, T.C., James, S.W. & Seastedt, T.R. (1992). Soil invertebrate and plant responses to mowing and carbofuran application in a North American tallgrass prairie. *Plant and Soil*, **144**, 117–124.

Van der Putten, W.H. & Duyts, H. (1998). Plant parasitic nematodes in successionally dominant monocots from coastal foredunes. In *Nematode Communities in Northern Temperate Grassland Ecosystems* (Ed. by R.G.M. De Goede & T. Bongers), pp. 241–260. Focus, Giessen.

Van der Putten, W.H. & Peters, B.A.M. (1997). How soil-borne pathogens may affect plant competition. *Ecology*, **78**, 1785–1795.

Van der Putten, W.H. & Van der Stoel, C.D. (1998) Plant parasitic nematodes and spatio-temporal variation in natural vegetation. *Applied Soil Ecology* (in press).

Van der Putten, W.H., Van der Werf-Klein Breteler, J.T. & Van Dijk, C.

(1989). Colonization of the root zone of *Ammophila arenaria* by harmful soil organisms. *Plant and Soil*, **120**, 213–223.

Van der Putten, W.H., Maas, P.W.Th., van Gulik, W.J.M. & Brinkman, H. (1990). Characterization of soil organisms involved in the degeneration of *Ammophila arenaria*. *Soil Biology and Biochemistry*, **22**, 845–852.

Van der Putten, W.H., Van Dijk, C. & Peters, B.A.M. (1993). Plant-specific soil-borne diseases contribute to succession in foredune vegetation. *Nature*, **362**, 53–56.

Wallace, H.R. (1963). *The Biology of Plant Parasitic Nematodes*. Edward Arnold, London.

Wallace, H.R. (1987). Effects of nematode parasites on photosynthesis. In *Vistas on Nematology* (Ed. by J.A. Veech & D.W. Dickson), pp. 253–259. E.O. Painter Printing Co., DeLeon Springs, FL.

Wasilewska, L. (1970). Nematodes of the sand dunes in the Kampinos Forest. I. Species structure. *Ekologia Polska*, **18**, 429–443.

Wasilewska, L. (1971). Nematodes of the dunes in the Kampinos Forest. II. Community structure based on numbers of individuals, state of biomass and respiratory metabolism. *Ekologia Polska*, **19**, 651–688.

Wasilewska, L. (1995). Differences in development of soil nematode communities in single- and multi-species grass experimental treatments. *Applied Soil Ecology*, **2**, 53–64.

Weaver, T. & Smolik, J. (1987). Soil nematodes of northern Rocky Mountain ecosystems: genera and biomasses. *Great Basin Naturalist*, **47**, 473–479.

Wharton, D.A. (1986). *A Functional Biology of Nematodes*. Croom Helm, London.

White, J.H. (1953). Wind-borne dispersal of potato-root eelworm. *Nature*, **172**, 686–687.

Wilcox-Lee, D. & Loria, R. (1987). Effects of nematode parasitism on plant–water relationships. In *Vistas on Nematology* (Ed. by J.A. Veech & D.W. Dickson), pp. 260–266. E.O. Painter Printing Co., DeLeon Springs, FL.

Williams, L., Schotzko, D.J. & McCaffrey, J.P. (1992). Geostatistical description of the spatial distribution of *Limonius californicus* (Coleoptera: Elateridae) wireworms in the Northwestern United States, with comments on sampling. *Environmental Entomology*, **21**, 983–995.

Yeates, G.W. (1987). How plants affect nematodes. *Advances in Ecological Research*, **17**, 61–113.

Yeates, G.W., Watson, R.N. & Steele, K.W. (1985). Complementary distribution of *Meloidogyne*, *Heterodera*, and *Pratylenchus* (Nematoda: Tylenchida) in roots of white clover. In *Proceedings of the 4th Australian Conference on Grassland Invertebrate Ecology* (Ed. by R.B. Chapman), pp. 71–79. Caxton Press, Christchurch.

Yeates, G.W., Bongers, T., De Goede, R.G.M., Freckman, D.W. & Georgieva, S.S. (1993). Feeding habits in soil nematode families and genera. An outline for soil ecologists. *Journal of Nematology*, **25**, 315–331.

Zacheo, G. (1993). Nematodes as plant parasites. In *Nematode Interactions* (Ed. by M.W. Khan), pp. 1–25. Chapman & Hall, London.

Zoon, F.C. (1995). *Biotic and abiotic soil factors in the succession of sea buckthorn, Hippophaë rhamnoides L. in coastal sand dunes*. PhD thesis, Wageningen Agricultural University.

Zoon, F.C., Troelstra, S.R. & Maas, P.W.Th. (1993). Ecology of plant-feeding nematode fauna associated with sea buckthorn (*Hippophaë rhamnoides* L. ssp. *rhamnoides*) in different stages of dune succession. *Fundamental and Applied Nematology*, **16**, 247–258.

Chapter 8

Competition for light, plant-species replacement and herbivore abundance along productivity gradients

J. Huisman,[1,2] J.P. Grover,[3] R. van der Wal[4,5] and J. van Andel[2]

Summary

This chapter argues that patterns of herbivore abundance along productivity gradients depend on the kind of resource that plant species are competing for. Recently published plant–herbivore theory that considers competition for nutrients as its point of departure predicts a general increase in herbivore abundance with increasing productivity, as well as a change in plant species composition from good nutrient competitors at low productivity to good grazing tolerators at high productivity. We develop a model that combines plant–herbivore theory with a mechanistic model of competition for light, and show that competition for light may lead to other patterns than competition for nutrients. For example, suppose that tall plants of low palatability shade small plant species preferred by the herbivore. In this case, the herbivore may decrease in abundance or even disappear with increasing productivity. Furthermore, multiple stable states

1 Present address: Laboratory for Microbiology, University of Amsterdam, Nieuwe Achtergracht 127, 1018 WS Amsterdam, The Netherlands
2 Department of Plant Biology, University of Groningen, PO Box 14, 9750 AA Haren, The Netherlands
3 Department of Biology, University of Texas at Arlington, Box 19498, Arlington, TX 76019, USA
4 Zoological Laboratory, University of Groningen, PO Box 14, 9750 AA Haren, The Netherlands
5 Present address: Institute of Terrestrial Ecology, Hill of Brathens, Glassel, Banchory, Kincardineshire AB31 4BY, UK

may occur. In one state tall plants of low forage quality outshade small plant species, whereas in the other state small plant species of high forage quality support a large herbivore population. Field data from salt marshes grazed by intermediate-sized vertebrate herbivores like rabbits, hares and geese support these predictions.

Introduction

Competition between plant species affects the species composition of plant communities (e.g. Harper 1977), and hence the suitability of the plant community as food for herbivores (McNaughton 1984; Drent & Prins 1987; Fryxell 1991). Grazing, in turn, affects the outcome of plant competition (e.g. Berendse 1985; Furbish & Albano 1994; Silvertown *et al.* 1994). How do these interactions shape plant species composition and patterns of herbivory along productivity gradients?

Most theoretical studies on plant competition and plant–herbivore dynamics published to date have focused on plant growth limited by nutrient availability (e.g. Phillips 1974; Holt *et al.* 1994; Grover 1995; Leibold 1996). In general, these studies predict a change in plant species composition from good nutrient competitors at low productivity to grazing-tolerant or grazing-resistant plants at high productivity, as well as a general increase in herbivore grazing pressure (Leibold 1996). In line with this latter prediction, data gathered from both aquatic and terrestrial ecosystems show an increasing trend in the annual amount of plant material consumed and an increasing trend in herbivore biomass with increasing productivity (McNaughton *et al.* 1989; Cyr & Pace 1993). The high variability around these large-scale trends, however, may obscure local patterns not in accordance with the theory. In productive environments, in particular, plants compete not only for nutrients but also for light. In contrast to nutrient competition, competition for light is strongly asymmetric: tall plants shade small plants, but not vice versa. Hence, when small plant species are the preferred food source for a herbivore, the herbivore might be in trouble in productive environments.

In the context of competition for light, herbivores affect the proliferation of plant species in three qualitatively different ways. Firstly, herbivory implies plant biomass removal. In this way herbivores have a direct effect on the plants. Secondly, via biomass removal herbivores affect the shading pattern within vegetation. This may release small species from shading by taller species. Thirdly, apparent competition (*sensu* Holt 1977) will arise

240

if herbivores supported by consuming one plant species also attack other plant species. Outcomes of apparent competition depend strongly on the feeding preferences of the herbivore. The herbivore may consume all species, it may prefer to consume only the tall plants, or it may consume only the small plants. Since these feeding preferences affect both apparent competition and competition for light, one might expect the preferences of the herbivore to have a profound impact on the patterns of plant species replacement and herbivore abundance along productivity gradients.

In the present paper, we develop a simple model that combines standard plant–herbivore theory with a mechanistic model of competition for light. The model is used to analyse how above-ground plant competition affects patterns of plant-species replacement and herbivore abundance along a productivity gradient. A comparison with the nutrient-based models described above shows that competition for light may lead to different patterns of herbivory along productivity gradients than competition for nutrients. These patterns are illustrated with a field example from a salt marsh grazed by rabbits, hares and geese.

Approach

Numerous detailed simulation models have been developed that represent the dynamics of individual plants, or small groups of plants, interacting with each other under light-limited conditions (e.g. Botkin *et al.* 1972; Shugart 1984; Kohyama 1992; Hara 1993; Pacala *et al.* 1996). In these size-structured models, the location and height of each individual is tracked. Light availability and light absorption are related to plant height and leaf biomass, and height growth and mortality vary in response to the amount of light absorbed. Plants produce seeds that are dispersed to produce seedlings, and for each seedling height growth starts anew. An obvious drawback of this detailed approach is that the models are mathematically complicated and place stringent requirements on biological knowledge of the plants involved. In fact, writing such models and validating them is quite difficult. These difficulties are reflected by the wide variety of approaches taken by different theoreticians (compare, for example, Ford & Diggle 1981; Johnson *et al.* 1989; Prentice & Leemans 1990; Ryel *et al.* 1990; Yokozawa & Hara 1992; Sorrensen-Cothern *et al.* 1993; Pacala *et al.* 1996). Despite the major achievements of these various models, we are not convinced that anyone has yet written a size-structured model of plant competition that is both generally applicable and widely accepted.

Without such a basis, we do not believe that a complicated structured model of plant competition *plus* herbivory will present much of a unifying advance. Therefore, we have chosen a more simplistic and less ambitious approach. Our model is a caricature, and we deliberately make bold assumptions in the hope that what remains still has a grain of truth.

Model structure

The structure of the model is visualized in Fig. 8.1. We consider two plant species; a tall plant species and a small plant species. One might think of, say, a tall grass and a small herb. The tall plant species (referred to as species 1) receives light from above with an incident light intensity I_{in}. This incident light is partly absorbed by the canopy of species 1. Light that has not been absorbed leaves the canopy of species 1 with an intensity $I_{out,1}$, and this provides the incident light intensity for the small plant species (species 2). Species 2 partly absorbs the light it receives, and light that has not been absorbed by species 2 leaves its canopy with an intensity $I_{out,2}$. We make no assumptions about the vertical distribution of leaf biomass within the canopy, except that all the leaf mass of species 1 lies above the leaf mass of species 2, so that the canopies of the two species do not overlap. This ensures that competition for light is asymmetric, with species 2 never capable of casting shade on species 1. As a consequence of this assumption, the model does not consider the possibility that seedlings of the tall plant species are outshaded by adults of the smaller species. Therefore, the model probably applies best to herbaceous vegetations in which large seeds or clonal growth enable tall species to break through the shade cast by smaller grasses and herbs.

Various constraints and trade-offs are expected to be associated with variation in plant height. Because species 1 is tall, it needs to invest in stem

---

Figure 8.1 (*Opposite*) Illustration of the model structure. (a) The model considers a tall plant species (species 1) that shades a small plant species (species 2). The tall plant invests more biomass in stem than the small plant. One or both plant species are grazed upon by a herbivore. (b) The light profile. The incident light intensity I_{in} is partly absorbed by the canopy of species 1. The light intensity that penetrates through the canopy of species 1 ($I_{out,1}$) is the incident light intensity for species 2. Light that is not absorbed by species 2 reaches the soil surface with an intensity $I_{out,2}$. (c) The relationship between light and photosynthesis for a sun species and a shade species. Species 1 is a sun species, whereas species 2 is either a sun or a shade species.

biomass. Species 2, on the other hand, is small and shaded by species 1. Therefore, we assume that species 1 allocates more of its total biomass to stem and less to leaves than species 2. We also consider two different

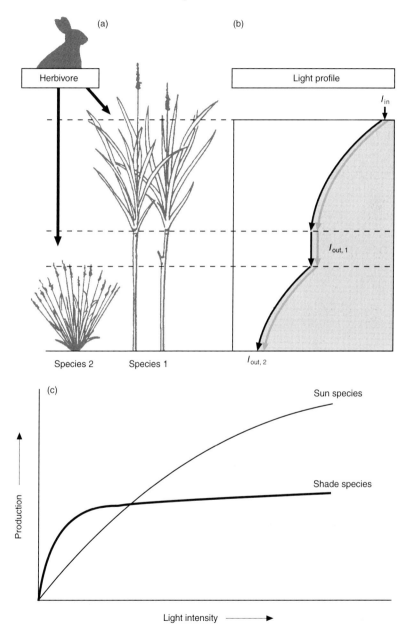

243

physiological strategies. Species may behave as sun species or shade species (e.g. Boardman 1977). Sun species have high maximum rates of photosynthesis but also suffer high biomass losses (by, for example, respiration, leaf abscission, root turnover) and have low rates of photosynthesis at low light intensities. In contrast, shade species have low maximum rates of photosynthesis but they also have low losses and are able to reach appreciable rates of photosynthesis at low light levels. Throughout the chapter we assume that species 1 is a sun species. Species 2 is either a sun species or a shade species. One or both plant species are grazed upon by a herbivore. We consider only above-ground herbivory, directed at leaves.

One plant species

This section describes the derivation and basic assumptions that underlie our mathematical formulation of light-limited growth. Our model formulation relies on many earlier studies of plant canopy photosynthesis (e.g. Monsi & Saeki 1953; Thornley & Johnson 1990), and extends recent work on phytoplankton competition for light (Huisman & Weissing 1994, 1995; Weissing & Huisman 1994) to terrestrial vegetation. First consider one plant species, and let W denote its total plant biomass. The rate of change of total plant biomass is governed by the balance between production and losses:

$$\frac{dW}{dt} = \text{production} - \text{losses}$$

$$= \int_0^z p(I(s))\omega_L(s)ds - \ell W$$

(1)

In this formulation, photosynthesis is restricted to the plant canopy. Vertical positions within the plant canopy are indicated by the depth s, where s ranges from 0 (top) to z (bottom). The notation $p(I(s))$ indicates that the specific rate of photosynthesis, p, is a function of light intensity I, and that the light intensity I is a function of depth s. The function $\omega_L(s)$ specifies the leaf biomass density at each depth s. Biomass losses are not restricted to the leaves, so the total plant biomass suffers a specific loss rate ℓ.

We make no assumptions about the distribution of leaf biomass density over depth. Hence our formulation applies to canopies of various shapes. The total leaf biomass, W_L, of the canopy is given by:

244

$$W_L = \int_0^z \omega_L(s)\,ds \tag{2}$$

It is convenient to relate total leaf biomass to total plant biomass by the leaf allocation parameter $A_L = W_L/W$. We assume that the specific rate of photosynthesis, $p(I)$, is an increasing but saturating function of light intensity I:

$$p(I) = \frac{p_{max}I}{h+I} \tag{3}$$

where p_{max} is the maximum rate of photosynthesis and h is the light intensity at which the rate of photosynthesis equals half its maximal rate. The light intensity I at depth s depends on the incident light intensity and on the cumulative amount of leaf biomass above depth s. More precisely, according to Lambert–Beer's law:

$$I(s) = I_{in}e^{-k\int_0^s \omega_L(\sigma)d\sigma} \tag{4}$$

where k is the light extinction coefficient of the leaves (which depends on pigment content, leaf angle, reflection of light, etc). Hence, the light intensity I_{out} that penetrates through the canopy is given by:

$$I_{out} = I_{in}e^{-kW_L} \tag{5}$$

This completes the formulation of our model for a single species.

The assumptions specified in Eqs 1–5 can be combined to arrive at the following dynamical system (Monsi & Saeki 1953; Huisman & Weissing 1994; Weissing & Huisman 1994):

$$\frac{dW}{dt} = \frac{p_{max}}{k}\ln\left(\frac{h+I_{in}}{h+I_{out}}\right) - \ell W \tag{6a}$$

$$I_{out} = I_{in}e^{-kA_L W} \tag{6b}$$

The model predicts that plant growth is only possible if the light supply exceeds the compensation point (Huisman & Weissing 1994):

$$I_{in} > I_c \tag{7}$$

where the compensation point is here defined not at the level of a single leaf but at the level of total plant biomass:

$$I_c = \frac{\ell h}{A_L p_{max} - \ell} \tag{8}$$

Figure 8.2 illustrates the dependence of plant growth on plant biomass when the light supply exceeds the compensation point. According to Eqs 6a,b, total production is an increasing, concave function of total plant biomass bounded by an upper asymptote (Fig. 8.2). The upper asymptote is caused by self-shading, and indicates that a limited energy supply (I_{in}) cannot result in an unlimited production. Total losses increase linearly with total biomass. The production and loss curve intersect at the equilibrium plant biomass W^*. Production exceeds losses and therefore plant biomass will increase when plant biomass is below W^*. Losses exceed production and therefore plant biomass will decrease when plant biomass is above W^*. As a consequence, independent of the initial conditions, in due course the plant biomass settles at the equilibrium plant biomass W^*. The light intensity that penetrates through the canopy of this equilibrium plant biomass will be indicated as I^*_{out}.

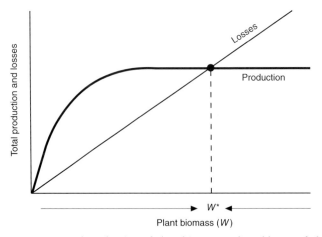

Figure 8.2 Total production of plant biomass and total losses of plant biomass as a function of plant biomass itself. Production and losses balance at the equilibrium plant biomass W^*. Redrawn from Huisman and Weissing (1994).

246

Two plant species

Now consider two plant species. Plant species 1 is tall and not shaded by the smaller plant species 2. Therefore, species 1 is not affected by species 2, and it will approach its biomass equilibrium W_1^* as described above. The light intensity $I_{out,1}^*$ penetrating through the canopy of species 1 provides the incident light intensity for species 2. Hence, species 2 is able to grow in the shade of species 1 if this incident light intensity exceeds its compensation point. If this condition is satisfied, species 2 will approach a stable equilibrium plant biomass W_2^*, analogous to the arguments given above. In other words, the two species may stably coexist if

$$I_{out,1}^* > I_{c,2} \tag{9}$$

Hence, coexistence is favoured if enough light penetrates through the canopy of species 1 and, from our definition of the compensation point (Eq. 8), if species 2 (i) has a high maximum rate of photosynthesis (high p_{max}), (ii) reaches appreciable rates of photosynthesis at low light (low h), (iii) has low biomass losses (low ℓ), and (iv) allocates most of its biomass to leaves (high A_L). Point (iv) is interesting, because it suggests that the natural trade-off between investing in stem and becoming tall, vs. investing in leaves and remaining small, actually promotes the condition required for coexistence.

When condition (9) is satisfied, we predict that two species coexist on a single limiting resource. This prediction differs from the standard theory of nutrient-based models. When species compete for a single limiting nutrient, theory predicts that at equilibrium the species able to grow at the lowest nutrient level should competitively displace all other species (the R^* rule, e.g. Armstrong & McGehee 1980; Tilman 1982). Our prediction also differs from the theory of competition for light in well-mixed aquatic environments. When phytoplankton species compete for light in a mixed water column, the species able to grow in the steepest light gradient should competitively displace all other species (the I_{out}^* rule, Huisman & Weissing 1994; Weissing & Huisman 1994; with recent support from phytoplankton competition experiments in Huisman *et al.* 1999). In our scenario, coexistence on a single resource results from the vertical patterning of the species in a unidirectional resource gradient. The small plant might have the right photosynthetic characteristics to be a better competitor for light in well-mixed aquatic environments, but in our terrestrial scenario it cannot displace the tall plant by competition for light simply because it does not shade the tall plant.

How is plant competition for light affected by environmental productivity? In order to investigate this question we incorporated the effect of various soil nutrient levels on plant growth by multiplying the production term of each plant species i with the Michaelis–Menten factor $N/(M_i + N)$, where N is the soil nutrient level and M_i is the species' half saturation constant for nutrient-limited growth. In order to keep our focus on light competition, we keep the soil nutrient level fixed, so that N is not a dynamic quantity and not affected by plant consumption. Essentially this means that we assume that a large pool of total soil nutrient buffers variation in the concentration of available nutrient in soil solution, e.g. through rapid exchange processes or mineralization. The dynamics now read:

$$\frac{dW_1}{dt} = \frac{N}{M_1 + N} \frac{p_{max,1}}{k_1} \ln\left(\frac{h_1 + I_{in}}{h_1 + I_{out,1}}\right) - \ell_1 W_1 \tag{10a}$$

$$I_{out,1} = I_{in} e^{-k_1 A_{L,1} W_1} \tag{10b}$$

$$\frac{dW_2}{dt} = \frac{N}{M_2 + N} \frac{p_{max,2}}{k_2} \ln\left(\frac{h_2 + I_{out,1}}{h_2 + I_{out,2}}\right) - \ell_2 W_2 \tag{10c}$$

$$I_{out,2} = I_{out,1} e^{-k_2 A_{L,2} W_2} \tag{10d}$$

Note that the species do not compete for soil nutrients, they only compete for light. However, according to this formulation, both light penetration through the canopy of species 1 and the compensation point of species 2 are a function of the soil nutrient level. And, therefore, coexistence of the two species depends on the soil nutrient level. This is illustrated in Fig. 8.3. This figure shows the equilibrium outcome of competition along a productivity gradient. Species 1 has the physiological characteristics of a sun species, whereas species 2 is a shade species. The parameter values of the species are given in Table 8.1. At low productivity, the leaf biomass of species 1 is low, and species 2 is able to grow in the shade cast by species 1. With increasing productivity, the leaf biomass of species 1 increases, light levels below the canopy of species 1 become fairly low, and species 2 is displaced. This demonstrates that the outcome of competition for light depends on the productivity of the system.

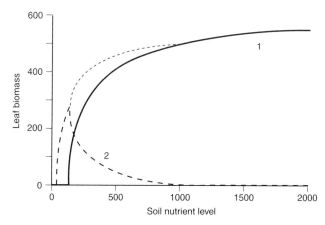

Figure 8.3 A tall plant species (1) and a small plant species (2) along a productivity gradient. The solid line indicates the leaf biomass of the tall plant, the dashed line indicates the leaf biomass of the small plant, and the thin dashed line indicates total leaf biomass. Parameter settings as in Table 8.1.

Table 8.1 Default parameter values used in the model simulations. The tall plant species (species 1) is a sun species. The small plant species (species 2) is a shade species, unless stated otherwise. The tall plant species invests more biomass in stem and hence less in leaves than the small plant species ($A_{L,1} = 0.4$, $A_{L,2} = 0.7$). In all simulations, the incident light intensity is $I_{in} = 400 \, \mu mol \, m^{-2} s^{-1}$, and the herbivore mortality rate is $m = 0.002 \, day^{-1}$.

Parameter	Meaning	Sun species	Shade species	Units
p_{max}	Maximum specific rate of photosynthesis	0.07	0.04	day^{-1}
h	Half saturation constant of photosynthesis	50	10	$\mu mol \, photons \, m^{-2} s^{-1}$
k	Light extinction coefficient of leaves	0.01	0.03	$m^2 g^{-1}$ leaf
ℓ	Specific loss rate of plant biomass	0.01	0.005	day^{-1}
M	Half saturation constant of nutrient limitation	200	200	$mg \, N \, kg^{-1}$ soil
a	Search rate of herbivore	0.002	0.002	$m^2 g^{-1}$ herbivore day^{-1}
c	Conversion of leaf biomass into herbivore biomass	0.01	0.01	g herbivore g^{-1} leaf

One plant species and a herbivore

Before we proceed to introduce a herbivore in our two-species plant community, let us first consider the interactions between one plant species and a herbivore. Since we are interested in competition for light, we consider above-ground herbivory only. We adopt the general model structure that has been used before in many plant–herbivore models, as well as other predator–prey theory (e.g. Rosenzweig 1973; Oksanen *et al.* 1981; DeAngelis 1992; van de Koppel *et al.* 1996; Crawley 1997). Let W and H denote plant biomass and herbivore biomass per unit area, respectively. The rate of change of plants and herbivores is represented by the differential equations:

$$\frac{dH}{dt} = g(W)H \tag{11a}$$

$$\frac{dW}{dt} = f(W) - b(W)H \tag{11b}$$

where $f(W)$ describes plant growth as a function of plant biomass, $b(W)$ is the specific rate of consumption of plant biomass by the herbivore (the functional response), and $g(W)$ is the specific growth rate of the herbivore population (the numerical response). To avoid some of the intricate non-equilibrium dynamics that may arise from saturating herbivore responses, we keep our model as simple as possible and assume a linear functional and numerical response. Hence, in the context of light limitation and above-ground herbivory we arrive at the following plant–herbivore model:

$$\frac{dH}{dt} = caA_L WH - mH \tag{12a}$$

$$\frac{dW}{dt} = \frac{N}{M+N} \frac{P_{max}}{k} \ln\left(\frac{h+I_{in}}{h+I_{out}}\right) - \ell W - a A_L WH \tag{12b}$$

$$I_{out} = I_{in} e^{-kA_L W} \tag{12c}$$

where a is the search rate of the herbivore (*sensu* Holling 1959), c is the conversion efficiency of leaf biomass into herbivore biomass, and m is the specific mortality rate of the herbivore. Note that our plant–herbivore system is just a specific formulation of the general model (Eqs 11a,b).

The model predicts that when leaf biomass is sufficiently high, the herbivore population will increase. An increased herbivore population consumes more leaves and thus reduces leaf biomass. The herbivore population is at equilibrium when it has reduced leaf biomass to the level:

$$W_L^* = \frac{m}{ac} \tag{13}$$

This equation states that the equilibrium leaf biomass does not depend on any growth features of the plant, it is only determined by parameters of the herbivore. In other words, once productivity is high enough for the herbivore to invade, it is the herbivore population that increases with productivity while leaf biomass is grazed down to a constant level W_L^*, independent of productivity. This pattern is illustrated in Fig. 8.4, and is known as 'top-down' control. Top-down control is also apparent in most other plant–herbivore models that have adopted the general structure of Eqs 11a,b (e.g. Oksanen *et al.* 1981; but see van de Koppel *et al.* 1996).

Figure 8.4 considers only one plant species, and sets the standard with which we shall compare the results derived in the remainder of this chapter. What patterns of plant biomass and herbivory should we expect when the herbivore grazes upon two plant species competing for light?

Two plant species and a herbivore

Combining the two previous sections, we now develop our full model of two plant species and a herbivore. To recapitulate, the plants compete for light. Plant species 1 is tall, and shades the small plant species 2. The herbivore may graze upon one or both of these plant species, and herbivory is only above-ground. The model reads:

$$\frac{dH}{dt} = c_1 a_1 A_{L,1} W_1 H + c_2 a_2 A_{L,2} W_2 H - mH \tag{14a}$$

$$\frac{dW_1}{dt} = \frac{N}{M_1 + N} \frac{p_{max,1}}{k_1} \ln\left(\frac{h_1 + I_{in}}{h_1 + I_{out,1}}\right) - \ell_1 W_1 - a_1 A_{L,1} W_1 H \tag{14b}$$

$$I_{out,1} = I_{in} e^{-k_1 A_{L,1} W_1} \tag{14c}$$

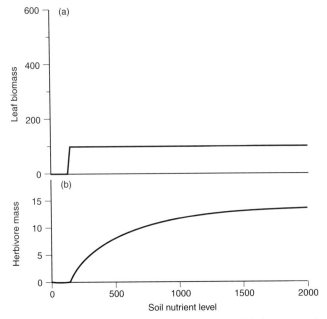

Figure 8.4 One plant species (a) and a herbivore (b) along a productivity gradient. Parameter settings as in Table 8.1, using species 1 as the plant.

$$\frac{dW_2}{dt} = \frac{N}{M_2 + N} \frac{p_{\max,2}}{k_2} \ln\left(\frac{h_2 + I_{\text{out,1}}}{h_2 + I_{\text{out,2}}}\right) - \ell_2 W_2 - a_2 A_{L,2} W_2 H \tag{14d}$$

$$I_{\text{out,2}} = I_{\text{out,1}} e^{-k_2 A_{L,2} W_2} \tag{14e}$$

To illustrate the rich behaviour of this model, we shall investigate a variety of scenarios. We introduce inedible plants, compare selective vs. non-selective herbivores, consider aspects of the food quality of the plants, and discuss the possibility for multiple stable states. Because competition for light is asymmetric, in all these scenarios it is important to realize which plant species is affected in which way. Grazing of the tall plant species has a totally different effect on the plant-herbivore dynamics than grazing of the small plant species.

Inedible plants
Suppose that one of the plant species is inedible. Now it matters which species is inedible. Firstly, suppose that the tall species (species 1) is

consumed by the herbivore ($a_1 > 0$, $c_1 > 0$) while the small species (species 2) is inedible ($a_2 = c_2 = 0$). The resulting patterns of plant biomass and herbivore biomass are shown in Fig. 8.5a. In this case, only species 1 is directly affected by the herbivore and the herbivore is affected only by species 1. Hence, species 1 and the herbivore behave as in the one plant, one herbivore model (compare Fig. 8.4 and Fig. 8.5a). The herbivore increases with productivity while it maintains the leaf biomass of species 1 at a constant level

$$W_{L,1}^* = \frac{m}{c_1 a_1} \tag{15}$$

Because species 1 is maintained at a low and constant level by the herbivore, species 2 receives more light, and increases in abundance along the productivity gradient. As a consequence, total leaf biomass (summed over species 1 and 2) and herbivore biomass both increase along the productivity gradient.

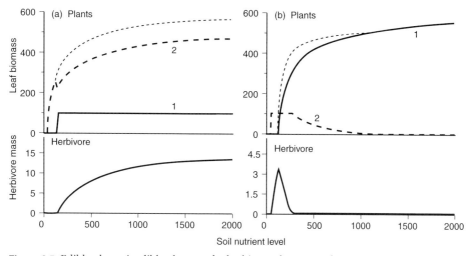

Figure 8.5 Edible plants, inedible plants and a herbivore along a productivity gradient. (a) The tall plant species is edible, and the small plant species is inedible ($a_2 = c_2 = 0$). (b) The tall plant species is inedible, and the small plant species is edible ($a_1 = c_1 = 0$). In the top panels, the solid lines indicate the leaf biomass of the tall plant, the thick dashed lines indicate the leaf biomass of the small plant, and the thin dashed lines indicate total leaf biomass. Default parameters, see Table 8.1.

Alternatively, the tall species may be inedible $(a_1 = c_1 = 0)$ while the small species is consumed by the herbivore $(a_2 > 0, c_2 > 0)$. This situation is completely different (Fig. 8.5b). Species 1 is not affected by the herbivore and not affected by species 2. As a consequence, species 1 behaves just as in monoculture or in the two-plant model, and increases with productivity (compare Fig. 8.3 and Fig. 8.5b). Because the herbivore is affected only by species 2, when present the herbivore maintains the leaf biomass of species 2 at the constant level

$$W_{L,2}^{*} = \frac{m}{c_2 a_2} \tag{16}$$

However, with increasing productivity species 2 is more shaded by species 1 and is not able to maintain a leaf biomass $W_{L,2}^{*}$. As soon as the leaf biomass of species 2 is reduced below $W_{L,2}^{*}$, the herbivore is no longer able to persist (compare, again, Fig. 8.3 with Fig. 8.5b). Here we thus have an example where total plant biomass increases with productivity, whereas the herbivore is only present at low productivity. In essence what happens is that plant species 1 has an indirect effect on the herbivore through outcompeting its food species (plant species 2) at high productivity levels.

A non-selective herbivore

Suppose that the herbivore consumes both plant species in a non-selective way, and that both plant species are of equal food quality (i.e. $a_1 = a_2 > 0$ and $c_1 = c_2 > 0$). In this case, the two plant species are equal from the perspective of the herbivore. According to Eq. 14a, the herbivore population is at equilibrium when it has reduced the total leaf biomass summed over both species to a constant level. In other words, total leaf biomass remains constant along a productivity gradient, and the herbivore population increases with productivity (Fig. 8.6; compare with Fig. 8.4). Within the plant community, however, the balance shifts. At low productivity, species 2 (the shade species) dominates. With increasing productivity, species 1 (the sun species) gradually takes over. This is a case of apparent competition. Due to grazing, total leaf biomass is so low that species 2 does not really suffer from shading by species 1. Because species 1 has a higher maximum photosynthetic rate, however, at high productivity species 1 is able to support a higher herbivore grazing pressure. Species 2, with a much lower maximum production rate, cannot withstand this high level of herbivory and disappears.

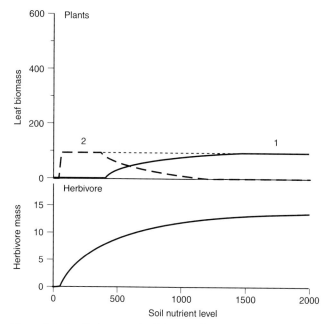

Figure 8.6 A tall plant species (1), a small plant species (2) and a non-selective herbivore along a productivity gradient. In the top panel, the solid line indicates the leaf biomass of the tall plant, the thick dashed line indicates the leaf biomass of the small plant and the thin dashed line indicates total leaf biomass. Default parameters, see Table 8.1.

A selective herbivore

The herbivore may prefer one of the species over the other species. Again, it matters which species is preferred. Suppose that species 1 is preferred and species 2 less preferred ($a_1 > a_2 > 0$). Figure 8.7a shows that species 1 becomes extremely rare, while the herbivore and the less-preferred species 2 both increase with increasing productivity. Apparent competition plays an important role. Species 2 supports a herbivore population, which in turn consumes most of species 1. Note that the pattern differs from the case where species 2 is inedible (compare Fig. 8.7a with Fig. 8.5a). When species 2 is inedible, the herbivore relies on species 1. As a consequence, species 1 cannot disappear due to herbivory because then the herbivore would disappear too. In contrast, when species 2 is less preferred but still eaten, the herbivore can do without species 1 and is thus able to deplete its own favourite dish. This is exactly what happens in Fig. 8.7a. Interestingly, the

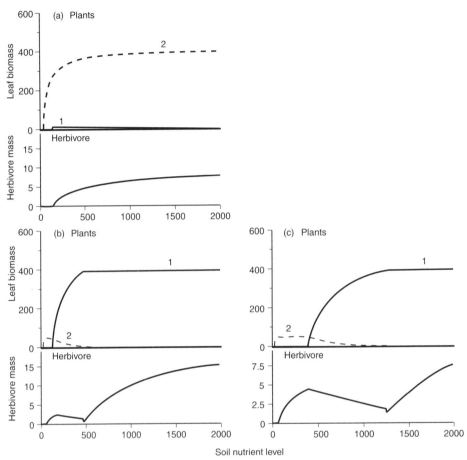

Figure 8.7 A preferred plant species, a less-preferred plant species and a herbivore along a productivity gradient. (a) The tall plant is preferred and the small plant is less preferred ($a_1 = 0.004$, $a_2 = 0.0005$). (b) The tall plant is less preferred, and the small plant is preferred ($a_1 = 0.0005$, $a_2 = 0.004$). (c) As in (b), but now the tall plant also has a higher nutrient requirement ($M_1 = 500$). In the top panels, the solid lines indicate the leaf biomass of the tall plant, and the dashed lines indicate the leaf biomass of the small plant. Default parameters, see Table 8.1.

herbivore population reaches a higher biomass when plant species 2 is not eaten than when it is eaten with low preference (compare the herbivore biomass in Fig. 8.5a and Fig. 8.7a).

256

When species 2 is the preferred species while species 1 is less preferred ($a_2 > a_1 > 0$), species 2 is at a real disadvantage. It suffers from both competition by shading and apparent competition via the herbivore. As a consequence, species 2 is only able to survive at the low-productivity end of the gradient, where species 1 is rare (Fig. 8.7b). Herbivore biomass shows a strange pattern along the productivity gradient: it first increases then decreases and then increases again. To be convinced that this herbivore pattern is no numerical artefact, we raised the nutrient requirements of species 1 by enlarging the parameter M_1, so that species 1 needs a higher soil nutrient content in order to invade. Now species 2 is more abundant at the low-productivity end of the gradient because it suffers less from shading, and the herbivore pattern becomes even more pronounced (Fig. 8.7c). Basically what happens is that, because of the preferences of the herbivore, at any given productivity species 2 alone is able to support a higher herbivore population than species 1 alone. Hence, when species 1 replaces species 2 (by shading) the herbivore population goes down. When species 2 is fully replaced, however, we are only left with species 1. That is, we are back at our one plant, one herbivore model, which says that the leaf biomass of species 1 should remain constant whereas herbivore biomass should increase with increasing productivity.

One might perhaps suggest that the herbivore pattern in Figs 8.7b,c is unlikely in reality because the model assumes static preferences whereas in reality herbivores are much more flexible in their feeding behaviour. However, the mechanism that produces the herbivore pattern is that plant species 2 alone gives a higher herbivore mass than plant species 1 alone. If this difference is large enough, even the most flexible herbivore cannot prevent a decrease in herbivore abundance when plant species 1 replaces plant species 2. At the extreme, if plant species 1 does not support a herbivore population at all, the herbivore will simply disappear with increasing productivity (as in Fig. 8.5b).

Food quality and priority effects

Besides differences in the preferences of the herbivore (encapsulated in the parameter a_i), the plant species may differ in their quality as a food source for the herbivore (the parameter c_i). We will show that differences in food quality may give rise to two alternative stable states. In other words, we will show that the same environment may support two completely different communities depending on which species become dominant first. Suppose that both plant species are sun species (see Table 8.1), and that species 2 offers a higher food quality than species 1 ($c_2 > c_1 > 0$). These two assumptions imply that, when grown alone, species 2 is able to support a much higher herbivore population than species 1. Now, when species 1 is dominant first it casts much shade on species 2, and species 2 is unable to grow (Fig. 8.8a). Note the pattern of herbivore biomass along the productivity gradient — the explanation is similar to that for Fig. 8.7 given above. When species 2 is dominant first, however, it supports a large herbivore population and species 1 is unable to grow because of the large herbivore grazing pressure (Fig. 8.8b). Hence, multiple stable states have arisen in quite a natural way. Either a dense and tall vegetation develops with plants of low forage quality, or a short vegetation of high-quality species supports a high herbivore population that in turn prevents the establishment of tall, low-quality plants. One aspect is essential here. The herbivore population must be able to prevent the establishment of the tall plant species, either by still eating the tall plant species (even though it is of low quality) or by effectively

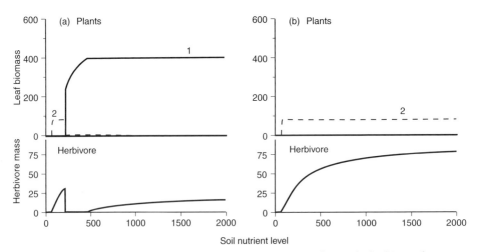

Figure 8.8 Two plant species of different forage quality, and a herbivore along a productivity gradient. Both plant species are sun species. The tall plant is low-quality food, and the small plant is high-quality food ($c_1 = 0.01$, $c_2 = 0.05$, $a_1 = a_2 = 0.0005$). This leads to two stable states. (a) In one state, the tall plant dominates and outshades the small plant. (b) In the other state, the small plant dominates and supports a high herbivore population. In the top panels, the solid lines indicate the leaf biomass of the tall plant, and the dashed lines indicate the leaf biomass of the small plant. Default parameters, see Table 8.1.

removing the tall plant species in some other way (e.g. trampling). If the tall plant species is hardly removed by the herbivore population (i.e. when a_1 is much lower than a_2), the model predicts that the tall plant species can easily invade a short vegetation and that multiple stable states do not exist.

Comparison with nutrient-based models

There is a large body of theory examining plant–herbivore dynamics assuming that plant growth is limited by nutrient availability (e.g. Phillips 1974; Leibold 1996; Grover 1997). The simplest of these models ignore the species composition of trophic levels, thus making no distinction between plant species (e.g. Oksanen *et al.* 1981). This leads to relatively simple patterns of abundance along a productivity (nutrient supply) gradient. At the lowest productivities, even plants cannot persist, but once productivity is high enough to support a plant population, its abundance first increases with productivity. When plant abundance is high enough to support a herbivore population, further increases in productivity will give only an increase in herbivore abundance. Plants remain at a density controlled by herbivores (as in Fig. 8.4). If higher trophic levels are added to this plant–herbivore food chain, then top-down control shifts to the highest trophic level. Depending on the number of trophic levels, plant and herbivore abundance may be constant or increasing through some productivity ranges. But, in these food chain models, plant and herbivore abundance are never decreasing functions of productivity.

More elaborate nutrient-based models allow for more than one plant species. One simple approach is to imagine an edible and an inedible plant species (Grover 1995). In such models, the inedible plant must be an inferior competitor for the nutrient, otherwise the edible plant and its herbivore are always competitively excluded. Inedible plants satisfying this assumption are only present at high productivity. At lower productivity, only the edible plants and the herbivore occur, and their abundances vary with productivity as in the simple food chain models. But, once productivity is high enough to allow invasion of inedible plants, any further increases in productivity serve only to increase inedible plant abundance, while the edible plant and the herbivore remain at a constant level. Some inedible plant species interfere with the ability of herbivores to consume their edible food, in which case abundances of all species increase with productivity once inedible plants are present. However, regardless of such complexities in the herbivore functional response, when one plant species is inedible, herbivore abundance is never a decreasing function of productivity.

When nutrient-based models of two edible plants and one herbivore are constructed, additional phenomena are possible (Holt *et al.* 1994; Leibold 1996; Grover 1997). The species that is the superior nutrient competitor dominates at low productivity, while the species more resistant to grazing dominates at high productivity. Under certain conditions, the two plant species can coexist in some intermediate range of productivities. Priority effects at intermediate productivity are also possible. Thus, the general pattern along a productivity gradient is that one plant species has unimodal abundance, with a peak at low productivity, while abundance of the other plant species increases with productivity as it replaces the first. At very low and very high productivities, herbivore abundance increases with productivity. Through the intermediate range of productivity where both plants are present, the response of herbivore abundance to productivity depends on its functional response. If this is linear (as we have assumed in this chapter), then herbivore abundance is constant in this productivity range. If the functional response is non-linear, then we conjecture that either increases or decreases are possible.

This survey of nutrient-based theory thus shows that the usual prediction is that herbivore abundance is a strictly non-decreasing function of productivity. Decreases arise only through complications of the herbivore functional response, and are thus a 'top-down' phenomenon imposed by the consumer. Other theory also portrays negative herbivore responses to productivity as top-down phenomena. For example, Abrams (1993) constructed several simple food web models with two non-competing basal (plant) species, and showed that responses to productivity depended on the details of higher trophic structure. Van de Koppel *et al.* (1996) posed yet another example of negative herbivore responses to productivity arising in a top-down way. In their model, plants are aggregated in a single trophic level, as in Oksanen *et al.* (1981), but they assumed that either the herbivore's functional or functional and numerical responses decrease with increasing plant biomass when plant biomass is high. This may lead to a collapse of the herbivore population at high productivity.

In contrast to nutrient-based theories, our model of light-based competition among plants suggests that herbivores could decrease with increasing productivity, as a consequence of a 'bottom-up' effect. This effect arises most strongly when only the small plant is edible, and importantly, when the taller plant would replace the small plant at high productivity even in the absence of herbivory. Herbivory may modify this pattern of species replacement, but the bottom-up effect of this inevitable plant species replacement is that herbivores simply have no food at higher productivity, and they are thus forced to disappear (see Fig. 8.5b). In contrast, when plants compete for a single nutrient, there is no species replacement in the absence of herbivory—a single plant species is the superior competitor at all nutrient supplies (Armstrong & McGehee 1980; Tilman 1982). If this superior nutrient competitor is edible, then inedible plants may invade at high productivity, but they can never replace the edible plants. The herbivores still have a food supply, and thus do not decrease in abundance. These contrasting patterns between light- and nutrient-based models are also present, but appear less forcefully when both plants are edible and differ in their preference for herbivores. This illustrates the profound importance of the mechanisms of plant competition for the propagation of 'bottom-up' and 'top-down' effects in lower portions of food webs.

A field example

Rabbits, hares and geese along a salt-marsh productivity gradient

In contrast to most nutrient-based plant–herbivore models, our model of competition for light predicts that herbivore populations may decrease with increasing productivity. This occurs when tall plant species are inedible, less preferred or of lower quality compared with the smaller plant species. Are these patterns also observed in the field? Here we present an example from a salt-marsh ecosystem grazed by rabbits, hares and geese.

Because the net tidal movements are directed eastwards and there deposit new sediment, the salt marsh of the Waddensea island of Schiermonnikoog, in The Netherlands, gradually spreads towards the east. As a consequence, various stages of salt-marsh development are situated next to each other, with the youngest successional stages at the far east end of the island, and the older stages, ageing over a 100 years, at the western part of the salt marsh (De Leeuw *et al.* 1993; Olff *et al.* 1997). The youngest stages are characterized by a sparse vegetation and a very low amount of total nitrogen in the soil. Fertilizer experiments in this area showed that nitrogen is a major factor limiting the production of plant biomass (H.J. van Wijnen, unpublished results). Both soil nitrogen and the nitrogen mineralization rate increase with successional age (Bakker *et al.* 1994; Olff *et al.* 1997). Hence this successional sequence can be viewed as a natural productivity gradient from low productivity in the youngest stages to high productivity in the older stages. The salt marsh is grazed by several vertebrate herbivores. Rabbits (*Oryctolagus cuniculus*) and hares (*Lepus europaeus*) are present year-round. Barnacle geese (*Branta leucopsis*) use the salt marsh in winter and early spring, and brent geese (*Branta bernicla bernicla*) are mainly present in April and May.

Figure 8.9a shows the pattern of plant species replacement along the successional gradient at the higher parts of the salt marsh (the plant species composition is different at the lower salt marsh, which is more frequently inundated by sea water; see Olff *et al.* 1997). Total plant biomass increases

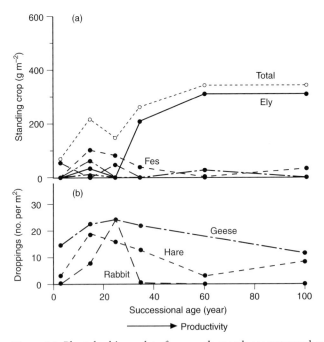

Figure 8.9 Plant–herbivore data from a salt-marsh ecosystem where productivity increases with successional age. (a) Above-ground standing crop of the six most abundant plant species, as well as total above-ground standing crop (total, thin dashed line), in relation to successional age. The species are *Elymus athericus* (Ely, solid line), *Festuca rubra* (Fes, dashed line), and the remaining four species *Agrostis stolonifera*, *Artemisia maritima*, *Elymus farctus* and *Plantago maritima*. (b) Number of droppings of rabbits, hares and geese (barnacle goose and brent goose combined) collected over 1 year in relation to successional age.

with successional age and, thus, with productivity. The younger stages at the higher salt marsh are vegetated with small to intermediate-sized grasses and herbs like *Festuca rubra*, *Agrostis stolonifera*, *Artemisia maritima* and *Plantago maritima*. Some of these species are preferred by the herbivores (e.g. *Plantago*, *Festuca*; Ydenberg & Prins 1981; Prop & Deerenberg 1991), whereas others are less frequently eaten (e.g. *Artemisia*). Within a time span of about 40 years, however, all these smaller species are replaced by a dense stand of the tall-grass species *Elymus athericus*. Competition for light in *Elymus*-dominated stands is likely to occur. We measured light absorption by *Elymus* at four different sites using a Licor LI-191SA line quantum sensor (400–700-nm range, in μmol photons m^{-2} s^{-1}). Only about 7% of the incident light intensity penetrated through the *Elymus* canopies. In shading experiments with natural sun light, seedlings of *Festuca*, *Artemisia* and several other species had very poor performance when light levels dropped below 20% of the incident light. Hence, shading by *Elymus* stands should have a major impact, virtually preventing growth of all smaller plant species beneath its canopy. In addition, *Elymus* is not a preferred species for any of the grazing herbivores. *Elymus* is a tall grass that produces a lot of structural tissue in its stem and stiff leaves. This might enable *Elymus* to become a good light competitor, but structural tissues are usually difficult to digest and reduce the quality of the plants as a food source for herbivores. Furthermore, vegetations dominated by *Elymus* often have a large amount of standing dead material (van de Koppel *et al.* 1996), making fresh plant parts more difficult to obtain for herbivores. Grazing pressure by hares, rabbits and geese is maximal in the successional stages of intermediate productivity, while the grazing pressure is relatively low in the older and more productive parts of the salt marsh dominated by *Elymus* (Fig. 8.9b; see also van de Koppel *et al.* 1996; Olff *et al.* 1997; Chapter 9). Hence, here we have an example of small edible plants dominating at low productivity but being displaced by a tall inedible or less-preferred plant species at high productivity, with a consequent decline in herbivore abundance. Qualitatively, at least, this pattern is in line with the expectations based upon our model observations.

Another prediction of the model is the possibility for multiple stable states. The model predicts multiple stable states when the tall plant species is of a low forage quality and able to develop a dense canopy outshading smaller plant species, while the small plant species offers a high forage quality and is able to support a high herbivore grazing pressure. Something similar to this seems to have happened in the herbivore removal experiments of Bazely and Jefferies (1986). They studied grazing by lesser snow geese (*Anser caerulescens caerulescens*) on a Canadian salt marsh. The geese maintained a short *Carex–Puccinellia* community, which produced relatively large quantities of high-quality forage. Cessation of grazing by means of herbivore exclosures led to an increase of above-ground plant biomass, changed the species composition of the plant community, increased the amount of litter and reduced the overall quality of the forage. The vegetation remained only poorly grazed when the fence was removed again. This suggests that herbivore-mediated multiple stable states indeed occur in this salt-marsh system, by the same mechanisms as proposed by our model.

Discussion

The results presented in this paper suggest that the mechanisms of plant competition have a major impact on the patterns of herbivory along productivity gradients. Competition for light leads to other patterns than competition for nutrients. For example, in contrast to most nutrient-based models (e.g. Oksanen *et al.* 1981; Grover 1995; Leibold 1996), our light-based model predicts that herbivores may decrease in abundance with increasing productivity. This occurs when tall plant species of low palatability domi-nate over smaller but more palatable plant species. Furthermore, under certain circumstances, multiple stable states may occur: one state where tall plants of low forage quality outshade small plants, and another state where small plants of high forage quality support a large herbivore population. The field data from the salt-marsh ecosystems that we have discussed seem to support these predictions.

The predictions are based on a model that is kept as simple as possible, and there is no doubt that this simple model neglects many important aspects of reality. For example, in reality many plants grow from seed. This means that each individual has to grow through the shade cast by other plants, from the lowest light level at the bottom of the habitat to the light level available at its mature adult height. In contrast, the model developed here is analytically tractable but assumes that the tall plant species is never shaded by smaller plant species. Hence our assumptions on light competition are probably restricted to herbaceous vegetations in which tall species have either large seeds or are capable of clonal growth since these traits may enable tall species to break through the shade cast by smaller plants. Furthermore, the model does not consider competition for nutrients, as we treated the soil nutrient level as a fixed parameter instead of a dynamic variable. The theory presented here is therefore most applicable in situations where competition for light is of overriding importance. We made this choice deliberately since we were here interested in effects of competition for light only. It is relatively straightforward, however, to introduce a dynamic nutrient (e.g. Huisman & Weissing 1995), and an important next step would be to combine the nutrient-based and light-based approaches. Also, we assumed fairly simple herbivore foraging behaviours. In reality, herbivores could discriminate between young and older plant parts, or they could change their foraging behaviour in response to the presence of other herbivores. More elaborate foraging models could thus be posed, but they would be more difficult to analyse and are beyond the scope of this chapter. Other major simplifications of the model include the linear functional response of the herbivore, the absence of plant dispersal, and neglect of both the horizontal patchiness in vegetation and the spatial movements of the herbivores. We see no reason, however, why more complicated models of plant competition and herbivore behaviour would necessarily produce monotonic trends of herbivore abundance along productivity gradients, or would exclude the possibility of multiple stable states. It seems more reasonable to assume that inclusion of additional complexities will mainly tend to alter the quantitative patterns predicted by the model while many of the qualitative trends will remain intact. The accordance between theory and data shows that our model assumptions as well as its predictions are, at the very least, biologically plausible.

In our model, the soil nutrient level operates as a factor that reduces the production of the plants. Other factors that reduce plant production (like salinity, pH and waterlogging) can be treated the same way, but with the highest stress level corresponding to the lowest soil nutrient level. Thus, the analysis presented here is not restricted to productivity gradients caused by variation in soil nutrient level, but is more general. It should in fact apply to any productivity gradient, provided that plant competition remains restricted to competition for light.

Recently, Huisman and Olff (1998) extended the model to multiple herbivore species. This revealed that generalist herbivores suppress plant competition for light and thereby facilitate selective herbivores that prefer small plant species. Field data support this (Huisman & Olff 1998).

Competition among plants is not a simple process. It has very different consequences depending on whether soil resources or light are involved. We have shown that different mechanisms produce different patterns of plant-species replacement and herbivore abundance. This illustrates that without explicit consideration of these mechanisms, ecologists will be in a poor position to predict how vegetation changes along productivity (or other) gradients, even in the absence of herbivory, and will be even less able to understand how herbivores modify these patterns. If our goal is to predict and understand community organization along major environmental gradients, a careful analysis of the mechanisms of species interaction will be required.

Acknowledgements

We thank Harm van Wijnen for his help during the field work at Schiermonnikoog, and Han Olff, Johan van de Koppel and the anonymous referees for their detailed comments on the manuscript. We are grateful to Dick Visser for drawing Fig. 8.1. James P. Grover acknowledges the support of NSF grant no. DEB-9418096. This research was made possible by a travelling grant to Jef Huisman, provided by the Center of Ecological and Evolutionary Studies (CEES), University of Groningen.

References

Abrams, P.A. (1993). Effect of increased productivity on the abundances of trophic levels. *American Naturalist*, **141**, 351–371.

Armstrong, R.A. & McGehee, R. (1980). Competitive exclusion. *American Naturalist*, **115**, 151–170.

Bakker, J.P., Olff, H., Loonen, M., Hazekamp, A., van Hooff, E. & Kats, R. (1994). Plant species composition, plant production, compartmentation of nitrogen and geese grazing in a successional gradient on salt marshes. *Acta Botanica Neerlandica*, **43**, 217.

Bazely, D.R. & Jefferies, R.L. (1986). Changes in the composition and standing crop of salt-marsh communities in response to the removal of a grazer. *Journal of Ecology*, **74**, 693–706.

Berendse, F. (1985). The effect of grazing on the outcome of competition between plant species with different nutrient requirements. *Oikos*, **44**, 35–39.

Boardman, N.K. (1977). Comparative photosynthesis of sun and shade plants. *Annual Review of Plant Physiology*, **28**, 355–377.

Botkin, D.B., Janak, J.F. & Wallis, J.R. (1972). Some ecological consequences of a computer model of forest growth. *Journal of Ecology*, **60**, 849–873.

Crawley, M.J. (1997). Plant–herbivore dynamics. In *Plant Ecology* (Ed. by M.J. Crawley), pp. 401–474. Blackwell Science Ltd, Oxford.

Cyr, H. & Pace, M.L. (1993). Magnitude and patterns of herbivory in aquatic and terrestrial ecosystems. *Nature*, **361**, 148–150.

DeAngelis, D.L. (1992). *Dynamics of Nutrient Cycling and Food Webs.* Chapman & Hall, London.

De Leeuw, J., De Munck, W., Olff, H. & Bakker, J.P. (1993). Does zonation reflect the succession of salt-marsh vegetation? A comparison of an estuarine and a coastal bar island marsh in The Netherlands. *Acta Botanica Neerlandica*, **42**, 435–445.

Drent, R.H. & Prins, H.H.T. (1987). The herbivore as a prisoner of its food supply. In *Disturbance in Grasslands* (Ed. by J. van Andel, J.P. Bakker & R.W. Snaydon), pp. 131–148. Junk Publishers, Dordrecht.

Ford, E.D. & Diggle, P.J. (1981). Competition for light in a plant monoculture modelled as a spatial stochastic process. *Annals of Botany*, **48**, 481–500.

Fryxell, J.M. (1991). Forage quality and aggregation by large herbivores. *American Naturalist*, **138**, 478–498.

Furbish, C.E. & Albano, M. (1994). Selective herbivory and plant community structure in a mid-atlantic salt marsh. *Ecology*, **75**, 1015–1022.

Grover, J.P. (1995). Competition, herbivory, and enrichment: nutrient-based models for edible and inedible plants. *American Naturalist*, **145**, 746–774.

Grover, J.P. (1997). *Resource Competition.* Chapman & Hall, London.

Hara, T. (1993). Effects of variation in individual growth on plant species coexistence. *Journal of Vegetation Science*, **4**, 409–416.

Harper, J.L. (1977). *Population Biology of Plants.* Academic Press, London.

Holling, C.S. (1959). The components of predation as revealed by a study of small mammal predation of the European pine sawfly. *Canadian Entomologist*, **91**, 293–320.

Holt, R.D. (1977). Predation, apparent competition, and the structure of prey communities. *Theoretical Population Biology*, **12**, 197–229.

Holt, R.D., Grover, J.P. & Tilman, D. (1994). Simple rules for interspecific dominance in systems with exploitative and apparent competition. *American Naturalist*, **144**, 741–771.

Huisman, J. & Olff, H. (1998) Competition and facilitation in multispecies plant-herbivore systems of productive environments. *Ecology Letters*, **1**, 25–29.

Huisman, J. & Weissing, F.J. (1994). Light-limited growth and competition for light in well-mixed aquatic environments: an elementary model. *Ecology*, **75**, 507–520.

Huisman, J. & Weissing, F.J. (1995). Competition for nutrients and light in

267

a mixed water column: a theoretical analysis. *American Naturalist*, **146**, 536–564.

Huisman, J., Jonker, R.R., Zonneveld, C. & Weissing, F.J. (1999). Competition for light between phytoplankton species: experimental tests of mechanistic theory. *Ecology* (in press).

Johnson, I.R., Parsons, A.J. & Ludlow, M.M. (1989). Modelling photosynthesis in monocultures and mixtures. *Australian Journal of Plant Physiology*, **16**, 501–516.

Kohyama, T. (1992). Size-structured multi-species model of rain forest trees. *Functional Ecology*, **6**, 206–212.

Leibold, M.A. (1996). A graphical model of keystone predators in food webs: trophic regulation of abundance, incidence, and diversity patterns in communities. *American Naturalist*, **147**, 784–812.

McNaughton, S.J. (1984). Grazing lawns: animals in herds, plant form, and coevolution. *American Naturalist*, **124**, 863–886.

McNaughton, S.J., Oesterheld, M., Frank, D.A. & Williams, K.J. (1989). Ecosystem-level patterns of primary productivity and herbivory in terrestrial habitats. *Nature*, **341**, 142–144.

Monsi, M. & Saeki, T. (1953). Über den Lichtfaktor in den Pflanzen-gesellschaften und seine Bedeutung für die Stoffproduktion. *Japanese Journal of Botany*, **14**, 22–52.

Oksanen, L., Fretwell, S.D., Arruda, J. & Niemela, P. (1981). Exploitation ecosystems in gradients of primary productivity. *American Naturalist*, **118**, 240–261.

Olff, H., De Leeuw, J., Bakker, J.P., Platerink, R.J., van Wijnen, H.J. & De Munck, W. (1997). Vegetation succession and herbivory in a salt marsh: changes induced by sea level rise and silt deposition along an elevational gradient. *Journal of Ecology*, **85**, 799–814.

Pacala, S.W., Canham, C.D., Saponara, J., Silander, J.A., Jr., Kobe, R.K. & Ribbens, E. (1996). Forest models defined by field measurements: estimation, error analysis and dynamics. *Ecological Monographs*, **66**, 1–43.

Phillips, O.M. (1974). The equilibrium and stability of simple marine biological systems. II. Herbivores. *Archiv für Hydrobiologie*, **73**, 310–333.

Prentice, I.C. & Leemans, R. (1990). Pattern and process and the dynamics of forest structure: a simulation approach. *Journal of Ecology*, **78**, 340–355.

Prop, J. & Deerenberg, C. (1991). Spring staging in brent geese *Branta bernicla*: feeding constraints and the impact of diet on the accumulation of body reserves. *Oecologia*, **87**, 19–28.

Rosenzweig, M.L. (1973). Exploitation in three trophic levels. *American Naturalist*, **107**, 275–294.

Ryel, R., Barnes, P.W., Beyschlag, W., Caldwell, M.M. & Flint, S.D. (1990).

Plant competition for light analyzed with a multispecies canopy model. I. Model development and influence of enhanced UV-B conditions on photosynthesis in mixed wheat and wild oat canopies. *Oecologia*, **82**, 304–310.

Shugart, H.H. (1984). *A Theory of Forest Dynamics*. Springer-Verlag, New York.

Silvertown, J., Lines, C.E.M. & Dale, M.P. (1994). Spatial competition between grasses — rates of mutual invasion between four species and the interaction with grazing. *Journal of Ecology*, **82**, 31–38.

Sorrensen-Cothern, K.A., Ford, E.D. & Sprugel, D.G. (1993). A model of competition incorporating plasticity through modular foliage and crown development. *Ecological Monographs*, **63**, 277–304.

Thornley, J.H.M. & Johnson, I.R. (1990). *Plant and Crop Modelling: A Mathematical Approach to Plant and Crop Physiology*. Oxford University Press, Oxford.

Tilman, D. (1982). *Resource Competition and Community Structure*. Princeton University Press, Princeton.

van de Koppel, J., Huisman, J., van der Wal, R. & Olff, H. (1996). Patterns of herbivory along a productivity gradient: an empirical and theoretical investigation. *Ecology*, **77**, 736–745.

Weissing, F.J. & Huisman, J. (1994). Growth and competition in a light gradient. *Journal of Theoretical Biology*, **168**, 323–336.

Ydenberg, R.C. & Prins, H.H.T. (1981). Spring grazing and the manipulation of food quality by Barnacle Geese. *Journal of Applied Ecology*, **18**, 443–453.

Yokozawa, M. & Hara, T. (1992). A canopy photosynthesis model for the dynamics of size structure and self-thinning in plant populations. *Annals of Botany*, **70**, 305–316.

Chapter 9

Cyclic grazing in vertebrates and the manipulation of the food resource

R.H. Drent[1] and R. van der Wal[1]

Summary

Two facets of the putative power of herbivores to manipulate their food supply according to the ideas of McNaughton are reviewed with special reference to a young salt marsh in the Dutch Wadden Sea. A general feature of vegetations described as 'grazing lawns' seems to be that return visits by the herbivore enhance the quality of food on offer. In the salt marsh we studied, intensive and repeated grazing by staging geese in spring protracted the period of high nutritive status of the food plants, effectively extending the exploitation window for the herbivores by 7 weeks. Detailed quantification of the visit interval in relation to regrowth of previously grazed ramets indicate that the geese time their grazing cycle in order to maximize the intake of nutritious regrowth. However, the observed visit interval tends to fall on the early side of the model outcome for intake maximization, suggesting that the competitive interaction between flocks on the marsh needs further attention. In the longer term, vertebrate herbivory on the marsh extends the period of persistence of the younger successional stages of the vegetation supporting the heaviest level of exploitation. In our island study, the year-round presence of the brown hare, *Lepus europaeus*, retarded establishment of the dwarf shrub *Atriplex* on the young marsh, and thus facilitated access by the geese to the palatable species otherwise screened by the bushy *Atriplex*. Exclosures placed on the marsh at sites of different age up to 40 years, combined with experimental removal or addition of *Atriplex* plants, allow the conclusion that hares

1 Zoological Laboratory, University of Groningen, Kerklaan 30, 9751 NN Haren, The Netherlands

facilitate grazing by geese and extend the goose exploitation period by 25 years. Why the herbivores fail to arrest the onward march of plant succession on the marsh is unresolved but may be caused by ongoing changes in the nutrient status of the soil. The grazing lawns are thus not permanent in our system, but the general features of the plant–herbivore interaction agree with the framework hypothesized by McNaughton.

Introduction

To Sam McNaughton (1979) we owe the provocative idea that herbivores are able to manipulate their food supply. In the short term, herding animals can exert massive grazing pressures and return periodically to harvest the nutritious early growth phases of vegetation kept in a state of high above-ground productivity of readily digestible material. In the longer term, the grazers (this may involve a whole guild) may positively influence the sustainability of a productive assemblage of palatable plants in what he termed a 'grazing lawn' (McNaughton 1984). McNaughton reasoned that the result of herbivory was a plant growth response packing 'productive, nutritious, and herbivore-sought tissues into a small volume near the soil surface' and drew attention to the distinction between evolutionary time (allowing genetic change) and ecological time (adaptive responses we can investigate today). To allow persistence of the lawns McNaughton theorized that sufficient tissue of the grazed plants must enjoy the protection of a spatial refuge contributing to the stability of grazing ecosystems. This chapter addresses two rather general questions about plant–herbivore interactions on the ecological time scale, covering both within- and between-season events.

1 How general is the finding that grazing enhances the *quality* of food on offer during later visits compared with previously ungrazed areas (implying that grazing should be cyclic)?

2 What evidence is there that grazing gives herbivores a long-term benefit by favouring the local persistence of the food plants?

The most controversial aspect of this 'grazing optimization' hypothesis is the suggestion that above-ground plant production is higher under the impact of grazing than in ungrazed controls (see Belsky 1986 for a critical review). As is discussed elsewhere in this volume (Chapter 10) this 'overcompensation' occurs only under rather special circumstances, and is linked to nutrient inputs from herbivore droppings. The crux of the grazing optimization hypothesis (McNaughton 1985) is that grazing directly (e.g.

by increasing nutrient input) or indirectly (by selectively removing competitors or reducing overshading) enhances the persistence of the food plants at the grazing site. Although at first sight one might dismiss the interaction by the sweeping statement 'herbivory is bad for the plants that get eaten and good for the ones that don't' (Crawley 1997) we must consider whether herbivory is in some cases the price the plant pays for persistence *in situ*.

From the herbivore point of view there is an advantage to renewed use of previously grazed 'lawns' as the enhanced quality of the regrowth (higher protein levels, lower concentration of structural carbohydrates inhibiting digestibility) provides a greater net yield (digestible nutrients) per bite than from lightly grazed vegetation (McNaughton 1984). A salient advantage of exploiting grazed vegetations resides in the more favourable ratio of live/dead standing material, enhancing the quality of the mouthful of the hard-pressed herbivore that must spend most of the available time in harvesting its daily ration. All these aspects have been quantified in a study of grazing lawns adjoining the 'dog towns' in which the colonial herbivore *Cynomys ludovicianus* (black-tailed prairie dog) lives. In the prairie grasslands of South Dakota Coppock *et al.* (1983a,b) showed that the lawns adjoining well-established colonies (3–8 years' residence) provided graminoids with a higher crude protein content, higher *in vitro* digestibility of dry matter and a more favourable live/dead ratio in the above-ground standing crop than more recently (2 years) colonized sites, which in turn were superior to the ungrazed surroundings. The long-term grazing regime exerts selection on the plant population, and Holland *et al.* (1992) in following up the prairie dog story reported that production was greater with grazing and maintained at higher grazing intensities on the lawns near the colony in comparison to ecotypes in ungrazed situations.

The concept of 'grazing optimization' leads logically to defining the intensity of offtake and grazing interval that would maximize the return to the herbivore in terms of net intake. The key to understanding why intake maximization of the herbivore entails a decrease in the potential standing crop of the food plant is the realization that the lifespan of an edible leaf in graminoids is a matter of days, and harvest rates depend on a trade-off between waiting for regrowth to occur and cropping leaves before senescence sets in. Parsons *et al.* (1983) have studied above-ground production in *Lolium perenne* pastures under a continuous grazing regime by sheep, combining measurements of the physiological basis for plant production with the herbivore offtake. Their general model (Fig. 9.1) is

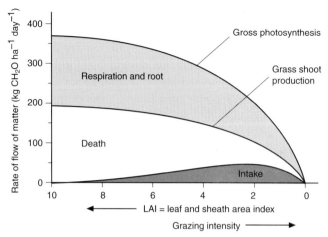

Figure 9.1 Physiological limitation to production of *Lolium perenne* under continuous grazing. The diagram shows the relationship between the uptake and loss of matter, and the animal intake that can be achieved in swards maintained at each of a range of the leaf area index (LAI) by adjustment of the grazing intensity (sheep stocking rate). From Parsons *et al.* (1983).

instructive as it relates total herbivore offtake to an index of grazing pressure, the leaf area index (LAI in the figure). The maximum yield to the sheep is attained at a LAI close to 2, where the optimal mix of consumption/senescence of the leaf crop occurs, at somewhat less than half of the potential gross photosynthetic production of the sward. This is a very illuminating study as it demonstrates that even without considering nutrient inputs from the sheep there is a sharp increase in the harvest to the herbivores if they can maintain a reasonably heavy grazing intensity. There is an analogy here with yield models for exploited animal populations, where the 'fishing intensity' must strike a balance between allowing further growth and hence relaxing pressure, or harvesting individuals that would (in a statistical sense) die anyway. To our knowledge there is no study of comparable detail for a natural system with wild herbivores, but Wilmshurst *et al.* (1995) come close with their experimental approach with penned wapiti *Cervus elaphus* allowed to choose between grass (*Poa, Bromus*) patches previously manipulated by mowing. Also in this case, net energy intake of the wapiti peaked at intermediate values of grass biomass, and herbivore preference tended to coincide. The data imply a relatively young sward age for maximal harvest (the trade-off between forage quality and abundance we have met with before) and a herd grazing pattern maintaining an even-aged sward and returning at the optimal interval. Hobbs and Swift (1988) point out that phenologically young grass in wapiti diets is roughly twice as digestible as mature or senescent grass, and in their model output the expected intake rate of digestible dry matter is always higher when feeding on grazed as distinct from ungrazed grassland, except at sites of very low productivity.

From the plant point of view, the enhancement of the quality of the above-ground biomass by means of repeated grazing episodes may have important repercussions by shortening the time for microbial decomposition for the dead plant material in comparison with the more refractive material usual in ungrazed situations. This acceleration of the mineralization loop has been evaluated by Pastor and Naiman (1992) and deserves recognition as one of the long-term impacts of herbivory (cf. Frank & McNaughton 1993 and Frank & Evans 1997). Herbivore effects are not all positive, however, and may cause decline or disappearance of the palatable species, with the result that the litter quality declines also (Pastor & Cohen 1997). These authors conclude that in the boreal forests of North America 'increased dominance by unbrowsed, highly defended, slow-growing evergreens and depression of soil nitrogen availability seems to be the common response of plant communities to browsers and grazers'. We will thus restrict ourselves to the graminoid systems for which the concept of an equilibrium between grazers and grazing lawns was originally envisaged.

In the following section we will explore the two leading questions inherent to the grazing optimization hypothesis (short-term and long-term benefits for the herbivores) against the background of our own studies of vertebrate herbivory on the salt marsh of Schiermonnikoog. In common with the tundra systems discussed by Oksanen *et al.* (1983, 1997), on our island predators are virtually absent, and the vertebrate herbivores are at least locally resource limited. In keeping with Oksanen's ideas, the plant cover of the young marsh is subjected to intense natural grazing.

Cyclic grazing in geese

We have studied herbivory on the island of Schiermonnikoog off the northern coast of The Netherlands (53°30′N, 6°18′E) since 1971. Most of the work reported here was undertaken in a 550-ha tract of marsh and dune, where spring grazing is concentrated on the lower marsh and concerns approximately 1500 brent geese, *Branta bernicla*, from mid-March to the close of May when the birds depart for their breeding area. This intensive grazing period is preceded by exploitation by *c.* 4000 barnacle geese *Branta leucopsis* staging on the upper marsh from mid-February until mid-April, most, however, departing by late March. Based on repeated census in November, approximately 300–600 hares (*Lepus europaeus*) are resident in this part of the island. Domestic stock have been excluded from this part of the marsh for the past 40 years and the plant associations studied on the lower marsh do not owe their origin to cattle grazing in the past, as virtually the entire study area at the eastern end of the island was bare of vegetation at the time of cattle exclosure. Further to the west a tract of salt marsh has been fenced, and here cattle graze in the summer months (May through to September) in addition to the geese, hares and locally some rabbits (*Oryctalagus cuniculus*). There are no voles (*Microtus*) on the island, so compared with the mainland the herbivore assemblage is impoverished.

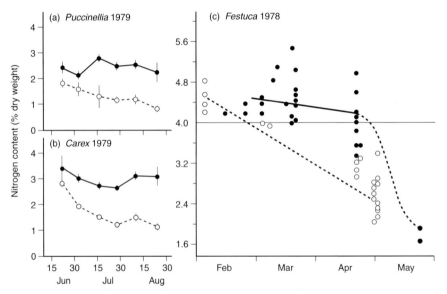

Figure 9.2 Seasonal shifts in total nitrogen content (% dry weight) of shoots of (a) *Puccinellia phryganodes*, (b) *Carex subspathacea* and (c) *Festuca rubra* from grazed (●) and ungrazed (○) sites. The 1979 data refer to snow goose grazing on the Hudson Bay salt marsh (Cargill & Jefferies 1984), the 1978 data to barnacle goose grazing on the Schiermonnikoog salt marsh. From Ydenberg and Prins (1981). Included is the lower threshold for goose utilization at this time of year (4% nitrogen content, see text).

Comparison of a grazed sward with exclosed controls shows that grazing enhances the quality of the plants (nitrogen content) on offer during subsequent visits (Fig. 9.2) (Ydenberg & Prins 1981; Cargill & Jefferies 1984). The herbivore thus maintains the vegetation in a state of rapid growth, delaying the decline in food quality associated with plant maturation. However, the data for barnacle geese grazing on *Festuca rubra* suggest that although the herbivores can usefully extend the period of exploitation, they are unable to prevent the decline in food quality. Since the lower threshold for *Festuca* utilization by the barnacle goose has been estimated as 4% nitrogen (J. Prop, unpublished data) we can quantify the extension of the staging period under the influence of repeat grazing as approximately 50 days, effectively more than doubling the spring exploitation phase. This same tactic is employed by the barnacle geese at the next staging site further north, where again *Festuca rubra* is the main forage species. Two further examples are included in Fig. 9.2 showing the advantage of return visits to the herbivore.

Cyclic grazing in brent geese

Visitation patterns of brent geese to specific parts of the salt marsh were quantified from observation towers from mid-April onwards prior to departure of the geese for the breeding grounds. At this time the geese feed mainly on the lower salt marsh, and hourly (or half-hourly) counts were done throughout the daylight period in quarter hectare blocks in the 30-ha intensive study plot. In between the counts the observers attempted to identify all colour-ringed individuals visiting the plot (coded leg rings can be deciphered at up to 300 m using × 35 telescopes).

The original observations indicated that blocks with a mosaic vegetation including *Plantago maritima* and/or *Triglochin maritima* in addition to the other main forage plants, the graminoids *Festuca rubra* and *Puccinellia maritima* tended to be visited every 4–5 days (Prins *et al.* 1980). Marked individual *Plantago* plants showed traces of regrazing, and subsequent observations were collected to test the idea that the timing of the grazing cycle was adjusted to maximize the yield to the geese on the short term, the flock returning to the area to profit from regrowth in an even-aged stand.

The rosette growth form and thick leaves of *Plantago* allows the geese to take large bites giving a higher intake rate than the alternatives. Brent geese selectively feed on the *Plantago* in preference to the grasses that predominate in their diet at this season. The lower threshold value of

Plantago grazing corresponds to the biomass providing an intake rate equal to the background grasses (mainly *Puccinellia*) and the majority of *Plantago* rosettes were grazed down to this level during a single passage of a goose flock (Prop & Loonen 1986; Drent & Prins 1987). When the visitation interval was related to a vegetation map of the grid, a pattern emerged (Fig. 9.3). The lower margins of the study plot were visited most frequently and here the brent mainly fed on *Puccinellia*. By monitoring the fates of individual leaves on grazed and ungrazed *Puccinellia* tillers, Prop (1991) demonstrated that the geese could maximize their bite size and hence their intake rate by revisiting at an interval of 2 days. This is indeed what the geese did, the observed grazing intensity allowing full regrowth of the vegetation in the interim.

On the raised 'islands' in the vegetation mosaic adjoining the *Puccinellia* sward, patches of *Festuca rubra* of several square metres grade into gullies with a sparse growth of *Puccinellia*. This transitional zone contains clumps of *Plantago maritima* and less frequently *Triglochin maritima*, which, with its fleshy leaves, also offers a high intake rate. In this 'island' vegetation

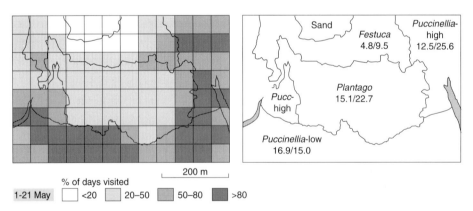

% of days visited

1-21 May   □ <20   ▨ 20–50   ▨ 50–80   ■ >80

Figure 9.3 Visitation pattern by brent geese to a salt-marsh study plot (50×50 m grid) at Schiermonnikoog, based on continuous observation in the daylight hours from 1 to 21 May 1986 (left) and vegetation map (right), the different numbers show the cumulative grazing pressure as measured by the brent geese droppings accumulated per square metre in the springs of 1985 and 1986, respectively. When the two panels are superimposed it appears that the low *Puccinellia* belts are visited most frequently, followed by the high *Puccinellia* and the *Plantago* 'island' area in the centre where the observation tower was located. Adapted from Prop (1991).

278

with the highly preferred *Plantago* plants the counts revealed a visitation interval of 4–8 days reminiscent of the earlier observations. Prop (1991) constructed a model to estimate optimal return times, based on the relationships between the intake rate of the geese and the size of *Plantago* rosettes, which flattens out at higher *Plantago* biomass, reaching a point where the geese cannot effectively increase harvest rate. Return times are optimal when the individual *Plantago* rosettes have recovered to the biomass level corresponding to the hump in the intake rate. The inclusion of a spatial constraint completes the recipe for the optimal grazing interval. This constraint is based on the finding that the flock, recognizable by individually marked geese and utilizing the *Plantago* resource, has a finite number of choices amounting to a total exploitation zone equivalent to four times the study plot. Predictions of this new model are shown in Fig. 9.4 along with a summary of the frequency of visitation intervals to this vegetation (assembled over several years). Clearly the optimal grazing regime embraces a rather wide plateau of 3–8 days, with the empirical observations skewed towards early harvest (see below).

In keeping with the finding that the visitation pattern is related to exploitation of specific plants, the marsh area originally employed as the study plot in 1978 was again studied in 1994 when the area had become

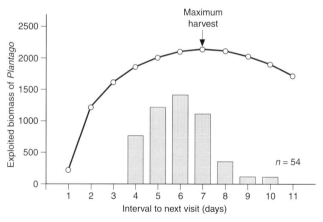

Figure 9.4 Visitation interval (line and circles) by brent geese to the *Plantago* zone shown in Fig. 9.3 as predicted by the brent goose intake maximization model incorporating the constraint of flock feeding area (see text). The frequency diagram depicts the frequency of flock visits to individual hectare blocs as actually observed in the years 1982–1997.

less attractive to the geese because of succession towards a dominance of non-edible plant species (Olff *et al.* 1997). When the 1994 counts are restricted to the blocks that still harboured *Plantago* the same pattern emerged again (see Fig. 9.5, where the 1978 and 1994 data are compared). Functionally, the overriding role of *Plantago* in setting the pace of local exploitation is understandable from the key role this food plant plays in contributing towards the accumulation of body reserves during spring (Prop & Deerenberg 1991) as a prerequisite for successful migration and subsequent reproduction on the tundras 5000 km away.

Rowcliffe *et al.* (1995) have recently found evidence for cyclic grazing in brent geese during the winter exploitation of *Puccinellia maritima* swards on the salt marshes adjoining the main agricultural feeding grounds in

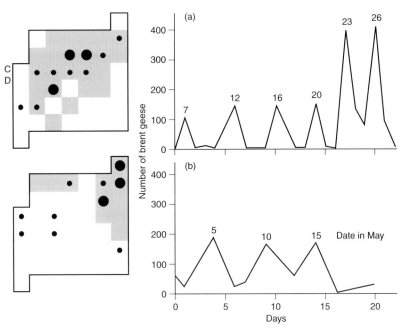

Figure 9.5 Visitation pattern by brent geese to an older salt-marsh plot 2 km west of the area depicted in Fig. 9.3, observed in (a) 1978 (from Prins *et al.* 1980) and (b) again in 1994. A cyclical pattern was maintained in the count sectors, which still provided *Plantago* feeding opportunities to the geese. Extent of vegetation with *Plantago* shown by shading on left (units are hectare blocs) where dots represent grazing minutes accumulated. Goose visitation in (b) restricted to shaded zone only (11 ha) and in (a) to the rows C and D (9 ha with *Plantago*).

their study area (the Norfolk coast). On the basis of near-daily counts, six of the 12 frequently grazed parts of the salt marsh were shown to be visited in a cyclic pattern, at intervals of 7–15 days. The advantage of this behaviour is presumed to reside in an enhanced quantity or quality in the food harvested on second and subsequent visits, but intake rates and diet selectivity were not studied. Why the sward as a whole deserves renewed exploitation at weekly or fortnightly intervals is as yet unresolved, as the measurements of elongation rates of individual tillers following grazing do not provide a simple explanation.

This winter study is an interesting contrast to the late spring exploitation of *Puccinellia* in our study site, where the brent geese return at 2-day intervals (Prop 1991) as has been mentioned above. Cyclic grazing thus occurs in both sites where it has been looked for, but the interval differs at different seasons in similar vegetations, and differs at the same season when adjacent types are exploited for different food plants. These findings implicate the dependence of the intake rate on the regrowth of leaves of high digestibility as the motor of the cycle, as has been argued for ungulate regrazing patterns in the Serengeti plains (see summary by Fryxell 1995). In both cases the difficulty is to explain why the flock (or herd) maintains its cohesion when it would pay the individual to 'jump the gun' and return to the previously grazed swards slightly ahead of the bulk of its competitors to reap the benefits of regrowth without the penalty of interference and depletion by group members. Fryxell (1995) considers the advantage of avoiding predation by group action as a counterbalancing selective force. To this can be added the further advantages of tapping the accumulated wisdom in the group in relation not only to predator avoidance (e.g. choice of safe roosting or sleeping areas) but regarding also information on alternative feeding sites, sources of water or specific nutrients. In our case of the geese it can be argued that imparting this knowledge to the juveniles still consorting with their parents at this time lends selective advantage to group cohesion, and it is certainly true that vigilant behaviour interferes with feeding when small groups are visually separated from the main flock (Inglis & Lazarus 1981).

Cyclic grazing in other vertebrates

It is logical to turn first to the extensive grassland systems of East Africa where the grazing optimization hypothesis orginated. McNaughton (1979) noted that wildebeest, *Connochaetes taurinus*, tended to revisit the Serengeti grassland units after regular intervals, harvesting the high-quality fresh growth resulting from the previous herd-grazing episode. The frequency of 'rotational passage grazing' seems to depend on rain behind the herd (or sufficient residual soil water) to generate new growth (McNaughton & Banyikwa 1995) but this has not been quantified in more detail.

In the nearby Lake Manyara National Park, Prins (1996) collected a large data set ($n = 212$) on intervals of visitation by African buffalo, *Syncerus caffer*, to patches of the grass *Cynodon dactylon* (patch size about 20 ha). The mean interval was 5.1 days (Fig. 9.6), which, on the basis of clipping experiments, allows harvest of regrowth in this rapidly growing grass, but in the judgement of Prins the bulk of the visits occur too soon to maximize the potential regrowth (median interval was only 3 days). Prins suggests that this short interval has been selected for in relation to the risk that the grazed patch would be exploited by the major food competitor, the elephant. By returning early, the buffalo profit from regrowth at a growth stage still insufficient for elephants to harvest.

The maturational stage of the vegetation determines both quality and quantity. Growth enhances biomass but with increasing age, forage quality declines and the ideal compromise is to regraze the patch to maximize the net energy or nutrient intake. In a recent manipulative study on wapiti, *Cervus elaphus*, in Canada, Wilmshurst *et al.* (1995) showed that the patch type preferred by the wapiti was indeed close to the predicted optimum. In this case, characteristics of the patch were manipulated by staggered mowing schedules to simulate previous grazing episodes, and a direct test of naturally occurring visitation intervals was not made. Nevertheless, the experiment is a powerful test of the ability of the grazer to discriminate between competing patches and select the stage providing the highest return in terms of energy intake.

We have been unable to find quantification of the regrazing cycle in caribou or domestic reindeer, but the evidence from long-term exclosures does not support the notion of an equilibrium between grazers and their preferred foods in the short term (see below).

There is some observational evidence for regrazing at intervals of more than 1 year (early-season grazing by muskox, *Ovibos moschatus*, on *Oxytropis viscida*, Mulder & Harmsen 1995) and a recent suggestion of a hypothetical mechanism that would generate grazing cycles of several years in the lemming, *Lemmus lemmus*. Seldal *et al.* (1994) have theorized that following gnawing wounds, *Carex bigelowii* and *Eriphorum angustifolium* produce a proteinase inhibitor blocking the digestive function of trypsin

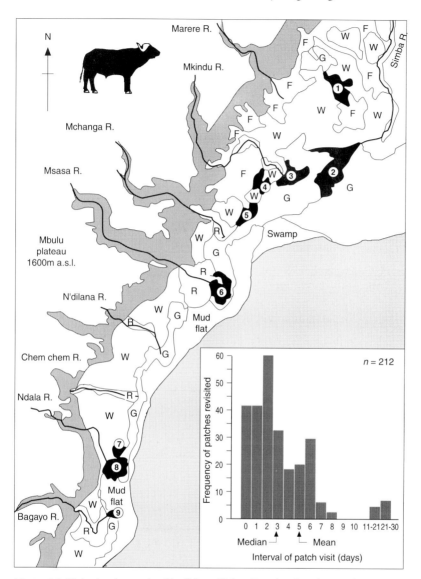

Figure 9.6 Visitation intervals of buffalo utilizing *Cynodon dactylon* patches identified in the map of the Lake Manyara region (1 = 9 ha, 2 = 49 ha, 3 = 25 ha, 4 = 13 ha, 5 = 15 ha, 6 = 33 ha, 7 = 8 ha, 8 = 18 ha, 9 = 4 ha). The escarpment is indicated by stippling, and vegetation codes are R = riverine bush, F = forest, W = woodland, G = grassland. Dimensions of the map are 11 × 15 km. Adapted from Prins (1996).

in the herbivore gut. Repeated grazing would thus cause a population decline in the herbivore, allowing the plants a period of recovery during which they might profit from the beneficial effects of prior grazing on the nutrient pool. Whatever the precise cause, it is certainly true that many small herbivorous rodents experience regular cycles in numbers, causing pulses of grazing and recovery in their target plants. As an adjunct to periodically high rodent numbers, Danell *et al.* (1981) showed by tracing year of establishment from growth rings that willows (*Salix* spp.) are only likely to meet conditions favourable to germination in the disturbed soil conditions following a population peak of the local rodents with associated large-scale burrowing.

Grazing equilibria in the long term

Central to the hypothesis of grazing optimization as propounded by McNaughton (1979, 1984, 1985) was the idea that recurring grazing gave rise to and served to maintain 'grazing lawns' providing a sustainable yield throughout the growth season to the foraging animals. For the Serengeti system where the hypothesis was developed with wildebeest herds as the crown witness, Sinclair (1995) confirms that despite sixfold increases in wildebeest numbers in the period 1963–77 (recovery from the rinderpest disease outbreak introduced accidentally with domestic stock) there is no evidence of an increase in unpalatable dicot species and, indeed, no changes in vegetation plots examined over a decade were detectable. These findings are compatible with the suggestion that wildebeest numbers are regulated by food shortage in the dry season, but that they have abundant food on the plains in the wet season. In this part of the range no deletrious effects on the vegetation have been detected and, indeed, Belsky (1992) showed convincingly by following exclosed vegetation plots over a 5-year period during high wildebeest numbers that grazing itself was essential to persistence of the system. In the absence of intensive grazing the 'grazing lawns' were rapidly transformed into rough vegetations with dense foliage, in which the short-statured species favoured by the grazers disappeared.

The rather special case of the Serengeti wildebeest (returning to carrying capacity from a disease-induced perturbation lasting almost a century) should caution us against accepting plant–herbivore equilibria as a matter of course. A good test of the equilibrium model would be the caribou, *Rangifer tarandus*, herds of northern Canada where in the presence of wolves and nowadays at least the virtual absence of people some stable relationship between grazer and vegetation could be expected. Manseau *et al.* (1996) have investigated this interaction by sampling vegetation plots with and without recent grazing in the summer range of a population of some 650 000 caribou in northern Quebec and Labrador. It proved difficult to locate ungrazed 'controls', as over 99% of the range showed signs of caribou usage. For the shrub tundra, the lichen mat was absent in grazed sites, and ground previously occupied by lichens was either bare, covered by fragments of dead lichens and mosses or recolonized by early succession lichen species not utilized by the caribou. In stands of dwarf birch grazed by caribou both ground cover and leaf biomass of *Betula glandulosa* was significantly lower than that in ungrazed sites, and above-ground productivity of vascular plants at these grazed stands was 25% lower than that of the ungrazed controls, primarily due to lower output of *Betula* itself. Even in the presence of wolf populations not subject to human control the caribou can depress plant biomass in the summer range to levels below $50 \, g \, m^{-2}$, a level below which food intake is expected to decline significantly. The authors' conclusion that vegetation on the summer range is not resilient to grazing and trampling and does not recover rapidly when grazing pressure is reduced seems inescapable. On the winter range, compacted snow and ice may hinder cratering by the caribou and act as refuges for the vegetation (White *et al.* 1981). From earlier work on domestic reindeer herds introduced to Arctic islands in Alaska (reviewed by Leader-Williams 1988) it is evident that overgrazed lichen stands may require 50 years or more to recover. In northern Fennoscandia, where the tundra is under persistent heavy grazing by reindeer, the vegetation appears to be converging towards a dominance by mosses (Oksanen 1983; Oksanen & Virtanen 1995).

Marshalling our own observations from the past 25 years, the geese have been unable to arrest vegetation change despite their intensive utilization of the lower marsh. Starting from bare sand, the Schiermonnikoog marsh accumulates a clay layer on the sand base and is colonized by plants (Olff *et al.* 1997). Brent geese prefer the early successional stages and the highest goose usage is observed in the 30-year span between year 10 and year 40 (Van de Koppel *et al.* 1996). In the first 10 years a *Salicornia* sp. and a *Puccinellia* sward develops, heavily utilized by the birds. As plant succession progresses, *Puccinellia* remains in the creeks dissecting the system, whereas on the higher 'islands' with lower flooding frequency *Festuca* takes over, leading to reduced goose grazing pressure (Fig. 9.7). Eventually the higher marsh

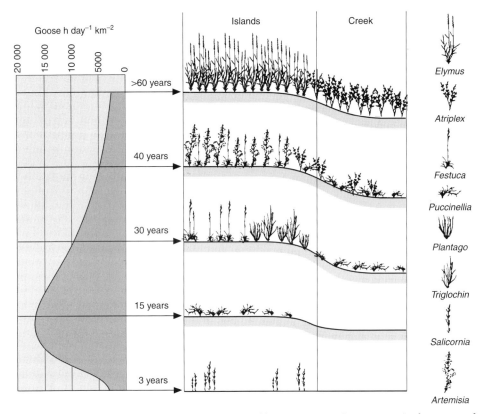

Figure 9.7 Changes in vegetation and brent goose grazing pressure in the course of salt-marsh succession at Schiermonnikoog. Depressions change to an *Atriplex* marsh, with very little *Puccinellia* or other food plants left. The slightly higher 'islands' turn from a *Puccinellia* sward into *Festuca*, after which *Elymus* invades and the geese vanish.

parts are invaded by *Elymus*, which forms dense stands, totally unsuitable for goose grazing. Meanwhile, the creek systems become overgrown by *Atriplex portulacoides* within 40 years. Our goose counts (Fig. 9.8) illustrate the trend of waning goose visitation as succession proceeds (we have been forced to shift our goose towers ever eastwards towards the end of the island experiencing accretion). Over time, geese abandoned older marsh areas after invasion of *Atriplex* and *Artemisia maritima*, while newly formed marsh vegetations were readily exploited. This is not to say that the herbivores acting in unison have no impact whatever on the rate of change,

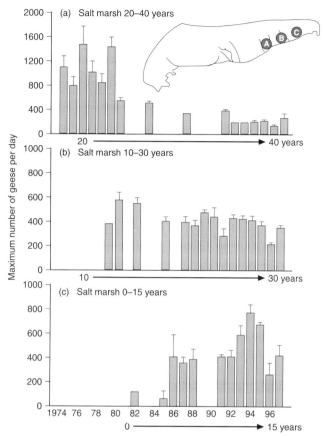

Figure 9.8 Eastwards shift of the prime grazing area for brent geese at Schiermonnikoog as shown by counts during the spring grazing peak (not all areas were counted in all years). The counts reflect gradual vegetation succession towards less palatable plants (the island is growing at the eastern end, ages of salt-marsh areas in years given; the series embraces stages in the first 40 years of plant colonization).

exclosure studies have revealed. In 1991 and 1994 a series of exclosures were placed in the prime goose area, allowing entry by hares but excluding geese, or excluding all herbivores (a variant allowing entry by geese but excluding hares has defeated our ingenuity so far).

A striking effect of excluding all herbivores is the rapid expansion of *Atriplex* that forms a high canopy whereas in both control and goose-free plots, cover of these bushy plants remained low (Fig. 9.9). From analysis

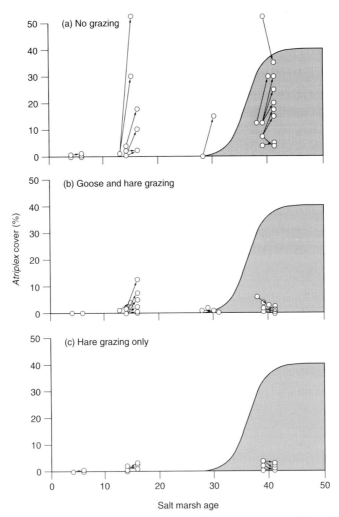

Figure 9.9 Exclosure experiments on the Schiermonnikoog salt marsh at the level of the intensive study plot (young successional stage) illustrating the runaway process of succession in the absence of herbivores (a). The grey area depicts the natural abundance of *Atriplex portulacoides*, whereas the open symbols show the trajectory (arrows) followed after exclusion of all herbivores. The panels (b) and (c) give *Atriplex* cover in adjacent plots where both geese and hare grazing occurred (b), or with only hare grazing. Natural *Atriplex* abundance was reconstructed on the basis of multiple relevees all along the time axes. From van der Wal (1998).

of hare droppings we found that *Atriplex* is an important component of the diet, particularly in the period of vegetative scarcity in winter and early spring when the hare gnaw off the woody stems completely. Small-scale experiments with individual plants of *Atriplex* transplanted into the outer marsh revealed that particularly in young plants not protected by wire cages mortality rate was twice as high relative to ungrazed plants. This inhibition of *Atriplex* in the early colonization phase is a potent means of resetting the successional clock, and depends solely on the hare. Taking the data presented in Fig. 9.8, we conclude that on account of hare grazing in winter and early spring when the *Atriplex* stems are highly susceptible to damage, the marsh remains virtually free of *Atriplex* and hence open to exploitation by brent geese for at least 25 years longer than in the absence of hare. How sensitive brent geese are for the encroachment by *Atriplex* was revealed by a small-scale experiment. When *Atriplex* bushes were transplanted to the younger marsh frequented by the geese, grazing pressure in the immediate vicinity ($4\,m^2$ including the $1\,m^2$ occupied by the bush transplants) was significantly reduced compared with adjacent controls. The converse treatment, removing *Atriplex* on the older marsh by clipping square-metre plots similar to the bare patches caused by hare grazing in early spring, resulted in a higher grazing intensity by the geese (compared with uncut control *Atriplex* patches nearby). Outcome of these manipulations is illustrated in Fig. 9.10.

Geese exploit the marsh only during the period of graminoid abundance, but these experiments show that hare grazing earlier in the season has an important facilitating effect on goose usage. Indeed, extrapolating from detailed vegetation mapping of the marsh zone currently exploited by brent geese in spring allows the conclusion that in the absence of hare grazing almost half of this core area would already have entered the phase of *Atriplex* encroachment, thus excluding the geese. This is not the full story, since there is a competitive interaction between the brent goose, barnacle goose and hare on the *Festuca* greens on the upper marsh, but our exclosures have not yet elucidated how this affects persistence of the grazing patches.

The conclusion from this work on Schiermonnikoog is that the marsh succession at the higher elevations marches relentlessly on to reach an end phase dominated by the grass *Elymus* that is unpalatable to the small native herbivores. At the intermediate elevations that form the core grazing area for the brent geese the vegetation retains its features of a grazing lawn only if severe hare grazing in the off season prevents the encroachment by the

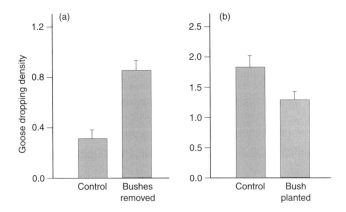

Figure 9.10 Brent goose dropping density (no. per m² per week) as a measure of goose visitation in experimental plots (see text) where *Atriplex* bushes were either removed (a) or planted (b) in the last week of March and goose droppings counted until the close of May. Removal experiments were performed in older marsh where goose grazing was low, whereas bushes were planted in a core feeding area on the younger marsh. From van der Wal (1998).

bushy *Atriplex*. The lower margins of the marsh, regularly inundated and locally still accreting, support pockets of *Puccinellia* allowing exploitation by geese, but these can persist only by the aid of abiotic influences. These trends so unfavourable to the geese can be reversed by cattle grazing, as has been shown in the western part of the marsh. Here, the *Elymus*, *Atriplex* and *Artemisia* are removed by the cattle, and *Festuca* and *Puccinellia* restored locally, with *Plantago* and *Triglochin* patches along the foreshore (Olff *et al.* 1997). From the evolutionary point of view this raises the question whether in prehistoric times a native ungulate was a member of the grazing community on the salt marsh. This is rather an academic question, since the dynamic state of the coastline before large-scale coastal defences were erected by humans about 1000 years ago must have guaranteed very extensive areas of young marsh. Food conditions must have been favourable for exploitation by small herbivores such as the barnacle goose, to say nothing of the extensive beds of eelgrass *Zostera* which, in former times, formed the staple food of the brent goose in its wintering area, as is still true in some parts of the range.

The closest parallel to this work is the study of the salt marsh on the Hudson Bay (Manitoba, Canada) where the lesser snow geese *Anser caerulescens caerulescens* of the La Perouse Bay colony feed. By erecting exclosures, Bazely and Jefferies (1986) demonstrated that removing the geese (the only important grazers in this system) accelerated plant succession. Indeed, intense grazing of the salt-marsh swards dominated by *Puccinellia phryganoides* and *Carex subspathacea* by the parent geese with their goslings enhances the growth of *Puccinellia* through the accelerated provision of nutrients via goose droppings. Moreover, the bare micropatches that result from goose grazing are colonized by cyanobacteria that fix nitrogen and hence also make a contribution to the nutrient pool (as summarized by Bazely & Jefferies 1997). The story is complicated by the early but intensive spring grubbing by snow geese, many of them on passage to more northerly colonies. That this grubbing results in a deterioration of the marsh habitat is now well documented (Kotanen & Jefferies 1997). The current mismatch between goose numbers and foraging habitat in this subarctic area results from unintentional supplementary feeding of wintering geese on a continental scale due to changes in agricultural practices. This complication should not detract from the observation that quite apart from the grubbing syndrome the conventional non-destructive summer grazing of the geese effectively arrests plant succession at an early stage.

Gauthier *et al.* (1995 and personal communication) have undertaken similar exclosure experiments at Bylot Island in the High Arctic of Canada, in this case in a tundra area exploited by greater snow geese (*Anser caerulescens atlantica*). Although the geese were found to remove a high proportion of the above-ground production, there is no evidence for overcompensatory growth. Indeed, Beaulieu *et al.* (1996) undertook experimental grazing trials with hand-raised goslings and have shown that neither a single nor repeated grazing treatment affected net above-ground primary production of the two graminoids tested (a grass, *Dupontia fisheri* and a sedge, *Eriophorum scheuchzeri*). Although both plant species were able to maintain production at a level similar to that of ungrazed controls, this was only possible at the cost of a reduction in below-ground reserves (as evidenced by lower contents of non-structural carbohydrates in the underground parts of the grazed plants following the trials). At the Bylot study site, input of nutrients via the droppings failed to have a positive effect on the graminoids, most likely because the nutrients were pre-empted by the moss carpet in which the food plants are embedded. Parallel results have been obtained at Ny Aalesund (on the west coast of Spitsbergen) where again the nutrients leaching from the goose pellets (in this case of barnacle geese) do not enhance graminoid production, but increase the moss carpet instead (Bakker & Loonen 1998).

291

Turning now to another goose brood-rearing area, Mulder *et al.* (1996) have tried to tease apart the factors determining the distribution of arrow-grass *Triglochin palustris* in the extensive salt marshes of the Yukon–Kuskokwim delta of south-western Alaska. At that study site, *Triglochin* is a preferred forage species by both *Branta bernicla nigricans* and *B. canadensis minima*, especially important for goslings where this plant often dominates the diet. Transplant experiments combined with fertilization treatments were designed to discover why the arrowgrass is restricted to such a narrow vertical range in the marsh. Paradoxically, fertilization (NPK) had a negative effect on the transplanted *Triglochin*, although the same treatment resulted in increased productivity of graminoids in the same setting. Arrowgrass plants occur singly, however, and Mulder and associates speculate that neighbour effects intervene, and they suggest that the upper limits in the marsh are set by a combination of light competition and the highly selective foraging by geese, effectively weeding *Triglochin* out. Further work is called for to elucidate how arrowgrass maintains a toehold in the lower marsh despite the heavy grazing. Clearly this patchy plant does not fit the simplified view of a happy commensalism with the grazer, triggered by the nutrient loop.

Conclusions

Despite the unabated interest in the interactions between vertebrate herbivores and their food plants it is somewhat dismaying to reflect that many of the questions posed at the 10th Symposium of the British Ecological Society, entitled *Animal Populations in Relation to their Food Resources*, and held in Aberdeen almost 30 years ago (Watson 1970), are still being pursued today. The generalization that many ungulates are adapted to early successional stages and are unable to arrest vegetation change despite heavy grazing pressures is echoed today (Davidson 1993). At the same time, the possibility is considered that under conditions of pasture heterogeneity patches of preferred species may attract consistent grazing and cycles of regrowth thus promoting vegetation change beneficial to the grazer, at least locally. This idea is reminiscent of the 'hot spots' discussed in the recent reappraisal of the work in the Serengeti (McNaughton & Banyikwa 1995). This does not mean that our science is at a standstill but rather that relatively few reports on really long-term studies on vegetation change in relation to utilization by wild herbivores have come to our attention in the interim, and that these do not lend themselves to simple abstraction. Work in the Serengeti including fence experiments (Belsky 1992) register loss in diversity when herbivores were excluded. By contrast, exclusion of lagomorphs (jackrabbits and cottontails) from creosote bush semidesert in New Mexico over a 50-year period failed to register changes that might be interpreted as successional (Gibbens *et al.* 1993). The most striking effects were increases in coverage of plant species known to be preferred forage species.

The painstaking efforts of Jefferies and his group working in the Hudson Bay salt marshes have uncovered an indisputable case of plant growth stimulation under the influence of herbivory by lesser snow geese, powered by nutrient inputs through the rapidly leaching droppings. When this work was replicated at the High Arctic site Bylot Island with the closely related greater snow goose, this feedback loop was not found, as the nutrient inputs did not reach the graminoids the geese fed upon, but were pre-empted by the ubiquitous moss carpet (Gauthier 1995; Beaulieu *et al.* 1996).

One facet of the original idea of how grazing herbivores 'manipulate' their food plants for their future benefit has been upheld, and this concerns the direction of the impact of the grazers on plant succession. In a recent review, Davidson (1993) collates results from 13 situations where herbivores (ranging from tortoises to elephants, but including geese) exploit graminoids, and in all cases the impact of the grazers was to reverse or halt plant succession. This finding contrasts with the vertebrate browsers (e.g. moose, *Alces alces*) where, according to Davidson's analysis, herbivory tends to hasten succession. Grazers are thus in this sense manipulators, and browsers followers of plant succession.

Following vegetation change is a tall order as has been detailed by Belsky (1992). Adding the complexity of possible control by herbivores leaves no alternative but to experiment within the constraints of the system studied, and above all to maintain the effort through the years. Our own attempts to understand plant–goose interactions in the early successional stages of the Schiermonnikoog salt marsh have dispelled notions of sweeping generalizations and tend rather to shrink the questions into smaller steps within our powers of discrimination. As we have seen, on the time scale of one season, cyclic grazing confers a measurable benefit to the geese in greatly extending the period of high nutritive status of the forage plant (barnacle geese × *Festuca*, see Fig. 9.2). Within the season, grazing cycle length seems adjusted to maximize intake for the herbivore (brent geese × *Plantago,* see Fig. 9.4). In our system the persistence of the plants vital to the grazers is not ensured by continued herbivory and ongoing vegetation succession eventually excludes the favoured plants and the herbivores that depend on them (Fig. 9.8). Although the geese are seasonal (spring) grazers, the resident hares, with more efficient digestive abilities, can survive by exploiting more woody vegetation in the off-season when they presumably bridge periods of deficit by drawing on body reserves. This winter and early spring hare grazing is a powerful factor retarding vegetation succession, as their heavy utilization of *Atriplex* at that time maintains the lower marsh in a grassy state and protracts the phase of goose utilization by several decades (see Fig. 9.9). Despite these facilitation effects the herbivores cannot prevent the vegetation at a specific site on the marsh from leaving them behind on the longer term. On the scale of the island, accretion at the eastern end has so far compensated for the onward march of plant succession, and the herbivores survive by being mobile and shifting along with the favoured vegetation belts. This opportunistic strategy is a logical attribute for herbivores inhabiting the highly dynamic coastal marsh systems, and their versatility has enabled the three species studied to survive in the human-modified agricultural systems of the present day (Fig. 9.11).

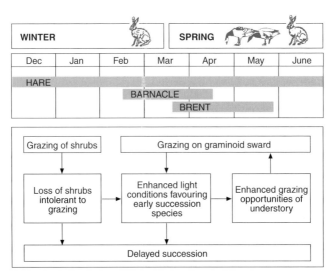

Figure 9.11 Overview of the major effects of vertebrate herbivores (the hare and two goose species) on plant succession on the Schiermonnikoog marsh.

Finally, three points on research strategies deserve attention. Firstly, exclosure techniques though they can be traced back at least as far as Darwin (as pointed out by McNaughton, 1984) are still underused. We agree with Jefferies *et al.* (1994) that a circumpolar network would be extremely valuable, and feel that the original plots should be large enough to allow slices to be returned to free herbivore access after a given time period to test the reversibility of vegetation change. Given the time scale of vegetation change, setting up the network presumes a long-term commitment by the participants. Secondly, experimentation with tame herbivores in pen situations on the natural vegetation have great potential, and could well be combined with the long-term exclosure work. Thirdly, a better integration with work on abiotic factors determining plant distribution and production would help to define the role of herbivores in enhancing persistence of their target plants. The techniques seem to be in place (e.g. Thornton *et al.* 1993) but a close-knit project embracing the physiology and ecology of both plant and herbivore *in situ* has not yet come about.

Acknowledgements

We thank the management authority (Natuurmonumenten) for permission to work on Schiermonnikoog, recall the many observers who helped in the field and record our debt to the plant ecologists (especially Jan Bakker and Han Olff) in our team effort on the marsh. Joost Tinbergen carried out the model computations depicted in Fig. 9.4. The manuscript benefitted greatly from the critical commentary by Kate Lessells and Bob Jefferies.

References

Bakker, C. & Loonen, M.J.J.E. (1998). The influence of goose grazing on the growth of *Poa arctica*: overestimation of overcompensation. *Oikos*, **82**, 459–466.

Bazely, D.R. & Jefferies, R.L. (1986). Changes in the composition and standing crop of salt-marsh communities in response to the removal of a grazer. *Journal of Ecology*, **74**, 693–706.

Bazely, D.R. & Jefferies, R.L. (1997). Trophic interactions in arctic ecosystems and the occurrence of a terrestrial trophic cascade. In *Ecology of Arctic Environments* (Ed. by S.J. Woodin & M. Marquiss), pp. 183–207. Blackwell Science Ltd, Oxford.

Beaulieu, J., Gauthier, G. & Rochefort, L. (1996). The growth response of graminoid plants to goose grazing in a High Arctic environment. *Journal of Ecology*, **84**, 905–914.

Bélanger, L. & Bédard, J. (1994). Role of ice scouring and goose grubbing in marsh plant dynamics. *Journal of Ecology*, **82**, 437–445.

Belsky, A.J. (1986). Does herbivory benefit plants? A review of the evidence. *American Naturalist*, **127**, 870–892.

Belsky, A.J. (1992). Effects of grazing, competition, disturbance and fire on species composition and diversity in grassland communities. *Journal of Vegetation Science*, **3**, 187–200.

Bergström, R. & Danell, K. (1995). Effects of simulated summer browsing by moose on leaf and shoot biomass of birch, *Betula pendula. Oikos*, **72**, 132–138.

Cargill, S.M. & Jefferies, R.L. (1984). The effects of grazing by lesser snow geese on the vegetation of a sub-arctic salt marsh. *Journal of Applied Ecology*, **21**, 669–686.

Coppock, D.L., Detling, J.K., Ellis, J.E. & Dyer, M.I. (1983a). Plant–herbivore interactions in a North American mixed-grass prairie I. Effects of black-tailed prairie dogs on intraseasonal aboveground biomass and nutrient dynamics and plant species diversity. *Oecologia*, **56**, 1–9.

Coppock, D.L., Ellis, J.E., Detling, J.K. & Dyer, M.I. (1983b). II. Responses of bison to modification of vegetation by prairie dogs. *Oecologia*, **56**, 10–15.

Crawley, M.J. (1983). *Herbivory, the Dynamics of Animal–Plant Interactions.* Blackwell Scientific Publications, Oxford.

Crawley, M.J. (1997). Plant–herbivore dynamics. In *Plant Ecology* (Ed. by M.J. Crawley), pp. 401–474. Blackwell Science Ltd, Oxford.

Danell, K., Ericson, L. & Jakobsson, K. (1981). A method for describing former fluctuations of voles. *Journal of Wildlife Management*, **45**, 1018–1021.

Davidson, D.W. (1993). The effects of herbivory and granivory on terrestrial plant succession. *Oikos,* **68**, 23–35.

Drent, R.H. & Prins, H.H.T. (1987). The herbivore as prisoner of its food supply. In *Disturbance in Grasslands* (Ed. by J. van Andel, J.P. Bakker & R.W. Snaydon), pp. 131–148. Junk, Dordrecht.

Ebbinge, B. (1992). Regulation of numbers of dark-bellied brent geese *Branta bernicla* on spring staging sites. *Ardea*, **80**, 203–228.

Frank, D.A. & Evans, R.D. (1997). Effects of native grazers on grassland N cycling in Yellowstone National Park. *Ecology*, **78**, 2238–2248.

Frank, D.A. & McNaughton, S.J. (1993). Evidence for the promotion of above-ground grassland production by native large herbivores in Yellowstone National Park. *Oecologia*, **96**, 157–161.

Fryxell, J.M. (1995). Aggregation and migration by grazing ungulates in relation to resources and predators. In *Serengeti II: Dynamics, Management, and Conservation of an Ecosystem* (Ed. by A.R.E. Sinclair & P. Arcese), pp. 257–273. Chicago University Press, Chicago.

Gauthier, G., Hughes, R.J., Reed, A., Beaulieu, J. & Rochefort, L. (1995). Effect of grazing by greater snow geese on the production of graminoids at an Arctic site (Bylot Island, NWT, Canada). *Journal of Ecology*, **83**, 653–664.

Gibbens, R.P., Havstad, K.M., Billheimer, D.D. & Herbel, C.H. (1993). Creosotebush vegetation after 50 years of lagomorph exclusion. *Oecologia*, **94**, 210–217.

Hobbs, N.T. & Swift, D.M. (1988). Grazing in herds: when are nutritional benefits realized? *American Naturalist*, **131**, 760–764.

Holland, E.A., Parton, W.J., Detling, J.K. & Coppock, D.L. (1992). Physiological responses of plant populations to herbivory and their consequences for ecosystem nutrient flow. *American Naturalist*, **140**, 685–706.

Inglis, I.R. & Lazarus, J. (1981). Vigilance and flock size in Brent Geese: the edge effect. *Zeitschrift für Tierpsychologie*, **57**, 193–200.

Jefferies, R.L., Klein, D.R. & Shaver, G.R. (1994). Vertebrate herbivores and northern plant communities: reciprocal influences and responses. *Oikos*, **71**, 193–206.

Kotanen, P.M. & Jefferies, R.L. (1997). Long-term destruction of wetland vegetation by lesser snow geese. *Ecoscience*, **4**, 1895–1898.

Leader-Williams, N. (1988). *Reindeer on South Georgia: the Ecology of an Introduced Population*. Cambridge University Press, Cambridge.

Manseau, M., Huot, J. & Crete, M. (1996). Effects of summer grazing by caribou on composition and productivity of vegetation: community and landscape level. *Journal of Ecology*, **84**, 503–513.

McNaughton, S.J. (1979). Grassland–herbivore dynamics. In *Serengeti: Dynamics of an Ecosystem* (Ed. by A.R.E. Sinclair & M. Northon-Griffiths), pp. 46–81. Chicago University Press, Chicago.

McNaughton, S.J. (1984). Grazing lawns: animals in herds, plant form, and coevolution. *American Naturalist*, 124, 863–886.

McNaughton, S.J. (1985). Ecology of a grazing ecosystem: the Serengeti. *Ecological Monographs*, **55**, 259–294.

McNaughton, S.J. & Banyikwa, F.F. (1995). Plant communities and herbivory. In *Serengeti II: Dynamics, Management, and Conservation of an Ecosystem* (Ed. by A.R.E. Sinclair & P. Arcese), pp. 49–70. Chicago University Press, Chicago.

Mulder, C.P.H. & Harmsen, R. (1995). The effect of muskox herbivory on growth and reproduction in an arctic legume. *Arctic and Alpine Research*, **27**, 44–53.

Mulder, C.P.H., Ruess, R.W. & Sedinger, J.S. (1996). Effects of environmental manipulations on *Triglochin palustris:* implications for the role of goose herbivory in controlling its distribution. *Journal of Ecology*, **84**, 267–278.

Oksanen, L. (1983). Trophic exploitation and arctic phytomass patterns. *American Naturalist*, **122**, 45–52.

Oksanen, L. & Virtanen, R. (1995). Topographic, altitudinal and regional patterns in continental and suboceanic heath vegetation of northernmost Fennoscandia. *Acta Botanica Fennica*, **15**, 1–80.

Oksanen, L., Aunapuu, M., Oksanen, T., Schneider, M., Ekerholm, P., Lundberg, P.A., Armuluk, T., Aruoja, V. & Bondestad, L. (1997). Outlines of food webs in a low Arctic tundra landscape in relation to three theories on trophic dynamics. In *Multitrophic Interactions in Terrestrial Systems* (Ed. by A.C. Gange & V.K. Brown), pp. 351-373. Blackwell Science Ltd, Oxford.

Olff, H., Bakker, J.P., Platerink, R.J., Van Wijnen, H. & De Munck, W. (1997). Vegetation succession and herbivory on a salt marsh. *Journal of Ecology*, **88**, 799–814.

Parsons, A.J., Leafe E.L., Collett, B., Penning, P.D. & Lewis, J. (1983). The physiology of grass production under grazing II. Photosynthesis, crop growth and animal intake of continuously-grazed swards. *Journal of Applied Ecology*, **20**, 127–139.

Pastor, J. & Cohen, Y. (1997). Herbivores, the functional diversity of plant species, and the cycling of nutrients in ecosystems. *Theoretical Population Biology*, **51**, 165–179.

Pastor, J. & Naiman, J. (1992). Selective foraging and ecosystem processes in boreal forests. *American Naturalist*, **139**, 690–705.

Prins, H.H.T. (1996). *Ecology and Behaviour of the African Buffalo: Social Inequality and Decision Making.* Chapman & Hall, London.

Prins, H.H.T., Ydenberg, R.C. & Drent, R.H. (1980). The interaction of brent geese (*Branta bernicla*) and sea plantain (*Plantago maritima*) during spring staging: field observations and experiments. *Acta Botanica Neerlandica*, **29**, 585–596.

Prop, J. (1991). Food exploitation patterns by brent geese *Branta bernicla* during spring staging. *Ardea*, **79**, 331–342.

Prop, J. & Deerenberg, C. (1991). Spring staging in brent geese *Branta bernicla*: feeding constraints and the impact of diet on the accumulation of body reserves. *Oecologia*, **87**, 19–28.

Prop, J. & Loonen, M.J.J.E. (1986). Goose flocks and food exploitation: the importance of being first. In *Acta XIX Congressus Internationalis Ornithologici*, Ottawa, Canada, pp. 1878–1887.

Rowcliffe, J.M., Watkinson, A.R., Sutherland, W.J. & Vickery, J.A. (1995).

Cyclic winter grazing patterns in brent geese and the regrowth of salt-marsh grass. *Functional Ecology*, 9, 931–941.

Seldal, T., Andersen, K.-J. & Högstedt, G. (1994). Grazing-induced proteinase inhibitors a possible cause for lemming population cycles. *Oikos*, 70, 3–11.

Sinclair, A.R.E. (1995). Equilibria in plant–herbivore interactions. In *Serengeti II: Dynamics, Management, and Conservation of an Ecosystem* (Ed. by A.R.E. Sinclair & P. Arcese), pp. 91–113. Chicago University Press, Chicago.

Thornton, B., Millard, P., Duff, E.I. & Buckland, S.T. (1993). The relative contribution of remobilization and root uptake in supplying nitrogen after defoliation for regrowth of laminae in four grass species. *New Phytologist*, 124, 689–694.

van de Koppel, J., Huisman, J., van der Wal, R. & Olff, H. (1996). Patterns of herbivory along a productivity gradient: an empirical and theoretical investigation. *Ecology,* 77, 736–745.

van der Wal, C.F.R. (1998). *Defending the Marsh: herbivores in a dynamic coastal environment*. PhD thesis, University of Groningen.

Watson, A. (Ed.) (1970). *Animal Populations in Relation to their Food Resources*. Blackwell Scientific Publications, Oxford.

White, R.G., Bunnell, F.L., Garre, E., Skogland, T. & Hubert, B. (1981). Ungulates on Arctic ranges. In *Tundra Ecosystems: A Comparative Analysis* (Ed. by L.C. Bliss, O.W. Heal & J.J. Moore), pp. 397–483. Cambridge University Press, Cambridge.

Wilmshurst, J.F., Fryxell, J.M. & Hudson, R.J. (1995). Forage quality and patch choice by wapiti (*Cervus elaphus*). *Behavioural Ecology*, 6, 209–217.

Ydenberg, R.C. & Prins, H.H.T. (1981). Spring grazing and the manipulation of food quality by barnacle geese. *Journal of Applied Ecology*, 18, 443–453.

Chapter 10

Herbivores, nutrients and trophic cascades in terrestrial environments

R.L. Jefferies[1]

Summary

Although consumption and digestion of plants by herbivores can result in increased nutrient turnover in grazed pastures, soil microbial activity and edaphic conditions may constrain decomposition processes and produce a nutrient bottleneck that limits plant growth. As a result, herbivores are frequently 'prisoners of their food supply', but herbivore mobility ensures that foragers are able to exploit high-quality forage when and where it occurs, although this may incur predation costs, particularly for small mammals. Return times of herbivores to preferred vegetation patches may be of the order of days or years, depending on rates of regrowth. The grazing episodes reinforce the maintenance of vegetational heterogeneity. Overcompensation of above-ground primary production in response to defoliation is unusual and is dependent on soil conditions and the presence of particular traits in both the herbivore and the forage species. A given herbivore can act as a keystone species, but its effects on vegetation are density dependent, and as the numbers increase the spatial and temporal scales of the effects change in a discontinuous manner. Where simple linear food chains prevail, increased numbers of herbivores may override bottom-up controls and this can lead to terrestrial trophic cascades. Examples of trophic cascades are examined, and collectively they indicate that both biotic and abiotic processes operate in tandem to bring about discontinuities in community and alternative vegetation states. The effects are non-linear and are rarely predictable a priori.

1 Department of Botany, University of Toronto, 25 Willcocks Street, Toronto, Ontario, Canada M5S 3B2

Introduction

Herbivores in their roles as consumers of vegetation and promoters of nutrient cycling affect both species assemblages and amounts of available biomass in plant communities (Huntly 1991). The effects not only depend on strong trophic interactions, including interactions with micro-organisms, but they also trigger non-trophic abiotic processes that alter resources contributing to changes in vegetation states characterized by different species assemblages (*sensu* Noy Meir 1975; Westoby 1979; Westoby *et al.* 1989). Herbivores/consumers and their associated gut flora act as transducers of different chemical states of essential elements for biological activity. The animals often are highly selective in their foraging choices, and this together with digestion, assimilation and excretion result in a different elemental composition in faecal material compared with that of ingested material (cf. Sterner 1986). For example, the urine of large herbivores is rich in nitrogen, but devoid of phosphorus. The structure of combined forms of carbon, nitrogen and phosphorus also change as ingested material passes through the gut (Van Soest 1994). However, compounds, such as lignin, and some secondary chemicals, either are not broken down, or else adversely affect rates of digestion (Bryant & Kuropat 1980; Mattson 1980; Bryant *et al.* 1983; Crawley 1983). These same compounds also influence the rate of decay of plant litter, and consequently the rate of release of limiting nutrients by decomposers (Melillo *et al.* 1982; Wedin & Pastor 1993). In contrast, the litter of palatable food plants, which are largely devoid of these compounds, is often rapidly mineralized by microorganisms (Hobbie 1992). Hence, these organisms in the gut of animals and in the soil effectively act as pacemakers of 'nutrient pumps' activated by the foraging of herbivores and detritivores. The outcome of these biogeochemical processess on soil nutrient availability, plant growth and the subsequent changes in plant species assemblages can be considerable (Vitousek 1985; Coley *et al.* 1985; Wedin & Tilman 1990).

Fluxes of nutrients from these herbivore/consumer 'pumps' contribute to ecosystem processes controlling primary productivity, plant assemblages and succession. However, these fluxes cannot be examined in isolation from related events occurring in the environment. Edaphic conditions may slow rates of microbial activity. In addition, an increased number of herbivores may override the regulatory bottom-up controls of resources and ecosystem processes imposed by the interactions of herbivores with soil organisms. Of particular interest are situations where foraging in terrestrial systems can result in strong top-down interactions originating in upper trophic levels that give rise to 'trophic cascades' that promote herbivore abundance (*sensu* Paine 1969; Carpenter *et al.* 1985). As discussed later, these may affect adversely primary producers and lead to alternative vegetation states brought about by 'runaway consumption' of existing vegetation when increasing numbers of herbivores go unchecked. The cascades, in addition, frequently produce non-trophic changes in edaphic properties, which are intimately linked to establishment of alternative vegetational states.

The strength of these biotic and abiotic links can be described by positive and negative feedback processes that ultimately control the fluxes of resources that are made available for plant and microorganism growth (DeAngelis *et al.* 1986; DeAngelis 1992; Wilson & Agnew 1992). Usually in

Table 10.1 Characteristics of trophic cascades and food webs with reference to terrestrial environments. The list is not exhaustive and not all characteristics may be represented in a given situation. Based, in part, on Strong (1992) and Polis and Strong (1996).

Trophic cascade	Food webs
Linear food chains	Reticulate food web
Distinct trophic levels	Absence of distinct trophic levels
No omnivory	Omnivory
Species diversity low	Species diversity high
Functional groups absent, or poorly represented	Presence of functional groups
'Runaway' consumption, particularly of plants	No 'runaway' consumption
Little accumulation of plant litter	Plant litter accumulates
Positive feedbacks sustained	Positive feedbacks transient, negative feedbacks sustained
Weak defences of primary producers	Strong defences of primary producers
Coupled non-trophic interactions reinforce biotic effects	Coupled non-trophic interactions reinforce biotic effects
Threshold responses leading to an alternative vegetation state and habitat change	
Recovery to original vegetation state long term, or else change is irreversible	
Cascade often the result of anthropogenic causes	

most terrestrial systems a series of coupled feedbacks operate in tandem, in which the effects of strong, but transient, interactions initiated by upper-level trophic organisms are weakened, and negative feedbacks act to dampen the effects of the strong interactions (McQueen 1990; DeMelo *et al.* 1992), so that trophic cascades and runaway consumption of plants by herbivores do not occur. Reticulate food webs consisting of many species predominate in terrestrial environments, in which distinct trophic levels are absent because of omnivory. In addition, different member species of a functional group within the reticulate network are capable of subsuming the functional role of a declining species in buffering top-down effects via these negative feedbacks (McQueen 1990; Power 1992; Strong 1992; Polis & Strong 1996) (Table 10.1). Simple linear food chains with distinct trophic levels, each represented by one or perhaps two species, are uncommon.

Initially in this chapter the dual effects of bottom-up processes that either promote or constrain resource availability are examined. The latter limit primary production, so that herbivores are very much 'prisoners of their food supply' (Drent & Prins 1987) as a result of density-dependent processes. Animal mobility and the presence of vegetational heterogeneity allow herbivores to exploit food pulses at different spatial and temporal scales. In later sections the effects of increasing numbers of herbivores that result in the loss of bottom-up controls are discussed in relation to trophic cacades. A number of case studies are examined. The effects are dependent on both the population sizes of herbivores and the spatial scales over

---

Table 10.2 Possible causes of declines in net nitrogen mineralization rates in soils.

---

1 Changes in the physical and chemical properties of plant litter that result in slower rates of litter breakdown (e.g. lignification, secondary chemical compounds)

2 Increased C : N ratios of plant litter, a consequence of higher amounts of structural biomass and maintenance tissue relative to photosynthetic tissue, results in net immobilization of nitrogen

3 Woody litter has a low surface area compared with fragmented leaf litter

4 There is efficient retranslocation of nutrients from senescing leaves into the plant that results in a high C : N ratio in leaf litter

5 Declining amounts of available soil nitrogen lead to a lower leaf area index

6 Build up of soil organic matter and reductions in bulk density reduce amounts of available nitrogen per unit volume of soil

7 Organic material in ageing soil casts produced by soil invertebrates is unavailable to microorganisms

---

which animals forage. As the effects describe the actions of a single species, they impinge on the keystone species concept, which is also discussed in this chapter. Many of the interactions involving herbivores are coupled to nitrogen transfer within ecosystems. The availability of this element, which often limits primary production, particularly where the C : N ratio is high (see below) (Table 10.2), is largely regulated by biological processes, hence its availability can be used as an index of the strength of interactions between organisms.

Nutrient pumps, microbial pacemakers and vegetational change

Individual animal species can have effects on resource availability that are disproportionately large in relation to their biomass or abundance (Sterner 1986, 1990; Huntly & Inouye 1988; Jefferies 1988a,b; Naiman *et al.* 1988; Whicker & Detling 1988). The effects reflect both the strength of trophic interactions between herbivores and other groups of organisms, and abiotic changes that result from these interactions. Within terrestrial environments vegetational processes give rise to spatial and temporal heterogeneity in plant assemblages (White & Pickett 1985) and in nutrient dynamics (Vitousek 1985) which are exploited by herbivores. Much of the heterogeneity reflects the outcome of differential growth responses of plants to variations in edaphic properties of soils (Belsky 1995), such as the availability of nitrogen and phosphorus (Chapin 1980), which are often the most limiting elements for plant growth in different environments (Vitousek 1985). The activities of larger herbivores (>1 kg), in particular, contribute to the maintenance of this heterogeneity, which include their effects on the rates of net mineralization of nitrogen, an indirect measure of primary productivity (Vitousek *et al.* 1989), and on the transfer of nutrients between sites. For example, in the latter case wildfowl can move large quantities of nutrients from farmland where they feed to managed wetlands where they roost (Post *et al.* 1998). At a site in New Mexico rates of nutrient deposition from faecal matter for an entire reserve (500 ha) may reach 300 kg N day^{-1} and in excess of 30 kg P day^{-1} (Post *et al.* 1998). These inputs promote the growth of cyanobacteria at the expense of green algae in drainage systems.

As indicated above, consumers influence the quality (e.g. low C : N ratio) and quantity of plant litter at different scales of space and time (Mills & Sinha 1971; Moore 1988; Wedin & Tilman 1990; Lavelle *et al.* 1995). At C : N ratios above 20–25, mineralization of soil and faecal nitrogen usually results in net immobilization of this element. These different ratios of $C:N$, $C:P$ and $N:P$ vary widely in both soils and water bodies, and their effects on species assemblages can be considerable, particularly in aquatic systems (Sterner 1986, 1990; Wedin & Tilman 1990; Pastor & Naiman 1992; Wedin & Pastor 1993; Vanni & Layne 1997; Vanni et al. 1997). In most soils only about 4% or less of the total nitrogen pool is in a labile state and is able to turn over rapidly (Hart & Gunther 1989; Nadelhoffer et al. 1992). This fraction is poorly characterized, but a large percentage is thought to consist of microbial biomass and non-recalcitrant plant material derived from tissues low in fibre and secondary chemicals in which the $C:N$ ratio is between 5 and 20. In spite of the presence of large amounts of organic nitrogen in soils, much of it is recalcitrant to decomposition processes. For example, the sequestration of nitrogen in tannin–protein complexes in dead beech leaves and roots accounts for 70% and 60%, respectively, of the total nitrogen in the different litter types (Toutain 1987). In addition, although faecal material from herbivores and soil invertebrates may have a favourable $C:N$ ratio resulting in the net mineralization of nitrogen, other factors, including abiotic processes, can be expected to limit the rates of mineralization (Table 10.2). Many autotrophic bacteria responsible for nitrification are slow-growing, even under favourable conditions (Sprent 1987). The energy yield of nitrification is low ($\Delta G_o' = -83.8\,\text{kcal}\,\text{mol}^{-1}$ of NH_4^+ converted to NO_3^-; Fenchel & Blackburn 1979), and the organisms are very sensitive to the periodic drying-out of soils which results in a marked decline in overall net rates of nitrogen mineralization. Hence, the organisms may be in dormant stages for long periods of time in spite of the presence of labile pools of nitrogen ('the sleeping beauty paradox': Jenkinson & Ladd 1981; Lavelle et al. 1995). Estimates of the turnover time of microbial biomass in the field compared with that in the laboratory are 1000 to 10 000 times slower (Clarholm & Rosswall 1980; Chaussod et al. 1988).

These constraints on microbial activity, in effect, create a nutrient bottleneck that tends to decouple the effect of consumers acting as a nutrient pump in accelerating the movement of nutrient fluxes from the soil to plants. In addition, although soil animals graze on micro-organisms, act as predators, restructure the soil and translocate substrates in promoting nutrient turnover (Bengtsson et al. 1995), some nutrients may be made unavailable by these activities. For example, earthworms and other macroinvertebrates sequester soil organic matter in casts (macroinvertebrates can digest 1000 tonnes of dry soil and plant litter per hectare annually; Lavelle *et al.* 1995). Initially, there is a sharp increase in the rate of mineralization during, and immediately after, digestion but the compact structure of ageing casts containing mineral nutrients and microbial biomass results in the blocking of mineralization on a time scale of months to years (Lavelle & Martin 1992). Hence, except for transient pulses of nutrients in faecal matter, these bottlenecks in rates of net mineralization can lead to a temporal decoupling between consumers, soil microbial activity and plant growth in maintaining fluxes of nutrients as a result of negative feedbacks.

The pattern of growth of forage plants in grasslands appears to be adapted to this decoupling of nutrient flow. Where large vertebrates forage in grasslands, most graminoid species are perennial, but above-ground growth is highly seasonal. Nutrients necessary to sustain short, rapid pulses of growth are drawn from below-ground storage organs and/or shoot bases, the plants do not appear to be highly dependent on available external supplies of nutrients during the period of rapid growth. The foraging strategy of large herbivores is to exploit the pulses of new growth in which the protein:biomass ratio is favourable and then to move to fresh pastures (Breman & de Wit 1983; McNaughton 1985; Prins 1996). The trickle of available nitrogen (and phosphorus) as a result of net mineralization is utilized by plants during the remainder of the season, which replenishes nutrient loss from defoliation. Although plants are not dependent on herbivores for growth, the latter, in contrast, are 'prisoners of their food supply' (Drent & Prins 1987) and in many situations the larger herbivores are strongly bottom-up controlled; there appears to be lack of good evidence of predator control. In a spatial and temporal context the herbivore moves to a different site to acquire high-quality forage when the supply of suitable forage is exhausted. However, in the case of small herbivores intense herbivory at 'safe sites' is often linked to predator avoidance. Movement of animals to the nearest supply of forage may result in predation, a situation that is exacerbated by the need of small herbivores for high-quality forage. Large herbivores, in contrast, can get by utilizing lower quality forage, at least in the short term. The occurrence of vegetational heterogeneity and mobility of herbivores ensure that the animals can exploit food resources during a narrow window of time when protein/biomass ratios are favourable at different sites. Hence, not only are the interactions between different herbivores and their forage plants strongly scale dependent in relation to both the size and mobility of the animals, but also the effect of the animals on soil nutrient fluxes and vegetational heterogeneity will be patch-size dependent. A consequence is that herbivory will tend to re-enforce existing spatial heterogeneity in vegetation patterns and soil characteristics. These 'hot spots' or 'nucleation sites' of forage plants with fast growth rates and high primary productivity are used frequently by herbivores year after year (McNaughton 1984; Jefferies 1988a,b; Mueller 1993). Often this leads to the development of grazing lawns *sensu* McNaughton (1984), which are probably more widespread than has been recognized to date. However, over time the sites act as a propagule trap; many fruits and seeds are brought to sites by animals (cf. Bullock & Primack 1977; Fischer *et al.* 1996), or the latter change site characteristics and seed distributions as a result of their overall disturbance activities (Johnston 1995). Selective foraging by herbivores may promote (Pastor *et al.* 1988; Pastor & Naiman 1992) or delay (Bazely & Jefferies 1986) successional processes.

Return times of herbivores to a particular patch may be either of the order of days (Drent & Prins 1987), or between seasons (Mulder & Harmsen 1995) or after a number of years have elapsed (Chernov 1985), depending on the regrowth potential of forage plants. The frequency of the return time is an index of the degree of coupling between herbivory, litter decomposition, microbial activity and plant growth. However, within these 'hot spots', although the interactions are strong, primary production is still largely a function of the physical conditions where the plants grow, as the supply of nutrients cannot be sustained in the short term following one or more defoliations. It is only in rare and unusual circumstances that above-ground biomass produced by compensatory growth of plants after defoliation exceeds that of ungrazed plants (cf. Belsky 1986, 1987). Conditions necessary to allow multiple defoliations to occur within the season include the colonial foraging behaviour of the herbivore that results in the deposition of substantial amounts of nitrogen-rich faecal material, which enables regrowth to occur in environments where turnover of soil nitrogen is limited (Hik *et al.* 1991). In addition, where regrowth occurs, grazing-tolerant plants characteristically have an indeterminate growth habit. Most root development takes place at, or just below, the soil surface, where nutrients can be readily absorbed from leachates of faecal material (Jefferies 1988a,b; Jefferies *et al.* 1994). More deep-rooted graminoid species are unable to respond as quickly, either because of the slow rate of movement of nutrients down the soil profile, or because phenological constraints limit growth. The amount of compensatory growth that is possible within a season is therefore a function of a number of variables, including the type and frequency of defoliation, the internal nutrient supplies in forage plants, the capacity of the soil to maintain nutrient fluxes to the root surfaces and the phenology of forage plants.

The effect of large mammals and herbivorous birds on vegetation and resource availability is also strongly dependent on the density of foraging animals. At a given density bottom-up processes, discussed above, may control forage production and the behaviour of the herbivore, as a result of density dependence (i.e. periodic use of a site, followed by migration to another site when the supply of forage is exhausted). In contrast, at higher densities top-down processes may dominate and replace bottom-up controls that may lead to the destruction of vegetation, as discussed later. This shift in dominance is strongly scale dependent, and reflects the depletion of forage over a large area brought about by increased numbers of herbivores. It is worth examining these interactions in relation to the keystone species concept before discussing trophic cacades, as they appear to relate directly to this concept.

Keystone species concept

A keystone species is defined currently as one 'whose impact on its community is large, and disproportionately large relative to its abundance' (Power *et al.* 1996). The term has been widely used to describe different modes of action of individual species on communities. Mills *et al.* (1993) recognize five major modes of action of species acting as keystone species, whereas Bond (1994) extends it to seven involving predators, herbivores, competitors, mutualists, earth movers, system processors and even abiotic agents. The effects of the individual species are both qualitative and quantitative, they may or may not involve trophic interactions, and a species may initiate a host of indirect effects on community assemblages. Mills *et al.* (1993) caution against the use of a term that is so wide ranging in meaning, except where the modes of action are defined carefully (see also Menge *et al.* 1994; Menge 1995).

The keystone concept is based on the premise that only one, or perhaps a few species, have a large effect on community structure and composition (Mills *et al.* 1993). Implicit in the paradigm is that keystone species may be expected to have strong interactions with other species if trophic relationships are involved, and that these interactions affect community structure. Hence, keystone species are given high community importance values (percentage of species lost from a community upon removal of a given species). Species–species interactions do not show a symmetric distribution, but a skewed distribution when keystone species are present (Mills *et al.* 1993). As yet, most of the above assertions remain untested and measures of interaction strengths are badly needed, particularly in relation to a related concept, the extended keystone hypothesis (Holling 1992). This states that in all terrestrial ecosystems a small number of plants, microorganisms and animals, together with associated abiotic processes, structure the landscape at different scales.

The definition of a keystone species (Power *et al.* 1996) hinges on numbers of individuals (or biomass) of a species relative to corresponding values for other species, and the strength of its effect as measured by the rate of change of community or ecosystem properties (e.g. nutrient cycling, species diversity, primary/secondary productivity). Keystone species are most likely (but not exclusively) to occur near the top of the food chain, because top predators have high *per capita* effects and low collective biomass compared with organisms at lower trophic levels (Power *et al.* 1996). The application of this concept, however, is sensitive to:

1 decreasing or increasing numbers of a species (little is said about increasing numbers) and the effects of the changes on trophic interactions;

2 whether the effects on other species are governed by negative or positive feedbacks; and

3 the resilience of the community/ecosystem properties to the changes in numbers of species.

Where changes in numbers result in sustained positive feedbacks, trophic interactions become unstable and trophic cascades result. Hence, the keystone concept is linked to this concept.

The difficulties outlined above may be illustrated with respect to interactions between lesser snow geese and coastal vegetation on the Hudson Bay coast (Jefferies 1988a,b; Bazely & Jefferies 1997), which are discussed in detail in the next section. The birds are the only major herbivore utilizing the vegetation, and their foraging activities are confined to 3 months during the snow-free season. At low densities of geese a positive feedback between the birds and the intertidal salt-marsh vegetation results in enhanced primary production in grazed vegetation compared with that of ungrazed vegetation, as a result of the rapid recycling of nutrients from faeces within the season (Cargill & Jefferies 1984; Bazely & Jefferies 1985). In addition, the intense grazing resets the successional clock each season, at least over a period of 10 years (Bazely & Jefferies

1986). Erection of exclosures leads to rapid changes in vegetation and a large increase in numbers of plant species. Hence, this bird may be regarded as a keystone species in maintaining community structure, but the effect is confined to intertidal salt marshes, and then to only a section of the marshes where nitrogen turnover is high because of the prevailing edaphic conditions (cf. Table 10.2) (Wilson & Jefferies 1996). The population of birds has increased dramatically in recent years and this has led to the destruction of vegetation in different habitats that include fresh-water sedge meadows and beach ridges. In the intertidal marshes a new feedback mechanism has replaced the former feedback that is associated with desertification of landscape (Srivastava & Jefferies 1996).

Again, the birds regulate community structure and strongly influence ecosystem properties. The communities are capable of recovery in the absence of goose foraging, but only over the long term. The species is a keystone species across a range of densities, but different mechanisms are involved depending on the density. In addition, spatial scales change from a local effect ($< 300 \, \text{m}$) to a regional change ($50 \, \text{km}$) in plant assemblages as the goose population increases. This involves increased mobility of geese as they forage in a variety of habitats and also temporal changes in their use of different plant assemblages. Not only are these herbivore and plant responses scale-dependent in relation to increasing numbers of geese, but also the desertification of landscape is a consequence of a trophic cascade, hence this concept and that of the keystone species are interconnected when the larger scales are examined.

Trophic cascades

Trophic cascades include situations where strong interaction effects of species with top-down dominance lead to 'runaway consumption and dominance through the food chain' (Strong 1992). They are characteristic of systems where linear food chains prevail. The cascade model assumes that a given species can be arranged a priori into a cascade, such that a given species can feed only on those species below it, and can itself be fed upon by only those species above it in the hierarchy (Lawton 1989). Often the biomass of primary producers is severely reduced when a herbivore species increases dramatically in the absence of control by a species at an upper trophic level. The trophic cascade model has been applied to aquatic ecosystems, particularly lakes, rather than terrestrial ecosystems (Kitchell & Carpenter 1992). Manipulations of piscivorous lake populations have been made in order to study trophic cascades (DeMelo *et al.* 1992). However, short-term, top-down manipulation of trophic systems can rarely be sustained, a large number of factors dampen the effect of initially strong top-down processes on lower trophic levels (Mills & Forney 1988; DeMelo *et al.* 1992; McQueen *et al.* 1992). In species-rich, highly differentiated, reticulate food webs characteristic of terrestrial environments the overall effects of downward-driven dominance are dissipated—the systems are buffered against these strong interactions (Lawton 1989; Strong 1992) (Table 10.1). The occurrence of sustained, chain-like responses by organisms at different trophic levels depends on positive feedback processes that bring about sudden and pronounced self-amplifying changes of matter: the system acts to reinforce change in the direction of the deviation (Maruyama 1963; DeAngelis *et al.* 1986). In the absence of negative feedbacks, chain-like responses associated with a major perturbation generate instability in the system. If the time delays are of sufficient length, the system moves to an alternative state more rapidly than negative feedbacks can respond, and a new equilibrium is established (cf. Oksanen 1990).

Although trophic cascades usually are considered to result from top-down effects (Hairston *et al.* 1960; Fretwell 1977; Oksanen *et al.* 1981; Power 1990, 1992), cascades may also operate from the bottom up (Hunter & Price 1992; Strong 1992). In aquatic ecosystems, such as the euphotic zone in lakes, nutrient cycling between primary producers and consumers does not directly involve decomposers, as in terrestrial systems. As such, the system is often temporarily depleted of nutrients and is particularly sensitive to nutrient additions. In two recent elegant studies, Vanni and Layne (1997) and Vanni *et al.* (1997) draw attention to the effects of fish and zooplankton not only acting as predators but also providing nutrients in faecal material for phytoplankton growth. In their studies, dinoflagellates and chrysophytes, in particular, responded to these nutrient inputs. $N:P$ ratios in food appeared to be important in determining the $N:P$ ratio of excreta of zooplankton (Sterner 1990; Vanni & Layne 1997). The fish also had strong effects on the rates of nutrient cycling of these two elements, that slowed the decline of phosphorus in the water, and the decrease in the total $N:P$ ratio resulted in increasing phytoplankton biomass (Vanni *et al.* 1997). Such studies not only indicate the complexities of top-down dominance, but also that individual species can have strong effects on both the physical and chemical characteristics of natural ecosystems and community assemblages—a conclusion that parallels results obtained in terrestrial environments (Huntly & Inouye 1988; Naiman *et al.* 1988; Wedin & Tilman 1990; Wedin & Pastor 1993).

There are few studies of trophic cascades in terrestrial ecosystems (e.g. Kajak *et al.* 1968; Spiller & Schoener 1990; Rosenheim *et al.* 1993; Marquis & Whelan 1994; Chase 1996). Distinct trophic levels are rare in nature and omnivory is widespread (Spiller & Schoener 1990; Power 1992), although omnivory does not necessarily preclude the occurrence of trophic cascades (Power 1990; Power *et al.* 1996). Where trophic cascades have been reported they share a number of common characteristics (Table 10.1). Most, but not all, reflect the influence of human agencies in either intentionally or unintentionally enforcing top-down dominance. This may involve the direct introduction of species representative of an upper trophic layer (cf. Smith & Steenkamp 1990, below), or an energy and nutrient subsidy that promotes the population growth of a top-level organism. Both types of manipulation, if sustained, have the potential to result in strong interactions, particularly where an introduced species is highly competitive. Although most of the case studies discussed below describe strong top-down effects, the processes that result in alternative vegetation states are driven by changes in the soil–microbial plant systems, which involve a large abiotic component.

The first example is the study by Smith and Steenkamp (1990) on Marion Island in the oceanic sub-Antarctic. The terrestrial environment is species-poor, but primary production is high. There are no macro-herbivores, so that most of the nutrients pass through the detritivore cycle. Most plant communities occur on soils that have low levels of available nutrients, and nutrient mineralization is the main bottleneck limiting primary production. However, rates of nutrient release mediated by free-living organisms alone are insufficient to account for the annual demand of nitrogen and phosphorus for vegetative growth. Larvae of a flightless moth (*Pringleophaga marioni*), as a result of digestion of plant litter, stimulate nitrogen mineralization 10-fold and phosphorus mineralization threefold in experimental systems. At present, the larvae process annually an estimated 1500 kg ha^{-1} of plant litter on the island (Crafford, in Smith & Steenkamp 1990). Marion Island supports a large population of introduced house mice (*Mus musculus*) that has increased in number in recent years as a result of the removal of feral cats and an interdecadal increase in temperature (Smith & Steenkamp 1990). The mice were introduced on the island before 1818 (Watkins & Cooper 1986), and between 83 and 94% of the diet of the mice is soil invertebrates (Rowe-Rowe *et al.* 1989). The mice eat 0.7% of the standing crop of these arthropods per day (Burger 1978). Crafford (in Smith & Steenkamp 1990) has estimated that the increasing mouse population reduces the annual turnover of litter by moth larvae from 2500 kg ha^{-1} to 1500 kg ha^{-1}. Smith and Steenkamp (1990) argue that the predation pressure on soil invertebrates has decreased, and will continue to decrease, rates of nutrient cycling causing further imbalances between primary production and decomposition. This will result in enhanced rates of peat accumulation, changes to the hydrological regime and changes in plant assemblages leading to alternative vegetation states (Gremmen 1981). The mice also eat seed heads of graminoid species; the sedge, *Uncinia compacta*, has almost been eradicated as a result of foraging by mice. Alterations in patterns of nutrient turnover have far-reaching effects on plant community structure that involve biotic and abiotic processes. This example of a trophic cascade indicates the dramatic effects that invasive organisms have on the functioning of terrestrial ecosystems where the structure of the food webs is relatively simple. Other examples indicate that altered vegetation states also occur when herbivores are introduced to predator-free islands. The introduction of reindeer to St Matthew Island (Klein 1968) led to dramatic changes in vegetation as the herbivore population increased.

The second example is taken from a collaborative project on changes in plant assemblages in a former prairie grassland in northern California (Strong *et al.* 1995; Maron & Connors 1996). The bush lupin, *Lupinus arboreus*, is thought to have invaded formerly heavily grazed grasslands when grazing ceased. In these areas there are now large stands of this nitrogen-fixing plant. However, periodic outbreaks of root-boring caterpillars of the ghost moth, *Hepialus californicus*, led to the death of bush lupins, and outbreaks of other invertebrate herbivores also contribute to their death. At one site an estimated 43 000 lupin bushes died at the end of the summer of 1991. Eventual re-establishment of bush lupin occurs because of an abundant and long-lived seed bank, but after germination and rapid growth, the new cohort of plants succumb to the effects of high population numbers of ghost moth caterpillars and other herbivores and the limit cycle is repeated. The outcome is that in this donor-controlled system (cf. Strong 1992) there is a mosaic of nitrogen-rich sites available to invasive weedy grasses, such as *Bromus diandrus*, that colonize the open nitrogen-rich dead lupin sites until lupin re-establish and the developing plant canopy shades the grasses. The repeated bouts of lupin germination, establishment and death convert a rich native plant community into a less diverse collection of introduced weeds (Maron & Connors 1996).

This donor-controlled system has created a bottom-up cascade where consumers and microorganisms subsequently act to release and redistribute nitrogen within these coastal prairies that results in alternative vegetation states. The cascade is also dependent, therefore, on top-down effects. In adjacent areas where the woody lupin bush has not invaded, total soil nitrogen and net rates of nitrogen mineralization are approximately half of corresponding values for sites where lupin is present, and populations of ghost moth caterpillars are low. Hence, the nitrogen-fixing plant promotes major changes in community assemblages but the changes are modulated by top-down processes.

The final two examples are where external energy and nutrient subsidies have had dramatic effects on populations of herbivores resulting in trophic cascades that have led to alternative vegetation states. The ultimate causes of the change, as in previous examples, are modifications to soil–microbial–plant systems involving positive feedbacks that are strongly linked to abiotic processes.

Population growth rates of species of Arctic breeding geese that winter in Europe and North America in terrestrial ecosystems have increased significantly over the last 30 years (Fig. 10.1) (Abraham *et al.* 1996; Bazely & Jefferies 1997). Although a number of factors have contributed to the increase, including decreased hunting pressure, the creation of reserves and refuges and interdecadal climate variation, changes in agriculture resulting in increased forage availability on both wintering and southern staging grounds appear to be a significant cause of the increase (Abraham *et al.* 1996). Populations of geese that winter in marine habitats have not increased significantly in numbers in recent decades, at least in North America (K.F. Abraham, personal communication, 1996). The increased use of agricultural resources by species of Arctic geese represents a significant energy and nutrient subsidy to the birds, and biomanipulations of Arctic breeding goose populations on a massive scale (Abraham *et al.* 1996; Bazely & Jefferies 1997).

The mid-continent population of the lesser snow goose (*Anser caerulescens caerulescens*) is at least 5 million birds. The birds breed in coastal salt- and fresh-water marshes around the shores of James Bay, Hudson Bay and south-west Baffin Island (Abraham *et al.* 1996). The population size is increasing at a rate of approximately 7% per annum (Cooke *et al.* 1995). Within coastal salt marshes, species richness is low, and

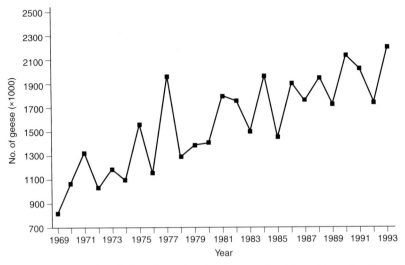

Figure 10.1 An index of abundance showing the growth of the mid-continent population of lesser snow geese from 1969 to 1993 based on mid-winter counts. Data provided by the Mississippi and Central Flyway Waterfowl Councils' Technical Sections. Modified from Abraham *et al.* (1996).

with the exception of the geese there are no other major herbivores, hence the trophic relationships can be represented as a trophic ladder as a first approximation. The system has shifted to strong top-down dominance resulting from increased numbers of these herbivores. In early spring just prior to above-ground plant growth, both staging and nesting geese grub for roots and rhizomes of the preferred salt-marsh graminoids, on which family groups also feed in the post-hatch period in summer. This initiates a positive feedback process that has led to the destruction of salt-marsh swards (Fig. 10.2) (Jefferies 1988a,b; Kerbes *et al.* 1990; Iacobelli & Jefferies 1991; Srivastava & Jefferies 1996). The removal of vegetation by geese creates bare patches of sediment with only a few remaining tillers of salt-marsh graminoids. In summer increased rates of evaporation from these patches render the upper layers of the sediments hypersaline, therefore detrimental to regrowth of forage graminoids (Srivastava & Jefferies 1995a,b, 1996). The salinities may be three times greater than that of oceanic sea water (Iacobelli & Jefferies 1991). The hypersaline conditions depress the growth of the salt-marsh graminoids, which fail to recolonize patches of bare, hypersaline soil, at least in the short term (Srivastava &

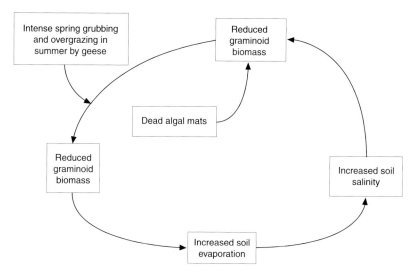

Figure 10.2 Positive feedback involving both trophic and non-trophic interactions that leads to the destruction of grazing swards and desertification of the soil system in coastal salt marsh of Hudson Bay. From Srivastava and Jefferies (1996).

Jefferies 1995b; T. Handa unpublished data). In time, an alternative state of bare mud is reached, where plant growth is severely restricted. In these denuded sites not only is the salinity high and plant growth limited, but also rates of soil mineralization are significantly lower compared with those beneath intact swards (Wilson & Jefferies 1996). In addition, in summer family groups of lesser snow geese heavily graze remaining vegetation, which exacerbates the effect of the degenerative positive feedback with negative consequences for plant growth (Srivastava & Jefferies 1996) and gosling development (Cooch *et al.* 1991; Francis *et al.* 1992). In particular, geese grub or heavily graze vegetation characteristic of early successional stages where inputs of nutrients (especially nitrogen) are high. As this source of vegetation has been depleted the birds forage in less preferred sites in which inputs and turnover of nitrogen are less (Wilson & Jefferies 1996). Where hypersaline conditions prevail the vegetation is unlikely to re-establish in the foreseeable future; some exclosures have been devoid of perennial plants for 12 years (R.L. Jefferies, unpublished data). At these sites the alternative vegetational state is represented by the annual species, *Salicornia borealis*, which is tolerant of high salinities. The plant is not eaten by the geese; 50% of its weight consists of salts.

The positive feedback initiated by lesser snow geese that has led to loss of vegetation is an example of self-amplifying changes, which are resulting in the destruction of existing plant communities and soils; rapid recovery is not possible. Although the geese act as the trigger for this series of changes, the major transformations are abiotic and involve modifications to soil properties that adversely affect plant growth.

The final example is the occurrence of a trophic cascade in sub-Saharan Africa, particularly the Sahel region. Landscape pattern in arid and semi-arid savannas of this region is determined primarily by soil moisture and edaphic properties of soils (Breman & de Wit 1983; Belsky 1995). In spite of the known ability of large herbivores to alter the structure and dynamics of plant communities (Cumming 1982; McNaughton & Georgiadis 1986; Belsky 1989; Coughenour 1991) disturbances by large herbivores (and fire) are secondary in importance to soil moisture and edaphic conditions in regulating landscape dynamics (Belsky 1995). In pristine grassland communities in national parks revegetation of patches created by animal disturbances is by plant species that are already abundant in surrounding communities (Belsky 1995). The pattern of revegetation is longer and involves alternative states only where fire and elephants create grassland–woodland mosaics (Sinclair 1995), or where disturbances affect soil moisture and edaphic properties (Belsky 1995), as discussed below.

As with lesser snow geese the effects of large mammals are intensified where the animals are compressed into high densities in either small game reserves, or where watering holes have been constructed for wildlife and livestock. In the southern Sahel grazing is the principal use of the land and humans are the top predators in this pastoral system (Breman & Cissé 1977; Sinclair & Fryxell 1985).

Fires play a minor role in the dynamics of the Sahel rangelands, because amounts of biomass are low and spatially scattered north of the 400-mm isohyet (Rietkerk et al. 1996). Human populations and arid ecosystems can be described as a simple four-compartment system: humans, herbivores, vegetation and soil (Graetz 1991). Pastoral societies in the Sahel until recently carried out a nomadic strategy very similar to the large ungulates in coping with seasonal rainfall associated with the movement of the intertropical convergence zone (ITCZ) and with the periodic droughts when the ITCZ failed to bring rains to the Sahel (Breman & de Wit 1983). The nomadic pastoral system was stabilized by negative feedbacks between vegetation, grazers and the human population (Graetz 1991), in which the ecological processes that characterized the feedbacks were energy and nutrient flow. The nomadic system was destabilized by the impairment of these feedbacks brought about by settlement, cultivation and 'overgrazing' after World War II (Breman & de Wit 1983; Sinclair & Fryxell 1985; Le Houérou 1989). As in the case of the destruction of the coastal salt marshes of Hudson Bay, the long-term consequence of sustained and intense grazing was indirect in action (Graetz 1991). Initially, much of the available plant growth was eaten during the wet season which left little food for the 9-month dry season (Warren & Maizels 1977). Annual herbs, particularly around wells, replaced perennial plants, and unpalatable trees and shrubs spread (Warren & Maizels 1977; Sinclair & Fryxell 1985). The increased exposure of bare soil devoid of plant litter had profound consequences. As the surface was bared, rain infiltration decreased and runoff increased that caused sheet and gully erosion (Prins 1989). The loss of litter also facilitated erosion and redistribution of soil by wind (Warren & Maizels 1977; Noble & Tongway 1986). Both processes were the outcome of positive feedbacks associated with high numbers of animals, and both altered the spatial patterns of nutrient movement and water flow in soils. Very small changes in loss of plant cover at, or about, a threshold can be expected to result in disproportionately large increases in soil erosion and gullying (Fig. 10.3) (Graetz 1991; Rietkerk *et al.* 1996).

Another feedback was the change in microclimate that occurred as plant cover was reduced (Graetz 1991). The soil surface temperature and the evaporative demands placed on plants were amplified by increases in the bare areas of soil that led to lack of plant establishment or poor growth. The end result of these processes was desertification — a man-made disturbance of these arid ecosystems where the adverse effects were exacerbated in years of drought, that brought about the impoverishment of these terrestrial ecosystems (Dregne 1983).

The last two studies on semi-arid grasslands and Arctic salt marshes relate catastrophic events to plant–soil interactions. The effects triggered by top-down dominance of herbivores are accentuated by short-term climatic conditions (i.e. late springs or droughts) and lead to multiple stable states. It is the indirect effects of lack of vegetational cover that bring about the collapse of these two systems associated with discontinuous processes. In the Sahel subsequent removal of grazers, or an occurrence of wet years after unprecedented drought rarely have enabled the former plant assemblages to re-establish (Breman & de Wit 1983; Sinclair & Fryxell 1985; Le Houérou 1989) (plots protected in 1984 in Arctic salt marshes are still largely devoid of vegetation where hypersaline conditions prevail).

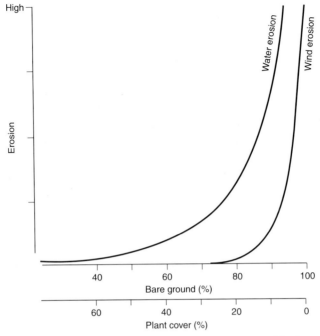

Figure 10.3 The relationship between the erosion potential of water and wind as a function of the relative proportion of plant cover. From Marshall (1973), in Graetz (1991).

Recently, Rietkerk *et al.* (1996) have interpreted the vegetation dynamics of the Sahel in relation to cusp catastrophe theory. They have proposed a hypothesis (as yet untested) that communities of annual grasses (a vegetational state) do not persist, because in drought years the seed bank is ultimately eliminated. Periodically, threshold values are reached that initiate discontinuous changes. This may result in alternative vegetative states of perennial grasses replacing annual herbs and vice versa, where the intermediate state of annual grasses is bypassed. The significance of the approach is that it places the descriptive studies in a theoretical framework and results in testable hypotheses.

Conclusions

The availability of nutrients for plant growth in faecal matter of herbivores, and consumers in general, is limited by the ability of the microbial gut flora to process ingested plant material. A functionally similar soil microflora is involved in the mineralization of faecal material, soil casts and plant litter. However, the relatively poor growth of the soil microflora when drought and low temperatures occur, and their inability to access nutrient sources within bound soil casts of animals effectively regulates nutrient availability for plant growth, in spite of large amounts of soil turnover by consumers. Edaphic conditions largely determine sites where high rates of net primary production can be sustained — the activities of herbivores/consumers only reinforce the suitability of these sites as prime foraging areas. Their role as agents facilitating nutrient transfer is constrained by microbial activity, the presence of unpalatable species and need for suitable edaphic conditions for plant growth. Bottom-up control ultimately regulates net primary production, and density-dependent processes result in both large and small herbivores exploiting the occurrence of vegetational heterogeneity in space and time in order to meet their food requirements. Predation may limit movement of the latter group of animals.

Where herbivore numbers are increasing, often as a result of external energy and nutrient subsidies and lack of predation, top-down dominance replaces bottom-up controls. These trophic cascades, which are rare in terrestrial ecosystems, occur where simple linear food chains exist and where species diversity is low. In these situations destruction of vegetation, soil degradation and alternative stable states are all likely to occur and involve both biotic and abiotic events. Although these trophic cascades are driven usually by top-down dominance (but not exclusively) their enactment involves changes to bottom-up processes. The occurrence of a cascade is scale dependent both in relation to numbers of herbivores and in a spatial and temporal context. Its effects on primary production and changes to the physical environment are non-linear and rarely predictable a priori.

Acknowledgements
I thank Peter Kotanen, Han Olff, Herbert Prins, David Wedin and two anonymous reviewers for their constructive criticisms. Their suggestions greatly improved the manuscript. The Natural Sciences and Engineering Research Council provided financial support for this work, part of which was written while on sabbatical leave at the Institute of Entomology in London. Mrs C. Siu kindly typed the manuscript.

321

References

Abraham, K.F., Jefferies, R.L., Rockwell, R.F. & MacInnes, C.D. (1996). Why are there so many white geese in North America? In *Seventh International Waterfowl Symposium* (Ed. by J.T. Ratti), pp. 79–92. Ducks Unlimited, Memphis, TN.

Bazely, D.R. & Jefferies, R.L. (1985). Goose faeces: a source of nitrogen for plant growth in a grazed salt-marsh. *Journal of Applied Ecology*, 22, 693–703.

Bazely, D.R. & Jefferies, R.L. (1986). Changes in the composition and standing crop of salt marsh communities in response to removal of a grazer. *Journal of Ecology*, 22, 693–703.

Bazely, D.R. & Jefferies, R.L. (1997). Trophic interactions in arctic ecosystems and the occurrence of a terrestrial trophic cascade. In *Ecology of Arctic Environments* (Ed. by S.J. Woodin & M. Marquiss), pp. 183–208. Blackwell Science Ltd, Oxford.

Belsky, A.J. (1986). Does herbivory benefit plants? *American Naturalist*, 127, 870–892.

Belsky, A.J. (1987). The effects of grazing: confounding of ecosystems, communities, organisms. *American Naturalist*, 129, 777–783.

Belsky, A.J. (1989). Landscape patterns in a semi arid ecosystem in East Africa. *Journal of Arid Environments*, 17, 265–270.

Belsky, A.J. (1995). Spatial and temporal landscape patterns in arid and semi arid African savannas. In *Mosaic Landscapes and Ecological Processes* (Ed. by L. Hansson, L. Fabrig & G. Merriam), pp. 31–56. Chapman & Hall, London.

Bengtsson, J., Zheng, W.D., Ågren, G.I. & Persson, T. (1995). Food webs in soil: an interface between population and ecosystem ecology. In *Linking Species and Ecosystems* (Ed. by C.G. Jones & J.H. Lawton), pp. 141–150. Chapman & Hall, London.

Bond, W.J. (1994). Keystone species. In *Biodiversity and Ecosystem Function* (Ed. by E.D. Schulze & H.A. Mooney), pp. 237–253. Springer-Verlag, Berlin.

Breman, H. & Cissé, A.M. (1977). Dynamics of Sahelian pastures in relation to drought and grazing. *Oecologia*, 28, 301–315.

Breman, H. & de Wit, C.T. (1983). Rangeland productivity and exploitation in the Sahel. *Science*, 221, 1341–1347.

Bryant, J.P. & Kuropat, P.J. (1980). Selection of winter forage by subarctic browsing vertebrates: the role of plant chemistry. *Annual Reviews of Ecology and Systematics*, 11, 261–285.

Bryant, J.P., Chapin, F.S., III & Klein, D.R. (1983). Carbon/nutrient balance of boreal plants in relation to vertebrate herbivory. *Oikos*, 40, 357–368.

Bullock, S.H. & Primack, R.B. (1977). Comparative experimental study of seed dispersal on animals. *Ecology*, **58**, 681–686.

Burger, A.E. (1978). Terrestrial invertebrates: a food resource for birds at Marion Island. *South African Journal of Antarctic Research*, **8**, 87–99.

Cargill, S.M. & Jefferies, R.L. (1984). The effects of grazing by lesser snow geese on the vegetation of a sub-arctic salt-marsh. *Journal of Applied Ecology*, **21**, 669–686.

Carpenter, S.R., Kitchell, J.R. & Hodgson, J.R. (1985). Cascading trophic interactions and lake productivity. *BioScience*, **35**, 634–639.

Chapin, F.S., III (1980). The mineral nutrition of wild plants. *Annual Review of Ecology and Systematics*, **11**, 233–260.

Chase, J.M. (1996). Abiotic controls of trophic cascades in a simple grassland food chain. *Oikos*, **77**, 495–506.

Chaussod, R., Houot, S., Guiraud, G. & Hétier, J.M. (1988). Size and turnover of microbial biomass in agricultural soils: laboratory and field measurements. In *Nitrogen Efficiency in Agricultural Soils* (Ed. by K.A. Smith & D.S. Jenkinson), pp. 321–326. Elsevier, London.

Chernov, Yu I. (1985). *The Living Tundra*. Cambridge University Press, Cambridge.

Clarholm, M. & Rosswall, T. (1980). Biomass and turnover of bacteria in a forest soil and a peat. *Soil Biology and Biochemistry*, **12**, 49–57.

Coley, P.D., Bryant, J.P. & Chapin, F.S., III (1985). Resource availability and plant anti herbivore defense. *Science*, **230**, 895–899.

Cooch, E.G., Lank, D.B., Rockwell, R.F. & Cooke, F. (1991). Long-term decline in body size in a snow goose population: evidence of environmental degradation? *Journal of Animal Ecology*, **60**, 483–496.

Cooke, F., Rockwell, R.F. & Lank, D.B. (1995). *The Snow Geese of La Pérouse Bay. Natural Selection in the Wild*. Oxford University Press, New York.

Coughenour, M.B. (1991). Spatial components of plant herbivore interactions in pastoral, ranching and native ungulate ecosystems. *Journal of Range Management*, **44**, 530–542.

Crawley, M.J. (1983). *Herbivory*. Blackwell Scientific Publications, Oxford.

Cumming, D.H.M. (1982). The influence of large herbivores on savanna structure in Africa. In *Ecology of Tropical Savannas* (Ed. by B.J. Huntley & B.H. Walker), pp. 215–245. Springer-Verlag, Berlin.

DeAngelis, D.L. (1992). *Dynamics of Nutrient Cycling and Food Webs*. Chapman & Hall, London.

DeAngelis, D.L., Post, W.M. & Travis, C.C. (1986). *Positive Feedback in Natural Systems*. Springer-Verlag, Berlin.

DeMelo, R., France, R. & McQueen, D.J. (1992). Biomanipulation: Hit or myth? *Limnology and Oceanography*, **37**, 192–207.

Dregne, H.E. (1983). *Desertification of Arid Lands*. Harwood, New York.

Drent, R.H. & Prins, H.H.T. (1987). The herbivore as prisoner of its food supply. In *Disturbance in Grasslands, Causes, Effects and Processes* (Ed. by J. van Andel, J.P. Bakker & R.W. Snaydon), pp. 131–148. Kluwer Academic Publishers, Dordrecht.

Fenchel, T. & Blackburn, T.H. (1979). *Bacteria and Mineral Cycling*. Academic Press, New York.

Fischer, S.F., Poschlod, P. & Beinlich, B. (1996). Experimental studies on the dispersal of plants and animals on sheep in calcareous grasslands. *Journal of Applied Ecology*, **33**, 1206–1222.

Francis, C.M., Richards, M.H., Cooke, F. & Rockwell, R.F. (1992). Long term changes in survival rates of lesser snow geese. *Ecology*, **73**, 1346–1362.

Fretwell, S.D. (1977). The regulation of plant communities by food chains exploiting them. *Perspectives in Biology and Medicine*, **20**, 169–185.

Graetz, R.D. (1991). Desertification: a tale of two feedbacks. In *Ecosystem Experiments, Scope, 45* (Ed. by H.A. Mooney, E. Medina, D.W. Schindler, E.D. Schulze & B.H. Walker). John Wiley & Sons, New York.

Gremmen, N.J.M. (1981). The vegetation of the subantarctic islands, Marion and Prince Edwards. *Geobotany*, **3**, 1–149.

Hairston, N.G., Smith, F.E. & Slobodkin, L.B. (1960). Community structure, population control and competition. *American Naturalist*, **94**, 421–425.

Hart, S.C. & Gunther, A.J. (1989). *In situ* estimates of annual net mineralization and nitrification in a subarctic watershed. *Oecologia*, **80**, 284–288.

Hik, D.S., Sadul, H.A. & Jefferies, R.L. (1991). Effects of the timing of multiple grazings by geese on net above-ground primary production of swards of *Puccinellia phryganodes*. *Journal of Ecology*, **79**, 715–730.

Hobbie, S.E. (1992). Effects of plant species on nutrient cycling. *Trends in Ecology and Evolution*, **7**, 336–339.

Holling, C.S. (1992). Cross scale morphology, geometry and dynamics of ecosystems. *Ecological Monographs*, **62**, 447–502.

Hunter, M.D. & Price, P.W. (1992). Playing chutes and ladders: heterogeneity and the relative roles of bottom up and top down forces in natural communities. *Ecology*, **73**, 724–732.

Huntly, N.J. (1991). Herbivores and the dynamics of communities and ecosystems. *Annual Review of Ecology and Systematics*, **22**, 477–504.

Huntly, N.J. & Inouye, R. (1988). Pocket gophers in ecosystems: patterns and mechanisms. *BioScience*, **38**, 786–793.

Iacobelli, A. & Jefferies, R.L. (1991). Inverse salinity gradients in coastal marshes and the death of stands of *Salix*: the effects of grubbing by geese. *Journal of Ecology*, **79**, 61–73.

Jefferies, R.L. (1988a). Pattern and process in arctic coastal vegetation in response to foraging by lesser snow geese. In *Plant Form and Vegetation Structure* (Ed. by M.J.A. Werger, P.J.M. van der Art, H.J. During & J.T.A. Verhoeven), pp. 381–300. S.P.B. Academic Publishing, The Hague.

Jefferies, R.L. (1988b). Vegetational mosaics, plant animal interactions and resources for plant growth. In *Plant Evolutionary Biology* (Ed. by L.D. Gottlieb & S.K. Jain), pp. 341–369. Chapman & Hall, London.

Jefferies, R.L., Klein, D.R. & Shaver, G.R. (1994). Vertebrate herbivores and northern plant communities: reciprocal influences and responses. *Oikos*, **71**, 193–206.

Jenkinson, D.S. & Ladd, J.N. (1981). Microbial biomass in soil: measurement and turnover. In *Soil Biochemistry* (Ed. by J.N. Ladd & E.A. Paul), pp. 415–417. Academic Press, New York.

Johnston, C.A. (1995). Effects of animals on landscape pattern. In *Mosaic Landscapes and Ecological Patterns* (Ed. by L. Hansson, L. Fahrig & G. Merriam), pp. 57–79. Chapman & Hall, London.

Kajak, A., Andzejewska, L. & Wojcik, Z. (1968). The role of spiders in the decrease of damages caused by *Acridoidea* on meadows experimental investigations. *Ekologia Polska, Series A*, **16**, 755–764.

Kerbes, R.H., Kotanen, P.M. & Jefferies, R.L. (1990). Destruction of wetland habitats by lesser snow geese: a keystone species on the west coast of Hudson Bay. *Journal of Applied Ecology*, **27**, 242–258.

Kitchell, J.F. & Carpenter, S.R. (1992). Cascading trophic interactions. In *The Trophic Cascade in Lakes* (Ed. by S.R. Carpenter & J.F. Kitchell), pp. 1–14. Cambridge University Press, Cambridge.

Klein, D.R. (1968). The introduction, increase, and crash of reindeer on St. Matthew Island. *Journal of Wildlife Management*, **32**, 350–367.

Lavelle, P. & Martin, A. (1992). Small scale and large scale effects of endogenic earth worms on soil organic matter dynamics in soils of the humid tropics. *Soil Biology and Biochemistry*, **24**, 1491–1498.

Lavelle, P., Lattund, C., Trigo, D. & Barois, I. (1995). Mutualism and biodiversity in soils. In *The Significance and Regulation of Biodiversity* (Ed. by H.P. Collins, G.P. Robertson & M.J. Klug), pp. 23–33. Kluwer Academic Publishers, Dordrecht.

Lawton, J.H. (1989). Food webs. In *Ecological Concepts* (Ed. by J.M. Cherrett), pp. 43–78. Blackwell Scientific Publications, Oxford.

Le Houérou, H.N. (1989). *The Grazing Land Ecosystems of the African Sahel.* Springer-Verlag, Berlin.

Maron, J.L. & Connors, P.G. (1996). A native nitrogen fixing shrub facilitates weed invasion. *Oecologia*, **105**, 302–312.

Marquis, R.J. & Whelan, C.J. (1994). Insectivorous birds increase growth of white oak through consumption of leaf chewing insects. *Ecology*, **75**, 2007–2014.

Marshall, J.K. (1973). Drought, land use and soil erosion. In *Drought* (Ed. by J. Lovett), pp. 55–80. Angus & Robertson, Sydney.

Maruyama, M. (1963). The second cybernetics: Deviation amplifying mutual causal processes. *American Scientist*, **51**, 164–179.

Mattson, W.R., Jr. (1980). Herbivory in relation to plant nitrogen content. *Annual Review of Ecology and Systematics*, **11**, 116–119.

McNaughton, S.J. (1984). Grazing lawns: animals in herds, plant form, and coevolution. *American Naturalist*, **124**, 863–886.

McNaughton, S.J. (1985). Ecology of a grazing ecosystem: the Serengeti. *Ecological Monographs*, **55**, 259–294.

McNaughton, S.J. & Georgiadis, N.J. (1986). Ecology of African grazing and browsing mammals. *Annual Review of Ecology and Systematics*, **17**, 39–65.

McQueen, D.J. (1990). Manipulating lake community structure: where do we go from here? *Freshwater Biology*, **23**, 613–620.

McQueen, D.J., Mills, E.L., Forney, J.L., Johannes, M.R.S. & Post, J.R. (1992). Trophic level relationships in pelagic food webs: comparisons derived from long term data sets for Oneida lake, New York (USA) and Lake St. George, Ontario (Canada). *Canadian Journal of Fisheries and Aquatic Sciences*, **49**, 1588–1596.

Melillo, J.M., Aber, J.D. & Muratore, J.F. (1982). Nitrogen and liguin control of hardwood leaf litter decomposition dynamics. *Ecology*, **63**, 621–626.

Menge, B.E. (1995). Indirect effects in marine rocky intertidal interaction webs: patterns and importance. *Ecological Monographs*, **65**, 21–74.

Menge, B.E., Berlow, E.L., Blanchette, C.A., Navarette, S.A. & Yamada, S.B. (1994). The keystone concept: variation in the interaction strength in a rocky intertidal habitat. *Ecological Monographs*, **64**, 249–286.

Mills, E.L. & Forney, J.L. (1988). Trophic dynamics and development of pelagic food webs. In *Complex Interactions in Lake Communities* (Ed. by S.R. Carpenter), pp. 14–27. Springer-Verlag, Berlin.

Mills, J.T. & Sinha, R.N. (1971). Interactions between a spring tail *Hypogastrura tullbergi*, and soil borne fungi. *Journal of Economic Entomology*, **64**, 398–401.

Mills, S.L., Soulé, M.E. & Doak, D.F. (1993). The keystone species concept in ecology and conservation. *BioScience*, **43**, 219–224.

Moore, J.C. (1988). The influence of microarthropods on symbiotic and non symbiotic mutualism in detrital based below ground food webs. *Agriculture, Ecosystems and Environment*, **24**, 147–159.

Mueller, F.P. (1993). *Herbivore plant soil interactions in the Boreal Forest:*

selective winter feeding by spruce grouse. PhD thesis, University of British Columbia, Vancouver, Canada.

Mulder, C.P.H. & Harmsen, R. (1995). The effect of muskox herbivory on growth and reproduction in an arctic legume. *Arctic and Alpine Research*, **27**, 44–53.

Nadelhoffer, K.J., Giblin, A.E., Shaver, G.R. & Linkins, A.E. (1992). Microbial processes and plant nutrient availability in arctic soils. In *Arctic Ecosystems in a Changing Climate, an Ecophysiological Perspective* (Ed. by F.S. Chapin, R.L. Jefferies, J.F. Reynolds, G.R. Shaver & J. Svoboda), pp. 281–300. Academic Press, New York.

Naiman, R.J., Johnston, C.A. & Kelley, J.C. (1988). Alteration of North American streams by beaver. *BioScience*, **38**, 753–761.

Noble, J.C. & Tongway, D.J. (1986). Herbivores in arid and semi arid rangelands. In *Australian Soils: the Human Impact* (Ed. by J.D. Russell & R.F. Isbell), pp. 243–271. University of Queensland Press, St Lucia.

Noy Meir, I. (1975). Stability of grazing systems: an application of predator prey graphs. *Journal of Ecology*, **63**, 459–481.

Oksanen, L., Fretwell, S.D., Arruda, J. & Niemela, P. (1981). Exploitation ecosystems in gradients of primary productivity. *American Naturalist*, **118**, 240–261.

Oksanen, T. (1990). Exploitation ecosystems in heterogeneous habitat complexes. *Evolutionary Ecology*, **4**, 220–234.

Paine, R.T. (1969). A note on trophic complexity and community stability. *American Naturalist*, **103**, 91–93.

Pastor, J. & Naiman, R.J. (1992). Selection foraging and ecosystem processes in boreal forests. *American Naturalist*, **139**, 690–705.

Pastor, J., Naiman, R.J., Dewey, B. & McInnes, P. (1988). Moose, microbes and the boreal forest. *BioScience*, **38**, 770–777.

Polis, G.A. & Strong, D.R. (1996). Food web complexity and community dynamics. *American Naturalist*, **147**, 813–846.

Post, D.M., Taylor, J.P., Kitchell, J.F., Olson, M.H., Schindler, D.E. & Herwig, B.R. (1998). The role of migratory wildfowl as nutrient vectors in a managed wetland. *Conservation Biology*, 910–920.

Power, M.E. (1990). Effects of fish in river food webs. *Science*, **250**, 811–814.

Power, M.E. (1992). Top down and bottom up forces in food webs: do plants have primacy? *Ecology*, **73**, 733–746.

Power, M.E., Tilman, D., Estes, J.A., Menge, B.A., Bond, W.J., Mills, L.S., Daily, G., Castilla, J.C., Lubchenco, J. & Paine, R.T. (1996). Challenges in the quest for keystones. *BioScience*, **46**, 609–620.

Prins, H.H.T. (1989). East African grazing lands: overgrazed or stably degraded? In *Nature Management and Eastlands Development* (Ed. by W.D. Verwey), pp. 281–305. IOS, Amsterdam.

Prins, H.H.T. (1996). *Ecology and Behaviour of the African Buffalo: Social Inequality and Decision Making.* Chapman & Hall, London.

Rietkerk, M., Ketner, P., Stroosnÿder, L. & Prins, H.H.T. (1996). Sahelian rangeland development: a catastrophe? *Journal of Range Management*, **49**, 512–519.

Rosenheim, J.A., Wilhoit, L.R. & Armer, C.A. (1993). The influence of intraguild predation among generalist insect predators on the suppression of an herbivore population. *Oecologia*, **96**, 439–449.

Rowe-Rowe, D.T., Green, B. & Crafford, J.E. (1989). Estimated impact of feral house mice on sub Antarctic invertebrates at Marion Island. *Polar Biology*, **9**, 457–460.

Sinclair, A.R.E. (1995). Equilibria in plant–herbivore interactions. In *Serengeti II: Dynamics, Management, and Conservation of an Ecosystem* (Ed. by A.R.E. Sinclair & P. Arcese), pp. 91–113. University of Chicago Press, Chicago.

Sinclair, A.R.E. & Fryxell, J.M. (1985). The Sahel of Africa: ecology of a disaster. *Canadian Journal of Zoology*, **63**, 987–994.

Smith, V.R. & Steenkamp, M. (1990). Climate change and its ecological implications at a sub antarctic island. *Oecologia*, **85**, 14–24.

Spiller, D.A. & Schoener, T.W. (1990). A terrestrial field experiment showing the impact of eliminating top predators on foliage damage. *Nature*, **347**, 469–472.

Sprent, J.I. (1987). *The Ecology of the Nitrogen Cycle.* Cambridge University Press, Cambridge.

Srivastava, D.S. & Jefferies, R.L. (1995a). The effects of salinity on the leaf and shoot demography of two arctic forage species. *Journal of Ecology*, **83**, 421–430.

Srivastava, D.S. & Jefferies, R.L. (1995b). Mosaics of vegetation and soil salinity: a consequence of goose foraging in an arctic salt marsh. *Canadian Journal of Botany*, **73**, 75–85.

Srivastava, D.S. & Jefferies, R.L. (1996). A positive feedback: herbivory, plant growth, salinity and the desertification of an arctic salt marsh. *Journal of Ecology*, **84**, 31–42.

Sterner, R.W. (1986). Herbivore direct and indirect effects on algal populations. *Science*, **231**, 605–607.

Sterner, R.W. (1990). The ratio of nitrogen to phosphorous resupplied by herbivores: zooplankton and the algal competitive arena. *American Naturalist*, **136**, 209–229.

Strong, D.R. (1992). Are trophic cascades all wet? Differentiation and donor control in species ecosystems. *Ecology*, **73**, 747–754.

Strong, D.R., Maron, J.L., Harrison, S., Connors, P.G., Jefferies, R.L. & Whipple, A. (1995). High mortality, fluctuations in numbers and heavy subterranean insect herbivory in bush lupine *Lupinus arboreus*. *Oecologia*, **104**, 85–92.

Toutain, F. (1987). Les Litières: Sièges de systèmes interactifs etmoteurs de ces interactions. *Revue Ecologie et Biologie du Sol*, **24**, 231–242.

Van Soest, P.J. (1994). *Nutritional Ecology of the Ruminant*, 2nd edn. Cornell University Press, New York.

Vanni, M.J. & Layne, C.D. (1997). Nutrient recycling and herbivory as mechanisms in the "top down" effect of fish on algae in lakes. *Ecology*, **78**, 21–40.

Vanni, M.J., Layne, C.D. & Arnott, S.E. (1997). "Top down" trophic interactions in lakes: effects of fish on nutrient dynamics. *Ecology*, **78**, 1–20.

Vitousek, P.M. (1985). Community turnover and ecosystem nutrient dynamics. In *The Ecology of Natural Disturbance and Patch Dynamics* (Ed. by S.T.A. Pickett & P.S. White), pp. 325–333. Academic Press, New York.

Vitousek, P.M., Matson, P.A. & van Cleve, K. (1989). Nitrogen availability and nitrification during succession: primary, secondary, and old field seres. In *Ecology and Arable Land* (Ed. by M. Clarholm & L. Bergström), pp. 161–171. Kluwer Academic Publishers, Dordrecht.

Warren, A. & Maizels, J.K. (1977). Ecological change and desertification. In *Desertification* (Ed. by United Nations), pp. 171–260. Pergamon Press, Oxford.

Watkins, B.P. & Cooper, J. (1986). Introduction, present status and control of alien species at the Prince Edward islands. *South African Journal of Antarctic Research*, **16**, 86–94.

Wedin, D.A. & Pastor, J. (1993). Nitrogen mineralization dynamics in grass monocultures. *Oecologia*, **96**, 186–192.

Wedin, D.A. & Tilman, D. (1990). Species effects on nitrogen cycling: a test with perennial grasses. *Oecologia*, **96**, 186–192.

Wedin, D.A. & Tilman, D. (1993). Competition among grasses along a nitrogen gradient: initial conditions and mechanisms of competition. *Ecological Monographs*, **63**, 199–229.

Westoby, M. (1979/80). Elements of a theory of vegetation dynamics in arid rangelands. *Israeli Journal of Botany*, **28**, 169–194.

Westoby, M., Walker, B. & Noy Meir, I. (1989). Opportunistic management for rangelands not at equilibrium. *Journal of Range Management*, **42**, 266–274.

Whicker, A.D. & Detling, J.K. (1988). Ecological consequences of prairie dog disturbances. *BioScience*, **38**, 778–785.

White, P.S. & Pickett, S.T.A. (1985). Natural disturbance and patch dynamics: an introduction. In *The Ecology and Natural Disturbance and*

Patch Dynamics (Ed. by S.T.A. Pickett & P.S. White), pp. 3–13. Academic Press, New York.

Wilson, D.J. & Jefferies, R.L. (1996). Nitrogen mineralization, plant growth and goose herbivory in an Arctic coastal ecosystem. *Journal of Ecology*, **84**, 841–851.

Wilson, J.B. & Agnew, A.D.Q. (1992). Positive-feedback switches in plant communities. *Advances in Ecological Research*, **23**, 236–336.

Part 3

Scaling up from food intake to community dynamics in vertebrate herbivores

Introductory remarks

H.H.T. Prins[1] and G.R. Iason[2]

This session reviews our ability to explain abundance of vertebrate herbivores by building upwards from processes of food acquisition and nutrient extraction to population and community level responses. It is particularly apposite to consider this spectrum as the implications of recent advances in our theoretical understanding of the functional response of herbivores (Spalinger & Hobbs 1992), for higher ecological levels, have yet to be fully explored. Central to this session is the appreciation that energy and nutrients are the major driving forces for population dynamics and that the surplus or deficits in relation to the animal's requirements determine whether young will be produced or whether starvation occurs. Superimposed on this, predation may play a role, and the differential ability of individuals both within and between species to gather energy and nutrients from the environment may determine whether there are competitive interactions that shape the structure of the assemblage of herbivores.

Iason and Van Wieren review different fermentation strategies that have evolved in mammalian herbivores to cope with food that is of low quality (low digestible energy content or low nutrient content) as compared with the food of carnivores. They show that within species the digestion coefficient for a given forage varies considerably as a result of acclimation to the diet, indicating that assumed constraints due to digestive limitation must be considered as flexible. Iason and Van Wieren cite evidence that fibre content of the food of ruminants is not only a major determinant of digestibility but that it also negatively influences the rate of intake, and therefore also the functional response, via the greater requirement for chewing fibrous diets. Hence, food quality and food quantity are intricately linked in herbivores.

1 Department of Environmental Sciences, Wageningen Agricultural University, Bornsesteeg 69, 6708 PD Wageningen, The Netherlands
2 Macaulay Land Use Research Institute, Craigiebuckler, Aberdeen AB15 8QH, UK

Fryxell takes us one step further into the labyrinth, by showing that the functional response curve should not be viewed as a univariate response to changing resource abundance. Not only are food abundance and quality of importance (Chapters 11 and 13), but behavioural reactions to the size and spatial distribution of food items may play a role too. Although he focuses on the beaver, the insights gained from his work are, we think, generally applicable.

Although simplicity in modelling is a worthy goal (Chapters 17 and 19) because of the complexity of natural systems, it seems likely that a plethora of details must be elucidated before we can achieve a mechanistic understanding of herbivore population densities and their fluctuations. Many of these details are brought together and explicitly integrated in the model of Illius and Gordon, which predicts from first principles the periodic winter mortality of Soay sheep, which is a predominant feature of their population dynamics. The choice of free-living Soay sheep as their model animal is instructive, as the relevant detailed information on energy expenditure, digestibility, and food selection could be derived from agricultural science. This provides the richness of biological detail required to enable Illius and Gordon to make the important step of successfully modelling the relationship between food intake and the numerical response of herbivores.

Klaassen's chapter, however, shows that while Illius and Gordon use age/sex classes in their model, intra-individual differences may be necessary to incorporate in the next generation of population models. His examples from ornithology show that the basal metabolic rate, which is perhaps the single most important keystone of modern animal ecology, is not a fixed parameter but varies in relation to cold exposure, and the proportion of fat and intestinal organs in relation to body mass. Unquestionably, these parameters oscillate over the year and Klaassen's work shows that physiologically based models could potentially become even richer in detail than that of Illius and Gordon and that as in the case of digestion (see Chapter 11) single modelled parameter values for metabolic processes are unlikely to suffice.

Finally, Krebs et al. provide even more complex cases where different herbivore species interact through competition and predation. Although populations still show a numerical response to changes in food abundance and food quality, their work centres on the thesis that herbivores are limited (not regulated) by predation. One may question whether this is generally true, and Krebs et al. point out that migration may offer a means to escape from limitation by predators. They have limited their aim to large effects, i.e. 'unless one can understand doubling (or halving) numbers in one generation by the change of an ecological factor, the effect is too small to be well understood yet'. This is quite humbling, because it shows that at the community level, our degree of understanding of interactions between different herbivores, their food supply and predators or diseases is still limited (see also Prins & Olff 1997).

The way ahead

1 Metabolic, digestive and ingestive processes are relatively well understood in ruminants compared with other groups but in any case are only relevant to population and community processes when considered in relation to the free-living animals' requirements for maintenance, growth and reproduction. The success of Illius and Gordon's model could be extrapolated to other animals if equivalent detailed information was available.

2 Ultimately, to understand numerical responses, the models should include the effects of plant quality variation over the year, feedback between consumption and plant quality, food choices of herbivores and spatial decision rules in relation to distribution and quality of forage.

3 The established link between metabolism, nutrition and at least some population processes is now clear, but our ability to extend these principles to communities remains poor. The way in which predators and/or competition influence time–energy budgets of herbivores may provide a route of feedback between the nutritional and community levels.

If these steps are undertaken, we will finally be able to understand the whole causal web between animal physiology, food abundance, food quality and community ecology of assemblages of coexisting herbivore species. This will enable us to manage herbivores in a truly scientific way, and to answer the ultimate question of animal ecology, namely, what determines animal numbers?

References

Prins, H.H.T. & Olff, H. (1997). Species richness of African grazer assemblages: towards a functional explanation. In *Dynamics of Tropical Communities* (Ed. by D.M. Newbery, H.H.T. Prins & N.D. Brown), pp. 449–490. Blackwell Science Ltd, Oxford.

Spalinger, D.E. & Hobbs, N.T. (1992). Mechanisms of foraging in mammalian herbivores: new models of functional response. *American Naturalist*, **140**, 325–348.

Chapter 11

Digestive and ingestive adaptations of mammalian herbivores to low-quality forage

G.R. Iason[1] and S.E. Van Wieren[2]

Summary

For herbivores utilizing fermentation strategies, digestibility depends on the retention time of plant cell wall (fibre) particles in the digestive tract. Clearance of indigestible plant components from the gastrointestinal tract limits the ability of herbivores to process large quantities of food. This leads to a positive relationship between digestibility and intake rate of different diets. Allometric studies across species suggest that metabolic requirements increase with body mass$^{0.75}$ whereas gut capacity scales with body mass$^{1.0}$. Larger animals should thus have a more favourable ratio of gut capacity and digesta retention time to mass specific requirements, and hence be able to tolerate lower quality forage.

We review digestive strategies that alleviate the passage rate constraint. Some hindgut fermenting herbivores have a separation mechanism that accelerates excretion of larger indigestible particles and selectively retains smaller and soluble particles in the caecum. This facilitates a negative relationship between digestibility and intake.

Foraging strategies of ruminants have long been classified along a continuum from concentrate selectors (browsers) to bulk and roughage feeders, which eat mainly grasses (grazers). Recent studies have attributed several measures of digestive function to body mass whilst finding a surprising lack of digestive differences between browsers and grazers. We

1 Macaulay Land Use Research Institute, Craigiebuckler, Aberdeen AB15 8QH, UK
2 Department of Environmental Sciences, Wageningen Agricultural University, Bornsesteeg 69, 6708 PD Wageningen, The Netherlands

compiled data from 278 digestion trials using 20 ruminant species, to test the prediction that fibre digestibility by grazers was greater than that by browsers. The negative effect of the concentration of the poorly digestible plant structural component lignin, explained the largest proportion of variation in cell-wall digestibility, followed by the predicted grazer–browser difference and a weak positive effect of body mass. Digestibility of cell wall by browsers was relatively lower at higher lignin concentrations indicating that they digest poor-quality diets less well than grazers. Although these cell-wall digestive effects are as predicted, the grazer–browser classification currently lacks a mechanistic digestive basis.

We tested the plasticity of digestion within a species, using mountain hares (*Lepus timidus*). These are typical of intermediate feeders that switch between a graminoid-eating (grazer) strategy in summer, to browsing in winter. During a 2-week period of dietary changeover of captive hares from grass to natural woody browse forages, growth rate was reduced. Browse-adapted hares digested woody browse significantly better than non-adapted hares. A dietary switch involves costs and represents a nutritional bottleneck. The marked digestive adaptation suggests that digestive constraints to foraging behaviour must be considered as flexible.

Whilst herbivore body size and qualitative aspects of forage plants are important (relatively) fixed constraints, flexibility of ingestive behaviour (chewing and ruminating) interacts with digestive processes to provide alternative strategies for meeting requirements. More intense chewing or diet selection increases diet digestibility, but results in a negative relationship with intake rate because of the greater time required. This may lead to interactions between nutrition and antipredator behaviour.

The intimate relationship between ingestion and digestion suggests an important role of digestive limitation in the functional responses of mammalian herbivores and mechanisms underlying population and community processes.

Introduction

The abundance of green plant material in the biosphere led Hairston *et al.* (1960) to suggest that herbivore populations are seldom food limited. However, in reaching this conclusion, they failed to acknowledge that in comparison to the diets of other organisms, most of the biomass available to herbivores is generally of very low nutritional quality. This is particularly true of mammalian herbivores whose relatively large size precludes the fine selection of plant cells or plant parts that may be achieved by invertebrates (Crawley 1989). In comparison to animal tissue, plant material is low in nitrogen and high in recalcitrant fibrous material, which is only slowly digested. Herbivores can circumvent this problem by selecting a diet higher in quality than the average of the available plant material and/or by evolving digestive strategies that permit nutrient extraction from the poor-quality forage. Harvesting of sufficient nutrients by herbivores requires long periods of activity and hence high energy expenditures (Robbins 1983).

The nutrients available to herbivores from plant material are the product of the breakdown of the plant's chemical structures by the herbivores' digestive processes; the determinants of quality of plant material to herbivores are hence characteristics of both plants and herbivores. Plant material ingested by herbivores can be viewed most simply as polysaccharide cell walls surrounding soluble cell contents present in vacuoles (Van Soest 1982). While most components of the cell contents are readily available to herbivores, the cell-wall components consist mainly of cellulose and hemicellulose, which cannot be degraded directly by animals because they lack cellulase enzymes. However, part of the cell-wall constituents can be digested in mammalian herbivores via fermentation by populations of symbiotic microorganisms residing in their gastrointestinal tracts, while other cell-wall components and tissues are refractory even to fermentative digestion (Table 11.1). Fermentation is a slow process requiring long gastrointestinal retention times of food for effective degradation. Digestion theory states that animals maximize their fitness by maximizing the long-term rate of energy or nutrient release from their food. Plotting net energy

Table 11.1 Potential digestibility of major plant constituents. Adapted from Jones and Wilson (1987).

	Cell contents	Cell wall
Completely digestible	Sugars, protein, non-protein nitrogen, lipid, organic acids, soluble minerals	Unlignified cellulose, hemicellulose, pectin
Partially digestible		Partially lignified cellulose and hemicellulose
Indigestible	Pigments, tannins	Lignin, highly lignified cellulose and hemicellulose, silica, cuticle

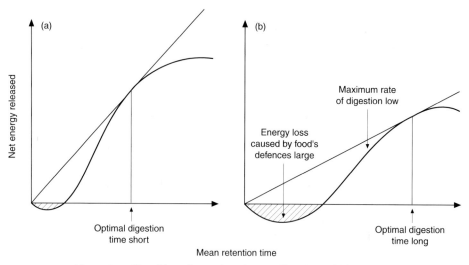

Figure 11.1 The effect of retention time on digestion of (a) a high-quality and (b) a low-quality diet. Curves show expected time course of energy release from food and the straight lines define the maximum rate of digestion. From Sibley (1981).

released against retention time of different feeds shows that the optimal retention time of low-quality foods is longer than that of high-quality foods (Fig. 11.1) (Sibley 1981), and this applies strongly to plant cell-wall material in the gut of herbivores. This difference also applies to different components of food; the release of the cell contents is relatively rapid and animals, such as geese, that specialize on this component of plants, only retain food for the short periods required to effect this release, although the overall quality of the diet as determined by the extent of digestion (digestibility) is low. The first aim of this work is to review both the plant determinants of digestibility, and the digestive strategies of mammalian herbivores that permit utilization of low-quality, fibrous forage; we also consider how these strategies interact with food intake and ingestive behaviour.

Our second aim is to investigate the relative importance of herbivore body mass, and ecophysiological digestive strategy as determinants of cell-wall digestibility by ruminants. The ruminants are a group of herbivores that are highly specialized to ferment plant materials. For over 25 years comparative ruminant nutrition has been dominated by the classification

340

of species along a continuum from bulk roughage feeders (grazers) to concentrate selectors (browsers) (Hofmann 1973, 1989). The classification is on the grounds that the observed gastrointestinal morphology permits different effectiveness of digestive processing of the two types of diet dominated by either graminoids, or woody plants (browse) and forbs, respectively, which have very different physical and chemical characteristics. However, many facets of digestive function that segregate grazers and browsers according to the Hofmann classification could also be predicted on the basis of species' body masses, identified as the main determinant of several aspects of digestive function in recent studies (Gordon & Illius 1994; Robbins *et al.* 1995). We attempt to distinguish these possibilities presenting new evidence using an interspecific, comparative approach.

In addition to the interspecific variation we also consider intraspecific plasticity in digestibility, in the context of how herbivores cope with seasonal variation in food quality. Specifically, we experimentally investigate the extent of digestive adaptation between the summer and winter diets within a single small herbivore species, the mountain hare (*Lepus timidus*), which switches between the grazer and browser strategy on a seasonal basis.

Plant determinants of digestibility

Lignin is a polyphenolic component of cell walls that imparts mechanical strength to the plant tissue. It is itself indigestible and limits the digestibility of plant cell wall (Swain 1979). It is the only major plant polymer whose subcomponents and structure are not precisely known, and the mechanisms by which lignin protects carbohydrates from degradation are not entirely understood, although various mechanisms have been proposed including effects on fermenting microbes and the formation of lignin–polysaccharide complexes (Van Soest 1982). Fermentation of plant samples by an inoculum of microorganisms under a set of standard *in vitro* conditions permits comparison of plant characteristics that determine digestibility, independently of variation due to herbivore digestion kinetics. Increasing lignin concentration in plant material has a strongly negative effect on animal's digestibility of cell walls (Fig. 11.2), and this primary effect of lignin on digestibility has also been demonstrated *in vivo* (Hanley *et al.* 1992).

With respect to herbivore nutrition, plants can be classified relative to their contents of potentially usable nutrients, an important distinction being

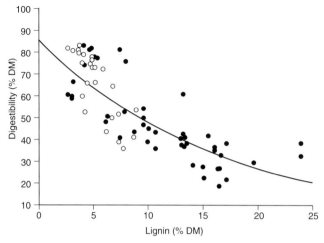

Figure 11.2 The relationship between lignin content and *in vitro* digestibility of dry matter (DM) in temperate grass (open circle) and browse (closed circle) species from an area of sandy soil in The Netherlands. *In vitro* digestibility represents maximum potential digestibility and was measured using a cattle rumen inoculum. $y = 84.06e^{-0.06x}$, $r^2 = 0.67$, $n = 70$.

the one between monocotylodons and dicotylodons. Figure 11.3 compares the mean chemical composition of a number of grass and woody browse species. This contrast is particularly relevant in view of the classification of herbivores along a continuum from grazers to browsers. For neutral detergent fibre (NDF), which represents total cell wall, lignin in NDF and condensed tannin content, all differences between grass and browse are significant ($P < 0.05$) (except NDF in winter). Grasses thus contain relatively more cell wall than woody plants but, grass cell walls are much less lignified than browse cell walls. This leads to a higher digestibility of grasses when compared with browse, the cell walls of which are potentially of very little value as substrate for fermentation (see Fig. 11.2). Nitrogen content in autumn and winter is higher in grasses, although for some grass species (not shown) nitrogen concentration can drop to low levels due to senescence. Herbivores have to cope with low plant nitrogen levels in many biotopes during certain periods and specific adaptations have evolved, which are discussed below.

In addition to fermentable cell walls and components of cell contents that can be used as nutrients by herbivores, plants also contain secondary

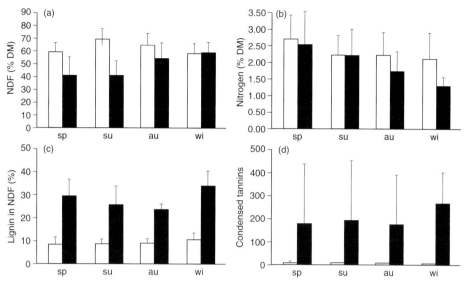

Figure 11.3 Seasonal variation in chemical composition of temperate grass (open bars) and browse (closed bars) species (Means with SD). Samples were hand plucked as being representative of bites taken by cattle and red deer. sp, spring; su, summer; au, autumn; wi, winter. (Condensed tannins are expressed as milligrams quebracho tannin equivalents per millilitre.)

metabolites such as tannins and other polyphenolics, with no known function within the plants, but which act as allelochemicals in interactions with other organisms including herbivores (Rhoades 1985). In general, woody plants contain higher levels of tannins and polyphenolics than do grasses (Fig. 11.3). We do not consider these so-called chemical defences here, but two main classes can be distinguished by their effects:

1 compounds that limit energy and protein digestibility, e.g. condensed tannins;

2 substances that have a toxic effect after absorption by the herbivore. Because the presence of secondary compounds lowers the nutritional value of plants to herbivores, animals have to take them into account while foraging. The options are avoiding them, or ingesting them and detoxifying them, both of which incur costs. The adaptations to ingestion and their associated costs are discussed by McArthur *et al.* (1991) and Foley and McArthur (1994).

Animal determinants of digestibility

The problem of herbivore body mass

In cross-species allometric comparisons the metabolic requirements of mammals has been shown to increases with body mass$^{0.75}$ (Nagy 1987). In contrast, the capacity of the gastrointestinal tract increases with body mass$^{1.0}$ (Parra 1978; Demment & Van Soest 1985). This means that smaller animals have higher mass-specific food requirements without an accompanying proportional increase in their gut capacity, which limits the volume of digesta retained and its rate of passage. It is hence considered paradoxical that some small mammals are able to utilize fibrous diets (Foley & Cork 1992; Justice & Smith 1992). Estimates for the lower size limit at which a fermentation strategy is viable are 15 kg for a ruminant and that a 1 kg mammal should be unable to meet maintenance energy requirements by eating a grass diet with more than 50% fibre (Cork 1994). We now consider some mechanisms that permit mammalian herbivores to digest low-quality fibrous forage, and approach and exceed the limits that we would expect from these allometric considerations.

Fermentation strategies, selective passage or retention of food particles, caecotrophy and urea recycling

Herbivores have developed special fermentation chambers that facilitate digestion of bulky food sources, permitting the long retention times of food particles required for the relatively slow process of cellulose fermentation. These chambers also provide the necessary anaerobic conditions, a pH of 6–7 and temperatures between 38 and 42°C required by the symbiotic microorganisms. Surface enlargement of these chambers is created by means of sacculation, papillation, folding and other mechanisms (see Hofmann 1973, 1988; Hume 1989). The major fermentation sites have been developed in either the foregut or the hindgut (caecum and colon). Foregut fermentation occurs in many taxonomic groups including primates and ungulates, with diverse sizes and feeding styles (Van Soest 1982). Within the group of foregut fermenters, the ruminants can be considered as the most developed and specialized, anatomically and morphologically, with respect to the capacity to utilize fibre. Of the extant ungulates consisting of c. 176 species, 145 are ruminants. Hindgut fermenters can be divided into colon fermenters and caecum fermenters, so called because of the digestive organ forming the main site of fermentation. Representatives of the first group are the equids and the pigs, while many rodents and the lagomorphs (rabbits and hares) are caecal fermenters. The endproducts of microbial fermentation are volatile fatty acids (VFA), which are readily absorbed and oxidized or converted to other energy sources, and CO_2, CH_4 and microbial matter, which can be partly digested and serve as an important source of amino acids. VFA produced by fermentation provide only 34% and 37% of the daily maintenance requirements of pigs and rabbits, respectively (Marty & Vernay 1984; Van Wieren 1996). This contrasts with the ruminants, in which VFA are the predominant source of energy.

Selective retention of poorly digestible particles can facilitate greater overall digestive decomposition. This should apply to all species digesting plant cell walls, but is theoretically most likely in larger species, which are less limited by their ratio of metabolic requirements to gut capacity than smaller herbivores for whom the retention of less digestible particles is not a viable option (Justice & Smith 1992). Retention can be promoted by various mechanisms and structures, including the omasum in ruminants, which acts as a sieve permitting only particles of a certain size to pass. In many hindgut fermenters selective retention is achieved by various anatomical devices and physical processes in the caecum or colon.

Several species of hindgut fermenter have a special caecal separation mechanism that helps deal with fibrous forage, not by prolonging retention but by accelerating excretion of fibre. Muscular contractions in the colon accomplish a separation of fibre particles from the non-fibre feed components, with peristaltic contractions rapidly moving less digestible, larger fibre particles through the colon for excretion in the so-called hard faeces. Antiperistaltic action moves fluids and small particles in a retrograde manner through the colon to the caecum, where they are retained for digestion and fermentation (e.g. microtine rodents: Hume *et al.* 1993; lagomorphs: Cheeke 1987; horses: Björnhag *et al.* 1984). A clear advantage of foregut fermentation is that the host has better opportunities to utilize essential vitamins and protein produced by the microorganisms. In contrast, hindgut fermenters have difficulty in capturing microbial protein and vitamins, which would be easily lost in the faeces. To minimize these losses many species of hindgut fermenter (especially the small rodents and lagomorphs) practise caecotrophy, whereby soft faeces are reingested (Hörnicke & Björnhag 1980). At intervals, the caecum contracts, and the caecal contents (soft faeces) are expelled through the colon and consumed directly from the anus. The excretion of the hard and soft faeces is subject to a circadian rhythm and in rabbits the separation mechanisms appear not to operate during formation of soft faeces (Jilge 1982).

Although the rumen, caecum and colon provide very similar environments for carbohydrate fermentation, differences in their function are apparent. The mechanisms for selective retention of poorly digestible food particles in large foregut fermenters and selective excretion of them in hindgut fermenters result in contrasting relationships between intake and digestibility. In the former, as digestibility decreases then so does intake, which is limited by the clearance of the indigestible material from the fermentation organ (Minson 1982). In contrast, in hindgut fermenters, intake increases as digestibility decreases over a broad range of digestibilities. The capacity of some small hindgut fermenters to increase intake

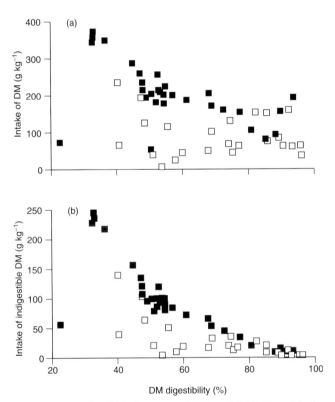

Figure 11.4 Intake of (a) dry matter (DM) and (b) indigestible dry matter on diets of different dry matter digestibility. Microtines (closed squares) are known to have digesta separation mechanisms whereas sciurids and murids (open squares) do not. From Cork (1994).

and thus throughput of food, as digestibility decreases, is undoubtedly contributed to by the separation mechanisms and caecotrophy, which, along with reductions in metabolic costs, help to reduce the constraints imposed by small body size (Fig. 11.4) (Cork 1994). However, several species such as members of the Sciuridae still utilize high-fibre diets in the absence of these mechanisms and this is probably facilitated by their larger body size (Hume *et al.* 1993).

Recycling of urea from catalytic processes, from the bloodstream back to the digestive tract for re-utilization by microbes is an important mechanism for conservation of nitrogen in many herbivores. In ruminants, plasma urea enters the rumen by two routes: with saliva and by diffusion through the rumen wall. With forage diets, 15–50% of the total urea recycled can follow the salivary route. From 23 to 92% of plasma urea is recycled to the digestive tract, with higher value associated with higher intakes. The amounts of nitrogen recycled to the rumen can be as high as 15 g day^{-1} for sheep and 60 g day^{-1} for cattle (Owens & Zinn 1988). Urea recycling is not confined to the rumen. It has also been established that significant urea recycling occurs to the hindgut of the horse (Prior *et al.* 1974), lagomorphs (Cheeke 1987) and other mammals.

Does body mass or ecophysiological digestive adaptation explain fibre digestibility by ruminants?

Twenty-five years ago Hofmann proposed the division of ruminants according to differences in anatomy of their digestive systems. These were associated with their contrasting foraging strategies, namely, grazers, which are bulk roughage feeders eating mainly grasses, concentrate selectors (or browsers), which selectively ingest forbs and parts of woody plants such as leaves, buds and twigs, and intermediate or mixed feeders (Hofmann & Stewart 1972; Hofmann 1973). The robust nature of Hofmann's categorization into feeding types is demonstrated in Fig. 11.5, which shows the average percentage of grass in the diet of 45 species of ruminant. The classification into browsers, grazers and intermediate feeders is based on Hofmann's categorization and not on diet composition, so no circular argument is involved in this analysis (Van Wieren 1996). The low variability in the percentage of grass in the diet of browsers as compared with the other two groups suggests that they are less flexible in tolerating large proportions of grass than some grazers are in tolerating large quantities of browse. Figure 11.5 also makes clear that small and large species occur in all groups.

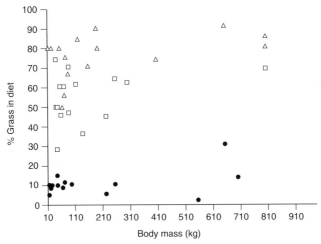

Figure 11.5 The percentage of grass in the diet of 45 ruminant species. (From Van Wieren 1996). (●) Browsers; (△) grazers; (□) intermediate feeders.

We have already demonstrated that grasses differ significantly from woody species in their composition, and it would be expected that within groups of herbivores, such as the ruminants, we would find species that are either specialized on the exploitation of grass or of browse. The grass specialists can be expected to have become adapted to the utilization of the fibre of highly fibrous forage, while the browse specialists have become adapted to the utilization of the cell contents of less fibrous forage. Associated with their different feeding niches, browsers and grazers differ in mouthparts and teeth, the browsers having smaller muzzle widths (Hofmann 1973; Janis & Ehrhardt 1988), which facilitate finer selection of plant parts (Gordon & Illius 1988).

Many anatomical characteristics associated with the digestive system have been described (Hofmann 1989), but only the main ones are summarized here. Grazers were reported to have a larger, more muscular rumen and smaller ostia between the rumen and the omasum than browsers. The simple rumenal structure and the small evenly papillated rumen wall of browsers were held to favour less selective retention and a more rapid passage of digesta to a generally larger hindgut, which would hence be expected to be the site of greater additional fermentation of cell-wall material than in the grazers. Saliva from the larger and presumably more productive salivary glands of browsers facilitates more rapid passage of digesta from the forestomach, buffers against the changes in acidity produced by the rapid fermentation rates of the cell soluble plant components ingested and may serve as a defence against the negative effects of dietary tannins (Hofmann 1989). In most respects, intermediate feeders are intermediate between browsers and grazers.

Despite a total of almost 200 citations of Hofmann's 1973 and 1989 publications, only recently have functional differences in digestion between grazers and browsers been tested. In comparative studies of ruminants, Gordon and Illius (1994) have found that neither fermentation rate, mass of rumen, hindgut nor total gastrointestinal tract contents differed as predicted between Hofmann's classes and nor did whole tract retention time of digesta. Similarly, Robbins *et al.* (1995) found no differences between Hofmann's classes in fibre digestibility nor in liquid flow rates from the rumen, although browser's salivary glands were larger than those of grazers. Although the detailed anatomical differences were claimed to be independent of body mass (Hofmann 1989), both studies found the major effects on all variables measured were attributable to herbivore body mass (BM), with the exception of fermentation rates in the caecum and energy content of VFA produced in the rumen (Gordon & Illius 1994); the net energy supplied by VFA from the rumen was, however, greater for grazers than for browsers after adjusting for body mass.

In order to test the relative importance of body mass and Hofmann's classes on digestibility of plant cell walls, and thereby extend the analysis of Robbins *et al.* (1995), we compiled the results of 278 feeding trials, although not all trials yielded results for all measurements. Digestibility data were available for 20 species of ruminant ranging in size from 11 to 816 kg. Of these, six were concentrate selectors (browsers): giraffe (*Giraffa camelopardus*), white-tailed, black-tailed and mule deer (*Odocoileus* spp.), moose (*Alces alces*), roe deer (*Capreolus capreolus*); five intermediate feeders: red deer or elk (*Cervus elaphus*) basasingha (*Cervus duvaucelli*), gemsbok (*Oryx gazella*), eland (*Taurotragus oryx*), domestic goat (*Capra hircus*); nine grass and roughage feeders (grazers): African buffalo (*Syncerus caffer*), gaur (*Bos gaurus*), bison (*Bison bison*), wisent (*Bison bonasus*), waterbuck (*Kobus ellipsiprymnus*), nilgai (*Boselaphus tragocamelus*), domestic cattle (*Bos taurus*), domestic sheep (*Ovis aries*) and bighorn sheep (*Ovis canadensis*). Because the data set was dominated by a few domestic species, namely, cattle, sheep, goats as well as some well-studied wild deer species, in regression analyses data points were weighted by the inverse of the frequency of occurrence of that species in the analysis being conducted.

NDF digestibility and body mass (BM) were transformed into natural logarithms. The dominant feed types were grass, browse and alfalfa (for more details see Van Wieren 1996).

Weighted regression of digestibility of NDF against BM showed only a weak positive relationship:

$$\log \text{NDF digestibility } (\%) = 3.35_{(SE\ 0.123)} + 0.104_{(SE\ 0.0234)} \log(BM) \qquad (1)$$
$$(F_{1,\ 178} = 19.56, P < 0.001; R^2 = 9.4\%)$$

In contrast, a much greater proportion of the total variation in log NDF digestibility was explained by an ANOVA using the three Hofmann classes (grazers, 4.05; browsers, 3.55; intermediate feeders, 3.90; SED = 0.043, $F_{2,\ 177} = 83.78, P < 0.001; R^2 = 67.4\%$).

Factors affecting NDF digestibility were analysed in a forward stepwise multiple regression analysis in which dietary lignin, BM, Hofmann classes and all interactions of these terms were entered as possible explanatory variables. The biggest single determinant of cell-wall (NDF) digestibility was the lignin content of the diet ingested (Fig. 11.6a; Table 11.2). This accords with the study of North American cervids (Hanley *et al.* 1992), which showed that NDF digestibility declined exponentially with lignin content of the cell wall. The next effects explaining variation in the stepwise regression

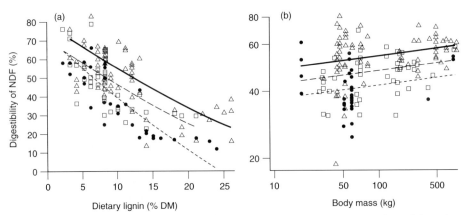

Figure 11.6 The effect of (a) dietary lignin and (b) body mass on digestibility of neutral detergent fibre (NDF), for grazers (open triangles), browsers (closed circles) and intermediate feeders (open squares). Exponential curves were fitted to raw data for dietary lignin and the lines for body mass are fitted after adjusting for all terms in the statistical analysis (see Table 11.2).

Table 11.2 The results of forward stepwise multiple regression of digestibility of neutral detergent fibre (NDF; ln transformed) with dietary lignin (dtLIG; % of dry matter), Hofmann classification (Hofmann: Grazer, Browser, Intermediate feeder, with coefficients expressed relative to grazers) and natural logarithm of body mass (logBM; kg) and all interactions as possible explanatory variables. Results are given for (a) main effects only and (b) the full model including the only significant interaction.

Explanatory variable*	Coefficient$_{(SE)}$	F_{df}	P	Cum R^2
(a) Main effects only				
Constant	$4.09_{(0.107)}$			
dtLIG	$-0.059_{(0.0042)}$	$302.0_{1,153}$	<0.001	50.2
Hofmann		$57.8_{2,153}$	<0.001	69.2
Browser	$-0.315_{(0.0395)}$			
Intermediate	$-0.161_{(0.326)}$			
logBM	$0.084_{(0.0159)}$	$27.8_{1,153}$	<0.001	73.8
(b) Main effects and interaction				
Constant	$3.84_{(0.119)}$			
dtLIG	$-0.036_{(0.0076)}$	$354.8_{1,151}$	<0.001	50.2
Hofmann	—	$67.9_{2,151}$	<0.001	69.2
Browser	$0.086_{(0.0920)}$			
Intermediate	$-0.101_{(0.0945)}$			
logBM	$0.093_{(0.0148)}$	$32.6_{1,151}$	<0.001	73.8
dtLIG × Hofmann	—	$14.4_{2,151}$	<0.001	77.4
Browser	$-0.042_{(0.0091)}$			
Intermediate	$0.005_{(0.0107)}$			

* See text for explanation of abbreviations.
F values are given with significance (*P*) and explained variation (R^2, in %).

(Genstat 5 1993) statistical model were the Hofmann classification followed by log(BM). The interaction between dietary lignin and Hofmann class on digestibility of cell wall was significant. Browsers digested NDF relatively poorly compared with grazers, on poor-quality (higher lignin) diets (Fig. 11.6a; Table 11.2).

Clearly the Hofmann classification explains more variation in NDF digestibility than does BM; both after adjusting for the effect of lignin concentration in the diet. Grazers digest plant cell-wall material to a greater extent than browsers do, and intermediate feeders are intermediate in their fibre digestion, all as predicted by Hofmann (1989). This result is in contrast to the findings of Robbins *et al.* (1995) who found no effect of Hofmann's classification on NDF digestibility, albeit with a much smaller data set. As well as being a suitable basis for prediction of diet composition and predicting the difference in NDF digestibility observed here, Hofmann's postulation is supported by the grazer–browser difference in salivary gland weights (Robbins *et al.* 1995) and net rumen energy production (Gordon & Illius 1994). However, although these results are predicted by the conventional wisdom derived largely from Hofmann's detailed investigations, they are very difficult to explain in mechanistic terms, in the light of the absence of differences between grazers and browsers in whole tract digesta retention time (Gordon & Illius 1994) or rumen liquid flow rates (Robbins *et al.* 1995). Both sets of authors acknowledge the paucity of the current data regarding passage rates, particularly concerning the distribution of the different classes across the range of body weights. In these data sets there is an absence of grazers below about 80 and 250 kg, respectively, and this problem of distribution of data also exists in Hofmann's anatomical data. Despite the emphasis on anatomical differences as being independent of body mass, a preliminary statistical analysis of Hofmann's (1973) data suggested that variation in several key characteristics of digestive anatomy could be explained better by body mass than by the Hofmann classification (Van Wieren 1996). In summary, Hofmann's classification predicts the superior fibre digestibility by grazers as compared with browsers, but evidence to support the mechanisms that he proposed to underlie the classification is currently lacking. Both the differential passage of diet components and the possible role of the hindgut in compensating for differences in retention and digestion in the foregut in ruminants may be areas of fruitful future research.

Intraspecific plasticity in digestibility associated with seasonal dietary changes

Climate is a major determinant of plant growth and primary productivity. Both are strongly affected by ambient environmental conditions (e.g. Bannister 1976), and senescent plant material contains a higher proportion of cell wall and less cell contents. Hence herbivores occupying seasonal environments must face fluctuations in the quality and quantity of their food supply. In order to prevent demand exceeding the supply during seasons of lower food quality and/or availability, herbivores can adopt or evolve one or more of several strategies. Here we consider seasonal adaptation in digestive efficiency of extraction of nutrients and energy as a strategy for coping with seasonal reduction in diet quality. We are, however, aware of the fact that as well as increasing the returns from poor-quality forage, herbivores also reduce their requirements during these seasons, e.g. via evolved life history strategies that avoid coincidence of the relatively demanding growth and reproductive processes with seasons of low food availability or quality. They also reduce their requirements for maintenance via seasonal metabolic and behavioural adjustments including hibernation, lowering metabolic rates whilst maintaining body temperature or inactivity. Migration to alternative sites is a further option to ensure that food requirements are met.

In seasonal environments, large seasonal shifts in diet ingested, which are driven by variation in food availability and quality, pose a challenge to animals' digestive systems. We might expect many of the morphological, physiological or behavioural adaptations to low-quality diets observed in cross-species comparisons to occur within species as diet quality changes with season. A reduction in diet quality in winter would be expected to lead to an associated increase in gut capacity, and seasonal morphological changes in the gastrointestinal tract of wild herbivorous mammals have been observed (e.g. rodents: Derting & Noakes 1995; snowshoe hares: Smith *et al.* 1980; chamois: Hofmann 1989). Field studies that measure changes in gut morphology do so against a background of seasonal changes in energy demands which would tend to reduce any effects that might be due to diet quality *per se*. For example, the higher food requirements and intake associated with seasonal reproduction may itself lead to greater masses of gut tissue in the summer (Hammond & Diamond 1994), and winter hibernation strategies would reduce the need for greater digestive efficiency in winter (Derting & Noakes 1995). Experimental evidence has shown that in response to lower quality of diet or greater energy requirements, animals' digestive ability increases and gastrointestinal morphology adapts by hypertrophy, to increase their capacity (Gross *et al.* 1985; Bozinovic 1995; Hammond & Wunder 1995). However, few reports of variation in digestive function, typical of that occurring seasonally, use natural forages (e.g. Batzli *et al.* 1994). In an experimental study we quantify the extent of phenotypic digestive adaptation to a dietary shift typical of that which occurs seasonally in the mountain or Arctic hare (*Lepus timidus*) from summer graminoid-dominated to a winter browse-dominated diet. We also consider the broader implications of the seasonal dietary change.

Mountain hares do not hibernate and have winter body reserves that can sustain a maximum of about 4 days without feeding (Thomas 1987). Despite having relatively low heat loss due to high thermal insulation of their winter coat, and adopting postures that conserve heat whilst resting and feeding (Wang *et al.* 1973), these animals must effectively procure a continuous food supply in winter. It is therefore important for them to be able to utilize the poor-quality woody browse forage that is available. Along the continuum of foraging strategies from bulk feeding grazers to highly selective browsers (Van Soest 1982), the mountain hare is considered to be a selective forager that utilizes woody browse in the winter but its diet consists of more grasses and forbs, when available in the summer (Flux 1970; Iason & Waterman 1988). We consider this seasonal switching of diet type to be a form of intermediate feeding (*sensu* Hofmann), but one which demands a very dynamic set of digestive changes, compared with the adoption of simultaneous intake of graminoid and dicotyledonous plants.

Eight young mountain hares were hand-reared to weaning when they were offered a diet of *ad libitum* commercial sheep pellets (Godfor-Notfor; Lantmannen, Orebro, Sweden) and timothy grass (*Phleum pratense*), for 3 weeks. Then over a period of 2 weeks, for four of the leverets plus two non-breeding adults, the grass was withdrawn totally and the pellets reduced to an amount that provided 50% of estimated maintenance requirements. These were replaced by an *ad libitum* supply of freshly harvested browse (the 'Browser' group). The browse plants were silver birch (*Betula pendula*) twigs cut at 5 mm diameter, shoots of ling heather (*Calluna vulgaris*) and stems of raspberry (*Rubus idaeus*). The experiment was conducted in autumn and residual leaf material remained on the raspberry stems. The remaining four leverets plus two further adults (the 'Grazer' group), were offered *ad libitum* fresh timothy grass, the same quantity of pelleted food as the 'Browser' group. Both groups were offered a small supplement (10 g) of bilberry (*Vaccinium myrtilus*). After the 2-week change-over period (days 44–58 in Fig. 11.8), the two groups were maintained on these diets for 5 more weeks. Hares were individually housed and water was available throughout. Four members of each group were then subjected to digestibility trials of each of the food types. Groups were balanced for adult and subadult hares. Trials were conducted in large cages fitted with separators to permit separate collection of urine and faeces (Pehrson 1981). Food offered and refused was collected and analysed separately for total nitrogen and NDF. The diets of the two groups represent summer and winter diets, respectively. Because the aim was to measure the extent of the seasonal digestive responses, it was necessary to prevent longer-term digestive adaptation taking place during the course of the digestion experiments. Digestion trials were hence limited in duration to a 2-day initial period on each single food species diet, followed by 2 days of intake measurements and collection of faeces. Hares were returned to their initial graze or browse diets for at least 3 days between trials. Two of the initial four animals, one from each group, lost weight during the digestibility trials, one of which refused to eat;

both were removed from the experiment. Data were hence analysed using a two-way, forage type *x* group analysis, fitted using restricted maximum likelihood methodology, which deals with unbalanced data sets by treating individual animals as a random factor (Genstat 5 1993). Main effects and the interaction between forage type and group were tested using the Wald statistic and where the interaction effects were significant, then tests between groups within forage type were made using the SED appropriate to the comparison, in the least significant difference test (Snedecor & Cochran 1980). After the experiment, the hares were returned to their respective original browse and graminoid diets for a further 4 weeks prior to slaughter, when their parotid salivary glands and caecum were dissected and weighed; the caeca were flushed and blotted prior to weighing.

There were significant differences between the forage types offered in the dry matter intake, dry matter digestibility (DMDIG), total nitrogen digestibility (NDIG) and the proportion of total nitrogen in the ingested diets ($P < 0.001$). There was no overall difference in DMDIG between groups but the forage type *x* group interaction was significant ($P < 0.001$) and this was due to superior digestibility of two of the three browse forages by the Browser group (Fig. 11.7a). The Browser group had significantly higher NDIG than the Grazer group ($P < 0.001$) and again this was due particularly to significantly higher values on the three browse forages (Fig. 11.7b). Intake of dry matter did not differ between the groups and nor was the group *x* forage type interaction significant ($P > 0.05$; Fig. 11.7c). The proportion of total nitrogen in the diet ingested did not differ consistently between groups, but the group *x* forage type interaction was significant ($P < 0.01$) due to selection of a higher nitrogen diet by the Browser group on the silver birch twigs (Fig. 11.7d), and the Grazer group on the raspberry. The NDF concentration in the diet ingested was not influenced by the group *x* forage type interaction. The parotid salivary glands of the Browser group were significantly larger than those of the Grazer group (Browser group: 1.03 g kg^{-1} BM, SD 0.105; Grazer group: 0.284 g kg^{-1} BM, SD 0.071; $t = 13.1$, df 7, $P < 0.001$) but there was no difference in the empty caecal weights (Browser group: 18.3 g kg^{-1}BW, SD 2.28; Grazer group: 15.27 g kg^{-1} BW, SE 2.67; $t = 1.90$, df 7, $P < 0.1$).

The effect of the dietary change on the growth of the leverets is apparent from the growth curves of the eight leverets during the course of the experiment (Fig. 11.8). Each curve is smoothed by fitting splines with 4 degrees of freedom to each hare. The precise birth dates of the hares, which were all caught wild, were unknown. However, although the hares were

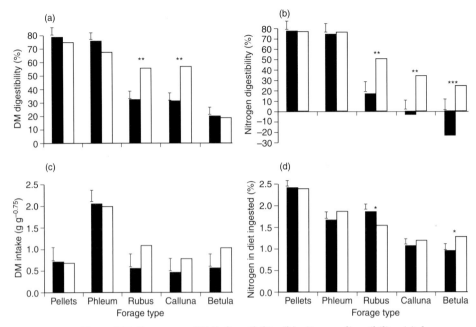

Figure 11.7 Dry matter (DM) digestibility, (b) nitrogen digestibility, (c) dry matter intake and (d) percentage of nitrogen in diet ingested of five diets by mountain hares either pre-adapted to woody browse ('Browser' group) (open bars) or not pre-adapted ('Grazer' group) (closed bars). Means and SED for each Grazer–Browser comparison.

clearly different ages, all were within the weight range over which growth is approximately linear (Pehrson & Lindlöf 1984), with the exception of the first few measurements of the smallest hare (No. 7). During the final 4 weeks of separate diets prior to the digestibility trials (corresponding to days 68–98 in Fig. 11.8), there were no significant differences in growth rates between the groups (Browser group: 19.0 g day^{-1}, SD 1.55; Grazer group: 20.9 g day^{-1}, SD 4.24; $t = 0.84$, df 7, NS).

Ecological and adaptive significance of seasonal digestive plasticity

The digestive constraints that are thought to influence animals' foraging choices have, until recently, been considered in theoretical studies to result from relatively inflexible plant and animal characteristics (see Owen-Smith 1994). This is surprising given the evidence that animals' digestive systems can adapt to increase their capacity in response to lower quality of diet

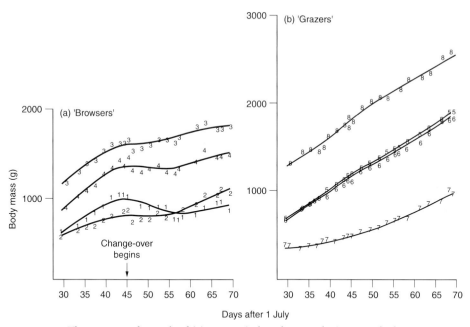

Figure 11.8 The patterns of growth of (a) mountain hare leverets during a gradual dietary shift from a diet of 50% grass/50% pelleted food to a 50% browse/50% pelleted diet ($N=4$) and (b) mountain hare leverets fed continuously on grass (50%) and pelleted food (50%). Curves were fitted to body mass data for each hare individually (labelled 1–8) using smoothing splines with 4 degrees of freedom.

or greater energy requirements (Gross *et al.* 1985; Green & Millar 1987). However, this experiment demonstrates the profound adaptive responses in digestive function that can be achieved by a herbivore whose foraging strategy involves flexibility between seasons. Although mountain hares are generally unable to maintain themselves on diets consisting of a single browse species (Pehrson 1981), the browse-adapted hares' digestion of the browse forages was superior to that of the unadapted grazers on the same diets. This difference could not be explained by a higher diet quality selected by the Browser group. The period of digestive adaptation provided a clear digestive benefit to the Browser group. The poor or negative digestibilities of nitrogen by the grazers on the browse diets probably reflect losses of endogenous and microbial protein from their unadapted population of gut microorganisms.

The pattern of growth of the hares showed that the gradual dietary

,ge resulted in a marked slowing of growth and in some cases a loss of
;ht. There was no subsequent compensatory growth leading to recovery
n this weight loss but once hares were established on their respective
gɪɑze and browse diets (days 68–98), their growth was similar, suggesting
that both diets were adequate to maintain growth. The results presented
here, combined with the observation that Scottish mountain hares increase
their condition during the winter (Flux 1970), suggest that the lower quality
of winter forage may not subject the hares to performance-limiting nutritional stress. Instead, we speculate that the period of dietary change-
over represents the main nutritional bottleneck for hare populations. This
suggestion is supported by the time of dietary change in Scottish mountain
hares being coincident with one of the main periods of leveret mortality
(Iason 1989). The dietary change imposed in this experiment was considerably more gradual than that which may affect wild hares whose
available food quality could change rapidly for the worse with a heavy
autumn snowfall. Because of the cost of dietary change in terms of reduced
growth of leverets, there would be selective pressure for intermediate feeders
to implement their autumn dietary change prior to it being enforced by
a sudden food shortage. If this prediction were borne out, the implication
is that diets selected should not always be those that are nutritionally most
favourable in the short term.

We found no effect of the dietary change independent of reproductive
status on the wet caecum mass of mountain hares, but nonetheless found
differences in digestive ability. The advantages of increased caecum size
in winter are hypothesized to be greater retention of food particles for
cellulolysis (Bozinovic 1995). However, the size and capacity of the main
organ of fermentation is only one of many measures of gut morphology and
physiology that may change with shifts in diet composition and food intake
(Karasov & Diamond 1988). We suggest that if the physiological characteristics that facilitate browsing on lignified plant tissue of low average
quality are the colonic particle separation mechanism and caecotrophy,
then a seasonal change in such functions may facilitate a seasonal shift to
browsing without increasing caecum size. The expected difference in caecum
dimensions between seasons may have been masked by the enlargement of
the caecum of the Grazer group, due to the extremely high intakes of the
Timothy diet, which, considering its digestibility, would have been expected
to lead to lower intake (Cork 1994).

Animals that eat tannin-rich diets are thought to produce salivary tannin-
binding proteins, which preferentially bind with and nullify the normal

protein binding and digestion inhibiting effects of the tannins (Robbins *et al.* 1991). The enlargement of the salivary glands of the Browser group and their superior protein digestion on the browse diets, are consistent with the finding that mountain hares produce such salivary proteins (Mole *et al.* 1990). The tannin-binding capacity of salivary gland protein extracts was also greater in the Browser group than the Grazer group (A.E. Hagerman & G.R. Iason, unpublished observations). This result on the induction of salivary adaptation in hares with changing diet confirms that the role of salivary glands is not restricted to the function of assisting digesta flow and buffering pH changes due to VFA production, as was originally thought for ruminants. Since most VFA would be produced in the hindgut of hares, somewhat distant from the site of salivary input to the digestive tract, then a protective function against inhibition of digestion by plant secondary metabolites is likely.

Interrelationships between buccal processing and digestion

Eating and chewing are important processes for herbivores, and occupy a large proportion of their time; a ruminant spends about 6–8 hours per day ruminating over and above the time spent ingesting its food. During chewing, food particles are degraded to small sizes, which leads to an increased surface area, which in turn facilitates microbial attack. Crushing of plant material also makes cellular contents more easily available for microbes or digestion by the animal's enzymes.

Recent studies have demonstrated that smaller animals can compensate for the expected lower digestive efficiency due to their body size by investing more effort in processing their food in order to facilitate digestion. In Nubian ibex (*Capra ibex nubiana*), despite the smaller size of females than males (26 vs. 62 kg) their digestibility of the same forage was the same as males. The females are considered to compensate by a greater chewing effort, making more chews per gram of food at both the stages of feeding and ruminating (Fig. 11.9) (Gross *et al.* 1995). Similarly, in a comparison of sheep (66 kg) and goats (44 kg), the smaller goats fed for a longer time and reduced particle sizes during feeding to a greater extent than sheep. Faster reduction of particle sizes by goats would be expected to decrease rumen retention time but increase the effectiveness of microbial action. Although the sheep were slightly but not significantly more efficient at rumination, the goats had significantly higher fibre digestibility (Domingue *et al.* 1991). This is not consistent either with a body size explanation, grazer/browser

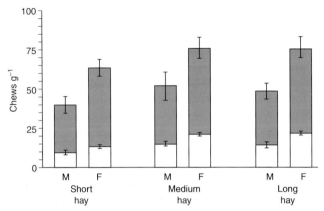

Figure 11.9 Differences in the chewing intensity (chews per gram) of male (M) and female (F) Nubian ibex during feeding (□) and ruminating (■) when offered diets of hay chopped to three different lengths; short, medium and long. Means ± SE. From Gross *et al.* (1995).

explanation (assuming that goats are more like browsers than sheep), nor with a simple digesta kinetics explanation. Further studies of the effectiveness of chewing would be needed to evaluate whether grazer–browser differences in ruminant's digestion of fibre could be attributed to differences in buccal processing.

A similar interaction between ingestive food processing and digestive efficiency has been recorded in seed-eating rodents, with the smaller species being more likely to remove the husks from seeds prior to ingestion than larger species or sexes (Khoklova *et al.* 1995).

Current models of the herbivore functional response to plant availability have demonstrated the short-term intake rate is limited by the clearance of ingested material from the mouth and that there is a tension between cropping of new plant material during ingestion and the process of chewing (Spalinger & Hobbs 1992). The efficiency of chewing is strongly negatively influenced by the NDF content of the diet (Shipley & Spalinger 1992), which, as we have seen, is also one of the main determinants of digestibility. Food processing in the mouth and digestive processing are clearly intimately linked.

Discussion

The ubiquitous distribution of herbivores confirms that they have successful adaptations to the general low quality of their diets, especially by having a large digestive capacity. Although this is theoretically closely related to their feeding niche, current insight of mechanisms of digestive function cannot explain the browser–grazer differences in feeding niches of ruminants, nor the apparent superiority of grazers at digesting plant cell walls. The hypothesized differences in digestive function that were thought to underlie the differences in feeding niche between grazers and browsers are also consistent in most cases with differences expected purely on the basis of body size (Gordon & Illius 1996). However, the occurrence of animals of different feeding niches in a broad range of body sizes still requires explanation. In contrast to Hofmann (1989) who emphasized that the feeding class difference was independent of body mass, we do not regard the browser–grazer classification and the allometric modelling of digestive function as being exclusive alternatives. From the work reviewed and presented here we conclude that allometric relations may provide the first order prediction of what to expect in terms of digestive functions such as passage rate but there remain many points at which differential passage and digestion of plant components may be effected, resulting in different outcomes for the animal. These mechanisms may be animal based or result from different chemical characteristics of contrasting diets (Gordon & Illius 1996). We suggest that further empirical studies are required, of digestive responses of herbivore feeding types across a range of body masses on controlled but contrasting forage types. Because of the hypothesized roles of saliva production, flow and composition in digesta kinetics (Robbins *et al.* 1995), these aspects require particular attention.

The variability about allometric-rule-based digestion is not unexpected. There is considerable known flexibility within herbivore species to digest the same forage to different extents as environmental conditions vary (e.g. rodents: Gross *et al.* 1985) or between seasons (e.g. ruminants: Iason *et al.* 1995). When a shift in diet composition is imposed by changing availability then herbivores can benefit from digestive adaptation in terms of increased nutrient and energy assimilation. This has been demonstrated here in mountain hares. If efficiency of any process can be increased such as chewing or digestion, then why is it not implemented continuously? Although the tissue of the digestive tract is considered costly to maintain (Webster 1981), the metabolic costs of adaptations such as growth of new tissue, and the costs in relation to other processes such as reproductive and growth requirements have not been tabled for any species. The best estimates available are for domestic sheep (Webster 1981). Until the costs as well as the benefits of a digestive or ingestive strategy are understood, the conditions under which strategies suspected to increase efficiency would be applied cannot be predicted. Such analyses would also facilitate the identification of periods of potential nutritional limitation of herbivore populations.

Theoretical studies of animal foraging at a range of scales including models of diet selection (e.g. Owen-Smith & Novellie 1982; Belovsky 1986; Schmitz *et al.* 1992) and patch selection (Fryxell 1991) include the animal's digestive processing capacity, along with rate of intake, as one of the poss-ible constraints (see Owen-Smith 1993). Other models consider digestive limitation to be the major feasible constraint (Verlinden & Wiley 1989). The flexibility in digestion described here means that such constraints will be difficult to parameterize and should be viewed as also being flexible rather than fixed.

It is not possible to consider food quality for herbivores independently of its quantity. Across communities there is generally a negative relationship between biomass available to herbivores and its average quality (Gordon 1989). We have identified several points during the ingestion and digestion processes at which food quality interacts with quantity. Firstly because of the digesta passage constraint from the foregut of ruminants, digestibility and intake are positively related. The specialization to accelerate passage of indigestible particles by small hindgut fermenters is essential for them to be able to meet their higher mass-specific maintenance requirements, but also facilitates a negative relationship between digestibility and intake in these species (see Fig. 11.4). This means that compensation for low diet quality by eating more is a feasible option for these animals. Secondly, there are likely to be trade-offs involved between nutritional quality of the diet and rate of intake and time spent searching for or chewing food (e.g. Gross *et al.* 1995). Where the time budget of an animal is influenced by its requirement for food of a minimum quality or in a certain state of subdivision, and this greater foraging time is spent at the expense of antipredator vigilance behaviour, then there is potential for interactions at the population level between diet quality and predation rate (Prins & Iason 1988; Illius & Fitzgibbon 1994). Recent work on the snowshoe hare (*Lepus americanus*) favours an interaction between predation and nutrition as the most likely cause of population limitation (Krebs *et al.* 1995).

The functional response of herbivores to plant abundance is considered fundamental to the population dynamics of herbivores and to their effects on plant communities (Crawley 1983; Barlow 1987; Chapter 13). Most recent studies have attempted to explain herbivore functional responses in terms of short-term ingestive behaviour (Spalinger & Hobbs 1992). One study has shown that large ruminants ingest a lower quality, higher fibre diet in response to lower density of browse forage availability (Shipley & Spalinger 1995). We suggest that this foraging solution is only sustainable within a short time scale and that in the longer term the digestive limitation imposed by clearance of the fibrous material from the rumen would be likely to limit intake and influence foraging choices. Few studies of wild herbivores have attempted to explain intake rate in terms of functional differences in digestive characteristics of food plant species. A large proportion of variation in voluntary intake of tree species (Fryxell *et al.* 1994) and dietary preferences by beavers (*Castor canadensis*) was explained by gut retention. This further emphasizes the intimate relationship between ingestive and digestive processes and endorses the role of the digestive processes in the functional responses and mechanistic explanations of population and community patterns.

Acknowledgements

We thank Iain Gordon for help, advice and comments at all stages of preparation of this manuscript, David Hirst of Biometrics and Statistics Scotland for statistical advice and Ake Pehrson for advice on hare husbandry. Han Olff, Rudi Drent, Andrew Illius and an anonymous referee provided helpful comments. GRI was supported by the Scottish Office Agriculture, Environment and Fisheries Department and a Royal Society European Exchange Fellowship.

References

Bannister, P. (1976). *Introduction to Physiological Plant Ecology*. Blackwell Scientific Publications, Oxford.

Barlow, N.D. (1987). Pastures, pests and productivity: simple grazing models with two herbivores. *New Zealand Journal of Ecology*, **10**, 43–55.

Batzli, G.O., Broussard, A.D. & Oliver, R.J. (1994). The integrated processing response in herbivorous small mammals. In *The Digestive System in Mammals: Food, Form and Function* (Ed. by D.J. Chivers & P. Langer), pp. 324–336. Cambridge University Press, Cambridge.

Belovsky, G.E. (1986). Optimal foraging and community structure: implications for a guild of generalist grassland herbivores. *Oecologia*, **70**, 35–52.

Björnhag, G., Sperger, I. & Holtenius, K. (1984). A separation mechanism in the large intestine of equines. *Canadian Journal of Animal Science*, **64** (Suppl.), 89–90.

Bozinovic, F. 1995. Nutritional energetics and digestive responses of an herbivorous rodent (*Octodon degus*) to different levels of dietary fibres. *Journal of Mammalogy*, **76**, 627–637.

Cheeke, P.R. (1987). *Rabbit Feeding and Nutrition*. Academic Press, Orlando.

Cork, S.J. (1994). Digestive constraints on dietary scope in small and moderately-small mammals: how much do we really understand? In *The Digestive System in Mammals: Food, Form and Function* (Ed. by D.J. Chivers & P. Langer), pp. 337–369. Cambridge University Press, Cambridge.

Crawley, M.J. (1983). *Herbivory: The Dynamics of Plant–Animal Interactions*. Blackwell Scientific Publications, Oxford.

Crawley, M.J. (1989). The relative importance of vertebrate and invertebrate herbivores in plant population dynamics. In *Insect Plant Interactions* (Ed. by E. Bernays), pp. 41–57. CRC Press, Boca Raton.

Demment, M.L. & Van Soest, P.J. (1985). A nutritional explanation for body-size patterns of ruminant and non-ruminant herbivores. *American Naturalist*, **125**, 641–675.

Derting, T.L. & Noakes, E.G. (1995). Seasonal changes in gut capacity in the white-footed mouse (*Peromyscus leucopus*) and meadow vole (*Microtus pennsylvanicus*). *Canadian Journal of Zoology*, **73**, 243–252.

Domingue, B.M.F., Dellow, D.W. & Barry, T.N. (1991). The efficiency of chewing during eating and ruminating in goats and sheep. *British Journal of Nutrition*, **65**, 355–363.

Doucet, C.M. & Fryxell, J.M. (1993). The effect of nutritional quality on forage preference by beavers. *Oikos*, **67**, 201–208.

Flux, J.E.C. (1970). The life history of the mountain hare (*Lepus timidus*) in Scotland. *Journal of Zoology*, **161**, 75–123.

Foley, W.J. & Cork, S.J. (1992). Use of fibrous diets by small herbivores: how far can the rules be bent? *Trends in Ecology and Evolution*, **7**, 159–162.

Foley, W.J. & McArthur, C. (1994). The effects and costs of allelochemicals for mammalian herbivores: an ecological perspective. In *The Digestive System in Mammals: Food, Form and Function* (Ed. by D.J. Chivers & P. Langer), pp. 370–391. Cambridge University Press, Cambridge.

Fryxell, J.M. (1991). Forage quality and aggregation by large herbivores. *American Naturalist*, **138**, 478–498.

Fryxell, J.M., Vamosi, S.M., Walter, R.A. & Doucet, C.M. (1994). Retention time and the functional response of beavers. *Oikos*, **71**, 207–214.

Genstat 5 (1993). *Genstat-5, Release 3, Reference Manual*. Genstat Committee. Clarendon Press, Oxford.

Gordon, I.J. (1989). Vegetation community selection by ungulates on Rhum: The food supply. *Journal of Applied Ecology*, **26**, 35–51.

Gordon, I.J. & Illius, A.W. (1988). Incisor arcade structure and diet selection in ruminants. *Functional Ecology*, **2**, 15–22.

Gordon, I.J. & Illius, A.W. (1994). The functional significance of the browser–grazer dichotomy in African ruminants. *Oecologia*, **98**, 167–175.

Gordon, I.J. & Illius, A.W. (1996). The nutritional ecology of African ruminants — a reinterpretation. *Journal of Animal Ecology*, **65**, 18–28.

Green, D.A. & Millar, J.S. (1987). Changes in gut dimensions and capacity of *Peromyscus maniculatus* relative to diet quality and energy needs. *Canadian Journal of Zoology*, **65**, 2159–2162.

Gross, J.E., Wang, Z. & Wunder, B.A. (1985). Effects of food quality and energy needs: changes in gut morphology and capacity of *Microtus ochrogaster*. *Journal of Mammalogy*, **66**, 661–667.

Gross, J.E., Demment, M.W., Alkon, P.A. & Kotzman, M. (1995). Feeding and chewing behaviours of Nubian ibex: compensation for sex-related differences in body size. *Functional Ecology*, **9**, 385–393.

Hairston, N.G., Smith, F.E. & Slobodkin, L.B. (1960). Community structure, population control and competition. *American Naturalist*, **94**, 421–425.

Hammond, K. & Diamond, J. (1994). Limits to dietary nutrient uptake in lactating mice. *Physiological Zoology*, **67**, 282–303.

Hammond, K.A. & Wunder, B.A. (1995). Effect of cold temperatures on the morphology of the gastro-intestinal tract of two Microtine Rodents. *Journal of Mammalogy*, **76**, 232–239.

Hanley, T.A., Robbins, C.T., Hagerman, A.E. & McArthur, C. (1992). Predicting digestible protein and digestible dry matter in tannin-containing forages consumed by ruminants. *Ecology*, **73**, 537–541.

Hofmann, R.R. (1973). *The Ruminant Stomach*. East African Literature Bureau, Nairobi.

Hofmann, R.R. (1988). Anatomy of the gastro-intestinal tract. In *The Ruminant Animal. Digestive Physiology and Nutrition* (Ed. by D.C. Church), pp. 14–43. Prentice-Hall, Englewood Cliffs, NJ.

Hofmann, R.R. (1989). Evolutionary steps of ecophysiological adaptation and diversification of ruminants: a comparative view of their digestive system. *Oecologia*, **78**, 443–457.

Hofmann, R.R. & Stewart, D.R.M. (1972). Grazer or browser: a classification based on the stomach structure and feeding habits of East African ruminants. *Mammalia*, **36**, 226–240.

Hörnicke, H. & Björnhag, G. (1980). Coprophagy and related strategies for digesta utilization. In *Digestive Physiology and Metabolism in Ruminants* (Ed. by Y. Ruckebush & P. Thivend), pp. 707–730. MTP Press, Lancaster.

Hume, I.D. (1989). Optimal digestive strategies in mammalian herbivores. *Physiological Zoology*, **62**, 1145–1163.

Hume, I.D., Morgan, K.R. & Kenagy, G.J. (1993). Digesta retention and digestive performances in sciurid and microtine rodents — effects of morphology and body size. *Physiological Zoology*, **66**, 396–411.

Iason, G.R. (1989). Body size and mortality of mountain hares (*Lepus timidus*). *Journal of Zoology (London)*, **219**, 676–680.

Iason, G.R. & Waterman, P.G. (1988). Avoidance of plant phenolics by juvenile and reproducing female mountain hares in summer. *Functional Ecology*, **2**, 433–440.

Iason, G.R., Sim, D.A. & Foreman, E. (1995). Seasonal changes in intake and digestion of chopped timothy hay by three breeds of sheep. *Journal of Agricultural Science*, **125**, 273–280.

Illius, A.W. & Fitzgibbon, C. (1994). Costs of vigilance in foraging ungulates. *Animal Behaviour*, **47**, 481–484.

Janis, C.M. & Ehrhardt, D. (1988). Correlation of relative muzzle width and relative incisor width with dietary preference in ungulates. *Zoological Journal of the Linnean Society*, **92**, 267–284.

Jilge, B. (1982). Rate of movement of marker particles in the digestive tract of the rabbit. *Laboratory Animals*, **16**, 7–11.

Jones, D.I.H. & Wilson, A. (1987). Nutritive quality of forage. In *The Nutrition of Herbivores* (Ed. by J.B. Hacker & J.H. Ternouth), pp. 65–90. Academic Press, Sydney.

Justice, K.E. & Smith, F.A. (1992). A model of dietary fibre utilization by small mammalian herbivores with empirical results for *Neotona*. *American Naturalist*, **139**, 398–416.

Karasov, W.H. & Diamond, J.M. (1988). Interplay between physiology and ecology in digestion. *BioScience*, **38**, 602–611.

Khoklova, I.S., Degen, A.A. & Kam, M. (1995). Body size, gender, seed husking and energy requirements in two species of desert gerbelline rodents. *Functional Ecology*, **9**, 720–724.

Krebs, C.J., Boutin, S., Boonstra, R., Sinclair, A.R.E., Smith, J.N.M., Dale, M.R.T., Martin, K. & Turkington, R. (1995). Impact of food and predation on the snowshoe hare cycle. *Science*, **269**, 1112–1114.

Marty, J. & Vernay, M. (1984). Absorption and metabolism of the volatile fatty acids in the hindgut of the rabbit. *British Journal of Nutrition*, **51**, 265–277.

McArthur, C., Hagerman, A.E. & Robbins, C.T. (1991). Physiological strategies of mammalian herbivores against plant defences. In *Plant Defences Against Mammalian Herbivory* (Ed. by R.T. Palo & C.T. Robbins), pp. 103–114. CRC Press, Boca Raton.

Minson, D.J. (1982). Effects of chemical and physical composition of herbage eaten upon intake. In *Nutritional Limits to Animal Production for Pasture* (Ed. by J.B. Hacker), pp. 167–182. CAB, Slough.

Mole, S., Butler, L.G. & Iason, G.R. (1990). Defense against dietary tannins in herbivores: a survey of proline-rich salivary proteins. *Biochemical Systematics and Ecology*, **18**, 287–293.

Nagy, K.A. (1987). Field metabolic rates and food requirement scaling in mammals and birds. *Ecological Monographs*, **57**, 111–128.

Owens, F.N. & Zinn, R. (1988). Protein metabolism of ruminant animals. In *Digestive Physiology and Nutrition of the Ruminant* (Ed. by D.C. Church), pp. 227–249. Prentice-Hall, New Jersey.

Owen-Smith, N. (1993). Evaluating optimal diet models for an African browsing ruminant, the kudu: how constraining are the assumed constraints? *Evolutionary Ecology*, 7, 499–524.

Owen-Smith, N. (1994). Foraging responses of kudus to seasonal changes in food resources: elasticity in constraints. *Ecology*, **75**, 1050–1062.

Owen-Smith, N. & Novellie, P. (1982). What should a clever ungulate eat? *American Naturalist*, **119**, 151–178.

Parra, R. (1978). Comparison of foregut and hindgut fermentation in herbivores. In *The Ecology of Boreal Folivores* (Ed. by G.G. Montgomery), pp. 205–230. Smithsonian Institution, Washington, DC.

Pehrson, A. (1981). Winter food consumption and digestibility in caged mountain hares. In *Proceedings of the World Lagomorph Conference*. (Ed. by K. Myers & C.D. MacInnes), pp. 732–742. IUCN, Gland, Switzerland.

Pehrson, S. & Lindlof, B. (1984). Impact of winter nutrition on reproduction in captive mountain hares (*Lepus timidus*). *Journal of Zoology (London)*, **204**, 201–209.

Prins, H.H.T. & Iason, G.R. (1988). Dangerous lions and nonchalant buffalo. *Behaviour*, **108**, 262–269.

Prior, R.L., Hintz, H.F., Lowe, J.E. & Visek, W.J. (1974). Urea recycling and metabolism in ponies. *Journal of Animal Science*, **38**, 565–571.

Rhoades, D.F. (1985). Offensive–defensive interactions between herbivores and plants: their relevance in herbivore population dynamics and ecological theory. *American Naturalist*, **125**, 205–235.

Robbins, C.T. (1983). *Wildlife Feeding and Nutrition*. Academic Press, Orlando.

Robbins, C.T., Hagerman, A.E., Austin, P.J., McArthur, C. & Hanley, T.A. (1991). Variation in mammalian physiological responses to a condensed tannin and its ecological implications. *Journal of Mammalogy*, 72, 480–486.

Robbins, C.T., Spalinger, D.E. & Van Hoven, W. (1995). Adaptation of ruminants to browse and grass diets: are anatomical-based browser–grazer interpretations valid? *Oecologia*, 103, 208–213.

Schmitz, O.J., Hik, D.S. & Sinclair, A.R.E. (1992). Plant chemical defence and twig selection by snowshoe hare: an optimal foraging perspective. *Oikos*, 65, 295–300.

Shipley, L.A. & Spalinger, D.C. (1992). Mechanics of browsing in dense food patches: Effects of plant and animal morphology on intake rate. *Canadian Journal of Zoology*, 70, 1743–1752.

Shipley, L.A. & Spalinger, D.E. (1995). Influence of size and density of browse patches on intake rates of foraging decision of young moose and white-tailed deer. *Oecologia*, 104, 112–121.

Sibley, R.M. (1981). Strategies of digestion and defecation. In *Physiological Ecology: An Evolutionary Approach to Resource Utilization* (Ed. by C. Townsend & P. Calow), pp. 109–139. Blackwell Scientific Publications, Oxford.

Smith, R.L., Hulbart, H.D.J. & Shoemaker, R.L. (1980). Seasonal changes in weights, caecal length and pancreatic function of showshoe hares. *Journal of Wildlife Management*, 44, 718–724.

Snedecor, G.W. & Cochran, W.G. (1980). *Statistical Methods*. Iowa State University Press, Iowa.

Spalinger, D.E. & Hobbs, N.T. (1992). Mechanisms of foraging in mammalian herbivores: new models of functional response. *American Naturalist*, 140, 325–348.

Swain, T. (1979). Tannins and lignins. In *Herbivores: Their Interaction with Secondary Plant Metabolites* (Ed. by R.A. Rosenthal & D.H. Janzen), pp. 657–683. Academic Press, New York.

Thomas, V.G. (1987). Similar winter energy strategies of grouse, hares and rabbits in northern biomes. *Oikos*, 50, 206–212.

Van Soest, P.J. (1982). *Nutritional Ecology of the Ruminant*. O & B Books, Corvallis, Oregon.

Van Wieren, S.E. (1996). *Digestive strategies in ruminants and nonruminants*. PhD thesis, Wageningen Agricultural University, Wageningen.

Verlinden, C. & Wiley, R.H. (1989). The constraints of digestive rate: an alternative model of diet selection. *Ecology*, 75, 1050–1062.

Wang, L.C.H., Jones, D.L., MacArthur, R.A. & Fuller, W.A. (1973). Adaptation to cold: energy metabolism in an atypical lagomorph, the arctic hare (*Lepus arcticus*). *Canadian Journal of Zoology*, **51**, 841–846.

Webster, A.J.F. (1981). Energy costs of digestion and metabolism in the gut. In *Digestive Physiology and Metabolism in Ruminants* (Ed. by Y. Ruckebush & P. Thivend), pp. 469–484. MTP Press, Lancaster.

Chapter 12

Functional responses to resource complexity: an experimental analysis of foraging by beavers

J. Fryxell[1]

Summary

Ecological theory has been based traditionally on univariate functional responses by consumers to changing resource abundance. The simplicity of this approach is clearly at odds with the complexity and variability of resources used by real consumers, particularly terrestrial herbivores, raising doubts about the utility of simple consumer–resource theory. In this chapter, I review the physiological and behavioural constraints on herbage intake acting at different scales of space and time, illustrated by our experimental work with beavers in deciduous forest environments. I then consider how rates of plant attack by beavers are influenced by behavioural responses to the size distribution of ramets, variation in nutritional quality among ramets, and the spatial distribution of ramets relative to the centre of beaver foraging territories, using paired experimental trials with manipulated changes in resource characteristics. Much of the variation in beaver functional and behavioural responses to resource complexity can be explained using optimality principles, offering hope that a robust theory of plant–herbivore dynamics may yet emerge, one that links fitness-maximizing behaviours with realistic environmental constraints.

1 Department of Zoology, University of Guelph, Guelph, Ontario, Canada, N1G 2W1

Introduction

Ecologists have long appreciated that processes affecting the rate of resource acquisition are fundamental in understanding predator–prey or competitive interactions (Lotka 1925; Volterra 1926). This is just as true of protists and algal species as it is terrestrial carnivores, herbivores and plants, so early work concentrated on development of a general theory of consumptive processes that might apply across different taxa and trophic levels. One of the great breakthroughs in ecology was Holling's (1959) demonstration that the rate of resource consumption in relation to resource density could be predicted from first principles (termed the functional response). By linking behavioural processes with patterns of resource use, Holling provided a coherent theoretical framework for predicting the outcome of ecological interactions.

Under most circumstances, Holling argued, one can visualize foraging as a process comprising two different and antagonistic behaviours: searching for prey and handling prey. Search is composed of consumer movement, perception and attack of potential resource items, whereas handling involves resource manipulation, dismemberment and feeding by the consumer. Under the generalized process envisaged by Holling these behaviours are antagonistic—time spent handling resources cannot be devoted to searching for new resources. It follows that the rate of resource consumption can be calculated as the reciprocal of the expected search time per item plus the expected handling time per item (Turelli *et al.* 1982; Mangel & Clark 1988). If resource items are homogeneously distributed at a constant density (N) in the environment and the consumer searches a constant area (a) per unit time, then the expected search time until a prey item is secured is $1/aN$. If we symbolize the expected handling time per resource item as h, then the rate of consumption (X) can be calculated as

$$X = \frac{1}{1/aN + h} = \frac{aN}{1 + ahN} \tag{1}$$

This formula is sometimes called the disk equation, in honour of its first empirical application to a human forager 'attacking' sandpaper disks stuck to a desktop (Holling 1959). This derivation produces a monotonically decelerating functional response that Holling christened the 'type II' curve. Patterns like it have been found in hundreds of experiments—particularly in simple environments with only a single type of prey.

Under other circumstances, one might expect to see a type I functional response, in which intake increases linearly with resource density or a type III functional response, in which intake initially accelerates with increasing resource density, but eventually decelerates and approaches an asymptote similar to that of the type II curve.

In the case of herbivores, search and handling are not always mutually exclusive. For example, a wildebeest grazing a lush carpet of grasses on the Serengeti Plain moves directly from bite to bite, without pausing to 'handle' those bites. Hence, in some circumstances, the most meaningful mechanistic trade-off for grazing herbivores is between cropping and masticating. Recent theory has utilized the time budget approach of the classic disk equation to derive functional responses incorporating these alternate trade-offs more applicable to grazing ungulates (Spalinger & Hobbs 1992; Farnsworth & Illius 1996). Nonetheless, the qualitative predictions often agree: consumption rates should increase curvilinearly with food abundance, where abundance might be measured as bite size or plant biomass per unit area rather than the density of individual prey per unit area.

Simple functional response models have proven themselves very useful in both understanding and predicting simple ecological relationships. Unfortunately, nature is rarely simple. Sometimes the consumer has a fixed territory or home range, so its pattern of search is far from random in space. Many foragers do not cease search to consume items, but rather process them as they look for new resources. This is particularly true of herbivores (Spalinger & Hobbs 1992), but it also applies to many carnivores that feed on extremely small prey, such as baleen whales attacking krill. Resource items are themselves unevenly distributed in space, doing further violence to the notion of random encounter. Moreover, few consumers use only a single resource type, but rather potential food items vary according to size, nutritional quality and species. All of these factors have the potential for altering rates of resource use.

We are only just beginning to appreciate ways of understanding behavioural strategies used to cope with such varied and different forms of resource complexity. My objective here is to show how one species, the beaver (*Castor canadensis*), responds to several forms of resource complexity and to develop behaviourally based foraging models that allow us to explore the implications of complexity in resource composition and spatial distribution.

Methodology

My coworkers and I have tried to tease apart the effects of resource complexity on plant–herbivore interactions through a series of controlled experimental analyses. It would not be terribly useful or practical to sieve through mountains of descriptive data to detect foraging patterns in relation to complex resource combinations. Nor does it make much sense to use a purely factorial experimental design to test possible relationships, because the number of potential combinations of factors to test in even the simplest of factorial designs would dwarf a typical research budget. The only logistically feasible approach is to compare foraging patterns in a less comprehensive set of experimental trials, using pairwise combinations of experimental conditions. One might then legitimately hope to develop a deep enough understanding to postulate a synthetic foraging model applicable to a wide range of conditions.

I have used such a protocol to study the foraging behaviour of captive beavers (*Castor canadensis*) presented with experimentally manipulated 'forests'. These experiments retain at least some of the native complexity inherent in real forest environments, all the while trying to control at least some of these factors. We have used a common methodology for all of our experiments, described in detail in earlier papers (Fryxell & Doucet 1993; Fryxell *et al.* 1994), so here I offer only a brief description.

Pairs of wild-caught beavers were kept in a 0.15-ha fenced enclosure with a natural overstory of mixed deciduous and coniferous trees, whose trunks were wrapped in wire mesh to prevent attack by beavers. We built a small, concrete-lined pond at one end of each enclosure, beside a plywood lodge large enough to house two adults. Animals were free to move around the enclosure at will, but were rarely found away from the 'lodge' or 'pond' except during foraging bouts occurring largely between dusk and dawn. Feeding by animals was restricted to saplings or branches pushed into the soil to anchor them. Beavers treat these experimental resources in a similar manner to that in the wild—cutting trees near ground level, pruning particularly large saplings, and then dragging whole saplings back to the pond or lodge before further dismembering the saplings and consuming the leaves, bark and twigs. We recorded the composition of the experimental forest each morning to determine which items had been removed overnight, thereby estimating daily rates of consumption per individual animal. On many nights, these indirect measurements were accompanied by direct nightly observation from a blind overlooking the dimly lit enclosures. The major variables changing across experimental treatments were species composition, size composition and the spatial distribution of saplings in the experimental enclosures. Each pairwise experiment is analysed using simple ANOVA, based on qualitative predictions derived from mechanistic foraging models described in the text. Although it is also possible to test such experiments using goodness-of-fit tests, we are here concerned more with assessing the qualitative effects of changing resource characteristics on foraging patterns than on an assessment of the power of specific optimal foraging models.

Size and species composition of resources

Variation in the size and species composition of potential food items of any consumer could affect rates of consumption, by affecting the rate of encounter during search and/or the handling time. For a herbivore, one might reasonably presume that rates of effective search are independent of species composition because prey are immobile and cannot directly avoid attack. On the other hand, handling time varies considerably among food items of different species. For a beaver, there are four significant components of handling time: the time required to cut down food items, the time devoted to dragging items back to the central place (provisioning), the time required for consumption of individual items, and the time required on average for digestive processing of items. Large trees obviously take much more time to cut down and to consume: in both cases time expenditure increases allometrically with tree diameter (Belovsky 1984; Fryxell & Doucet 1993). Provisioning time also varies with size of resources, because large items cannot be dragged intact back to the pond (Fryxell & Doucet 1993), but variation in this component of handling time is probably minor compared with other components. None of these components should vary for different species of plant prey.

The last component of handling time, time in the digestive tract, is somewhat controversial. There is evidence that herbivores often process herbage for a considerable time in the digestive tract (Warner 1981). Nonetheless, there is some disagreement whether processing time is an intrinsic property of different foods or whether processing time is mediated by the consumer itself. For example, it is often claimed that equids compensate for foods of low nutritional content by simply passing those foods more quickly through the digestive tract (Chapter 11). My coworkers and I have tested whether nutritional content of food items constrains intake in beavers by physiological experiments using a wide variety of natural food plants. Our results indicate that daily rates of intake are indeed constrained by processing time in the digestive tract. The *ad libitum* rate of intake declines hyperbolically with the average retention time of food particles in the digestive tract (Fryxell *et al.* 1994). Similar hyperbolic relations are observed when comparing different forage species or experimentally manipulated mixed diets. Mixed diets are retained in the digestive tract in direct proportion to their component species — implying that aggregate handling time is a simple additive process in beavers (Fryxell *et al.* 1994).

To further explore the influence of processing time on intake, we conducted a set of foraging experiments to measure functional responses towards two food types with widely differing retention time in the gut (Fryxell *et al.* 1994). In the first treatment, we presented beavers with a uniformly spaced 'forest' of saplings of trembling aspen (*Populus tremuloides*) of similar size, redistributing sapling placement each day. The second treatment was otherwise identical, except that we used speckled alder (*Alnus rugosa*) instead of trembling aspen saplings. Our earlier physiological studies had shown that aspen tissues had a mean retention time of 14 h, whereas alder tissues were retained 2.6 times longer (37 h) (Fryxell *et al.* 1994). If feeding rates in the field are constrained by interspecific variation in retention time among plant species, then one would predict maximum feeding rates between aspen and alder to differ the same way. These predictions were upheld: the maximum feeding rate per day recorded in the aspen trials was 2.5 times that in the alder functional response trials (Fig. 12.1). These results demonstrate that plant species traits have important consequences for rates of intake by beavers: rapidly digesting plants yield higher rates of maximum intake than slowly digesting plants. We know that this difference in intake rates is not simply a physiological response to compensate for differences in plant nutritional quality, because alders are less digestible and have lower energy content than do aspens (Fryxell & Doucet 1993). Beavers eat less of alders not because they can satisfy their minimal requirements in doing so, but because digestive constraints allow no other option. A beaver may starve to death in a forest of alders, even though it is eating as much as it can process each day.

Our results suggest that differences in retention time can influence forage preferences. The contingency model of optimal diet breadth (MacArthur & Pianka 1966; Charnov 1976; Stephens & Krebs 1986) predicts that consumers should try to improve rates of energy gain by expanding their diet to include resources of low profitability (energy content divided

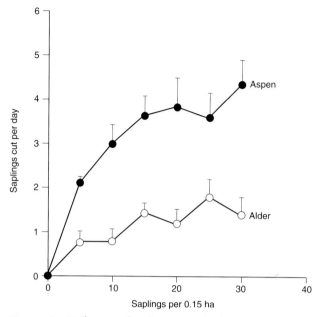

Figure 12.1 Daily rates of sapling cutting by beavers (mean $\pm$ SE) in relation to sapling density of monospecific stands of trembling aspens or speckled alders ($F_{1, 94} = 74.7$; $P < 0.001$). Data from Fryxell *et al.* (1994).

by handling time) when the probability of encounter with more highly profitable items is low.

We tested the contingency model of optimal diet choice by creating more complex forests of saplings composed of three different woody plant species (Fryxell & Doucet 1993). Of these three species, trembling aspen is highly profitable, speckled alder is of intermediate profitability, and red maple (*Acer rubrum*) has very low profitability. Our experimental forest was composed initially of equal mixtures of these three plant species and we allowed beavers to deplete this forest over time, thereby skewing resource availability through selective harvesting. This experimental design was intended to mimic the kinds of changes in resource availability that might occur following colonization of a virgin patch of forest. At various stages of forest depletion, we predicted the probability of each species being harvested by beavers according to a contingency model of diet selection, whereby beavers should choose a diet breadth (i) that maximizes net energy gain (E):

$$E_i = \frac{a \sum_{j=1}^{i} e_j N_j}{1 + a \sum_{j=1}^{i} h_j N_j} \qquad (2)$$

where h_j is the handling time and e_j is digestible energy content of a single food item j, and potential food items are ranked in descending order of profitability. We include handling time in the denominator, even though we recognize that a large fraction of that handling time occurs during daylight hours when beavers do not forage on shore. This is because the typical retention time in the digestive tract exceeds 12 h, so it can theoretically compete with other feeding activities.

Our results were consistent with the contingency model, in that the sequence of sapling depletion matched the rank order of sapling profitabilities: aspens were preferred to alders, which were preferred in turn to maples (Fig. 12.2). Therefore, the diet of beavers was primarily composed of aspens when aspens were as common as anything else, but beavers expanded their diet to include alders when aspens became rare. Dilution of the frequency of attack on aspens, which should be advantageous in enhancing energy gain, led to a change in the functional response on aspens. Instead of the monotonically decelerating (type II) functional response observed in the earlier trials with monospecific aspen stands (Fig. 12.1), the beaver functional response on aspen in mixed stands had a sigmoid shape, more like Holling's type III category (Fig. 12.3). This has interesting implications, because sigmoid functional responses can stabilize trophic interactions between consumers and their resources (Murdoch & Oaten 1975; Comins & Hassell 1976; Tansky 1978). The sigmoid response induced by adaptive diet choice is only stabilizing, however, when alternative prey are scarcely able to sustain consumers (Fryxell & Lundberg 1994). As explained earlier, this is likely to be the case for systems in which alder saplings are the most abundant alternative (Fryxell et al. 1994). These observations suggest that adaptive resource choice by beavers during depletion of its most preferred prey alters the beaver functional response in a manner that could stabilize trophic interactions.

Size variation among food items is expected to have similarly strong effects on rates of resource consumption, because the time required to consume and provision individual trees depends on their size. The mass of foliage, twigs and bark of saplings consumed by beavers is allometrically

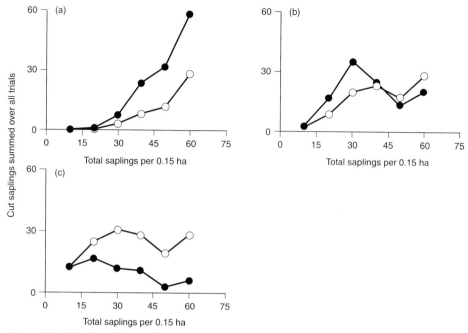

Figure 12.2 Number of cut saplings summed over all trials of (a) trembling aspen, (b) speckled alder or (c) red maple, shown as closed circles, in relation to that expected if animals were indifferent to plant species at various stages of sapling depletion in functional response experiments, shown as open circles (Kolmogorov–Smirnov tests showed significant deviation ($P<0.05$) from a null model at sapling densities of 20, 30, 40, 50 and 60, but not 10; Fryxell & Doucet 1993).

related to sapling linear dimensions, with a scaling coefficient of approximately 2.5 (Fryxell & Doucet 1993). Given a constant rate of feeding once a cut tree has been dragged to the pond, feeding time should similarly scale allometrically with tree size. The time required to cut down trees is similarly related to tree size, with an allometric coefficient of 2.6 (Fryxell & Doucet 1993; Walton 1994). There is a slight tendency for slower dragging of large saplings relative to small saplings, but this time investment is dwarfed by feeding and cutting time expenditures. If one combines these components of nutritional benefits vs. time costs, energetic profitability is a bell-shaped function of tree size, whose height declines with increasing distance from the pond. One can apply the contingency model of optimal

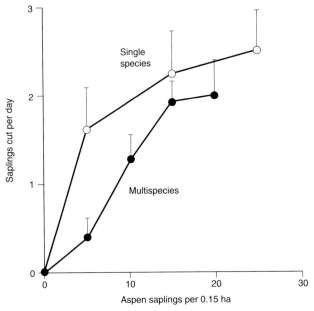

Figure 12.3 Daily rates of aspen sapling cutting by beavers (mean ± SE) in relation to the density of aspen saplings in monospecific stands or multispecies stands ($F_{2,109} = 40.2$, $P < 0.05$; Tukey *post hoc* test of treatment effect $P < 0.001$; Fryxell & Doucet 1993). Unlike Fig. 12.1, saplings were not redistributed uniformly every day to new locations.

energy gain (Eq. 2) to predict the size range of resources that is expected to be selected. In our case, the contingency model predicts that beavers ought to prefer trees of intermediate size (Schoener 1979; Jenkins 1980; Fryxell & Doucet 1993), as do linear programming models of similar structure (Belovsky 1984). Observational data from field studies supports both predictions (Jenkins 1980; Pinkowski 1983; Belovsky 1984; McGinley & Whitham 1985).

It is technically impossible to plant fully grown trees in our experimental enclosures, so we had to content ourselves with feeding trials with saplings ranging in diameter from 0.2 to 5.5 cm. Over this range, beavers are predicted to select the largest saplings available and avoid saplings of approximately 1 cm close to the pond and 2 cm at 50 m distance (Fryxell & Doucet 1993). During depletion experiments of the sort described earlier, we consistently found that beavers selected saplings of 4 and 5 cm diameter, avoided saplings of 1 cm diameter, and were indifferent to those of 2 or 3 cm diameter (Fig. 12.4). Our calculations suggested that sapling choice changed with resource availability in a manner consistent with optimal diet choice theory (Fryxell & Doucet 1993), but it is unlikely this hypothesis could be distinguished from a simpler model based on constant preferences ranked by energetic profitability.

If beavers harvested the optimal size range of saplings, then the mean size of selected food items would be lower than that available in the environment. Since maximum consumption rate is inversely related to prey size, because of differences in handling time, this implies that maximum cutting rates should be higher when beavers can exercise choice than when

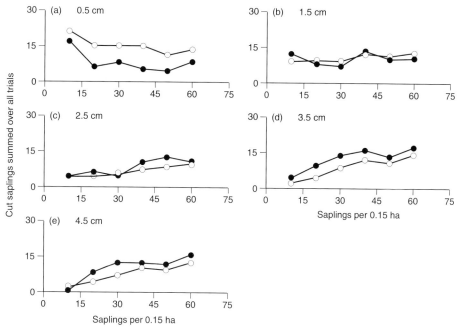

Figure 12.4 Number of cut saplings summed over all trials of aspen saplings of five different diameter classes, shown as closed circles, in relation to that expected if animals were indifferent to plant size at various stages of sapling depletion in functional response experiments, shown as open circles (Kolmogorov–Smirnov tests showed significant deviation from null model at sapling densities of 20, 30 and 40, $P < 0.05$; Fryxell & Doucet 1993).

they cannot. When prey density is very low, however, animals should harvest virtually everything available and functional responses should be little affected. We tested this prediction by comparing the functional response of beavers presented with identical-sized saplings of 2.5 cm diameter with a second experiment in which beavers fed on saplings averaging 2.5 cm diameter, but with wide size variability. The functional response of animals presented with wide size variation was considerably lower than that of animals presented with resources of constant size at high sapling densities, but the two functional response curves converged at low sapling densities (Fig. 12.5).

We do not fully understand the dynamical implications of such size selectivity. Preliminary size-structured models suggest minor changes in local stability due to adaptive food selection (Fryxell & Lundberg 1997). Uncertainty in our understanding of the dynamic implications of size-structured consumption derives from the inherent difficulty in

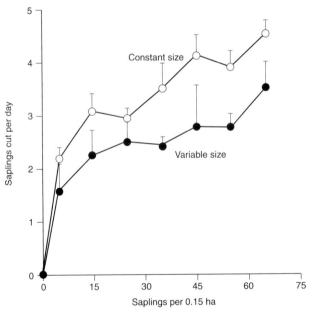

Figure 12.5 Daily rates of aspen sapling cutting by beavers (mean ± SE) in relation to aspen sapling density in single-size stands or multisize aspen stands ($F_{1,122} = 13.4$, $P < 0.05$; Fryxell & Doucet 1993).

mathematically analysing size-structured population models. Recent advances in individual-based modelling may soon rectify this imbalance (see Chapters 17 and 18). Until then, we have to content ourselves with an appreciation that size selection has demonstrable effects on the rate of consumption which may or may not have important ecological ramifications at the population level.

Spatial distribution of resources

The disk equation model assumes that all resources are equally likely to be encountered during a foraging bout. This assumption seems inappropriate for territorial foragers, such as beavers, because foraging bouts always begin at the centre of the territory before extending outwards (Schoener 1979; Jenkins 1980; Belovsky 1984). In this section, I consider an alternative foraging model predicting rates of resource use as a probability function of both distance from a central place and local resource density. Although I will develop the model with respect to beavers, it is sufficiently general that it could, in principle, be applied to other central-place foragers.

Let us assume that a foraging beaver cannot identify prospective food items from the central place and instead must seek out resources during discrete bouts of food searching behaviour. At first glance, this may seem a questionable assumption, because potential food items are often visible a great distance away from the pond. Our earlier sensory experiments have clearly demonstrated, however, that beavers use olfaction rather than vision to locate suitable prey (Doucet *et al.* 1994). Beavers require olfactory sampling from close range before plant identification can occur. This is perhaps not surprising, given that beavers forage on land almost exclusively at night. My coworkers and I have witnessed numerous previous experiments in large enclosures in which beavers passed within 5 m of preferred food plants, but seemingly failed to recognize them. Moreover, beavers under controlled experimental conditions often travel much further than would be strictly necessary before cutting down saplings (Fryxell 1992), which would not be expected if they could simply direct themselves to the nearest suitable sapling from the central place. We therefore assume that a foraging beaver travels at a constant velocity along a randomly orientated vector rooted at the central place, all the while searching a path of constant width on either side for suitable resources (Fig. 12.6). Once encountered, items are retrieved back to the central place for processing, consumption

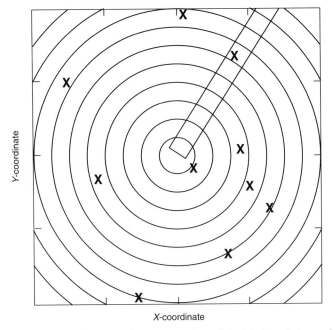

X-coordinate

Figure 12.6 Diagrammatic representation of the Markov chain model of central-place foraging. The forager chooses a search vector at random, then travels along that vector, searching a fixed radius on either side of the forager until a prey item is encountered. The search path in this example is shown by the rectangle orientated from the centre to the upper right corner of the figure. Potential prey items are symbolized by X. Each foraging bout consists of a series of failures in distance intervals 1 to $i-1$, culminating with success in distance interval i, in this example the seventh distance interval.

and digestion before the next foraging bout can begin. If no suitable resources are encountered within a maximum foraging radius from the central place, then the forager is assumed to return via the same route to the central place before initiating a new search. In this generalized foraging process, the probability that food encounter occurs at a given distance during a foraging bout is clearly conditional on the probability that no resources were successfully encountered earlier in the foraging bout. Such a pattern is often called a Markov chain in the theoretical literature (Stewart 1994). Hence, one can model central-place foraging as a Markov chain of successive search failures that is terminated by a successful encounter (Fryxell & Lundberg 1997).

384

More formally, assume a circular territory of constant radius. This circular territory can be thought of as comprising a series of concentric rings (labelled $i = 1, 2, ..., I$) of constant width. For convenience, I will refer to these concentric rings as distance intervals. For example, a circular territory with a radius of 100 m could be defined as comprising 100 distance intervals of 1 m width. Following standard probability theory (Scheaffer & Mendenhall 1975), the probability of a successful encounter during a single foraging bout occurs within distance interval i can be calculated by the probability that no prey are encountered as the beaver traverses distance intervals 1 to $i-1$ multiplied by the conditional probability that at least one acceptable food item falls within the forager's search path within distance interval i. Assuming that the local spatial distribution of resources is random, these probabilities of encounter should follow a Poisson distribution, the parameters of which are determined by the beaver's sensory radius (r), the width of each distance interval (w), and the local density of prey per unit area within distance interval i (N_i). Hence, the probability of failure to encounter any acceptable food items from the beginning of a bout until the forager reaches distance interval i can be calculated as:

$$F_i = \exp\left[-zrw \sum_{i=1}^{i-1} N_i\right] \qquad (3)$$

Note that F_i is the cumulative probability within a given bout that the forager does not encounter any acceptable food items in distance intervals $i = 1$ to $i-1$, hence the summation. The conditional probability of successful capture within distance interval i (given that the forager reaches that distance interval) is:

$$S_i = 1 - \exp(-zrwN_i). \qquad (4)$$

These distance-dependent probabilities of search success can be used to calculate the expected rate of success per foraging bout and the expected time expenditures in travel and handling food items. The ratio of success rate over time expenditure will be used to estimate the overall consumption rate by a single forager (i.e. its functional response).

The probability during any given foraging bout that a food item is found within distance interval i can therefore be calculated as S_iF_i. By definition, the expected value of any discrete random variable is $E(Z) = \Sigma\, zf(z)$, where z is the value of the random variable and $f(z)$ is the probability density function (Scheaffer & Mendenhall 1975). The expected foraging success per foraging bout is $\Sigma\, S_i F_i$, summed over the number of distance intervals ($i=1$ to I). Similarly, the expected handling time $= \Sigma\, hS_i F_i$, where h is the time it takes to handle a single food item. The expected time spent travelling per foraging bout can be calculated by considering the full range of potential outcomes: the forager can obtain an item in any of I distance intervals or it might fail to encounter any suitable items at all. Multiplying the probabilities of each of these outcomes by the time it takes to travel to and from distance interval i yields the following expression for expected travel time:

$$E\,(\text{travel time}) = \left(\frac{2w}{v}\right)\sum_{i=1}^{I} F_i \tag{5}$$

where v is forager travel velocity.

The functional response can be calculated by dividing the expected foraging success per bout by the expected travel time plus handling time per bout (Turelli *et al.* 1982; Mangel & Clark 1988):

$$X = \frac{\displaystyle\sum_{i=1}^{I} S_i F_i}{\displaystyle\sum_{i=1}^{I} \left(\frac{2w}{v} + hS_i\right) F_i} \tag{6}$$

The resource mortality rate in distance interval i can be calculated by $X_i = XS_i F_i / \Sigma S_i F_i$, where $S_i F_i / \Sigma S_i F_i$ is the proportion of all successful encounters that would be expected to occur in distance interval i. In order to model changes in local resource density, one then needs to divide the distance-specific mortality rate (X_i) by the area (A_i) of concentric ring i, where $A_i = \pi w^2 (2i-1)$.

The ecological implications of central place foraging can be most readily visualized by considering a hypothetical experiment in which a beaver colonizes a pristine area populated by randomly distributed plant ramets that do not rejuvenate over a short period of depletion. Subsequent changes in the spatial distribution of stems are strictly due to depletion by the forager. The Markov chain model implies that the risk of attack should decrease exponentially with distance as the beaver first begins to forage (Fryxell & Lundberg 1997). This leads to more rapid depletion of resources near the central place than farther away. As resources become more scarce

close to the central place, the forager has a lower probability of encountering a suitable food item early in a foraging bout and tends to move
further afield before successfully encountering a suitable ramet. Note that
this gradual extension of foraging area over time does not result from
an active change in search strategy on the part of the forager, but is
an inevitable consequence of increased probability of search failure when
plants become scarce close to the central place. This spatial skew in resource
distribution leads to an approximately normal distribution of *per capita*
predation risk with respect to distance.

This pattern of spatially biased resource use has interesting effects on
the functional response to changes in mean resource density and the
spatial distribution of resources (Fig. 12.7). Consumption rates should be
initially high, because ramets are common close to the central place and
expected search and retrieval times are short. Local depletion close to the
central place causes a sharp decline in consumption rates due to increased
search and retrieval times. The net result is a functional response that is

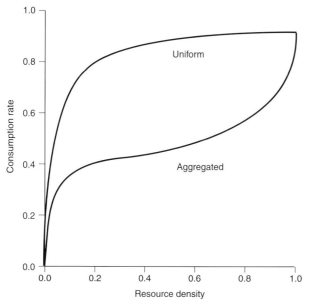

Figure 12.7 Predicted effect of central-place foraging on the rate of resource
consumption in relation to the mean density of non-renewing resources. The
expectation for a central-place forager with resources that are continually
randomized in space is provided for comparison.

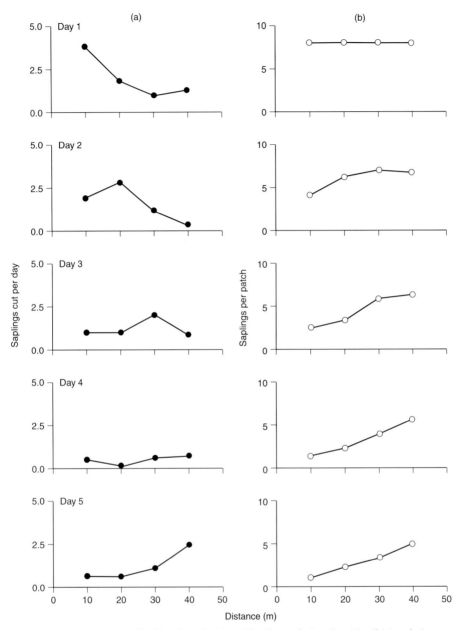

Figure 12.8 Daily risk of attack (a) and local population densities (b) in relation to distance to the central place for aspen saplings in experimental enclosures inhabited by captive beavers. Top panels correspond to the beginning of trials, when saplings were uniformly distributed among patches. Proceeding over time, nearby patches became more rapidly depleted than far patches, leading to a highly aggregated spatial distribution with highest sapling availability far from the central place.

388

depressed relative to that of a forager feeding on continually re-randomized resources.

To test the predictions of the Markov chain model, we first presented beavers in a 30×50 m experimental enclosure with an even spatial distribution of aspen saplings whose positions were redistributed every day for 6 days (Fryxell *et al.* 1994). We then presented the same animals with aspen saplings with an aggregated spatial distribution, clumped at distances of 10, 20, 30 and 40 m from the lodge, and allowed the study animals to continually deplete this experimental 'forest'.

The proportion of saplings remaining in each patch declined over time, with the rate of decline inversely related to distance from the central place (Fig. 12.8). As a result, the spatial distribution of saplings quickly became skewed, due to selective foraging near the pond. This skew in spatial distribution was maintained over the rest of the experiment. The frequency of sapling consumption was inversely related to distance at the beginning of the experiment, but quickly took on a hump-shaped distribution at successive stages of resource depletion (Fig. 12.8). Thereafter, a travelling wave of sapling consumption moved across the artificial forest, as predicted by the simple Markov chain model. This shifting pattern of resource use was obviously caused by beavers moving farther afield as saplings became progressively depleted near the central place.

During these same experimental trials, we recorded the number of saplings cut per day per individual beaver. Our results show a pronounced dip in the functional response of beavers foraging on resources with an aggregated spatial distribution (Fig. 12.9), relative to the functional response of beavers presented with uniformly distributed resources every day. Hence, the pattern of sapling exploitation by beavers in our experimental enclosures was at least qualitatively consistent with the Markov chain foraging model.

Simulation studies suggest that skewed resource distribution arising due to central place foraging can have a powerful stabilizing influence on trophic interactions (Fryxell & Lundberg 1997). The reason for this is fairly easy to see. Central-place foragers are less efficient at exploiting scarce resources, because those resources tend to be distributed far from the centre of territories. Hence, the rate of resource consumption is far less likely to outpace resource recruitment. Factors that reduce foraging efficiency at low resource densities are well known to have a stabilizing effect on consumer–resource dynamics (Rosenzweig & MacArthur 1963; Murdoch & Oaten 1975). The stabilizing effect of skewed resource distribution arising

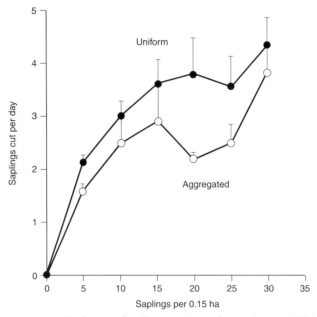

Figure 12.9 Daily rate of sapling cutting per beaver (mean ± SE) in relation to sapling density in trials with uniform sapling distribution, as in Fig. 12.1, or aggregated sapling distribution as described in Fig. 12.8. Rates of intake differed significantly between treatments ($F_{1,67} = 4.07$, $P < 0.05$).

from central-place foraging can be added to the numerous other stabilizing attributes of territoriality.

Discussion

There have been many previous studies of the adaptive basis of food selectivity and explicit tests of the contingency model of optimal diet breadth. The overwhelming majority of such experiments show patterns roughly consistent with the optimality theory predictions (Stephens & Krebs 1986), particularly that highly profitable food items tend to be selected most strongly and that diet expansion tends to occur as the frequency of encounter with highly ranked prey declines. In that sense, our behavioural studies of diet choice by beavers are not exceptional. Few empirical studies, however, have gone on to examine how patterns of selectivity affect the consumer functional response to changes in resource abundance (Murdoch

& Oaten 1975; Ranta & Nuutinen 1985; Colton 1987). Our results confirm the prediction that optimal diet choice can produce an inflection in the consumer functional response (Holt 1983; Gleeson & Wilson 1986; Persson & Diehl 1990; Fryxell & Lundberg 1994).

Such an inflection in the functional response is most likely to be stabilizing when alternative prey are so unprofitable that consumers could barely sustain themselves on them and when consumers exercise partial preferences rather than absolute diet preferences (Fryxell & Lundberg 1994). Both features are probably true of beavers: alternative resources, even when superabundant, are scarcely able to meet a beaver's metabolic needs let alone permit successful reproduction (Fryxell *et al.* 1994) and beavers show quite pronounced partial preferences for different plant species (see Fig. 12.2). Hence, it is conceivable that diet choice has a stabilizing influence on interactions between beavers and their plant resources. In this sense, complexity in plant community structure could have a demonstrable stabilizing influence on rates of resource use.

The depiction of central-place foraging as a Markov chain process has a number of interesting ecological implications that previously have not received much attention. Our foraging experiments with captive beavers were consistent with the Markov chain model. The mean distance to cut saplings increased as saplings became depleted close to the lodge, but the search time per unit area tended to always be negatively related to distance in beavers (Fryxell 1992). The Markov chain model is also consistent with a body of field observations showing that the frequency of cut trees or saplings tends to decline with increasing distance from water (Hall 1960; Jenkins 1980; Pinkowski 1983; Belovsky 1984; McGinley & Whitham 1985; Basey *et al.* 1988; Donkor 1993; Walton 1994). This is not to say that beavers always search along a vector as they move randomly away from their lodge, only that such an abstraction seems to predict the observed distributions of search time and prey attack better than null models that lack explicit travel time costs and distance-dependent probabilities of search. More detailed analyses of search pattern, well beyond the modest scope of this chapter, would be needed to discriminate between the Markov chain model and other models of adaptive space use (Andersson 1978; Morrison 1978; Getty 1981a; Bovet & Benhamou 1988, 1991; Pickup & Chewings 1988; Pickup 1994).

Inverse relationships between predation or activity rates and distance have been documented for other central-place foragers as well (Andersson 1981; Aronson & Givnish 1983; Getty 1981b; Huntly *et al.* 1986; Greig-Smith

1987; Martinsen *et al.* 1990; Pickup & Chewings 1988; Pickup 1994), and are circumstantially consistent with distance-specific patterns of seedling survival (Howe & Primack 1975) expected from distance-specific seed predators (Janzen 1970, 1971). Other factors, however, such as avoidance of predators (Covich 1976; Taylor 1988; Walton 1994; Basey & Jenkins 1995) or territorial competitors (Lewis & Murray 1993) offer alternative hypotheses for centralized patterns of resource use.

The Markov chain formulation implies that the daily rate of consumption should decline sharply as prey become depleted close to the central place, because of an increase in the time required for search and retrieval of individual food items. Our experimental functional response trials with beavers fairly convincingly suggest just such a reduction in foraging efficiency as woody plants became severely depleted close to the lodge. Such effects on predator efficiency could have a strongly stabilizing influence on predator–prey dynamics, because both the resource rate of decrease and the predator rate of increase are reduced once the spatial distribution of food has become highly skewed (Fryxell & Lundberg 1997).

The shape of the functional response of a central-place forager should be dependent on the relative magnitudes of the width of the search path and the velocity of search, rather than just the product of these parameters, as would be the case for the Holling's (1959) disk equation. When a central-place forager has a wide search path and travels at low velocity, prey depletion tends to be concentrated close to the starting point, which leads to pronounced depression in the functional response curve. On the other hand, a forager with a narrow search path and high travel velocity depresses resources less around the central place, which leads to less depression in the functional response. Hence, the stabilizing effect of central-place foraging should depend considerably on the perception radius and travel velocity of the consumer, as well as the maximum area searched around the central place.

In sum, our results indicate that rates of resource use are strongly affected by resource complexity. Interestingly, all of the forms of complexity considered in this series of experiments (variation in prey size, species and spatial distribution) had depressing effects on rates of resource use. In the absence of any further detail, that feature alone would tend to suggest greater likelihood of stability than that of a hypothetical situation with no variation amongst resources. Whether such behavioural modifications are responsible for the apparent stability of beaver populations (J. Fryxell, unpublished data) is as yet unanswered. One can conclude, however, that any meaningful model of beaver–plant dynamics must account for these features. Such a model should include, at the very least, elements of centrally biased search (such as the Markov chain model) combined with probabilities of prey acceptance once encountered based on energetic profitability (such as the contingency model of prey size and prey species breadth).

Acknowledgements

This work was funded through grants from the Natural Sciences and Engineering Research Council of Canada, Ontario Renewable Resources Research Grant Program and the Ontario Environmental Youth Corps. I thank Gord Lewer, Chris Doucet, Bea Beisner, Rob Waywell, Kim Cuddington, Russ Walton and Sue Stoddart for performing the beaver functional response experiments. John Wilmshurst, Carey Bergman, Sue Pennant, Ted Case, Kevin McCann, Peter Yodzis, Gord Hines and two anonymous reviewers made very constructive comments on a previous version of the manuscript.

References

Andersson, M. (1978). Optimal foraging area: size and allocation of search effort. *Theoretical Population Biology*, **13**, 397–409.

Andersson, M. (1981). Central place foraging in the whinchat, *Saxicola rubetra*. *Ecology*, **62**, 538–544.

Aronson, R.B. & Givnish, T.J. (1983). Optimal central-place foragers: a comparison with null hypotheses. *Ecology*, **64**, 395–399.

Basey, J.M. & Jenkins, S.H. (1995). Influences of predation risk and energy maximization on food selection by beavers (*Castor canadensis*). *Canadian Journal of Zoology*, **73**, 2197–2208.

Basey, J.M., Jenkins, S.H. & Busher, P.E. (1988). Optimal central-place foraging by beavers: tree-size selection in relation to defensive chemicals of quaking aspen. *Oecologia*, **76**, 278–282.

Belovsky, G.E. (1984). Summer diet optimization by beaver. *American Midland Naturalist*, **111**, 209–222.

Bovet, P. & Benhamou, S. (1988). Spatial analysis of animals' movements using a correlated random walk model. *Journal of Theoretical Biology*, **131**, 419–433.

Bovet, P. & Benhamou, S. (1991). Optimal sinuosity in central place foraging movements. *Animal Behaviour*, **42**, 57–62.

Charnov, E.L. (1976). Optimal foraging: attack strategy of a mantid.

American Naturalist, **110**, 141–151.

Colton, T.F. (1987). Extending functional response models to include a second prey type: an experimental test. *Ecology*, **68**, 900–912.

Comins, H.N. & Hassell, M.P. (1976). Predation in multi-prey communities. *Journal of Theoretical Biology*, **62**, 93–114.

Covich, A.P. (1976). Analyzing shapes of foraging areas: some ecological and economic theories. *Annual Review of Ecology and Systematics*, 7, 235–257.

Donkor, N.T. (1993). *Influence of beaver foraging and edaphic factors on the structure of boreal forest communities.* MSc thesis, University of Guelph, Guelph, Ontario.

Doucet, C.M., Walton, R.A. & Fryxell, J.M. (1994). Perceptual cues used by beavers foraging on woody plants. *Animal Behaviour*, **47**, 1482–1484.

Farnsworth, K.D. & Illius, A.W. (1996). Large grazers back in the fold: generalizing the prey model to incorporate mammalian herbivores. *Functional Ecology*, **10**, 678–680.

Fryxell, J.M. (1992). Space use by beavers in relation to resource abundance. *Oikos*, **64**, 474–478.

Fryxell, J.M. & Doucet, C.M. (1993). Diet choice and the functional response of beavers. *Ecology*, **74**, 1297–1306.

Fryxell, J.M. & Lundberg, P. (1994). Diet choice and predator–prey dynamics. *Evolutionary Ecology*, **8**, 407–421.

Fryxell, J.M. & Lundberg, P. (1997). *Individual Behavior and Community Dynamics.* Chapman & Hall, New York.

Fryxell, J.M., Vamosi, S.M., Walton, R.M. & Doucet, C.M. (1994). Retention time and the functional response of beavers. *Oikos*, **71**, 207–214.

Getty, T. (1981a). Analysis of central-place space-use patterns: the elastic disc revisited. *Ecology*, **62**, 907–914.

Getty, T. (1981b). Territorial behavior of eastern chipmunks (*Tamias striatus*): encounter avoidance and spatial time-sharing. *Ecology*, **62**, 915–921.

Gleeson, S.K. & Wilson, D.S. (1986). Equilibrium diet: optimal foraging and prey coexistence. *Oikos*, **46**, 139–144.

Greig-Smith, P.W. (1987). Bud-feeding by bullfinches: methods for spreading damage evenly within orchards. *Journal of Applied Ecology*, **24**, 49–62.

Hall, J.G. (1960). Willow and aspen in the ecology of beaver on Sagehen Creek, California. *Ecology*, **41**, 484–494.

Holling, C.S. (1959). Some characteristics of simple types of predation and parasitism. *Canadian Entomologist*, **91**, 385–398.

Holt, R.D. (1983). Optimal foraging and the form of the predator isocline. *American Naturalist*, **122**, 521–541.

Howe, H.F. & Primack, R.B. (1975). Differential seed dispersal by birds of the tree *Casearia nitidia* (Flacourtiaceae). *Biotropica*, 7, 278–283.

Huntly, N.J., Smith, A.T. & Ivins, B.L. (1986). Foraging behavior of the pika (*Ochotona princeps*) with comparisons of grazing versus haying. *Journal of Mammalogy*, 67, 139–148.

Janzen, D.H. (1970). Herbivores and the number of tree species in tropical forests. *American Naturalist*, 104, 501–528.

Janzen, D.H. (1971). Seed predation by animals. *Annual Review of Ecology and Systematics*, 2, 465–492.

Jenkins, S.H. (1980). A size–distance relation in food selection by beavers. *Ecology*, 61, 740–746.

Lewis, M.A. & Murray, J.D. (1993). Modelling territoriality and wolf–deer interactions. *Nature*, 366, 738–740.

Lotka, A.J. (1925). *Elements of Physical Biology*. Williams & Wilkins, Baltimore, MD.

MacArthur, R.H. & Pianka, E.R. (1966). On optimal use of a patchy environment. *American Naturalist*, 100, 603–609.

Mangel, M. & Clark, C.W. (1988). *Dynamic Modeling in Behavioral Ecology*. Princeton University Press, Princeton, NJ.

Martinsen, G.D., Cushman, J.H. & Whitham, T.G. (1990). Impact of pocket gopher disturbance on plant diversity in a shortgrass prairie community. *Oecologia*, 83, 132–138.

McGinley, M.A. & Whitham, T.G. (1985). Central place foraging by beavers (*Castor canadensis*): a test of foraging predictions and the impact of selective feeding on the growth form of cottonwoods (*Populus fremontii*). *Oecologia*, 66, 558–562.

Morrison, D.W. (1978). On the optimal strategy for refuging predators. *American Naturalist*, 112, 925–934.

Murdoch, W.W. & Oaten, A. (1975). Predation and population stability. *Advances in Ecological Research*, 9, 2–131.

Persson, L. & Diehl, S. (1990). Mechanistic individual-based approaches in the population/community ecology of fish. *Annales Zoologici Fennici*, 27, 165–182.

Pickup, G. (1994). Modelling patterns of defoliation by grazing animals in rangelands. *Journal of Applied Ecology*, 31, 231–246.

Pickup, G. & Chewings, V.H. (1988). Estimating the distribution of grazing and patterns of cattle movement in a large arid zone paddock: an approach using animal distribution models and Landsat imagery. *International Journal of Remote Sensing*, 9, 1469–1490.

Pinkowski, B. (1983). Foraging behavior of beavers (*Castor canadensis*) in North Dakota. *Journal of Mammalogy*, 64, 312–314.

Ranta, E. & Nuutinen, V. (1985). Foraging by the smooth newt (*Trituris vulgaris*) on zooplankton: functional responses and diet choice. *Journal of Animal Ecology*, **54**, 275–293.

Rosenzweig, M.L. & MacArthur, R.H. (1963). Graphical representation and stability conditions of predator–prey interactions. *American Naturalist*, **97**, 209–223.

Scheaffer, R.L. & Mendenhall, W. (1975). *Introduction to Probability: Theory and Applications*. Duxbury Press, North Scituate, MA.

Schoener, T.W. (1979). Generality of the size–distance relation in models of optimal foraging. *American Naturalist*, **114**, 902–914.

Spalinger, D.E. & Hobbs, N.T. (1992). Mechanisms of foraging in mammalian herbivores: new models of functional response. *American Naturalist*, **140**, 325–348.

Stephens, D.W. & Krebs, J.R. (1986). *Foraging Theory*. Princeton University Press, Princeton, NJ.

Stewart, W.J. (1994). *Introduction to the Numerical Solution of Markov Chains*. Princeton University Press, Princeton, NJ.

Tansky, M. (1978). Switching effect in predator–prey systems. *Journal of Theoretical Biology*, **70**, 263–271.

Taylor, R.J. (1988). Territory size and location in animals with refuges: influence of predation risk. *Evolutionary Ecology*, **2**, 95–101.

Turelli, M., Gillespie, J.H. & Schoener, T.W. (1982). The fallacy of the fallacy of the averages in ecological optimization theory. *American Naturalist*, **119**, 879–884.

Volterra, V. (1926). Fluctuations in the abundance of a species considered mathematically. *Nature*, **118**, 558–560.

Walton, R.A. (1994). *Predation risk, energy-maximization, and distance-dependent foraging by beavers*. MSc thesis, University of Guelph, Guelph, Ontario.

Warner, A.C.I. (1981). Rate of passage of digesta through the gut of mammals and birds. *Nutritional Abstracts Revue Series B*, **51**, 789–820.

Chapter 13

Scaling up from functional response to numerical response in vertebrate herbivores

A.W. Illius[1] and I.J. Gordon[1,2]

Summary

Considerable progress has been made in the last decade towards understanding the relationships between vertebrate herbivores and their food supply. Mechanistic approaches to analysing the constraints on food intake, and the consequences for population dynamics, are replacing the classical theoretical descriptions of predator–prey dynamics. The challenge of the former approach is to discover what our mechanistic understanding can reveal about process and pattern in plant–herbivore relationships. This chapter describes the modelling of the processes of food intake and diet selection, from the level of the individual bite, up to daily nutrient intake, metabolism, energy balance, reproduction and mortality, thus integrating the mechanisms underlying population dynamics. Two examples, of a temperate and a savanna grazing system, are used to show how far mechanistic modelling can be used to explain the relationship between vegetation and herbivore abundance and the physiological basis of overcompensatory population dynamics.

Introduction

In analysing trophic interactions and their population consequences, classical mathematical ecology aims to achieve a tractable analytical description of the phenomena of interest. This approach requires the introduction of

1 Institute of Cell, Animal and Population Biology, University of Edinburgh, West Mains Road, Edinburgh EH9 3JT, UK
2 Macaulay Land Use Research Institute, Craigiebuckler, Aberdeen AB15 8QH, UK

the minimum number of parameters consistent with capturing the main properties of the system's behaviour, in order that the parameter space can be investigated analytically or graphically (e.g. Rosenzweig & MacArthur 1963; Noy Meir 1975). Simplified representations of ecological interactions are justified, not only by the clarity with which they may reveal each parameter's effects, but also in the absence of sufficient knowledge of underlying mechanisms to build more detailed and, hence, more realistic models. However, considerable progress has been made in the last few decades towards a mechanistic understanding of the relationships between vertebrate herbivores and their food supply. Mechanistic approaches, addressing the physiological processes that relate food intake to its consequences for population dynamics, are now set to augment the classical theoretical descriptions of predator–prey dynamics. Instead of treating population dynamics explicitly, as, for example, in setting a parameter governing density dependence, mechanistic modelling defines the nature of plant–herbivore relationships implicitly, allowing quantitative hypotheses to be made and tested. The challenges for the mechanistic approach are to discover what further can be revealed about processes and patterns in plant–herbivore relationships; to show how far our mechanistic understanding can account for population and community phenomena, and to produce quantitative and prescriptive solutions to practical ecological problems.

This chapter sets out an approach to mechanistic modelling of mammalian herbivore population ecology, to show how the functional response can be scaled up to the numerical response. In order to substantiate the claim that such an approach is feasible, a detailed description of the modelling of the component processes is presented, using the St Kilda grazing system as an example. A second example, of a savanna system, is presented in less detail, to show how the approach can be generalized to other systems.

The unmanaged population of feral Soay sheep on the island of Hirta, in the St Kilda archipelago, off north-west Scotland, shows erratic population dynamics in a temperate environment. The sheep and their grazing system have been the subject of long-term study (Jewell et al. 1974; Clutton-Brock et al. 1991). Although climatic effects on sheep population dynamics may yet be revealed by longer-term study, the overriding influence appears to be the low or absent density dependence of fecundity, with the consequence that the sheep population periodically exceeds the number that can be supported by the island's winter food resources. This results in overcompensatory mortality (Grenfell *et al.* 1992): a sharply focused decline in sheep numbers from around 1500 to 700 in late winter. The purpose of modelling this grazing system in terms of the physiological processes governing the flow of energy and nitrogen from the vegetation and through the sheep population is to test whether the system can be adequately represented in this way, and to examine the causes of overcompensatory mortality.

In contrast to the relatively simple and temperate system of grass and sheep on Hirta, our second example is of tropical savanna grazing systems. Semi-arid savanna ecosystems are complex, in consisting of a range of vegetation and herbivore types, and are dominated by annual and seasonal variability in rainfall, which are the main source of variation in herbivore population size. A model of herbivore and vegetation dynamics is used to test whether secondary production can be predicted mechanistically from rainfall, and to consider the effect of vegetation structure on animal population dynamics.

Component processes

We are concerned here with describing the processes of food intake and diet selection, from the level of the individual bite, up to daily nutrient intake, metabolism, energy balance, reproduction and mortality, thus integrating the mechanisms underlying population dynamics of mammalian herbivores. These descriptions are then combined in a model that calculates the daily flux of material and energy through compartments representing the vegetation and the sheep population. A list of the state variables and parameters used in the model is given in Table 13.1. Further details of the model structure and of the vegetation submodel are given in the Appendix.

Herbage intake and nutrient supply

Daily nutrient intake is the product of intake rate while feeding, the nutrient content of the diet and the time spent feeding. Short-term nutrient intake rate is largely determined by the animal's physical responses to vegetation structure and abundance. The degree of selectivity exerted represents a trade-off between the nutrient content of the diet and the rate at which it can be eaten. The time spent feeding is driven by the animal's appetite, and is constrained by environmental factors such as day length and weather, and by animal factors such as fatigue, the ability to utilize or

Table 13.1 Definition of state variables and parameters.

	Units	Definition	Eq.
State variable			
L1, L2, L3	$kg\,ha^{-1}$	Live leaf: emerging, emerged, mature, respectively	
D	$kg\,ha^{-1}$	Dead leaf	
L	kg	Animal fat free mass	
F	kg	Animal fat mass	
N		Animal numbers in each sex and age class	
Parameter			
I_B	mm	Incisor arcade breadth	
H	mm	Sward surface height	
B_D	mm	Bite depth	
V_{max}		Maximum foraging velocity	1, 2
R_{max}		Maximum eating rate	3
h		Minimum handling time per bite in the absence of chewing	3
w		Width of search path	1, 2
D	m^{-2}	Density of bites	3
S		Bite mass	1, 2, 3
B	s^{-1}	Rate of biting	4
I_d	$kg\,day^{-1}$	Maximum daily dry matter intake	4
d	proportion	Digestibility of food	4
u_g	none	Degree of maturity in digestive tract capacity	4
W	kg	Animal weight	4
A	kg	Animal weight at maturity in body size	4
P_G	$kg\,day^{-1}$	Potential daily protein growth	5
u	none	Degree of maturity in body size	5
u_L	none	Degree of maturity in fat-free mass	5
W_b, W_{bt}	kg	Birth weight, birth weight given unrestricted fetal growth	7, 9
F_r	none	Fat mass as a fraction of the maximum possible fat mass	8, 11
$E_c(t)$	$MJ\,day^{-1}$	Energy deposition in the conceptus t days after conception	8
m	none	Reproductive (pregnancy) allocation parameter	8
$Y(n)$	kg	Daily milk yield in the nth week *post-partum*	10, 11
q	none	Reproductive (lactation) allocation parameter	11
σ_f	kg	Standard deviation of body fat	

store nutrients, and by digestive constraints (see Illius & Jessop 1996; Illius 1997; Chapter 11).

Intake rate
Both experimental and theoretical analyses of the functional response of mammalian herbivores show that bite mass is the variable exerting the greatest effect on intake rate (Hodgson 1985; Spalinger & Hobbs 1992). Grass swards consist of a three-dimensional array of plant tissues, and

bite mass varies with the volume of the sward that the animal can enclose in each bite and with the bulk density of the grazed horizon (Black & Kenney 1984; Illius & Gordon 1987; Burlison *et al.* 1991). Bite volume is determined by the horizontal area of the sward covered by each bite and the depth to which the incisors penetrate into the sward. Bite mass can be estimated by the model of Illius and Gordon (1987) from the animal's incisor arcade breadth and bite depth (governing bite volume) and from the height and biomass of the vegetation (governing the bulk density of the herbage enclosed in a bite). Incisor arcade breadth (I_B mm) in Soay sheep is given by the allometric relationship: $I_B = aW^{0.245}$, where W is body mass (kg) a is 12.28 for females and 12.03 for males ($r^2 = 64$; residual SD $= 1.6$ mm; Illius *et al.* 1995). Bite depth of Soay sheep grazing *Agrostis–Festuca* swards was found to be closely related to sward surface height (H mm) by the following expression: $B_D = 4.45 + 0.4 H$ ($r^2 = 0.94$, residual SD $= 0.4$; A.W. Illius *et al.*, unpublished observations). The relationships between sward surface height and sward biomass for the three vegetation types were taken from Armstrong *et al.* (1997).

   The determinants of the rate of biting were greatly clarified by the work of Spalinger and Hobbs (1992), who recognized that bite rate could be limited either by the animal's need to search for its next bite or by its need to chew ingested herbage. They derived equations for the functional response under conditions where intake rate is limited by one of three processes: rate of encounter with cryptic food items (Process 1); rate of encounter with apparent items (Process 2); and rate at which food can be chewed and swallowed (Process 3). Given an animal with maximum foraging velocity V_{max}, maximum eating rate R_{max}, minimum handling time per bite in the absence of chewing h and width of search path w, foraging on plants offering bites at density D (per m^2), and offering a bite of mass S, then the rate of biting B can, by adapting Spalinger and Hobbs (1992) slightly, be written as:

Process 1:
$$B_1 = \frac{V_{max} w D}{(1 + n V_{max} w D)} \tag{1}$$

Process 2:
$$B_2 = \frac{V_{max} \sqrt{D}}{(1 + n V_{max} \sqrt{D})} \tag{2}$$

Process 3:
$$B_3 = \frac{R_{max}}{(S + R_{max} h)} \tag{3}$$

Intake rate is simply the product of B and S. In Eqs 1 and 2, n is a coefficient describing the time lost to future encounters whilst pausing to take a bite, and can probably be approximated by h. The first two equations describe rate of biting as limited by encounter rate, and distinguish cases where (1) potential bites can only be detected at close range, perhaps because obscured by dead herbage, from (2) those where bites can be detected at a distance. Process 1 applies where the average distance between plants is greater than the detection distance, i.e. where $1/\sqrt{D} > w$. Equation 3 describes the case where encounters with large S are sufficiently frequent to cause bite rate to be limited by chewing rate. Thus, as bite mass becomes restricted, bite rate can increase up to a maximum, h. The actual rate of biting will be governed by whether encounter or chewing rate is slower. If bite mass and biomass density are known, then intake rate and distance moved during foraging can be calculated. V_{max} is an allometric function of animal weight (W kg) and is $0.5\,W^{0.13}\,\mathrm{m\,s^{-1}}$ (Pennycuick 1979). Search path width, w (m) can be assumed to scale with animal stature, and so with limb length, or with $W^{0.25}$ (Alexander 1977). Shipley et al. (1994) determined R_{max} ($\mathrm{mg\,s^{-1}}$) to be $12.3\,W^{0.69}$ and h (s) to be 0.9.

Plant community selection and nutrient content of the diet
On each day, the vegetation type selected to be grazed was assumed to be that offering the highest energy intake rate. Within a community, diet composition is determined by the vertical distribution of biomass in each leaf compartment, and by the effect of bite depth on the relative amounts removed from each leaf age (Fig. 13.1; see also Hodgson 1985). It was assumed that seasonal variation in nutrient content of the vegetation is primarily accounted for by the changing proportions of live and dead leaf. The organic matter digestibility of live and dead *Agrostis–Festuca* leaf was estimated from the data of Armstrong et al. (1986) to be 0.77 and 0.44, respectively, with values of 0.68 and 0.41 for live and dead *Molinia*. The metabolizable energy (ME) content of digested organic matter is $15.6\,\mathrm{MJ\,kg^{-1}}$ (ARC 1980). The nitrogen concentrations of *Agrostis–Festuca* live and dead leaf were taken to be 24.5 and $10.6\,\mathrm{g\,kg^{-1}}$ with 17.5 and $8.0\,\mathrm{g\,kg^{-1}}$ for *Molinia* (Milner & Gwynne 1974; Armstrong et al. 1986). Quantification of the metabolism and utilization of energy and nitrogen follows ARC (1980) and IDPW (1993).

Grazing time
Given the short-term rate of nutrient intake, daily intake is determined by

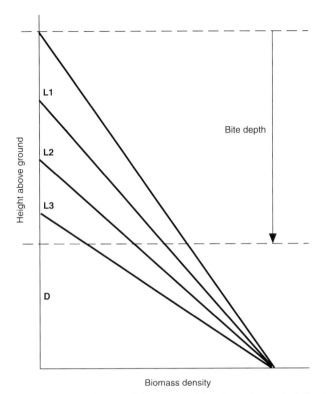

Figure 13.1 Prediction of diet composition from the vertical distribution of leaf (state variables L1, L2, L3, D) in the sward, and hence the relative amounts removed when the sward is grazed to a given depth.

the time spent grazing, which, assuming that animals seek to maximize their daily nutrient intake, will be the maximum allowed by environmental and other constraints. Additionally, digestive constraints may limit daily intake to the maximum daily turnover of digestive tract contents. Preliminary analysis of the diet quality selected by Soays suggested that digestive constraints are unlikely to apply, and this factor is therefore not included in the model. The setting of digestive constraints in cases where they might apply, such as in savanna ecosystems, is given in the next section. The settings of constraints on grazing time that are thought to be operative on Hirta are as follows: Soay sheep rarely graze at night (Grubb & Jewell 1974; Stevenson 1994); ruminants appear to have an upper limit on the number of bites that they will take in a day (Hodgson 1985); and intake will

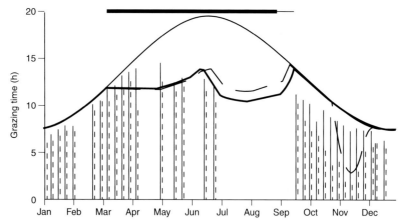

Figure 13.2 Seasonal pattern of observed (vertical lines; Grubb & Jewell 1974). and predicted grazing time in adult females (——) and males (- - - -). The sine curve shows day length, which constrains winter foraging. The thick horizontal line above indicates the period when grazing time was constrained to the time required for 40 000 bites day^{-1}, and the thin horizontal line indicates when appetite constrained grazing time. The reduction in grazing by rutting rams can be seen in November.

ultimately be constrained by the animal's ability to utilize or dispose of nutrients (see Illius & Jessop 1996). It was thus assumed that the grazing time of Soays is constrained to the minimum of:

1 the hours of daylight;
2 the time taken for 40 000 bites (estimated as the upper limit in a small-bodied ruminant: see Illius & Gordon 1987);
3 the time required to meet the animal's daily energy requirements (which are described below).

Rutting also reduces grazing time in rams, and this effect of male reproductive investment is discussed later.

The resulting predictions of grazing time are compared with observations of Grubb and Jewell (1974) in Fig. 13.2.

Digestive constraints
Daily intake of foods of low digestibility is constrained by the low rates of ruminal digestion and passage (e.g. Laredo & Minson 1973). The model of Illius and Gordon (1991, 1992) quantified the relationship between food

and animal characteristics and allows prediction of the maximal intake of a food in relation to its digestibility (*d* proportion) and the animal's size. Model output from a range of these inputs was summarized by regression to give the maximum daily intake (kg) under digestive constraints:

$$I_d = 0.09d^{1.1}A^{0.81}u_g \qquad \text{(grass)}$$
$$I_d = 0.13d^{1.35}A^{0.77}u_g \qquad \text{(browse)}$$

$$\text{where} \quad u_g = \left(\frac{W}{A}\right)^{0.75}$$

(4)

The term u_g scales gut capacity to body mass in immature animals (given W, the current weight and A, the mature weight), accounting for the relatively larger digesta load of weaned immature animals (B.G. Lowman, unpublished data). The prediction equations agree well with published observations at low digestibility, when digestive constraints would be expected to apply, but overestimates intake by up to 25% on higher digestibility diets, which experimental animals may not choose to eat to capacity (see Illius 1997).

Energy expenditure
The first call on the animal's energy intake is to supply its requirements for basal metabolism, activity and thermoregulation.

Maintenance and activity
Thermoneutral resting metabolism (MJ day^{-1}) is 0.3 $WA^{-0.27}$ (Taylor *et al.* 1981), where *A*, the mature mass, is 32 and 24 kg, respectively, for male and female Soay sheep (McClelland *et al.* 1976). Energy costs of foraging arise from maintenance of posture (4.22 $W^{0.735}$ J s^{-1} of foraging), from travel (15.8 $W^{0.589}$ J m^{-1}; Taylor *et al.* 1982) and from eating (1.54 times the resting metabolic rate; Graham 1964). Sheep were assumed to move 500 m during foraging and 500 m commuting between feeding and sheltering sites (I.R. Stevenson, personal communication).

Thermoregulation
Sheep, being warmer than their environment on St Kilda, continuously lose heat. To maintain body temperature, heat generated by maintenance energy expenditure, activity, and the waste heat of metabolism from growth and reproduction must either exceed the heat loss to the environment, or

additional energy expenditure must be incurred to make up the deficit. Heat production during metabolism is calculated according to ARC (1980). Heat loss to the environment was modelled according to the electrical analogue model (McArthur 1987) in which heat flows through the series of resistances posed by skin, coat and boundary layer. Radiative exchanges are considered separately. Parameter values were taken from Campbell *et al.* (1980), McArthur and Monteith (1980a,b), McArthur (1987), Monteith and Unsworth (1990) and Stevenson (1994). Environmental temperature and windspeed were taken to be, respectively, the daytime mean temperature and $5\,\mathrm{m\,s^{-1}}$ while foraging, and $2°C$ and $1.25\,\mathrm{ms^{-1}}$ while sheltering (Campbell 1974; Stevenson 1994). Webb and King (1984) showed that wetting an animal's coat reduces its resistance to sensible heat flux by a factor of 2, so the effect of the high rainfall on St Kilda was modelled by assuming that half of the coat of sheep is wet during half the period foraging. The surface area $(\mathrm{m^2})$ and trunk diameter (m) of Soay sheep ranging in size from juveniles to adults was described by the allometric functions: $0.14\,W^{0.565}$ and $0.085\,W^{0.376}$, respectively ($n = 14$, $r^2 = 0.98$ for both, residual c.v. = 6.0, 4.0%; A.W. Illius & G.A. Lincoln, unpublished data).

Energy deficits are met by metabolism of fat and protein, yielding respectively 39.3 and $13.5\,\mathrm{MJ\,kg^{-1}}$ (Blaxter 1989), after allowing for energy lost in urea and 4.6 MJ lost as heat of synthesis of urea. The protein proportion of energy loss was estimated from the data collated by Reeds and Fuller (1983) as $(0.12–0.19M)/(1-M)$ for $M < 0.63$, otherwise zero, where M is the energy intake expressed as a multiple of maintenance.

Growth and body composition

Two state variables are used to represent distinct aspects of growth and body composition. Fat-free mass (L kg) represents the animal's size and degree of maturity, to which most physiological functions are related. Fat mass (F kg) is the major energy store, and reflects both the physiological maturity of the animal and its history of energy balance.

Fat-free mass

Growth in fat-free mass can be described as a function of the animal's capacity for net protein synthesis, which in turn determines the deposition of minerals and water. The genetically standardized potential for protein growth in mammals can be described by:

$$P_G = 0.001565uA^{0.73}(4.475 - u^{0.65})\log_e\frac{1}{u} \qquad \text{kg day}^{-1} \qquad (5)$$

where u is the degree of maturity in body mass, given by W/A (Illius & Jessop 1995). This assumes that the mass of digesta is $0.178W$, the fat content of digesta-free mass is $0.184u^{0.65}$, and the protein content of fat-free mass is 0.23. The minerals and water deposited in proportion to protein amount to $0.23P_G$ and $3u^{-0.11}P_G$, respectively (Emmans & Fisher 1986), so that the potential increase in fat-free mass, nutrition permitting, can be calculated. In animals showing seasonal fluctuations in size and composition, it is preferable to relate potential growth to the degree of maturity in fat-free mass $u_L = L/A_L$, the current and mature fat-free mass, respectively. Since Soay sheep, like other highly seasonal mammals, show an annual metabolic cycle of growth and fattening (G.A. Lincoln, personal communication) it was assumed that P_G declines to zero at the winter solstice, and is maximal at the summer solstice:

$$P_G(t) = P_G\left(0.5 + 0.5\sin\left(2\pi\frac{t+273}{365}\right)\right) \qquad (6)$$

where t is the solar day. Fat-free mass is incremented daily from birth by the amount of growth less mobilization (below).

Fat mass
The maximum fat mass attainable at any body size by Soay sheep (F_{max}) was estimated from the data of McClelland *et al.* (1976) to be $aW^{2.0}$ kg, where a is 0.0099 for females and 0.0079 for males. It is assumed that if an animal has met all other energy requirements and has reached its maximum fatness, then it will stop eating due to metabolic constraints. Fat mass in lambs at birth was estimated from the formula of McDonald *et al.* (1979), adjusted to the fat mass usable for energy metabolism by subtracting the quantity of structural lipid in cell membrane and adding the energetic equivalent of glycogen (Mellor & Cockburn 1986) to be:

$$F = 0.0242W_b\left(\frac{W_b}{W_{bt}}\right)^{1.4} - 0.0082W_b \qquad (7)$$

where W_b is birth weight (kg) and $W_{bt} = 2.5$ kg is the expected birthweight following unrestricted fetal growth (see below).

407

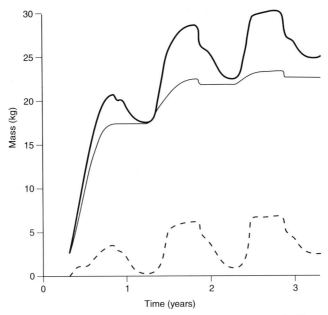

Figure 13.3 Predicted growth patterns in Soay rams over the first 3 years of life on Hirta. Fat mass (- - -), fat-free mass (————) sum to total live mass (——).

As an example, the predicted patterns of growth in fat mass and fat-free mass in males, and assuming a low population density throughout, are shown in Fig. 13.3. The predicted pattern of body-mass change with age and season is close to that observed (Doney *et al.* 1974), although the modelled maturation rate, which is based on the interspecific scaled mean, may be slightly higher than is exhibited by Soays, as suggested by the data of McClelland *et al.* (1976).

Fecundity and reproductive investment by females
Fecundity
In ungulates, the probability of conception, and hence the fecundity rate, is generally related to the animal's degree of maturity in body size and to its body reserves (Sæther 1997). Conception probability can therefore be modelled as a logistic function of animal mass. For example, the savanna model describes state-dependent fecundity in the simplest terms, where conception probability is 1 if fat mass exceeds 0.5 of the maximum adult fat

mass, otherwise zero. In the particular case of St Kilda, a different approach was followed, since there is virtually no observed density dependence or weight dependence in fecundity (Clutton-Brock *et al.* 1991, 1992), and hence no quantified physiological basis for predicting it. Since about 90% of females breed each year and only 10–20% produce twins, the rate of reproduction was approximated by assuming that every female produces a single lamb. This simplification approximates the observed fecundity.

Reproductive investment

This consists of the energy and protein costs of pregnancy, including the energetic cost of the additional weight carried, and lactation for a single offspring.

During pregnancy, the unit costs of energy and protein in the growing conceptus are described according to ARC (1980) and IDPW (1994). The principal identified causes of variation in lamb birth weight are the mother's age and the population size (Clutton-Brock *et al.* 1992), and these are likely to be due, respectively, to the degree of maturity in body size and adequacy of nutrition (see Gunn *et al.* 1986; Peart 1967). Therefore, unit costs (i.e. energy and protein use per kilogram of lamb at birth) were scaled by the degree of maturity in fat-free mass ($u_L = L/A_L$) and by a function of the current fat mass as a fraction of the maximum fat mass ($F_r = F/F_{max}$). Energy deposition in the conceptus, E_c (MJ), at t days after conception is estimated according to ARC (1980) and Blaxter (1989) as:

$$E_c(t) = 0.0433 u_L F_r^m e^{-0.00643t} 10^{(3.322 - 4.979 e^{-0.00643t})} \tag{8}$$

and lamb birth weight, at the end of the 150-day pregnancy as:

$$W_b = 5.72 E_c(150) \tag{9}$$

The parameter m determines reproductive allocation as a function of current nutritional status, and was evaluated empirically to be 0.25. As m tends to 0, F_r^m tends to 1 at all but very low values of F_r, favouring fetal growth over preservation of maternal body reserves. As m tends to 1, F_r^m tends to F_r and so fetal growth becomes more directly proportional to F_r. Equations 6 and 7 replicate the observed pattern of birth weight with age and population size, via its predicted effect on body condition (Fig. 13.4).

Taylor and Murray (1987) give peak milk output as $0.5A^{0.73}$MJ day^{-1}, and this is equal to 1.1 kg day^{-1} in mature Soays, assuming sheep's milk to contain 4.64 MJ kg^{-1} (Blaxter 1989). The lactation curve was described by the formula of Wood (1967):

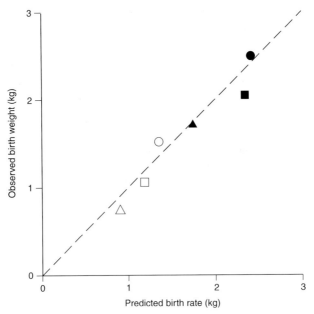

Figure 13.4 Comparison of observed (Clutton-Brock *et al.* 1992) and predicted birth weights of single lambs born to adult ewes, yearlings and ewe lambs (aged 4, 2 and 1 years, respectively; ●, ■ and ▲) at low population density ($N=200$ sheep, excluding neonates; solid symbols) and at high population density ($N=600$; open symbols).

$$Y(n) = an^b e^{-cn} \tag{10}$$

where $Y(n)$ is the daily milk yield (kg) in the nth week post-partum and *a, b* and *c* are constants. Data from lactation in Scottish blackface sheep (Doney *et al.* 1983) were scaled to the shorter lactation length of Soays (Taylor & Murray 1987) to obtain $b=0.8$ and $c=-0.343$ (residual SD = 10%). The effect of ewe body size on lactation potential was described by scaling milk yield to the degree of maturity in fat-free mass at parturition $(u_L(p) = L(p)/A_L)$, as for pregnancy. The effect of body condition on lactation (e.g. Doney *et al.* 1981) was described by a function of F_r:

$$Y(n) = u_L(p)F_r^q an^b e^{-cn} \tag{11}$$

where *q* determines reproductive allocation as a function of current nutritional status, as with *m* in the case of pregnancy. Malnutrition during lactation is known to reduce milk yield and lamb survival (Doney *et al.*

1981; Mellor & Cockburn 1986) implying a value for q of between 0.5 and 1, and a value of 0.68 was obtained from the data of Doney *et al.* (1981). They found that the milk yield of undernourished ewes in poor body condition was, on average, 73% of control milk yield, while substituting the reported values for body condition into Eq. 9 gives a predicted reduction averaging 69% ($F_{1,8} = 0.37$, ns). The coefficient *a* was estimated to be 2.0, on the assumption that the standard peak yield of 1.1 kg day^{-1} is achievable by a mature animal eating enough to maintain her fat reserves at half the maximum fatness. Finally, the yield of colostrum, which is an important source of energy to the newborn lamb, was estimated to be energetically equivalent to 2.4 times the milk yield predicted by Eq. 9 for the first day of lactation (Mellor & Cockburn 1986).

Reproductive investment by males

The cost associated with reproduction in males is assumed to be the loss in body condition during rutting. Grubb and Jewell (1974) showed that rams, whose grazing time is normally similar to that of ewes, restrict their period of grazing to about 50% of that of ewes during the rut in November. Rams spend commensurately more time moving around between groups of ewes (Stevenson 1994). Analysis of daily census data collected during the rut in the years from 1989 to 1992 (T.H. Clutton-Brock, unpublished data) showed that adult rams exhibit both earlier and more intense inappetance than younger ones, especially lambs. Over the whole rut, the average daily grazing time of male lambs, yearlings and adults was 0.81, 0.68 and 0.59 that of their female counterparts. The proportionate reductions in grazing time at the height of the rut were 0.54, 0.36 and 0.33. These age-specific patterns of reduced grazing time were used in the model to calculate the reduction in intake and the increase in locomotion by rutting males (see Fig. 13.2).

The mean proportionate loss of grazing time by rutting ram lambs, yearlings and adults (0.19, 0.32 and 0.41 of ewes' grazing time) was found to be linearly related to the degree of maturity of body weight and can be approximated by $0.43u$ ($u = 0.47$, 0.81 and 0.98 in ram lambs, yearlings and adults, respectively, estimated from data collected in November). The fact that male reproductive investment scales with physical maturity provides a strong justification for scaling female reproductive investment in a similar way (above).

Mortality

During winter, when intake is depressed by low vegetation biomass and when climatic heat loss is high, mortality occurs due to the exhaustion of body reserves. Mean body fat in each age and sex class is obtained daily from the calculated energy balance, and is assumed to be normally distributed with standard deviation σ_f. Mortality occurs in the proportion of animals in the tail of this distribution that projects below zero. Variance in body weight measured in the month prior to the crash in March 1992 was used to estimate σ_f within sex and age classes.

Model performance

Results presented in the previous section (see Figs 13.2–13.4) show that the model accounts for components of animal performance (grazing time, birth weight, body growth) for which observations have been made on the study population, and with acceptable accuracy. This section considers population phenomena, resulting from interaction of plant and animal components, in terms of the way population density, season and reproductive investment affect intake, gain and loss of body reserves, mortality and population dynamics.

Seasonal cycles of body composition

A demonstration of the predicted seasonal changes in fat mass in males and females over 2 years (at medium and high population density in years 1 and 2, respectively) is given in Fig. 13.5. The effects of reproductive investment and season can be separated by comparing the normal case (Fig. 13.5a) with the case where the components of reproductive investment in each sex (rutting, pregnancy, lactation) are set to zero (Fig. 13.5b). Loss of fat reserves due to food shortage and high environmental heat loss is predicted to occur in winter, and is reversed in spring. Comparing the predictions during the first year at medium population density with those subsequently at high population density shows that the higher density results in greater fat loss, due to greater competition and depletion of the vegetation (Fig. 13.5b). The predicted costs of reproduction are evident: rutting in males causes marked loss of fat in November, and the effects of pregnancy and especially lactation on female fat mass is observed in spring. Reproductive investment at high density, and hence low food intake, reduces mean fat reserves to nearly zero (implying about 50% mortality). Additionally, because male reproductive investment precedes that in females, and they

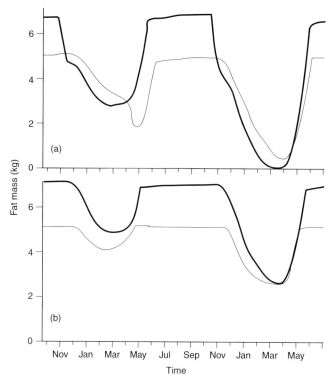

Figure 13.5 Changes in fat mass of males (———) and females (———) over 2 years: the first year at medium population density, the second at high density. (a) The normal case, with effects due to both reproductive investment and season. (b) Seasonal effects alone, after setting reproductive costs to zero.

are therefore first to exhaust their fat reserves, the resulting male mortality reduces competition for the vegetation and so diminishes the mortality of females.

Regardless of the population density, the model predicts that female fat mass can completely recover by the end of lactation, with the implication that fecundity will be unaffected by population density. This provides a physiological explanation of the virtual absence of density dependence in fecundity observed on Hirta (Clutton-Brock *et al.* 1991).

Differential mortality and population dynamics
The model's prediction of the proportions of each sex and age class

surviving a crash is in good agreement with the patterns observed on Hirta. These have been characterized by Grenfell *et al.* (1992) using the function:

$$S = \frac{d}{\left(1 + (aN)^b\right)} \tag{12}$$

where S is the proportion of any sex and age class surviving winter, d is density-independent mortality, a is the reciprocal of the threshold population size above which density-dependent mortality occurs, and b is the strength of density dependence and indicates overcompensation where $b > 1$ (Fig. 13.6a). N is the number of females, excluding lambs. We obtained comparable values of these parameters by analysing population changes predicted by the simulation model (Fig. 13.6b). The predictions do not differ significantly from those observed, although the population size at which overcompensatory mortality occurs is overestimated. Accordingly, the average population size is slightly overpredicted. The model does not account for the small amount of density-independent mortality, such as would occur through accidental deaths, so values for d are higher than observed. This will cause slight underestimation of the density-dependence parameter, b, but even so, the agreement between predicted and observed estimates is good, especially when comparing sex differences within

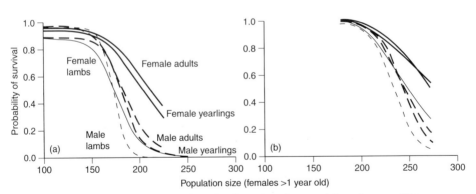

Figure 13.6 (a) Proportion of each age/sex class surviving winter on Hirta, 1985–1993, plotted in relation to population size in autumn. The lines are curves fitted to observations, using $S = d/(1 + (aN)^b)$, where N is the number of females older than 1 year. (b) Model predictions.

414

age classes. Sex differences in observed date of death are also accurately predicted by the model: the majority of male mortality occurs before that in females.

Given the lack of density-dependent fecundity and the periodic over-shoot in the population beyond winter food resources, overcompensatory mortality results in fluctuations of the modelled population. The effect is similar to that shown by the matrix model of Grenfell *et al.* (1992), which used the survival probabilities estimated by Eq. 12.

Plant–animal relationships in semi-arid savannas

We introduce this example to examine the predicted relationship between primary and secondary production over a gradient in mean annual rainfall, the main environmental determinant of plant and animal biomass in savannas. The example also illustrates the generalization of the animal model to different body sizes. The animal component is largely as described for the St Kilda model (see Appendix), and a complete description of the model is in preparation.

Relationship between mean annual rainfall and
ecological carrying capacity

The effect of mean annual rainfall on predicted ecological carrying capacity (defined as the long-term mean animal density in the absence of offtake and predation) is shown in Fig. 13.7. Twenty-year records of daily rainfall at eight sites in Botswana and Zimbabwe, along a SW–NE transect from 300 to 870 mm mean annual rainfall, were used as inputs to the model. The 20-year mean carrying capacities at each site were predicted by allowing herbivore numbers to be determined by the supply of vegetation (via food intake, birth and mortality), without animal offtake. Two conditions were simulated, with low ($100\,kg\,ha^{-1}$) and intermediate ($2000\,kg\,ha^{-1}$) starting values for the biomass of trees. In the latter case, predicted carrying capacity is reduced by tree encroachment, the main reason being that browse appears to offer animals lower rates of nutrient intake than grass. Predicted carrying capacity increases with rainfall, in broad agreement with the carrying capacities observed by Fritz and Duncan (1994). Note that the only parts of the model adjusted to match known output concerned the vegetation: grass yield was adjusted in relation to rainfall to match that observed by Dye and Walker (1987) at Bulawayo (see Appendix). The prediction of animal intake, growth, birth and mortality was not adjusted

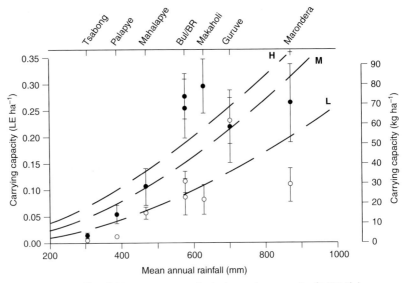

Figure 13.7 Predicted 20-year mean ecological carrying capacity ($\pm$ SEM) in relation to the rainfall at seven locations from south-west Botswana to north-east Zimbabwe (Tsabong, Palapye, Mahalapye, Bulawayo and Buffalo Range, Makaholi, Guruve, Marondera). Closed symbols: 100 kg ha^{-1} starting tree biomass; open symbols: 2000 kg ha^{-1} starting tree biomass. For comparison, the curves (right axis) show the carrying capacity observed by Fritz and Duncan (1994) in relation to rainfall at sites of high, medium or low soil nutrient availability (H, M, L, respectively). Livestock equivalents (LE) are defined in the Appendix.

to conform to expected *system* performance but were independently derived, being based on the principles outlined above.

Discussion

The examples show that mechanistic modelling can give acceptable predictions of herbivore carrying capacity in relation to primary production, and can also reveal some of the physiological mechanisms underlying population performance and community dynamics.

The St Kilda model reveals that the mechanisms underlying susceptibility to mortality are: energy intake relative to metabolic needs, growth rate, energy partition to body reserves and the timing and extent of reproductive investment. Age and sex differences in the operation of these mechanisms

416

determine the mean survival probability of each class, and, together with individual variation in energy balance, govern the degree of asymmetry of mortality across the population. Overcompensatory mortality arises because the period of reproductive investment of Soays is sufficiently brief, in accordance with their small body size, to allow them to regain body condition before the next breeding season, thereby removing an effect of density on fecundity (Clutton-Brock *et al.* 1997).

The results of the savanna model reported here show that rainfall, which is known to be the primary determinant of ungulate abundance, can be used to predict vegetation growth and hence animal population dynamics. Model predictions of carrying capacity increase roughly as the cube of rainfall, in comparison with observed carrying capacities varying with approximately the square of rainfall (Fritz & Duncan 1994). A doubling in mean annual rainfall from 300 to 600 mm would increase predicted carrying capacity 10-fold, from 0.012 to 0.12 (model) or, according to data of Fritz and Duncan (1994) fourfold, from 0.03 to 0.12. Natural systems are clearly sensitive to mean annual rainfall. The model overestimates sensitivity, mainly due to underprediction of carrying capacity at low rainfall, and sensitivity to bush encroachment in the absence of fire, which would tend to stabilize the vegetation of natural systems. Better understanding of vegetation dynamics is required, particularly in relation to the rainfall–growth responses of diverse vegetation types (e.g. forbs, ephemerals), which are important in low-rainfall regimes.

The savanna model also predicts some of the consequences for community dynamics of differences in animal body size and diversity of plant growth forms. The modelled dietary differences are accounted for by the fact that the nutritious growing tips and fallen fruits of browse are scarce, small in size and take longer to harvest, giving lower rates of energy intake to larger animals than they can obtain by grazing, supporting the predictions of Illius and Gordon (1987). The effect of this foraging strategy on the vegetation is partly determined by the assumption in the model that the grass/tree balance is determined by competition for water (Dye & Spear 1982). Removal of leaf tissue by herbivores limits transpiration ability and so allows greater access to water, and hence growth, by the less-defoliated plant type.

Analytical modelling provides phenomenological descriptions, and has undoubted power in exploring the framework of ecological interactions. It does not seek to explore the underlying mechanisms nor to give accurately quantified solutions. These can, on the other hand, be attempted by mechanistic modelling, as this paper has shown. For example, while the analytical model of Grenfell *et al.* (1992) demonstrated clearly that overcompensatory mortality could account for the erratic population dynamics of sheep on St Kilda, it could not provide any insight into the causes of it.

The price of the mechanistic approach is the greater complexity of the description, which requires a large number of parameters to be defined and given values, and which tends to obscure the parameters' relative importance. It is self-evident that mechanistic modelling is only viable if the greater amount of detail is sufficiently robust to realize the goal of an accurate description of the system. It is for that reason that we have laid out the details of our animal model. We would argue that the substantial literature on the physiology of domestic and wild ruminants allows a robust mathematical description, and that the many parameters can be defined with sufficient confidence. Furthermore, the principle we have applied is that the modelling of the system's components and their behaviour must be separated from the behaviour of the whole system. To investigate hypotheses about the way that physiological properties of animals and plants determine system dynamics, it is legitimate to specify those properties as accurately as possible, and then to examine the implications of the interaction between them. Provided that only the performance of individual components, *and not the whole system's response*, are modified by the modeller to accord with expectation, the investigation of the system's behaviour is a true test of the underlying assumptions. The ability of a model to give good predictions of component processes, as shown above, is therefore a prerequisite of scaling up from functional to numerical response in mechanistic terms.

Acknowledgement
Part of this work was supported by the UK Government's Department for International Development.

Appendix

Overview of the model of the St Kilda grazing system
The plant submodel describes grass growth, offtake by sheep and decomposition of herbage biomass on the 15 ha of *Holcus–Agrostis*, 16 ha of

Agrostis–Festuca and 11 ha of *Molinia* that comprise the Hirta study site. The plant submodel was adapted from Johnson and Thornley's (1983) model of vegetative grass growth. For each vegetation type, there are four state variables, representing the biomass of successive ages of leaf: new, expanding, mature and decaying (L1, L2, L3, D, respectively). The input to the new leaf compartment was taken to be the monthly average of daily net above-ground primary production under grazing, which was estimated on Hirta by D.R. Bazely (unpublished observations) in 1991 and 1992, augmented by data from Job and Taylor (1978). Maturation of live leaf and decay of dead leaf was modelled by a temperature-dependent first-order function (see Clutton-Brock *et al.* 1997).

The animal submodel consists of three state variables, representing animal numbers, fat-free body mass and fat mass, in each of three age classes (lambs, yearlings and adults) in each sex. The animal processes are described in terms of the daily flow of material from the vegetation (offtake), and the metabolism and expenditure of energy and protein in maintenance, activity, reproduction, growth of fat-free mass, depletion and repletion of fat reserves and heat loss to the environment. The model structure is of the 'super-individual' type (Scheffer *et al.* 1995), in that it describes separately each sex and age class. This has many of the advantages of an individual-based model, but without the large computational requirement.

The model has an iteration interval of 1 day, and values of the state variables are carried over to the following day. To examine differential mortality and population dynamics, simulations were run for 100 years, and results for the first 20 years discarded to remove any effect of the starting values given to state variables. Simulations were carried out without inter-annual climatic variation.

Overview of the model of a savanna grazing system

The vegetation component of the model is based on that of Dye (1983), who modelled grass growth in a semi-arid savanna in south-east Zimbabwe. His model was applied to growth in two vegetation components: perennial grasses and woody browse. The vegetation submodel predicts biomass production and allocation under competition between grass and trees for soil water. Trees were assumed to have the same rain-use efficiency as grasses, in the absence of clear evidence to the contrary. The phenology and allometric relationships between the plant parts of these components (Poupon 1976; Rutherford 1984; Dye & Walker 1987) was used to predict the daily growth of green leaf, stem and seed (grasses) and green leaf, twig, wood and fruit (trees). Literature estimates of tissue senescence, decomposition and invertebrate herbivory were included in the prediction of tissue flow from net photosynthesis through to loss from the system. The state variables were, for grasses: carbohydrate stores, green leaf, dead leaf, green stem plus seed, dead stem, fallen seed; and for trees: carbohydrate stores, green leaf, fallen leaf, current season's twig, wood, fruit, fallen fruit.

The animal component is the same as for the St Kilda model, with minor modifications. It allows several animal species to be included, each being specified by the mature size of each sex. Allometric functions of body mass are used to calculate the duration of pregnancy and lactation, in addition to the physiological and morphological variables (such as the maximum rate of protein deposition, maximum fat mass and incisor breadth) already described in this way. Two example species were used, approximating to the size of goats and cattle. To make comparisons between different animal species, it was assumed that animals are equivalent when compared on the basis of the metabolic weight of the mature animal, with immature animals rated in proportion to their degree of maturity. We therefore defined animals in relation to the metabolic weight of a mature bull, $A = 450$. Thus, one livestock equivalent (LE), is $450^{0.75} = 98 \, \text{kg}^{0.75}$. Given the range of ages and weights in the average herd, there are about 1.5 'average' cattle per LE and about 7.6 'average' goats.

As before, the model has an iteration interval of 1 day, using daily rainfall data, and values of the state variables are carried over to the following day. Since the balance of tree and grass biomass was predicted to vary widely over long runs, model runs of 20 years were used to generate results, after a run-in period of 5 years.

References

Agricultural Research Council (1980). *The Nutritional Requirements of Ruminant Livestock*. Commonwealth Agricultural Bureaux, Slough.

Alexander, R.McN. (1977). Allometry of the limbs of antelopes. *Journal of Zoology, London*, **183**, 125–146.

Armstrong, H.M., Gordon, I.J., Sibbald, A.R. Hutchings, N.J. & Milne, J.A. (1997). A model of grazing by sheep on hill systems in the U.K. I. The prediction of seasonal production, biomass and digestibility of vegetation. *Journal of Applied Ecology*, **34**, 166–185.

Armstrong, R.H., Common, T.G. & Smith, H.K. (1986). The voluntary intake and *in vivo* digestibility of herbage harvested from indigenous hill plant communities. *Grass and Forage Science*, **41**, 53–60.

Bircham, J.S. & Hodgson, J. (1983). The influence of sward conditions on rate of herbage growth and senescence in mixed swards under continuous stocking management. *Grass and Forage Science*, **38**, 323–332.

Black, J.L. & Kenney, P.A. (1984). Factors affecting diet selection by sheep. II. Height and density of pasture. *Australian Journal of Agricultural Research*, **35**, 565–578.

Blaxter, K. (1989). *Energy Metabolism in Animals and Man*. Cambridge University Press, Cambridge.

Burlison, A.J., Hodgson, J. & Illius A.W. (1991). Sward canopy structure and the bite dimensions and bite weight of grazing sheep. *Grass and Forage Science*, **46**, 29–38.

Campbell, G.S., McArthur, A.J. & Monteith, J.L. (1980). Windspeed dependence of heat and mass transfer through coats and clothing. *Boundary-Layer Meteorology*, **18**, 485–493.

Campbell, R.N. (1974). St Kilda and its sheep. In *Island Survivors* (Ed. by P.A. Jewell, C. Milner & J. Morton Boyd), pp. 8–35. Athlone Press, London.

Clutton-Brock, T.H., Price, O.F., Albon, S.D. & Jewell, P.A. (1991). Persistent instability and population regulation in Soay sheep. *Journal of Animal Ecology*, **60**, 593–608.

Clutton-Brock, T.H., Price, O.F., Albon, S.D. & Jewell, P.A. (1992). Early development and population fluctuations in Soay sheep. *Journal of Animal Ecology*, **61**, 381–396.

Clutton-Brock, T.H., Illius, A.W., Wilson, K., Grenfell, B.T., MacColl, A. & Albon, S.D. (1997). Stability and instability in ungulate populations: An empirical analysis. *American Naturalist*, **149**, 196–219.

Doney, J.M., Ryder, M.L., Gunn, R.G. & Grubb, P. (1974). Colour, conformation, affinities, fleece and patterns of inheritance in the Soay sheep. In *Island Survivors* (Ed. by P.A. Jewell, C. Milner & J. Morton Boyd), pp. 88–125. Athlone Press, London.

Doney, J.M., Gunn, R.G. Peart, J.N. & Smith, W.F. (1981). Effect of body condition and pasture type on herbage intake, performance during lactation and subsequent ovulation rate in Scottish Blackface ewes. *Animal Production*, **33**, 241–247.

Doney, J.M., Peart, J.N., Smith, W.F. & Sim, D.A. (1983). Lactation performance, herbage intake and lamb growth of Scottish Blackface and East Friesland × Scottish Blackface ewes grazing hill or improved pasture. *Animal Production*, **37**, 283–292.

Dye, P.J. (1983). *Prediction of variation in grass growth in a semi-arid*

421

induced grassland. PhD thesis, University of Witwatersrand, South Africa.

Dye, P.J. & Spear, P.T. (1982). The effects of bush clearing and rainfall variability on grass yield and composition in South-West Zimbabwe. *Zimbabwe Journal of Agricultural Research*, **20**, 103–118.

Dye, P.J. & Walker, B.H. (1987). Patterns of shoot growth in a semi-arid grassland in Zimbabwe. *Journal of Applied Ecology*, **24**, 633–644.

Emmans, G.C. & Fisher, C. (1986). Problems in nutritional theory. In *Nutrient Requirements of Poultry and Nutritional Research* (Ed. by C. Fisher & K.N. Borman), pp. 9–39. Butterworths, London.

Fritz, H. & Duncan, P. (1994). On the carrying capacity for large ungulates of African savanna ecosystems. *Proceedings of the Royal Society, London, Series B*, **256**, 77–82.

Graham, N. McC. (1964). Energy costs of feeding activities and energy expenditure of grazing sheep. *Australian Journal of Agricultural Research*, **15**, 969–973.

Grenfell, B.T., Price, O.F., Albon, S.D.A. & Clutton-Brock, T.H. (1992). Overcompensation and population cycles in an ungulate. *Nature*, **355**, 823–826.

Grubb, P. & Jewell, P.A. (1974). Movement, daily activity and home range of Soay sheep. In *Island Survivors* (Ed. by P.A. Jewell, C. Milner & J. Morton Boyd), pp. 160–194. Athlone Press, London.

Gunn, R.G., Doney, J.M., Smith, W.F. & Sim, W.A. (1986). Effects of age and its relation with body size on reproductive performance in Scottish Blackface ewes. *Animal Production*, **43**, 279–284.

Hodgson, J. (1985). The control of herbage intake in the grazing ruminant. *Proceedings of the Nutrition Society*, **44**, 339–346.

IDPW (1993). AFRC Technical Committee on Responses to Nutrients. Nutrient requirements of ruminant animals: protein. *Nutrition Abstracts and Reviews, Series B*.

Illius, A.W. (1997). Advances and retreats in specifying the constraints on intake in grazing ruminants. In *Proceedings of the XVIII International Grassland Congress*. Winnipeg, Canada.

Illius, A.W. & Gordon, I.J. (1987). The allometry of food intake in grazing ruminants. *Journal of Animal Ecology*, **56**, 989–999.

Illius, A.W. & Gordon, I.J. (1991). Prediction of intake and digestion in ruminants by a model of rumen kinetics integrating animal size and plant characteristics. *Journal of Agricultural Science, Cambridge*, **116**, 145–157.

Illius, A.W. & Gordon, I.J. (1992). Modelling the nutritional ecology of ungulate herbivores: Evolution of body size and competitive interactions. *Oecologia*, **89**, 428–434.

Illius, A.W. & Jessop, N.S. (1995). Modelling metabolic costs of allelochemical ingestion by foraging herbivores. *Journal of Chemical Ecology*, **21**, 693–719.

Illius, A.W. & Jessop, N.S. (1996). Metabolic constraints on voluntary intake in ruminants. *Journal of Animal Science*, **74**, 3052–3062.

Illius, A.W., Albon, S.D. Pemberton, J. Gordon, I.J. & Clutton-Brock, T.H. (1995). Selection for foraging efficiency during a population crash in Soay sheep. *Journal of Animal Ecology*, **64**, 481–492.

Jewell, P.A., Milner, C. & Morton Boyd, J. (Eds) (1974). *Island Survivors*. Athlone Press, London.

Job, D.A. & Taylor, J.A. (1978). The production, utilization and management of upland grazings on Plylimon, Wales. *Journal of Biogeography*, **5**, 173–191.

Johnson, I.R. & Thornley, J.H.M. (1983). Vegetative crop growth model incorporating leaf area expansion and senescence, and applied to grass. *Plant, Cell and Environment*, **6**, 721–729.

Laredo, M.A. & Minson, D.J. (1973). The voluntary intake, digestibility, and retention time by sheep of leaf and stem fractions of five grasses. *Australian Journal of Agricultural Research*, **24**, 875–888.

McArthur, A.J. (1987). Thermal interaction between animal and microclimate: a comprehensive model. *Journal of Theoretical Biology*, **126**, 203–238.

McArthur, A.J. & Monteith, J.L. (1980a). Air movement and heat loss from sheep. I. Boundary layer insulation of a model sheep, with and without fleece. *Proceedings of the Royal Society, London, Series B*, **209**, 187–208.

McArthur, A.J. & Monteith, J.L. (1980b). Air movement and heat loss from sheep. II. Thermal insulation of fleece in wind. *Proceedings of the Royal Society, London, Series B*, **209**, 209–217.

McClelland, T.H. Bonaiti, B. & Taylor, St C.S. (1976). Breed differences in body composition of equally mature sheep. *Animal Production*, **23**, 281–293.

McDonald, I., Robinson, J.J., Fraser, C. & Smart, R.I. (1979). Studies on reproduction in prolific ewes. 5. The accretion of nutrients in the foetuses and adnexa. *Journal of Agricultural Science, Cambridge*, **92**, 591–603.

Mellor, D.J. & Cockburn, F. (1986). A comparison of energy metabolism in the new-born infant, piglet and lamb. *Quarterly Review of Experimental Physiology*, **71**, 361–379.

Milner, C. & Gwynne, D. (1974). The Soay sheep and their food supply. In *Island Survivors* (Ed. by P.A. Jewell, C. Milner & J. Morton Boyd), pp. 160–194. Athlone Press, London.

Ministry of Agriculture, Fisheries & Food (1967). *Potential Transpiration*. Technical Bulletin No. 16. HMSO, London.

Monteith, J.L. & Unsworth, M.H. (1990). *Principles of Environmental Physics*, 2nd edn. Edward Arnold, London.

Noy Meir, I. (1975). Stability of grazing systems: an application of predator–prey graphs. *Journal of Ecology*, **63**, 459–481.

Peart, J.N. (1967). The effect of different levels of nutrition during late pregnancy on the subsequent milk production of Blackface ewes and on the growth of their lambs. *Journal of Agricultural Science, Cambridge*, **68**, 365–371.

Pennycuick, C.J. (1979). Energy costs of locomotion and the concept of "Foraging Radius". In *Serengeti: Dynamics of an Ecosystem* (Ed. by A.R.E. Sinclair & M. Norton-Griffiths), pp. 130–163. University of Chicago Press, Chicago.

Poupon, H. (1976). La biomasse et l'évolution da sa répartition au cours de la croissance d'*Acacia senegal* dans une savane sahelienne (Sénégal). *Bois et Forêts Tropicales*, **166**, 23–38.

Reeds, P.J. & Fuller, M.F. (1983). Nutrient intake and protein turnover. *Proceedings of the Nutrition Society*, **42**, 463–471.

Rosenzweig, M.L. & MacArthur, R.H. (1963). Graphical representation and stability condition of predator–prey interactions. *American Naturalist*, **97**, 209–223.

Rutherford, M.C. (1984). Relative allocation and seasonal phasing of growth of woody plant components in a South African savanna. *Progress in Biometeorology*, **3**, 200–221.

Sæther, B.-E. (1997). Environmental stochasticity and population dynamics of large herbivores: a search for mechanisms. *Trends in Ecology and Evolution*, **12**, 143–149.

Scheffer, M., Baveco, J.M., DeAngelis, D.L., Rose, K.A. & Van Nes, E.H. (1995). Super-individuals as a simple solution for modelling large populations on an individual basis. *Ecological Modelling*, **80**, 161–170.

Shipley, L.A., Gross, J.E., Spalinger, D.E., Hobbs, N.T. & Wunder, B.A. (1994). Scaling of functional response in mammalian herbivores: the allometry of food intake in heterogeneous environments. *American Naturalist*, **143**, 1055–1082.

Spalinger, D.E. & Hobbs, N.T. (1992). Mechanisms of foraging in mammalian herbivores: new models of functional response. *American Naturalist*, **140**, 325–348.

Stevenson, I.R. (1994). *Male-biased mortality in Soay sheep*. PhD dissertation, University of Cambridge.

Taylor, C.R., Heglund, N.C. & Maloiy, G.M.O. (1982). Energetics and mechanics of terrestrial locomotion. I. Metabolic energy consumption as a function of speed and body size in birds and mammals. *Journal of Experimental Biology*, **97**, 1–21.

Taylor, St C.S. & Murray, J.I. (1987). Genetic aspects of mammalian survival and growth in relation to body size. In *The Nutrition of Herbivores* (Ed. by J.B. Hacker & J.H. Ternouth), pp. 487–533. Academic Press, Sydney.

Taylor, St C.S., Turner, H.G. & Young, G.B. (1981). Genetic control of equilibrium maintenance efficiency in cattle. *Animal Production*, **33**, 179–194.

Webb, D.R. & King, J.R. (1984). Effects of wetting on insulation of bird and mammal coats. *Journal of Thermal Biology*, **9**, 189–191.

Wood, P.D.P. (1967). Algebraic model of the lactation curve in cattle. *Nature*, **216**, 164–165.

Chapter 14
Physiological flexibility and its impact on energy metabolism and foraging behaviour in birds

M. Klaassen[1]

Summary

Many aspects of foraging, for instance locomotion, social interactions, digestion, but also maintenance, require the expenditure of energy, and therefore the metabolism of food and energy reserves. In most current models of foraging behaviour, the individual's physiology is taken to be static. However, recent work shows that it may vary in response to changes in diet, food intake rate, predictability of food resources and temperature. In addition, during the annual cycle an individual may change its body composition in anticipation of reproduction, migration and wintering. Virtually all organs and reserve depots can be involved in these alterations. Focusing on (monogastric) birds, this potential for flexibility and its boundaries are reviewed with regard to their impact on energy expenditure. It is argued that these energy budget modifications may have distinguished consequences for foraging and habitat use.

Introduction

Most environments are highly variable in space and time. Meeting the demands of growth, reproduction and maintenance, under constantly changing circumstances, requires special adaptations. Besides innate adaptations, many organisms have the ability to respond to the ecological conditions encountered by physiological or morphological flexibility. Although not unlimited, the extent of this flexibility is often remarkable,

1 Netherlands Institute of Ecology, Centre for Limnology, Rijksstraatweg 6, 3631 AC Nieuwersluis, The Netherlands

enabling survival and reproduction in a range of different environmental conditions.

Life histories ultimately evolve to maximize lifetime reproductive success of an individual. However, in many long-living animal species it is often difficult if not impossible to evaluate an animal's behaviour in terms of reproductive success. Often, the rate at which energy can be extracted from the environment, and measures of economy of its use for life processes, are related to various fitness components. Therefore, energy has also become one of the major currencies to evaluate costs and rewards of foraging behaviour (see Chapter 13).

Reviewing 63 empirical tests of optimal diet, optimal patch use and central place foraging models on birds, Maurer (1996) concluded that the results from about one-quarter of the studies supported the models' predictions, one-half gave only qualitative support, and the rest did not match theoretical predictions. Several factors have been suggested for this high falsification rate (Maurer 1996), but here I would like to elaborate on the possible role of physiological flexibility (Piersma & Lindström 1997) in explaining the apparent conflict between data and theory. Physiological alterations in plants to herbivory have been widely recognized and proven to be of major importance for our understanding of plant–bird interactions (Ydenberg & Prins 1981; Cargill & Jefferies 1984; Bazely & Jefferies 1986; Sedinger & Raveling 1986; Madsen 1989). The scope of this paper is to evaluate the physiological flexibility of birds with special emphasis on their importance for individual energy budgets, challenging the tradition of assigning fixed values to the various components of an individual's energy budget.

In this Chapter, I will not deal with primary metabolic reactions, such as differences in thermoregulatory costs caused by changing operative temperatures or specific costs for digestion (heat increment of feeding or specific dynamic action; Blaxter 1989) associated with different diets (see Chapter 11). Instead, I will focus on the more profound secondary reactions impinging on the bird's metabolic machinery, the bird's internal composition. Much of this chapter will thus be allocated to changes in body composition and standard or basal metabolic rate (BMR) of birds. BMR is the metabolism of a resting, postabsorptive bird in the thermoneutral zone. BMR is not only a quantitatively important component of a bird's energy budget (Drent & Daan 1980; Daan *et al.* 1990; Bryant & Tatner 1991; Drent *et al.* 1992), but it may also provide information on a bird's functional capacities. Various ecological and behavioural correlates (e.g. climate,

428

habitat, parental investment, daily rhythm) with BMR exist in interspecific comparisons (Bennett & Harvey 1987; Daan *et al.* 1990; Klaassen & Drent 1991; Ricklefs *et al.* 1996). However, here I will concentrate on intra-individual changes in BMR; despite what terms like 'basal' and 'standard' may suggest, I will show that an individual's BMR is far from constant (see also Piersma & Lindström 1997).

I will begin by considering the energetic consequences of body mass changes. Just calling to mind the strong interannual and interindividual variations in body mass that we may observe in the field (see Chapter 13), a change in body mass may be one of the major physiological tools of an animal to respond to varying ecological conditions. Clearly, a drop in body mass may not always be adaptive, but merely a result of food shortage or illness. However, we will learn that there are a number of good reasons for animals to regulate their body mass, one of which is the energetic implications. Body-mass changes have an impact on BMR, but I will show that BMR is also sensitive to environmental changes without accompanying changes in body mass. I then turn to the possible relationship between body composition and BMR and the limits to metabolic flexibility. Finally, the potential consequences of alterations in the individual's energy budget for the bird's optimal strategies, its behaviour, are reviewed.

Changes in flight cost with body mass

Energy intake and expenditure may not always be in balance. During periods of food shortage or hard work, body stores provide both energy and nutrients. Although retaining body stores will reduce the risk of starvation it may also have some drawbacks. In their elaborate review, Witter and Cuthill (1993) identified six potential costs of body fuel stores: mass-dependent metabolic costs, mass-dependent predation risk, mass-dependent risk of injury, mass-dependent foraging costs, pathological costs and reproductive costs. Thus, the amount of body stores may be subject to a trade-off between various factors impinging on an animal's fitness (McNamara & Houston 1987; Alerstam & Lindström 1990; McNamara *et al.* 1990; Houston & McNamara 1993). Body reserves may serve as an energy source over small time scales, e.g. for shorebirds to overcome high tides and for diurnal birds to overcome nights. In those cases, body reserves mostly make up only a small proportion of the bird's total body mass. In birds preparing for longer periods of low and/or unpredictable food conditions massive fuel loads are deposited, e.g. in anticipation of the winter or periods of extreme energy expenditure such as migratory flight; migratory birds may more than double their body mass (Kvist *et al.* 1993; Piersma & Gill 1998). Clearly, the deposition of such large reserves will alter transport costs dramatically. However, the impact of intra-individual weight increase on flight costs has not yet been estimated empirically. Aerodynamic theory predicts that an individual's flight cost will rapidly increase with body mass: the energetic cost of flight at maximum speed is expected to increase in proportion with mass to the power 1.5 (Pennycuick 1975). The only indirect empirical support for this dramatic change in flight cost towards higher body mass comes from measurements of wing-beat frequency. According to aerodynamic theory, wing-beat frequency increases with flight cost and should increase intra-individually with the square root of body mass, which has been confirmed experimentally in two species of birds (Pennycuick *et al.* 1996). The expected intra-individual relationship between flight cost and body mass (exponent 1.5) is very different from the empirical interspecific allometric relationship that has an exponent of 0.81 (Norberg 1996). The difference between intra-individual predictions and interspecific findings may reside in wing shape, which changes interindividually in concert with body mass, whereas it remains unchanged intra-individually. Indeed, partly correcting for differences in wing shape by accounting for wing span, empirical flight cost estimates scale to body mass with the power 1.37 (Norberg 1996).

Changes in basal metabolic rate with body mass

An increase in body mass not only affects the costs of transport but also alters the BMR of a bird. Intra-individually, the allometric relationship between BMR and body mass was found to be 1.67 in the kestrel, *Falco tinnunculus* (Daan *et al.* 1989), 1.38 in the knot, *Calidris canutus* (Piersma *et al.* 1995), 1.26 in the redshank, *Tringa totanus* (Scott *et al.* 1996) and 1.25 in the thrush nightingale, *Luscinia luscinia* (Lindström *et al.*, in press). Interspecifically BMR scales with body mass with an exponent of 0.68 (Gavrilov & Dolnik 1985). Thus, as for flight costs, BMR increases with body mass more rapidly intra-individually than interspecifically. It should be noted that also intra-individual and intraspecific comparison may yield different results emphasizing the need to be aware of the level at which comparisons are made (Daan *et al.* 1989; see also Bennett & Harvey 1987). The reason for this different rate of change of BMR and body mass between species than within individuals is unclear. It is possible that with a change in body mass, the relative contribution of different organs differ at the interspecific and intra-individual levels. This interpretation is analogous to that explaining the differences in scaling of flight costs with body mass. Here, wing shape may vary between species but not within individuals.

Body mass may thus have a strong impact on a bird's energy budget. However, birds might compensate for this by changing their time allocation. Indeed, Bryant and Tatner (1991) found no clear effects of body mass on daily energy expenditure in a variety of bird species.

Changes in basal metabolic rate independent of changes in body mass

Although body mass is clearly important in determining the metabolism of a bird, it is not the sole factor. Lindström *et al.* (in press) used a thrush nightingale in their simulation of a long-distance autumn migration using a wind tunnel. Over a period of 11 weeks the bird made a single non-stop flight lasting 12–16 h about once every week, covering a total air distance of 6300 km. Before and after each flight session the bird had a body mass of, on average, 28 and 24 g, respectively. BMR was monitored during the recovery period following each flight session. In Fig. 14.1a the BMR of this thrush nightingale is plotted against body mass showing a strong correlation between the two, as expected. However, it was also observed that, independent of body mass, BMR in this individual slowly decreased over time (Fig. 14.1b). The origin of this change in BMR is unknown. Although the experimental protocol did not change during the course of the experiment the bird might still have been adjusting its metabolic machinery to the specific conditions it was undergoing. It is possible that the bird was in the process of changing its physiology in anticipation of the warmer climate it was expecting to meet at its imagined destination (see below). Whatever caused this change in the basal metabolism over time, it again illustrates the flexibility in the metabolism of birds.

Below the lower critical temperature the energy expenditure of endotherms increases with a decrease in ambient temperature. However, at temperatures above the lower critical temperature, i.e. within the thermoneutral zone, winter-acclimatized birds often exhibit higher basal metabolic rates than summer-acclimatized birds (Pohl & West 1973; Swanson 1991; Cooper & Swanson 1994). In these birds the higher BMR

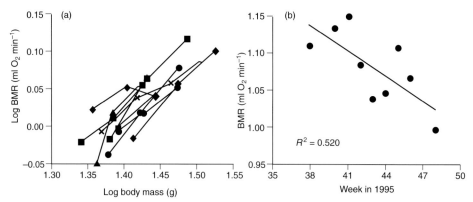

Figure 14.1 (a) Double logarithmic plot of basal metabolic rate (BMR) vs. body mass of an individual thrush nightingale going through various body-mass cycles sparked by weekly simulated migratory flights lasting 12–16 h. Following each flight, BMR was measured daily during 3 consecutive days; each of the nine series of three measurements is connected with lines. (b) Using regression lines drawn through the data points of each of the nine series, BMR at an average body mass of 26.6 g was calculated and depicted in relation to week number. From Lindström *et al.* (in press).

is associated with improved cold resistance. In summer, small passerines may become immediately hypothermic at an ambient temperature of approximately –60°C whereas in winter they can remain homothermic at that temperature (Dawson *et al.* 1983). At least in part, this better cold resistance in winter must be attributed to the bird's relatively new and thus denser and better insulating plumage. The coincidence of a better cold resistance and a higher BMR in winter compared with summer may also indicate a functional link between metabolic capacity and BMR. There are, however, cases where peak metabolic rate increased in winter while BMR remained stable (O'Conner 1996). Thus, it remains unclear whether seasonal changes in BMR truly contribute to increased thermogenic capacity or represent a separate acclimatization response. Irrespective of the ultimate or proximate cause, the fact remains that BMR is often found to change in response to, or in anticipation of, cold stress.

Until recently, alterations in BMR in relation to cold had exclusively been studied in birds experiencing large differences in ambient temperature between seasons. It was therefore interesting to study this aspect in garden warblers, *Sylvia borin*, long-distance migrants, which normally do not ex-

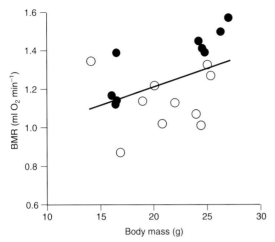

Figure 14.2 Basal metabolic rate (BMR) in cold (●) and warm (○) acclimatized garden warblers in relation to body mass. The two groups of 2-year-old hand-raised siblings, equally distributed over the two groups, were tested at 24°C after maintaining them at 3°C and 23°C, respectively, for 5 months under a 12:12 h light–dark cycle. BMR in the two groups was significantly different after correction for body mass (ANCOVA, $F_{1,16} = 9.78$, $P = 0.006$). From M. Klaassen, M. Oltrogge and L. Trost (unpublished observations).

perience great temperature differences between summer and winter. After maintaining two groups of birds, one at 5°C and the other at 23°C for a period of 5 months, the BMR of the birds of the two groups was measured. As in earlier experiments on high-latitude resident birds, cold-exposed garden warblers had a higher BMR (Fig. 14.2) (M. Klaassen, M. Oltrogge and L. Trost, unpublished observations). Thus, cold-acclimatization appears not to be a special feature of resident birds at temperate and higher latitudes. In addition, the seasonal changes in BMR may result from direct exposure to cold and are not necessarily under photoperiodic control. These experiments thus indicate that the flexibility of the metabolism is a basic trait of a bird, not a result of evolutionary adaptation in populations in cold areas.

Body compositional changes
BMR reflects the sum of the minimal metabolism of all organs in an

organism. This implies that changes in BMR as found in the cold-acclimatization experiments may be caused by changes in the function or size of one or more organs. Throughout the annual cycle the body composition of birds may change as illustrated for Bewick's swans, *Cygnus columbianus bewickii*, by comparing the sizes of various organs during winter and during moult in late summer (Fig. 14.3) (M. Klaassen, J.H. Beekman & P. Reffeltrath, unpublished observations). During moult, swans, like other waterfowl, are flightless and there is a tendency for breast muscles to be smaller during this period (see also Piersma (1988) for a similar result in moulting great crested grebes, *Podiceps cristatus*). On the other hand, gizzard and kidney are significantly larger during moult than during winter. This may be related to differences in diet between late summer and winter (tundra and aquatic vegetation vs. grass, respectively; Dirksen *et al.* 1991).

The deposited fuel mainly consists of fat, which contains the highest amount of energy per unit mass (Blaxter 1989). However, carcass analyses (Lindström & Piersma 1993) and studies on the nitrogen and energy balance of birds (Lindgård *et al.* 1992; Klaassen & Biebach 1994; Klaassen *et al.* 1997), indicate that body stores also contain proteins. These proteins

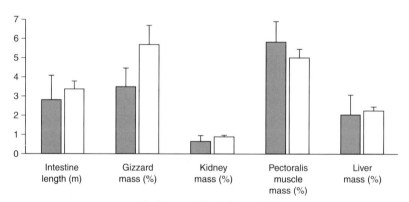

Body compositional characteristics

Figure 14.3 Body compositional characteristics of wintering (■) and moulting (□) Bewick's swans. The averages and SD of intestine length (m; Mann–Whitney U-test, $U_{6,6} = 16.0$, $P = 0.749$), gizzard mass (% of total body mass; $U_{7,6} = 1.0$, $P = 0.004$), kidney mass (% of total body mass; $U_{7,6} = 6.0$, $P = 0.032$), pectoralis muscle mass (% of total body mass; $U_{7,5} = 6.0$, $P = 0.062$) and liver mass (% of total body mass; $U_{6,6} = 6.0$, $P = 0.0547$) are presented. From M. Klaassen, P. Reffeltrath and J.H. Beekman (unpublished observations).

are probably necessary to support the extra mass acquired during fuel deposition. Increasing flight costs with increasing body mass need to be compensated for by the synthesis of muscle tissue. Indeed, in migratory birds muscle hypertrophy has been observed in association with increased body mass (Kendall *et al.* 1973; McLandress & Raveling 1981; Marsh 1984). Thus, the muscle content of a bird may change in conjunction with a change in fuel load.

The increased flight costs of a heavy bird may not only affect its muscle size but may trigger a cascade of changes involving tissues that are supporting the metabolism of the bird, such as liver, heart, blood, kidneys, gizzard, intestines, etc. (Piersma *et al.* 1996a,b). Disproportional reductions in heart and kidney (kestrels: Daan *et al.* 1989), intestine and gizzard (garden warbler: Hume & Biebach 1996) were found in intraspecific comparisons between starved (low body mass) and non-starved (high body mass) birds. These data are in line with the observation that starved birds may initially have a relatively low intake level when again placed under *ad libitum* food conditions (Ketterson & King 1977; Klaassen & Biebach 1994; Klaassen *et al.* 1997).

Differences in the quantity of food ingested may contribute to differences in gizzard size (Ankney 1977; Dykstra & Karasov 1992; Hammond *et al.* 1994; Secor *et al.* 1994). Adaptations in the gastrointestinal system resulting from changes in diet composition have also been reported (Karasov & Diamond 1988; Levey & Karasov 1989; Piersma *et al.* 1993, 1996a).

The picture emerges that body composition frequently changes. When depositing fat as fuel, which is presumably metabolically inert, protein is deposited in conjunction, which conceivably incurs a certain extra energetic cost for maintenance.

Body composition and basal metabolic rate

The breakdown of flight muscles during moult (see Fig. 14.3) (Piersma 1988) might be an energy-saving strategy if their maintenance incurs an energetic cost. Any increase in muscle mass to support the extra weight may have contributed to the elevated BMR of the fuelled bird. Apart from muscles, other organs important in supporting the metabolism of the bird and possibly having an impact on the level of BMR of fuel-depositing birds were mentioned above. There is some evidence from interspecific comparisons for their contribution to the level of BMR. Birds with a relatively high level of energy expenditure and a high BMR were found to have large livers and hearts (Daan *et al.* 1990; see Konarzewski & Diamond 1995 for comparable findings in mice).

There is a considerable body of literature on changes in the digestive tract, which mostly involve changes in digestive capacity. In case the capacity for digestion incurs a high maintenance cost in terms of energy or nutrient turnover it seems conceivable that changes in the digestive tract are regulated. *In vitro* energy metabolism of tissues from the gut are indeed among the highest of all body tissues (Field *et al.* 1939; Davies 1961; Itazawa & Oikawa 1986). Moreover, low BMRs were associated with a low digestive capacity in garden warblers (Klaassen & Biebach 1994). The BMR of slowly developing and conceivably poorly fed arctic tern, *Sterna paradisaea*, chicks was lower than that of rapidly growing conspecifics, taking body mass differences into account (Klaassen & Bech 1992). In mammals (Koong *et al.* 1983) and reptiles (Secor *et al.* 1994) positive relationships between nutritional status and BMR have been found. Thus, when feeding on a high plane of nutrition leading to an increase in digestive capacity, not only the energy requirements for digestion (referred to as specific dynamic action or heat increment of feeding; Blaxter 1989) will increase, but also there may be an effect on the BMR of the animal and flexible digestive capacity might therefore play a role in the economization of energy use.

BMR is expected to reflect the size and functioning of all body components but attempts to relate specific tissue metabolic rates to whole organism energy turnover have so far failed (reviewed in Blaxter 1989). However, more knowledge of tissue-specific flexibility in size may help explain inter- and intra-individual variation in BMR. Until now, the collection of data on body composition has mostly involved the sacrifice of the study object for carcass analysis. Data on the individual capacity for flexibility are therefore lacking. With the recent advent of echoscopy, computer tomography and nuclear magnetic resonance imaging, non-lethal measurement of organ size has become possible (Fig. 14.4). This opens a promising field of research into the relationships between environmental constraints, body composition and whole-body metabolism.

However, more knowledge on tissue-specific metabolism is also required. Cytochrome-c oxidase activity and oxygen consumption have been used to estimate tissue-specific metabolic intensities (Field *et al.* 1939; Davies 1961; Itazawa & Oikawa 1986). The results show much variation, which may result from fundamental differences between birds, mammals and fish. However, the possibility that tissue-specific metabolic activity may also vary intra-individually cannot be ruled out. In muscles of chicks such

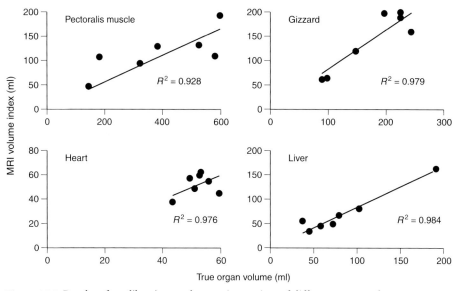

Figure 14.4 Results of a calibration study to estimate sizes of different organs of Bewick's swans using non-invasive nuclear magnetic resonance imaging (MRI). A volume index measured by MRI is plotted against volume measurements using water displacement after carcass dissection. Regression lines were forced through the origin. From Piersma and Klaassen (in press).

changes have been observed during development (Grav *et al.* 1988) and also in response to cold acclimatization (Barré *et al.* 1987). Lundgren and Kiessling (1986) showed a higher activity of cytochrome-c oxidase in migratory than in premigratory reed warblers, *Acrocephalus scirpaeus*. These changes were, however, not observed in other species of birds in relation to migration (Marsh 1981; Driedzic *et al.* 1993) and winter acclimatization (Marsh & Dawson 1982; Carey *et al.* 1989).

Limits to the bird's flexibility

The previous results suggest that birds have a remarkable ability to reorganize their physiology in an adaptive fashion to environmental constraints (see also Piersma & Lindström 1997). Clearly there are limits to this flexibility. Birds can only gain a certain amount of weight and still be able to fly (Hedenström & Alerstam 1992). However, only few of these physical limits have been experimentally explored. Some information is available on digestive limitations. This is probably because *ad libitum* food conditions may rarely occur under natural conditions. Digestive bottle-necks may become ecologically important in birds with extremely high energy expenditures, such as hummingbirds (e.g. Diamond *et al.* 1986), high growth or fuel deposition rates (e.g. Lindström 1991; Klaassen *et al.* 1997) or birds that feed on food containing much indigestible ballast (Kenward & Sibley 1977; Zwarts & Dirksen 1990). General theoretical explorations of the effect of digestive constraints on behavioural decisions in birds have commenced only recently (Bednekoff & Houston 1994).

Implications of flexibility in energy expenditure for optimality models

Animals are expected to manage their energy resources in an economic fashion. Optimal foraging theory assumes energy intake and expenditure are weighed against another in order to maximize net gain (Stephens & Krebs 1986). However, in most studies on avian foraging subtle accli-mations in the metabolism of the study species in response to changes in body mass, diet composition, plane of nutrition, cold acclimatization and other factors mentioned above, are not taken into account. However, these small physiological changes may have strong implications for a bird's behaviour. I will exemplify this with a study of stopover decisions during migration of birds (Klaassen & Lindström 1996). Lindström and Alerstam (1992) presented a model that predicts optimal departure fuel loads as a function of the rate of fuel deposition in time-minimizing migrants. Minimizing the time spent on migration may be an advantageous strategy for birds competing for resources at the breeding and wintering sites (Alerstam & Lindström 1990). A basic feature of the model is that the distance that can be covered per unit of fuel deposited diminishes with increasing fuel load. This is an effect of the increasing flight costs associated with increasing body mass (Pennycuick 1975). The main prediction of the time-minimization hypothesis is that the optimal amount of fuel upon departure from a given stopover site is determined by its rate of fuelling at that site, and by conditions expected further along the migration route (Alerstam & Lindström 1990). From a good stopover site where it attains a high fuel-deposition rate it should leave with a high fuel load; from a poor stopover site it should leave with a low fuel load. Lindström and Alerstam (1992) compared model predictions with empirical data on bluethroats, *Luscinia svecica*, that were given supplementary food at a stopover site (Fig.

438

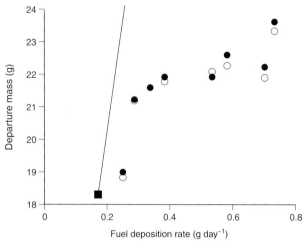

Figure 14.5 Departure mass (g) in relation to fuel deposition rate (g day^{-1}) for a natural population of bluethroats at a migratory stopover site in Sweden (■) and eight experimentally fed, but free-living individuals at the same site (●). The line depicts the predicted relationship of optimal departure mass in relation to fuel-deposition rate according to Lindström and Alerstam (1992). The two major assumptions on which that prediction is based are: (i) the migratory bird wants to migrate as fast as possible, i.e. it is a time minimizer; (ii) due to an increase in flight costs with body mass each unit of additional fuel stored has a lower additional return in terms of flight range. Lindström and Alerstam's prediction gives only qualitative support. Adding reasonable fuel-load associated energetic costs gives also quantitative support (○). From Klaassen and Lindström (1996).

14.5). They found that these birds left at much lower fuel loads than predicted. Klaassen and Lindström (1996) hypothesized that the difference between prediction and empirical data might be a result of extra resting metabolic and flight costs associated with an increase in fuel load during stopover. They developed a new version of the Lindström and Alerstam (1992) model taking fuel-load associated costs during stopover into account and showed that the observed departure fuel loads may be explained by taking changes in basal metabolic costs and transport costs with body mass into account.

The potentially large impact of body mass-dependent energetic costs on the behaviour of birds is becoming increasingly accepted in theoretical studies of avian foraging (DeBenedictis *et al.* 1978; McNamara *et al.* 1990; Houston 1992; Houston & McNamara 1993; Klaassen & Lindström 1996).

However, experimental measurements of these costs are still scarce. The importance of metabolic flexibility unrelated to body mass in birds is little recognized. Metabolic flexibility has not found its entrance in optimality models yet and empirical data are scarce, despite indications that this flexibility is ubiquitous (Piersma & Lindström 1997). Clearly, this is a field of research awaiting exploitation and with a great potential to illuminate our understanding of avian behaviour.

Acknowledgements

I thank Rudi Drent, Åke Lindström, Bart Nolet and Han Olff for providing helpful comments on an earlier version of the manuscript.

References

Alerstam, T. & Lindström, Å. (1990). Optimal bird migration: the relative importance of time, energy and safety. In *Bird Migration: The Physiology and Ecophysiology* (Ed. by E. Gwinner), pp. 331–351. Springer, New York.

Ankney, C.D. (1977). Feeding and digestive organ size in breeding Lesser Snow Geese. *The Auk*, **94**, 275–282.

Barré, H., Bailly, L. & Rouanet, J.L. (1987). Increased oxidative capacity in skeletal muscle from cold-acclimated ducklings: comparison with rats. *Comparative Biochemistry and Physiology B*, **88**, 519–522.

Bazely, D.R. & Jefferies, R.L. (1986). Changes in the composition and standing crop of salt-marsh communities in response to the removal of a grazer. *Journal of Ecology*, **74**, 693–706.

Bednekoff, P.A. & Houston, A.I. (1994). Avian daily foraging patterns: effects of digestive constraints and variability. *Evolutionary Ecology*, **8**, 36–52.

Bennett, P.M. & Harvey, P.H. (1987). Active and resting metabolism in birds: allometry, phylogeny and ecology. *Journal of Zoology, London*, **213**, 327–363.

Blaxter, K. (1989). *Energy Metabolism in Animals and Man*. Cambridge University Press, Cambridge.

Bryant, D.M. & Tatner, P. (1991). Intraspecific variation in avian energy expenditure: correlates and constraints. *Ibis*, **133**, 236–245.

Carey, C., Marsh, R.L., Bekoff, A., Johnston, R.M. & Olin, A.M. (1989). Enzyme activities in muscles of seasonally acclimatized house finches. In *Physiology of Cold Adaptation in Birds* (Ed. by C. Bech & R.E. Reinertsen), pp. 95–104. Plenum Press, New York.

Cargill, S.M. & Jefferies, R.L. (1984). The effects of grazing by lesser snow geese on the vegetation of a sub-Arctic salt marsh. *Journal of Applied Ecology*, **21**, 669–686.

Cooper, S.J. & Swanson, D.L. (1994). Seasonal acclimatization of thermoregulation in the black-capped chickadee. *The Condor*, **96**, 638–646.

Daan, S., Masman, D., Strijkstra, A. & Verhulst, S. (1989). Intraspecific allometry of basal metabolic rate: relations with body size, temperature, composition, and circadian phase in the kestrel, *Falco tinnunculus*. *Journal of Biological Rhythms*, **4**, 267–283.

Daan, S., Masman D. & Groenewold, A. (1990). Avian basal metabolic rates: their association with body composition and energy expenditure in nature. *American Journal of Physiology*, **259**, R333–R340.

Davies, M. (1961). On body size and tissue respiration. *Journal of Cellular and Comparative Physiology*, **57**, 135–147.

Dawson, W.R., Marsh, R.L. & Yacoe, M.E. (1983). Metabolic adjustments of small passerine birds for migration and cold. *American Journal of Physiology*, **245**, R755–R767.

DeBenedictis, P.A., Gill, F.B., Hainsworth, F.R., Pyke, G.H. & Wolf, L.L. (1978). Optimal meal size in hummingbirds. *American Naturalist*, **112**, 301–316.

Diamond, J., Karasov, W.H., Phan, D. & Carpenter, F.L. (1986). Digestive physiology is a determinant of foraging bout frequency in hummingbirds. *Nature*, **320**, 62–63.

Dirksen, S., Beekman, J.H. & Slagboom, T.H. (1991). Bewick's Swans *Cygnus columbianus bewickii* in The Netherlands: numbers, distribution and food choice during the wintering season. *Wildfowl*, **Suppl. 1**, 228–237.

Drent, R.H. & Daan, S. (1980). The prudent parent: energetic adjustments in avian breeding. *Ardea*, **68**, 225–252.

Drent, R.H., Klaassen, M. & Zwaan, B. (1992). Predictive growth budgets in terms and gulls. *Ardea*, **80**, 5–17.

Driedzic, W.R., Crowe, H.L., Hicklin, P.W. & Sephton, D.H. (1993). Adaptations in pectoralis muscle, heart mass, and energy metabolism during premigratory fattening in semipalmated sandpipers (*Calidris pusilla*). *Canadian Journal of Zoology*, **71**, 1602–1608.

Dykstra, C.R. & Karasov, W.H. (1992). Changes in gut structure and function of house wrens (*Troglodytes aedon*) in response to increased energy demands. *Physiological Zoology*, **65**, 422–442.

Field, J., Belding, H.S. & Martin, A.W. (1939). An analysis of the relation between basal metabolism and summated tissue respiration in the rat. *Journal of Cellular and Comparative Physiology*, **14**, 143–157.

Gavrilov, V.M. & Dolnik, V.R. (1985). Basal metabolic rate, thermoregulation and existence energy in birds: world data. In *Proceedings of the XVIII International Ornithological Congress*, Moscow pp. 421–466.

Grav, H.J., Borch-Iohnsen, B., Dahl, H.A., Gabrielsen, G.W. & Steen, J.B. (1988). Oxidative capacity of tissues contributing to thermogenesis in Eider (*Somateria mollissima*) ducklings: changes associated with hatching. *Journal of Comparative Physiology B*, 158, 513–518.

Hedenström, A. & Alerstam, T. (1992). Climbing performance of migrating birds as a basis for estimating limits for fuel-carrying capacity and muscle work. *Journal of Experimental Biology*, 164, 19–38.

Hammond, K.A., Konarzewski, M., Torres, R.M. & Diamond, J. (1994). Metabolic ceilings under a combination of peak energy demands. *Physiological Zoology*, 67, 1479–1506.

Houston, A.I. (1992). On the need for a sensitive analysis of optimization models, or, 'This simulation is not as the former'. *Oikos*, 63, 513–517.

Houston, A.I. & McNamara, J.M. (1993). A theoretical investigation of the fat reserves and mortality levels of small birds in winter. *Ornis Scandinavica*, 24, 205–219.

Hume, I.D. & Biebach, H. (1996). Digestive tract function in the long-distance migratory garden warbler, *Sylvia borin*. *Journal of Comparative Physiology B*, 166, 388–395.

Itazawa, Y. & Oikawa, S. (1986). A quantitative interpretation of the metabolism–size relationship in animals. *Experientia*, 42, 152–153.

Karasov, W.H. & Diamond, J. (1988). Interplay between physiology and ecology in digestion: intestinal nutrient transporters vary within and between species according to diet. *BioScience*, 38, 602–611.

Kendall, M.D., Ward, P. & Bacchus, S. (1973). A protein reserve in the pectoralis major flight muscle of *Quelea quelea*. *Ibis*, 115, 600–601.

Kenward, R.E. & Sibley, R.M. (1977). A woodpigeon (*Columba palumbus*) feeding preference explained by a digestive bottleneck. *Journal of Applied Ecology*, 14, 815–826.

Ketterson, E.D. & King, J.R. (1977). Metabolic and behavioral responses to fasting in the white-crowned sparrow (*Zonotrichia leucophrys*). *Physiological Zoology*, 50, 115–129.

Klaassen, M. & Bech, C. (1992). Resting and peak metabolic rates of Arctic tern nestlings and their relations to growth rate. *Physiological Zoology*, 65, 803–814.

Klaassen, M. & Biebach, H. (1994). Energetics of fattening and starvation in the long-distance migratory garden warbler, *Sylvia borin*, during the migratory phase. *Journal of Comparative Physiology B*, 164, 362–371.

Klaassen, M. & Drent, R. (1991). An analysis of hatchling resting metabolism: in search of ecological correlates that explain deviations from allometric relations. *The Condor*, **93**, 612–629.

Klaassen, M. & Lindström, Å. (1996). Departure fuel loads in time-minimizing migrating birds can be explained by the energy costs of being heavy. *Journal of Theoretical Biology*, **183**, 29–34.

Klaassen, M., Lindström, Å. & Zijlstra, R. (1997). Composition of fuel stores and digestive limitations to fuel deposition rate in the long-distance migratory Thrush Nightingale, *Luscinia luscinia*. *Physiological Zoology*, **70**, 125–133.

Koong, L.J., Nienaber, J.A. & Mersmann, H.J. (1983). Effects of plane of nutrition on organ size and fasting heat production in genetically obese and lean pigs. *Journal of Nutrition*, **113**, 1626–1631.

Konarzewski, M. & Diamond, J. (1995). Evolution of basal metabolic rate and organ masses in laboratory mice. *Evolution*, **49**, 1239–1248.

Kvist, A., Lindström, Å. & Tulp, I. (1993). Excessive migratory fattening in a captive bluethroat *Luscinia s. svecica*. *Ornis Svecica*, **3**, 161–163.

Levey, D.J. & Karasov, W.H. (1989). Digestive responses of temperate birds switched to fruit or insect diets. *The Auk*, **106**, 675–686.

Lindgård, K., Stokkan, K.A., Le Maho, Y. & Groscolas, R. (1992). Protein utilization during starvation in fat and lean Svalbard ptarmigan (*Lagopus mutus hyperboreus*). *Journal of Comparative Physiology B*, **162**, 607–613.

Lindström, Å. (1991). Maximum fat deposition rates in migrating birds. *Ornis Scandinavica*, **22**, 12–19.

Lindström, Å. & Alerstam, T. (1992). Optimal fat loads in migrating birds: a test of the time-minimization hypothesis. *American Naturalist*, **140**, 477–491.

Lindström, Å. & Piersma, T. (1993). Mass changes in migrating birds: the evidence for fat and protein storage re-examined. *Ibis*, **135**, 70–78.

Lindström, Å., Klaassen, M. & Kvist, A. (in press). Variation in basal metabolic rate and energy intake of a bird migrating in a wind-tunnel. *Functional Ecology*.

Lundgren, B.O. & Kiessling, K.H. (1986). Catabolic enzyme activities in the pectoralis muscle of premigratory and migratory juvenile reed warblers *Acrocephalus scirpaeus* (Herm.). *Oecologia*, **68**, 529–532.

Madsen, J. (1989). Spring feeding ecology of Brent geese *Branta bernicla*: effects of season and grazing on carrying capacity of salt marsh vegetation. *Danish Review of Game Biology*, **13**, 1–16.

Marsh, R.L. (1981). Catabolic enzyme activities in relation to premigratory fattening and muscle hypertrophy in the gray catbird (*Dumetella carolinensis*). *Journal of Comparative Physiology B*, **141**, 417–423.

Marsh, R.L. (1984). Adaptations of the gray catbird *Dumetella carolinensis* to long distance migration: flight muscle hypertrophy associated with elevated body mass. *Physiological Zoology*, **57**, 105–117.

Marsh, R.L. & Dawson, W.R. (1982). Substrate metabolism in seasonally acclimatized American goldfinches. *American Journal of Physiology*, **242**, R563.

Maurer, B.A. (1996). Energetics of avian foraging. In *Avian Energetics and Nutritional Ecology* (Ed. by C. Carey), pp. 250–279. Chapman & Hall, New York.

McLandress, M.R. & Raveling, D.G. (1981). Changes in diet and body composition of Canada geese before spring migration. *The Auk*, **98**, 65–79.

McNamara, J.M. & Houston, A.I. (1987). Starvation and predation as factors limiting population size. *Ecology*, **68**, 1515–1519.

McNamara, J.M., Houston, A.I. & Krebs, J.R. (1990). Why hoard? The economics of food storing in tits, *Parus* spp. *Behavioral Ecology*, **1**, 12–23.

Norberg, U.M. (1996). Energetics of flight. In *Avian Energetics and Nutritional Ecology* (Ed. by C. Carey), pp. 199–249. Chapman & Hall, New York.

O'Conner, T.P. (1996). Geographic variation in metabolic seasonal acclimatization in house finches. *The Condor*, **98**, 371–381.

Pennycuick, C.J. (1975). Mechanics of flight. In *Avian Biology*, Vol. 5 (Ed. by D.S. Farner & J.R. King), pp. 1–75. Academic Press, New York.

Pennycuick, C.J., Klaassen, M., Kvist, A. & Lindström, Å. (1996). Wingbeat frequency and the body drag anomaly: wind-tunnel observations on a thrush nightingale (*Luscinia luscinia*) and a teal (*Anas crecca*). *Journal of Experimental Biology*, **199**, 2757–2765.

Piersma, T. (1988). Breast muscle atrophy and constraints on foraging during the flightless period of wing moulting great crested grebes. *Ardea*, **76**, 96–106.

Piersma, T. & Gill, E., Jr. (1998). Gut's don't fly: small digestive organs in obese bar-tailed godwits. *The Auk*, **115**, 196–203.

Piersma, T. & Klaassen, M. (in press). Methods of studying the functional ecology of protein and organ dynamics in birds. In *Proceedings of the 22nd International Ornithological Congress* (Ed. by N. Adams & R. Slotow). University of Natal, Durban.

Piersma, T. & Lindström, Å. (1997). Rapid reversible changes in organ size as a component of adaptive behaviour. *Trends in Ecology and Evolution*, **12**, 134–138.

Piersma, T., Koolhaas, A. & Dekinga, A. (1993). Interactions between stomach structure and diet choice in shorebirds. *The Auk*, **110**, 552–564.

Piersma, T., Cadée, N. & Daan, S. (1995). Seasonality in basal metabolic rate and thermal conductance in a long-distance migrant shorebird, the knot. *Journal of Comparative Physiology B*, **165**, 37–45.

Piersma, T., Bruinzeel, L., Drent, R., Kersten, M., Van der Meer, J. & Wiersma, P. (1996a). Variability in basal metabolic rate of a long-distance migrant shorebird (red knot, *Calidris canutus*) reflects shifts in organ sizes. *Physiological Zoology*, **69**, 191–217.

Piersma, T., Everaarts, J.M. & J. Jukema (1996b). Build-up of red blood cells in refuelling bartailed godwits in relation to individual migratory quality. *The Condor*, **98**, 363–370.

Pohl, H. & West, G.C. (1973). Daily and seasonal variation in metabolic response to cold during rest and forced exercise in the common redpoll. *Comparative Biochemistry and Physiology*, **45A**, 851–867.

Ricklefs, R.E., Konarzewski, M. & Daan, S. (1996). The relationship between basal metabolic rate and daily energy expenditure in birds and mammals. *American Naturalist*, **147**, 1047–1071.

Scott, I., Mitchell, P.I. & Evans, P.R. (1996). How does variation in body composition affect the basal metabolic rate of birds? *Functional Ecology*, **10**, 307–313.

Secor, S.M., Stein, E.D. & Diamond, J. (1994). Rapid upregulation of snake intestine in response to feeding: a new model of intestinal adaptation. *American Journal of Physiology*, **266**, G695–G705.

Sedinger, J.S. & Raveling, D.G. (1986). Timing of nesting by Canada geese in relation to the phenology and availability of their food plants. *Journal of Animal Ecology*, **55**, 1083–1102.

Stephens, D.W. & Krebs, J.R. (1986). *Foraging Theory*. Princeton University Press, Princeton, NJ.

Swanson, D.L. (1991). Seasonal adjustments in metabolism and insulation in the dark-eyed junco. *The Condor*, **93**, 538–545.

Witter, M.S. & Cuthill, I.C. (1993). The ecological costs of avian fat storage. *Philosophical Transactions of the Royal Society of London, B*, **340**, 73–92.

Ydenberg, R.C. & Prins, H.H.T. (1981). Spring grazing and the manipulation of food quality by barnacle geese. *Journal of Applied Ecology*, **18**, 443–453.

Zwarts L. & Dirksen, S. (1990). Digestive bottleneck limits the increase in food intake of whimbrels preparing for spring migration from the Banc d'Arguin, Mauritania. *Ardea*, **78**, 257–278.

Chapter 15

Community dynamics of vertebrate herbivores: how can we untangle the web?

C.J. Krebs,[1] A.R.E. Sinclair,[1] R. Boonstra,[2] S. Boutin,[3] K. Martin[4] and J.N.M. Smith[1]

Summary

To answer the important question '*what determines the abundance of herbivores?*' we need to study vertebrate community dynamics. We can answer this question either by perturbation experiments or by observing the interactions between the main species in a system for long enough for the natural system variation to illuminate the nexus of interactions. We review our perturbation experiments on the Yukon boreal forest vertebrate community and compare our results with the observational analysis of the long-term dynamics of the Serengeti ungulate community of East Africa and arctic-breeding geese in Europe and North America. We conclude that all vertebrate herbivores are limited primarily in abundance by predation unless they have evolved an escape mechanism in space or time. Bird and ungulate migrations permit escape from predators in space. The properties of each community depend largely on the unique adaptations of the component species. The significant linkages in food webs tend to be vertical rather than horizontal and may be relatively few for each herbivore. This

1 Department of Zoology, University of British Columbia, Vancouver, BC V6T 1Z4, Canada
2 Department of Life Sciences, University of Toronto, Scarborough Campus, Scarborough, Ontario M1C 1A4, Canada
3 Department of Biological Sciences, University of Alberta, Edmonton, Alberta T6G 2E9, Canada
4 Department of Forest Sciences, University of British Columbia, Vancouver, BC V6T 1Z4, Canada

gives us some hope that experimental procedures of less than monumental complexity can untangle the web efficiently.

Introduction

Herbivores are animals that feed on living plants and play a central role in most ecological communities. To understand community composition we need to address the issue of what affects the abundance of herbivores. In this chapter we try to map the methods by which ecologists have tried to answer this apparently simple question and to discuss three case studies from polar to tropical habitats. We concentrate on communities dominated by mammals and birds. We have not tried to review all the literature on herbivore abundance in this paper but hope to provide a general framework for further studies of herbivore abundance.

Six main factors could possibly determine the abundance of a herbivore:

1 predators;
2 food supply;
3 cover;
4 disease;
5 parasites;
6 climatic conditions.

The three key decisions we must make before addressing the question of what limits herbivore abundance are to decide on:

1 the response variable;
2 the temporal scale of the response; and
3 the hierarchical level at which to address this issue.

The response variable we define as *average abundance*, and we will address the temporal scale of approximately one generation time. We recognize that one could define other response variables, such as rates of population change, and other temporal scales much shorter or longer than a generation time. We have picked these particular variables because we wish to consider management issues for herbivores, and to do so we require concrete variables that can be estimated (abundance) and a time frame that is within the scope of human management decisions. The third decision we needed to make is the hierarchical level to address. We restrict our discussion to comparisons within species in different communities and among species in the same community. We will not compare species between communities, e.g. to determine why the average abundance of white-tailed deer in Georgia is greater than that of barren ground caribou in northern Canada. This third decision is perhaps less clear than the other two, but we have made it in order to provide a series of hypotheses that are open to experimental manipulations of the type that wildlife and conservation managers might have available. Many global patterns in the abundance of herbivores can be correlated with differences in underlying vegetative productivity, but we argue that knowing these global correlations is not very useful to local managers.

Note that we are addressing the issue here of population limitation, not the problem of population regulation, which addresses related but different questions (Sinclair 1989; Krebs 1994). Population limitation is usually determined by manipulating the level of a suspected factor and measuring the resulting change in abundance. Some ecologists (e.g. Sinclair 1989) have argued that all factors simultaneously limit abundance. Our objective here is to determine those factors that have major effects on abundance. In this regard we will begin by assuming that unless we can double (or halve) numbers in one generation by changing the factor, the effect is too small to be studied by short-term field experiments. We are interested only in large effects.

For initial simplicity we will ignore the possibility of multiple stable states for herbivore populations (Chapter 8; Sinclair 1989) and thus potential situations in which one factor limits abundance at low density and a second factor limits abundance at high density. We ignore this possibility for practical reasons—in spite of the great attraction of multiple stable states to theoreticians, there are virtually no empirical examples of such systems that stand up to rigorous scrutiny.

We consider three alternative hypotheses about the action of predators, food supply, cover, disease, parasites and climate to determine abundance.
1 *A single factor limits average abundance.* On the face of it, this hypothesis would seem incorrect since all organisms have multiple requirements for existence. The focus here is on the practical question: for species *x* can we change average abundance by changing only one of these six factors?
2 *Several independent factors limit average abundance.* This is a particular form of the multifactor hypothesis of population limitation. As such it fits more with the views of many ecologists who have a philosophical dislike of single-factor explanations. The key point in this alternative is that the several factors that affect abundance operate sequentially and linearly. In practice this means, hypothetically, that if you change factor A and double numbers, and change factor B and triple numbers, you expect that if you change both factor A and factor B at the same time you will change

numbers by the simple multiple (2×3) or 6 times. Other models of how several independent factors operate could be postulated, but we begin with this simple one.

3 *Several interacting factors limit average abundance.* This is the most complex alternative hypothesis, since it is both multifactoral and postulates a mutual dependence between effects. In practice, an interactive explanation would be recognized by the fact that changing factor A and factor B at the same time does not result in their joint effect being predictable from their single effects. If this hypothesis applies to a community, interest may centre on exactly how the ecological interaction of factor A and factor B operates mechanistically.

A straw poll among ecologists would probably find most of them supporting hypotheses 2 and 3 above, the multifactor hypotheses. If this turns out to be the most frequent model for natural communities, it raises the multifactor dilemma that it is difficult to deal with more than three factors in any realistic large-scale, factorial experiment. There are two possible solutions to this dilemma. Firstly, we can hope that all factors operate independently (hypothesis 2 above). Secondly, we can hope that for systems with interactions only two or at most three factors show interactive effects (hypothesis 3).

Possible approaches

There are three possible approaches to answering the general question of what limits herbivore abundance in a particular system. The traditional approach is through single-species population dynamics. Each herbivore species can be analysed separately, and during the past 50 years a considerable amount of information has accumulated, particularly on locally abundant herbivores in temperate regions (e.g. Skogland 1985; Klenner & Krebs 1991). The second possible approach is through aggregating herbivores into a single trophic level and addressing questions at this level of aggregation (e.g. Abrams 1993; Hairston and Hairston 1993; Sinclair 1975). The third possible approach is to arrange species into food webs, which can range from single-species webs to highly aggregated webs in which the linkages are trophic levels (e.g. Messier 1995; Polis 1994). It is important to keep in mind the advantages and disadvantages of each of these three approaches to herbivores.

The single-species approach is highly traditional and has the advantage that it accommodates readily to experimental work. On the negative side, it assumes that all the critical factors affecting the species have been identified, and that unsuspected indirect effects are not central to changes in abundance. The single-species approach has low generality, and seems to require repeated observations on every single species that you wish to have information about. For a crisis discipline like conservation biology it may not be very useful.

The analysis of the dynamics of whole trophic levels has been one approach to deal with the great complexity of natural communities. There is a high degree of generality in this approach, since different communities with quite different complements of species can be compared. This trophic-level approach has often adopted energetics as the currency of exchange between levels and thus implicitly assumes that food energy is the central limiting factor in community structure. Because of the complexity of communities and the large amount of data needed to construct an energetic model, the community approach has typically been purely descriptive with no experimental orientation.

The third approach integrates species and trophic levels through food webs and is a hybrid of the first two approaches. In details for a single species, it can mimic the traditional single-species approach. But the food web approach requires a high level of ecological and natural history knowledge of a particular community, and for this reason it is unlikely to apply widely. By aggregating food webs many ecologists have attempted to make general conclusions about food web structure (Pimm *et al.* 1991). This third approach is the one we adopt here in an attempt to understand herbivore population limitation.

Case studies

We summarize studies on three communities both to illustrate how the general question of herbivore limitation can be addressed and to search for some generality that can be subjected to further testing.

Yukon boreal forest vertebrates

From 1986 to 1996 we analysed the vertebrate community dynamics of the boreal forest in the region of Kluane Lake, Yukon, Canada. The boreal forest vertebrate community in Canada is dominated by the 10-year cycle of snowshoe hares. Many separate aspects of our collaborative studies have been published already (e.g. Doyle & Smith 1994; Boutin *et al.* 1995; Krebs *et al.* 1995; Rohner 1995) and we wish to concentrate here on the

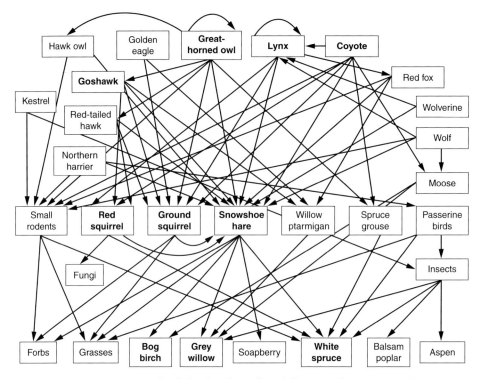

Figure 15.1 Food web for the Kluane boreal forest vertebrate community.

herbivores in this community and ask what determines their abundance. The food web for the Kluane region is shown in Fig. 15.1. There are 14 species of vertebrate herbivores in this system: snowshoe hare, red squirrel, arctic ground squirrel, seven species of mice and voles, moose, spruce grouse, ruffed grouse and willow ptarmigan. Figure 15.2 shows the average biomass pyramid for herbivores over the 10 years of study. Snowshoe hares, the two squirrels and moose together make up 92% of the total herbivore biomass. Here, we ask what the major factors are affecting the abundance of these herbivores.

We addressed this question experimentally with a series of large-scale manipulations (Krebs *et al.* 1995). The major manipulations were as follows.

1 *Predation and food supplies.* A two-factor experiment involved reducing predation pressure by mammalian predators by means of an electric fence

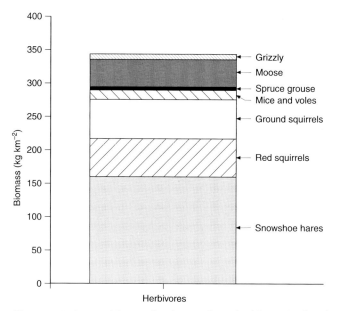

Figure 15.2 Average biomass for the vertebrate herbivores in the Kluane boreal forest, 1986–96.

around 1-km² blocks of forest and the addition of supplemental food in the form of commercial rabbit chow. Because of the large scale of the manipulation, only one experiment could be done of the combined predator reduction + food addition treatment and one experiment of the predation reduction treatment. Two areas with the food addition treatment and three control areas completed the unbalanced design.
2 *Nutrient addition.* To test for food limitation directly, we added nutrients to two areas in the form of commercial NPK fertilizer each year, and two control areas completed this design.
3 *Dominant herbivore removal.* Snowshoe hares are the most abundant herbivore in this community, and we eliminated hares from one 4-ha area on an unmanipulated site and one 4-ha area on a fertilized site to measure the impacts of snowshoe hare absence on the abundance of other herbivores.

We summarize the impact of these manipulations on the average abundance of each of the herbivore species by means of a density ratio:

$$\text{Density ratio of species} = \frac{\text{average density of species } x \text{ on type A areas}}{\text{average density of species } x \text{ on control areas}}$$
$$x \text{ for factor A}$$

If a particular treatment is having no effect on average abundance, the density ratio is expected to be 1.0. Table 15.1 summarizes the estimated density ratios for snowshoe hares, red squirrels, arctic ground squirrels, and mice and voles for the Kluane experiments. Two patterns emerge from Table 15.1. Snowshoe hare and arctic ground squirrel abundances show a mixed response to fertilization of the vegetation, are doubled by mammalian predator elimination, tripled by supplemental food addition and increased nine to 11 times by the combined treatment of predator reduction and food addition. These two herbivore species thus fit hypothesis 3 above in which both food and predation limit abundance and there is a positive interaction between these two factors.

In contrast, red squirrels, mice and voles are not affected either by the elimination of terrestrial predators, the addition of supplemental rabbit chow, or the combination treatment. These herbivores showed only one significant change to all these manipulations, a *reduction* in population abundance when the vegetation was fertilized.

It is important to distinguish in these experiments among the differential effects of fertilization and food addition. Fertilization increased the growth rate of grasses, herbs, shrubs and trees (Nams *et al.* 1993; Turkington *et al.* 1998; C.J. Krebs unpublished observations). But there were two negative effects of fertilization: mushroom production was depressed on fertilized areas, and some herb species (such as lupine *Lupinus arcticus*)

Table 15.1 Density ratios for the four major treatments at Kluane from 1989 to 1995. The expected density ratio is 1.0 if there is no effect of treatments on average population density.

Species	Factors manipulated			
	Fertilizer addition	Predator reduction	Food addition	Predator reduction and food addition
Snowshoe hare	1.30	2.00	3.13	9.67
Red squirrel	0.50	0.90	1.00	0.95
Arctic ground squirrel	0.50	2.00	4.00	10.0
Mice and voles	0.66	1.00	1.30	1.10

decreased as a result of fertilization (Graham 1994). The high-quality food represented by rabbit chow cannot be produced by boreal forest vegetation no matter what levels of nutrients are added (Turkington *et al.* 1998). The food addition experiment thus represents the experimental manipulation of the system to a level beyond what could ever occur in nature, and as such converts the system from a natural system into a nearly-agricultural one.

The reduction of red squirrel numbers on the fertilized areas probably follows from the elimination of the mushroom crop after fertilization. Mushrooms are important alternative food for red squirrels, particularly in years with no spruce seed production. We have no direct information on why mouse and vole numbers should have been reduced on the fertilizer grids. We infer that the increase in grasses at the possible expense of seed-bearing herbaceous species may have reduced the food supply of mice and voles, but more detailed experiments are needed to determine the exact mechanism. We suspect that grass seeds at Kluane are not good food for mice and voles.

At Kluane, most small rodents appear to be limited in abundance by food supply, and this food limitation results especially in poor overwinter survival (Gilbert & Krebs 1981, 1990; Schweiger & Boutin 1995). We believe this low overwinter survival is a key factor preventing the development of population cycles in the dominant species of Kluane small rodents. High overwinter survival is a necessary prerequisite for population growth in rodents, and food resources in the Kluane boreal forest are almost never sufficient to permit high overwinter survival for small rodents. The supplemental rabbit chow is not a good food supply for mice and voles and thus we observed no response by small rodents to the food addition experiments (Table 15.1).

We do not know the mechanism by which the interaction between food and predation is mediated in snowshoe hares and arctic ground squirrels. Since we have not replicated the combination treatment, we cannot reject the suggestion that the observed interaction is caused by an initial difference between the blocks, and is not caused by the treatments. We believe, however, that the size of the interaction is too large for this argument to be correct. If the interaction does operate, it may operate through the risk of predation (Hik 1995). If a herbivore's feeding habits are affected by predation risk, the elimination of mammalian predation might create behavioural or physiological effects that would not be seen in the single-factor manipulations. Boonstra and Singleton (1993) have suggested that predation risk may cause physiological stress that has long-term consequences for snowshoe hares during the decline and low phase of the

10-year cycle. The important point is that predation may have both direct and indirect effects that feed through to changes in abundance.

Little work has been done on the effects of parasites and diseases on boreal forest herbivores, and the limited view from this work is that neither of these factors is effective in limiting abundance. Keith *et al.* (1985) studied the helminth parasites of snowshoe hares in Alberta and concluded that none of the parasites they followed was an important mortality cause. Sovell (1993) used the vermicide ivermectin to manipulate the abundance of two pathogenic nematodes in snowshoe hares at Kluane. Total reproductive output increased about 30% on treated areas but the survival of young hares was not affected by maternal treatment. Sovell (1993) concluded that the effects of parasitism in snowshoe hares was compensatory to other forms of loss. Murray *et al.* (1997) found that nematode parasites could interact with predation to reduce hare survival, and that synergistic effects of parasitism and nutrition could affect hares if food became limiting.

For the Kluane vertebrate community we thus conclude that two of the major herbivore species (snowshoe hares and ground squirrels) are jointly limited in abundance by predation and food (hypothesis 3). By contrast, the third major herbivore (red squirrels) as well as voles and mice seem to be limited in abundance by food supply alone and thus fit hypothesis 1. Removing predators would thus be an effective management tool for snowshoe hares but would do nothing for red squirrel or vole abundance. Of the less abundant vertebrates, spruce grouse and ruffed grouse seem to be limited in abundance by predation alone as a byproduct of the predation associated with the hare cycle.

Ungulates of the Serengeti–Mara ecosystem
Studies of herbivores in the Serengeti–Mara ecosystem have made use of natural or unplanned perturbations to determine the limiting factors for ungulates. Research on ungulates began in the late 1950s (Grzimek & Grzimek 1960) and for a few of the 28 large herbivorous mammals such as wildebeest (*Connochaetes taurinus*) and African buffalo (*Syncerus caffer*) there has been a reasonably continuous record of numbers, while for others the data are more fragmented. The main findings from these studies, as reported by Sinclair (1979), Sinclair and Norton-Griffiths (1979), Packer (1990), Sinclair *et al.* (1985), Prins and Douglas-Hamilton (1990) and Fryxell *et al.* (1988), have been synthesized in Fryxell and Sinclair (1988), Fritz and Duncan (1993), Sinclair (1985) and Sinclair (1995). Since 1960 there have been three perturbations to the system, as outlined below.

Rinderpest

Rinderpest is a virus disease of cattle that can affect closely related ruminant species such as African buffalo and wildebeest. It does not, however, affect all ruminants (e.g. topi (*Damaliscus korrigum*) and kongoni (*Alcelaphus buselaphus*) were not noticeably affected) and non-ruminants such as Burchell's zebra (*Equus burchelli*) are unaffected. The disease is exotic to Africa, and its introduction and subsequent epizoosis in the 1890s is described elsewhere (Jarman 1974; Sinclair 1977; Sinclair & Norton-Griffiths 1982). In the early 1960s it disappeared from the wildlife coincident with a vaccination campaign in cattle.

The most important consequence of the rinderpest disappearance was a fivefold increase in the wildebeest population between 1963 and 1977 (Fig. 15.3). Buffalo numbers also increased fivefold, but their effects on the ecosystem were less marked than those produced by the wildebeest eruption (Sinclair 1979). After 1977 the wildebeest population levelled off, and it remained at approximately 1.3 million animals until 1993.

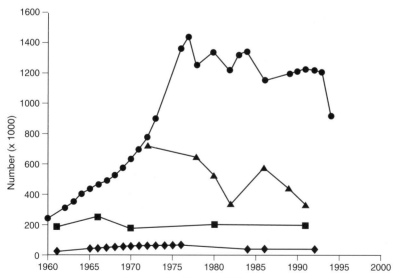

Figure 15.3 Population changes of wildebeest (●), zebra (■), Thomson's gazelle (▲) and African buffalo (◆) in the Serengeti ecosystem. Data from Dublin *et al.* (1990), Campbell and Borner (1995), Mduma (1996).

Changes in rainfall

The Serengeti ecosystem is unique because it is a grassland and savanna system in a rainfall regime that is unusually high but variable. Such a rainfall pattern occurs because the Serengeti sits at the convergence of three weather systems. The dominant system is the Indian Ocean monsoon that comes from the south-east and is responsible for the main rains in the March–June period. The other two weather systems come from Lake Victoria in the west, and the Mau highlands in Kenya to the north. Together they ensure that there is nearly always some rain occurring somewhere in the ecosystem each month. In general, there is a rainfall gradient from high (1200 mm year^{-1}) in the northern woodlands to low (500 mm year^{-1}) in the south-eastern plains.

There have been two periods of unusually high rainfall during this century, 1955–65 and 1971–76. In addition, there has been one severe drought from March 1993 to February 1994. Since rainfall is the main determinant of vegetation growth (Sinclair 1977; McNaughton 1979, 1983), these rainfall fluctuations have resulted in major differences in available food for the ungulates.

Poaching

A marked increase in settlement along the western boundaries of both the Serengeti National Park and Mara Reserve has been documented over the past 20 years (Dublin 1986; Campbell & Hofer 1995). In some areas the increase in the human population has approached 15% per year, mostly from immigration but partly from the 3% annual population growth rate. The human population increase combined with a large decrease in antipoaching patrols resulted in an invasion of northern and western Serengeti by poachers in 1977–88. The consequence of this was the complete extermination of black rhinoceros (*Diceros bicornis*) (which lost 52% of its population in 1 year alone, 1977), an 80% drop in elephant (*Loxodonta africana*) numbers and a 60% drop in buffalo population by 1980 (see Fig. 15.3) (Sinclair 1995). In northern Serengeti, large predators, particularly lions (*Panthera leo*), were also reduced (Packer 1990).

Limiting factors for Serengeti ungulates
Food supply

Although the wildebeest population increased following the disappearance of rinderpest in 1963, it did so in two stages, one during 1963–66, the other in 1970–77. The first was a direct result of rinderpest removal, and the population began to level out during 1967–69 as food supply was depleted. However, the second increase resulted from the high rainfall and, therefore, high available food supply during 1971–76 (Sinclair 1979; Sinclair & Norton-Griffiths 1982). Subsequently, the wildebeest levelled out as a result of density-dependent dry-season mortality caused by inadequate nutrition (Sinclair *et al.* 1985), and this is seen in Fig. 15.4, where the rate of population change is shown to be related to the dry-season food supply. In essence, during the dry season, grass growth cannot keep up with consumption, and animals ultimately die from intraspecific competition for food. This result was essentially similar to that for buffalo, which also increased following rinderpest release, and levelled out in the mid-1970s through lack of dry-season food (Sinclair 1977). This conclusion was confirmed by the drought of 1993, when numbers dropped to 900 000, a result that was predicted by a population model based on

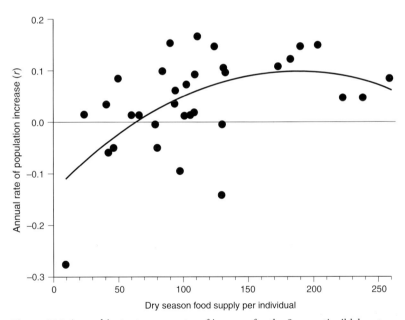

Figure 15.4 Annual instantaneous rates of increase for the Serengeti wildebeest population (r) as a function of dry-season food supply per individual (kg individual^{-1} km^{-2} dry-season grass). The fitted line is a second-degree polynomial ($r = -0.136 + 0.00252\ X - 0.00000678\ X^2$ $n = 33$, $r = 0.57$, $P = 0.003$). Data from Sinclair (1979) and Mduma (1996).

previously estimated density–food relationships (Mduma 1996; Mduma *et al.* 1997)

Thomson's gazelle (*Gazella thomsoni*) is a small (16-kg) antelope whose population follows the wildebeest migration. Studies by McNaughton (1977, 1979) suggested that during the late wet season wildebeest grazing created habitat and food for this gazelle species, thus supporting its population through 'facilitation'. However, in the dry season this effect disappears, and grazing by wildebeest is most likely having a negative effect on Thomson's gazelle numbers. The evidence comes from the decline in gazelle numbers as the wildebeest increased and levelled out — gazelle dropped from 720 000 in 1972 to 344 000 in 1991 (see Fig. 15.3) (Dublin *et al.* 1990; Campbell & Borner 1995).

Wildebeest, zebra and Thomson's gazelle are grazers that perform long-distance movements (*c.* 100–200 km across their home range) through the year, following the seasonal cycle of rainfall and grass growth — these species are *migrants*. Other grazers move only short distances (*c.* 10 km), up and down the soil catena (the soil gradient from ridge top to valley bottom — these are the *residents*. Populations that migrate have a reduced predation rate because no predator species can follow these populations throughout their annual cycle (Fryxell *et al.* 1988). In addition to escaping predators, these populations can also make use of ephemeral high-quality food supplies, such as those on the semi-arid Serengeti plains, which are not available to non-migrant species. Thus, migrants can maintain a higher density per unit area (of their annual geographic range, Fig. 15.5) than resident species of similar body weight. Where wildebeest are resident in western Serengeti their density is one-fifth that of the migrant population, and even in Ngorongoro Crater, with very much higher food productivity, resident wildebeest density is half that of migrants. It is notable that all large populations of ungulates, such as the American plains bison (*Bison bison*), Asian saiga antelope (*Saiga tatarica*), barren-ground caribou (*Rangifer tarandus*) and Sudanese white-eared kob (*Kobus kob*) are migrants (Fryxell & Sinclair 1988).

Evidence that food is limiting also comes from other studies of the relationship between population density and rainfall, since rainfall is an index of food supply in the Serengeti. This has been observed for buffalo populations in eastern Africa (Sinclair 1977) and for ungulate species in general throughout Africa (Fritz & Duncan 1993). In general, food limits the very large species such as elephant, rhinoceros, hippopotamus (*Hippopotamus amphibius*) and buffalo, as well as most migrant species.

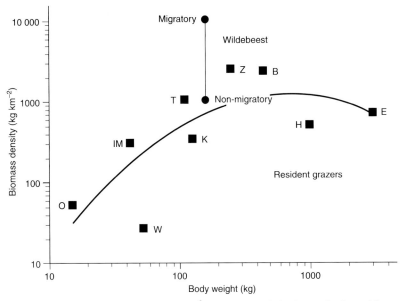

Figure 15.5 Population biomass (kg km^{-2}) increases with body weight for resident grazers (■) in the Serengeti. Note that the biomass is expressed per square kilometre of the entire annual geographic range for each population. The wildebeest population (●) that is resident (lower circle) can be compared with its migratory equivalent (upper circle). T, topi; K, kongoni; IM, impala; W, warthog; B, African buffalo; E, elephant; O, oribi; H, hippopotamus. Z, Burchell's zebra (a browser) is shown for comparison.

Predation

Although food supply is the main limiting factor for some ungulates in the Serengeti, this is not so for all species. At least some smaller resident ungulates appear to be limited more by predation. Evidence for limitation by predators comes from the responses of impala (*Aepyceros melampus*) and topi populations in northern Serengeti, where predators were reduced by poaching during 1977–88. Both populations increased nearly 10-fold following predator reduction compared with their population's pre-removal. In other areas of the Serengeti, where predators were not removed, neither species showed an increase (Sinclair 1995). This could not be due to a change in food supply because of simultaneous buffalo poaching, at least for impala. Measurements of impala mortality using telemetry have shown that nearly all mortality is caused by predation.

Other evidence for predator limitation comes indirectly from the pattern of habitat selection and niche partitioning. In general, ungulates separate out along four axes of the habitat, namely (i) a woodland–grassland vegetation structure, (ii) a gradient in the diet from pure dicot to pure monocot; (iii) the degree to which high-protein plant components (buds, leaves) or low-quality components (stems) are eaten; and (iv) the degree to which aridity in the environment is tolerated (Lamprey 1963; Gwynn & Bell 1968; Murray & Brown 1993). Despite species-specific differences, the degree of overlap along all axes remains high, sometimes as high as 80–90%. For example, eight ungulate species in the Serengeti overlapped with wildebeest in the latter's preferred habitat (green *Themeda* grassland less than 50 cm high) by 80% or more, and all except impala increased this overlap in the dry season, when wildebeest were known to be short of food (Sinclair 1985). This result is contrary to that predicted by the standard models of interspecific competition for food (Putman 1996), and it suggests that some other factor is limiting some of these ungulates below the level set by their food supply.

In addition to niche overlap, ungulate species are also found closer together than one would expect according to a random distribution. This positive association between species is so marked that one commonly sees mixed herds of five or six species of ungulates grazing together. The positive association is shown in both the wet and dry season, although slightly less so in the dry season. Such a pattern is predicted by the predation hypothesis but not by interspecific competition.

Zebra show some circumstantial evidence for predator limitation. First, although the population is partially migratory, its biomass is only marginally higher than that of resident species of similar body weight (see Fig. 15.5). More importantly, since zebra were not affected by rinderpest, the zebra population remained constant at 200 000 while wildebeest increased fivefold (see Fig. 15.3). This observation is contrary to predictions from either interspecific competition or facilitation (because the diets of zebra and wildebeest greatly overlap), but it is predicted by predator limitation (Sinclair 1985). Predation accounts for 60–74% of annual adult zebra mortality (Sinclair & Norton-Griffiths 1982) compared with only 25–30% for wildebeest and buffalo (Sinclair 1977). The difference in predation mortality between zebra and wildebeest is partly due to the greater availability of zebra to lions and other predators throughout the year. Some zebra do not migrate as far as the wildebeest but are scattered through the woodlands; and in these habitats zebra are eaten by lions three times more frequently than availability would predict (B.C.R. Bertram, personal communication in Sinclair 1985). Therefore, one might consider that when zebra are migrating only short distances, they are treated by predators as residents and, overall, they are predator limited.

Disease and parasites

The disappearance of rinderpest in 1963–64 and the subsequent fivefold increase in numbers of both wildebeest and buffalo is strong evidence that this exotic disease was able to hold both populations below the level of their food supply. However, because this disease is an exotic it may be expected to have a more marked, perhaps aberrant, affect on population numbers.

Indigenous diseases and parasites have a more subtle effect on populations. All ungulates suffer from a large number of parasites. In buffalo and wildebeest, at least 12 different blood diseases showed prevalence ranging from 16 to 96%, these including herpes virus, foot-and-mouth disease, anthrax, trypanosomiasis and theileriasis. At least 57 species of endoparasitic helminths and nematodes, and 17 species of ticks have been recorded with prevalence up to 80% (Sinclair 1977). Except in rinderpest and anthrax, these diseases and parasites become pathogenic only when the host is suffering from malnutrition, as indicated by very low bone-marrow fat content (Sinclair 1977). Therefore, indigenous diseases and parasites act synergistically with food supply to accelerate mortality. They do not act alone as a limiting factor.

Social behaviour

Social behaviour in the form of interference competition and dispersal does not appear as a dominant factor amongst the Serengeti ungulates and thus is not a proximate factor limiting local abundance. Many species are not territorial, and in these a dominance hierarchy amongst males does not always result in the exclusion of subordinates. Lower-ranking males are confined to the outside of the herd, where they may be more vulnerable to predation.

Some species such as impala and topi are territorial amongst males but not females, and the bachelor males form herds that associate loosely with the females (Jarman 1974). The very small ungulates such as oribi (*Ourebia ourebi*) and dik-dik (*Rhynchotragus kirki*) are both territorial and monogamous (Jarman 1974; Arcese *et al.* 1995). One might expect in these species that territorial exclusion and dispersal may play a role in population limitation. However, so few studies have been conducted on this aspect that this hypothesis remains only speculative. Jarman (1974) showed elegantly that social organization is closely associated with body size, food type and antipredator behaviour.

Several studies of ungulate populations in the Serengeti–Mara ecosystem show that no single limiting factor predominates over all grazing species. Some generalities are: (i) large grazers are food limited; (ii) migrant species are also food limited even if small in body size; (iii) smaller resident grazers tend to be predator limited; (iv) disease and parasites act synergistically with food; and (v) the very small species may be limited in density by social behaviour as well as predator limited. These generalizations should be treated as hypotheses to be tested with further research. We suggest that, amongst the Serengeti grazers, body size and migratory behaviour are major predictors for the type of limiting factor that predominates.

Arctic-breeding geese

There are almost no birds that are true grazers, as opposed to seed eaters, fruit eaters and omnivores. Geese are a notable exception, and much research has been carried out on their populations. For the last 30 years many goose populations in both Europe and North America have been increasing in abundance (Ebbinge 1992; Cooke *et al.* 1995). We review here the reasons postulated for these increases, which have become a large-scale management problem.

Brent geese wintering in western Europe have increased from about 20 000 birds in 1960 to over 250 000 in 1996 (Fig 15.6a) (B. Ebbinge, personal communication, 1997). Barnacle geese and white-fronted geese from western Europe have also increased in recent years, but the timing for these three species is not coincidental. Ebbinge (1992) argues that the reduction of hunting pressure is the most likely cause of these population increases. At present there is concern that the population of brent geese will become limited on the breeding grounds because of habitat saturation leading to intensive predation from arctic fox, herring gulls and snowy owls. Since these predators are primarily lemming predators, geese population limitation may emerge as an indirect effect of the lemming cycle.

Many geese populations in North America have similarly increased in numbers since the 1950s. Figure 15.6b shows the population growth of the snow goose colony at La Perouse Bay, Manitoba. The explanation of population increases in North American geese differs from those for European geese. Hunting pressure has not been a limiting factor in North

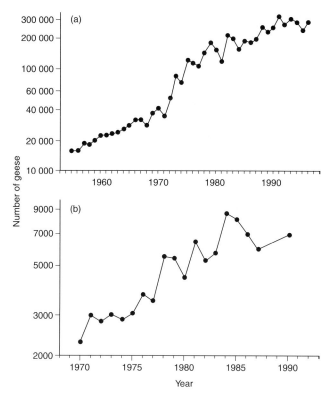

Figure 15.6 (a) Population growth in brent geese (*Branta bernicla bernicla*) that winter along the coasts of France, the UK and The Netherlands. Year refers to winter (1960 = 1960–61 winter). Data from Ebbinge (1992 and personal communication, 1997), (b) Population growth in lesser snow geese (*Anser caerulescens caerulescens*) breeding colony from La Perouse Bay, Manitoba. The population had been increasing at 8% per year from 1970 to 1985. Around 1990 the colony began spreading out into adjacent areas because of widespread habitat destruction. Data from Cooch and Cooke (1991) and D. Wilson (personal communication, 1997).

America during the past 50 years, and winter habitat is believed to have been limiting goose numbers before the intensification of agriculture during the past decades (see Chapter 10; Cooch & Cooke 1991). Fertilizer application to arable crops coupled with mechanized harvesting that leaves large volumes of waste grain in the fields has resulted in excessive food supplies on the wintering grounds. This agricultural subsidy has reduced winter mortality rates of geese, and the establishment of a comprehensive network of waterfowl refuges has also removed some of the hazards of migration and overwintering for geese.

Arctic-breeding geese seem thus to be food limited primarily, with human hunting pressure as a potential secondary limitation. There is no suggestion that natural predation plays anything other than a minor role in population limitation, and similarly disease, parasitism and social factors do not seem to limit numbers. The key food limitation appears to have been on the winter range, and this has been alleviated in recent years by agricultural subsidies from grain farming (Chapter 10). Whether this continuing agricultural subsidy on the winter range will transfer food limitation to the summer breeding grounds remains to be seen. Habitat destruction on the summer breeding range, as described by Hik *et al.* (1992) and Jefferies *et al.* (1994), may not limit population growth in the near future because birds can move to new areas.

Geese have relatively few antipredator adaptations, which implies that in evolutionary time they have not been limited by predation. Like many of the large ungulates of the Serengeti, adult geese are relatively immune to predation, and the only potential source of predator effects would be on egg predation or gosling predation. In some local populations these forms of predation could be important, but in the more global picture predators do not seem to limit population growth.

Discussion

The simple question of what limits the abundance of vertebrate herbivores does not have a simple answer. But over the past 40 years enough examples have been analysed to suggest a pattern that will answer this question for species not yet studied. We suggest the following outline as a hypothesis for future studies.

1 Vertebrate herbivores are normally limited in abundance by predators unless they can achieve escape in space or time.

2 Large mammals can assume active defence or simply become too large for predators to be effective, and these species normally become food limited.

3 Migratory species can escape predator limitation in space when predators are territorial or have limited movements.

4 Smaller species can evade predator limitation by escape in time (through hibernation for example) or occasionally by escape in space by behavioural adaptations.

5 Predator limitation may also be frustrated by excessive intraguild predation, or by predators killing other predators.

6 Disease and parasitism rarely limit herbivore abundance but may act synergistically with food limitation.

7 Small rodents may breed at a sufficiently high rate that predators cannot catch them, unless social limitation through spacing behaviour becomes limiting first.

Figure 15.7 shows schematically the possible interactions of predators and herbivores. In the simplest scenario, the herbivores and predators overlap completely in space and in time (Fig. 15.7a). Unless the herbivore has some effective behavioural adaptation to avoid predation, the limiting factor is expected to be predation and not food supply, and this is the typical situation for Kluane snowshoe hares and spruce grouse. In a more complex scenario (Fig. 15.7b), the herbivores have a refuge from their predators in time or space. Predators now face a season without the particular prey, and must migrate or become generalists to survive that season. This is the situation for migratory geese, migratory wildebeest in the Serengeti and for arctic ground squirrels at Kluane. These species are more likely to suffer food limitation. Much will depend on the particular community and the relative abundance of each species. Finally, in a third scenario herbivores

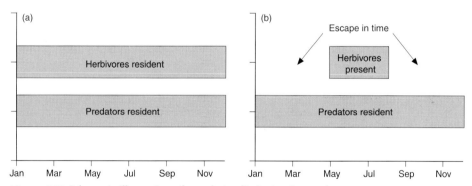

Figure 15.7 Schematic illustration of population limitation in vertebrate herbivores. (a) Complete overlap in space and time of herbivores and their predators. Predators can potentially control herbivore abundance. (b) Partial overlap in time of herbivores and predators, allowing the possible escape of the prey from predator limitation and subseqent limitation from food supplies. Migration is one strategy of escape, hibernation is another.

escape from possible predator limitation by virtue of their size (e.g. elephants), high reproductive rate (e.g. voles) or successful antipredator adaptations (e.g. red squirrel).

An alternative approach to a more predictive model for predator limitation would be to consider antipredator strategies in a behavioural context. Vertebrate prey species take two broad approaches to predators—be cryptic and avoid detection or take active defence. Cryptic species like snowshoe hares have only limited ability to avoid a predator once they are detected. Snowshoe hares cannot take active defence because they are unable to inflict any damage on a predator, so the stand-and-fight option is not available to them. By contrast, adult Canada geese can attack arctic fox and inflict significant damage. Similarly, lions will not attack elephants and wolves will not attack adult muskoxen. We might expect that the species that take active defence as a strategy will not be limited in abundance by predation, and we might then predict that they will be limited by food supplies, disease or climatic catastrophes.

While this general scenario can be applied to a new community of herbivores, we are concerned that its predictive value is low. This concern is partly based on our experience in the Kluane project in which we anticipated severe predator limitation on red squirrels once the main prey species (snowshoe hares) had declined, only to observe that red squirrels were completely unaffected by predators because of their behavioural adaptations against predators (Stuart-Smith & Boutin 1995). Our ability to generalize about the factors limiting herbivores is still more circumscribed than we would like.

There is an enormous literature on top-down and bottom-up control of communities, beginning with Hairston *et al.* (1960) and synthesized by Oksanen (1988). We cannot review this literature here except to point out that we differ from these approaches to understanding community organization because we do not believe that there are generic 'herbivores' and 'predators' that behave as units within the limits set by ecosystem productivity. We think for vertebrates that trophic interactions critically depend on species quirks, so that one cannot classify all the Serengeti ungulates as 'herbivores' without knowing more about their individual life history strategies. In our opinion these general models for ecosystem dynamics may not have predictive power for all terrestrial vertebrates and limited utility for management issues that arise, for example, with overabundant geese. While these models can provide a broad framework for generalization, they omit the devil in the details. We prefer the devil.

Acknowledgements

We thank Robert Bromley, David Hik and Deb Wilson for their assistance in reviewing this literature on herbivore limitation, and all the research assistants and postgraduate students of the Kluane Project who helped us gather the Yukon data. We thank Dennis Chitty for his comments on this paper.

References

Abrams, P.A. (1993). Effect of increased productivity on the abundances of trophic levels. *American Naturalist*, **141**, 351–371.

Arcese, P., Jongejan, G. & Sinclair, A.R.E. (1995). Behavioural flexibility in a small African antelope: group size and composition in the oribi (*Ourebia ourebi*, Bovidae). *Ethology*, **99**, 1–23.

Boonstra, R. & Singleton, G.R. (1993). Population declines in the snowshoe hare and the role of stress. *General and Comparative Endocrinology*, **91**, 126–143.

Boutin, S., Krebs C.J., Boonstra, R., Dale, M.R.T., Hannon, S.J., Martin, K., Sinclair, A.R.E., Smith, J.N.M., Turkington, R., Blower, M., Byrom, A., Doyle, F.I., Doyle, C., Hik, D., Hofer, L., Hubbs, A., Karels, T., Murray, D.L., Nams, V., O'Donoghue, M., Rohner, C. & Schweiger, S. (1995). Population changes of the vertebrate community during a snowshoe hare cycle in Canada's boreal forest. *Oikos*, **74**, 69–80.

Campbell, K. & Borner, M. (1995). Population trends and distribution of Serengeti herbivores: implications for management. In *Serengeti II: Dynamics, Management, and Conservation of an Ecosystem* (Ed. by A.R.E. Sinclair & P. Arcese), pp. 117–145. University of Chicago Press, Chicago.

Campbell, K. & Hofer, H. (1995). People and wildlife: spatial dynamics and zones of interaction. In *Serengeti II: Dynamics, Management, and Conservation of an Ecosystem* (Ed. by A.R.E. Sinclair & P. Arcese), pp. 534–570. University of Chicago Press, Chicago.

Cooch, E.G. & Cooke, F. (1991). Demographic changes in a snow goose population: biological and management implications. In *Bird Population Studies* (Ed. by C.M. Perrins *et al.*), pp. 168–169. Oxford University Press, Oxford.

Cooke, F., Rockwell, R.F. & Lank, D.B. (1995). *The Snow Geese of La Pérouse Bay. Natural Selection in the Wild*. Oxford University Press, New York.

Doyle, F.I. & Smith, J.N.M. (1994). Population responses of Northern Goshawks to the 10-year cycle in numbers of snowshoe hares. *Studies in Avian Biology*, **16**, 122–129.

Dublin, H.T. (1986). *Decline of the Mara woodlands: the role of fire and elephants*. PhD thesis, University of British Columbia, Vancouver.

Dublin, H.T., Sinclair, A.R.E., Boutin, S., Anderson, E., Jago, M. & Arcese, P. (1990). Does competition regulate ungulate populations? Further evidence from Serengeti, Tanzania. *Oecologia*, **82**, 283–288.

Ebbinge, B. (1992). *Population limitation in arctic-breeding geese*. PhD thesis, University of Groningen, Groningen.

Fritz, H. & Duncan, P. (1993). Large herbivores in rangelands. *Nature*, **364**, 292–293.

Fryxell, J.M. & Sinclair, A.R.E. (1988). Causes and consequences of migration by large herbivores. *Trends in Ecology and Evolution*, **3**, 237–241.

Fryxell, J.M., Greever, J. & Sinclair, A.R.E. (1988). Why are migratory ungulates so abundant? *American Naturalist*, **131**, 781–798.

Gilbert, B.S. & Krebs, C.J. (1981). Effects of extra food of *Peromyscus* and *Clethrionomys* populations in the southern Yukon. *Oecologia*, **51**, 326–331.

Gilbert, B.S. & Krebs, C.J. (1990). Population dynamics of *Clethrionomys* and *Peromyscus* in southwestern Yukon, 1973–1989. *Holarctic Ecology*, **14**, 250–259.

Graham, S.A. (1994). *The relative effect of clipping, neighbours, and fertilization on the population dynamics of Lupinus arcticus (Family Fabaceae)*. MSc. thesis, University of British Columbia, Vancouver.

Grzimek, B. & Grzimek, M. (1960). *Serengeti Shall Not Die*. Hamish Hamilton, London.

Gwynn, M.D. & Bell, R.H.V. (1968). Selection of vegetation components by grazing ungulates in the Serengeti National Park. *Nature*, **220**, 390–393.

Hairston, N.G., Smith, F.E. & Slobodkin, L.B. (1960). Community structure, population control, and competition. *American Naturalist*, **94**, 421–425.

Hairston, N.G., Jr. & Hairston, N.G., Sr. (1993). Cause–effect relationships in energy flow, trophic structure, and interspecific interactions. *American Naturalist*, **142**, 379–411.

Hik, D.S. (1995). Does risk of predation influence population dynamics? Evidence from the cyclic decline of snowshoe hares. *Wildlife Research*, **22**, 115–129.

Hik, D.S., Jefferies, R.L. & Sinclair, A.R.E. (1992). Foraging by geese, isostatic uplift and asymmetry in the development of salt-marsh plant communities. *Journal of Ecology*, **80**, 395–406.

Jarman, P.J. (1974). The social organisation of antelope in relation to their ecology. *Behaviour*, **48**, 215–267.

Jefferies, R.L., Klein, D.R. & Shaver, G.R. (1994). Vertebrate herbivores and northern plant communities: reciprocal influences and responses. *Oikos*, **71**, 193–206.

Keith, L.B., Cary, J.R., Yuill, T.M. & Keith, I.M. (1985). Prevalence of helminths in a cyclic snowshoe hare population. *Journal of Wildlife Diseases*, **21**, 233–253.

Klenner, W. & Krebs, C.J. (1991). Red squirrel population dynamics. I. The effect of supplemental flood on demography. *Journal of Animal Ecology*, **60**, 961–978.

Krebs, C.J. (1994). *Ecology: The Experimental Analysis of Distribution and Abundance*. Harper Collins, New York.

Krebs, C.J., Boutin, S., Boonstra, R., Sinclair, A.R.E., Smith, J.N.M., Dale, M.R.T., Martin, K. & Turkington, R. (1995). Impact of food and predation on the snowshoe hare cycle. *Science*, **269**, 1112–1115.

Lamprey, H.F. (1963). Ecological separation of the large mammal species in the Tarangire Game Reserve, Tanganyika. *East African Wildlife Journal*, **13**, 63–92.

McNaughton, S.J. (1977). Serengeti ungulates: feeding selectivity influences the effectiveness of plant defense guilds. *Science*, **199**, 806–807.

McNaughton, S.J. (1979). Grassland–herbivore dynamics. In *Serengeti: Dynamics of an Ecosystem* (Ed. by A.R.E. Sinclair & M. Norton-Griffiths), pp. 43–81. University of Chicago Press, Chicago.

McNaughton, S.J. (1983). Serengeti grassland ecology: the role of composite environmental factors and contingency in community organization. *Ecological Monographs*, **53**, 291–320.

Mduma, S.A.R. (1996). *Serengeti wildebeest population dynamics: regulation limitation and implications for harvesting*. PhD thesis, University of British Columbia, Vancouver.

Mduma, S.A.R., Hilborn, R. & Sinclair, A.R.E. (1997). Limits to exploitation of Serengeti wildebeest and implications for its management. In *Dynamics of Tropical Communities* (Ed. by D.M. Newbery, H.T. Prins & N. Brown), pp. 243–265. Blackwell Science Ltd, Oxford.

Messier, F. (1995). Trophic interactions in two northern wolf–ungulate systems. *Wildlife Research*, **22**, 131–146.

Murray, D.L., Cary, J.R. & Keith, L.B. (1997). Interactive effects of sublethal nematodes and nutritional status on snowshoe hare vulnerability to predation. *Journal of Animal Ecology*, **66**, 250–264.

Murray, M.G. & Brown, D. (1993). Niche separation of grazing ungulates in the Serengeti: an experimental test. *Journal of Animal Ecology*, **62**, 380–389.

Nams, V.O., Folkard, N.F.G. & Smith, J.N.M. (1993). Effects of nitrogen fertilization on several woody and nonwoody boreal forest species. *Canadian Journal of Botany*, **71**, 93–97.

Oksanen, L. (1988). Ecosystem organization: mutualism and cybernetics or plain Darwinian struggle for existence? *American Naturalist*, **131**, 424–444.

Packer, C. (1990). *Serengeti Lion Survey*. Game Department, Tanzania National Parks, Arusha, Tanzania.

Pimm, S.L., Lawton, J.H. & Cohen, J.E. (1991). Food web patterns and their consequences. *Nature*, **350**, 669–674.

Polis, G.A. (1994). Food webs, trophic cascades and community structure. *Australian Journal of Ecology*, **18**, 121–136.

Prins, H.H.T. & Douglas-Hamilton, I. (1990). Stability in a multi-species assemblage of large herbivores in East Africa. *Oecologia*, **83**, 392–400.

Putman, R.J. (1996). *Competition and Resource Partitioning in Temperate Ungulate Assemblies*. Chapman & Hall, London.

Rohner, C. (1995). Great horned owls and snowshoe hares: what causes the time lag in the numerical response of predators to cyclic prey? *Oikos*, **74**, 61–68.

Schweiger, S. & Boutin, S. (1995). The effects of winter food addition on the population dynamics of *Clethrionomys rutilus*. *Canadian Journal of Zoology*, **73**, 419–426.

Sinclair, A.R.E. (1975). The resource limitation of trophic levels in tropical grassland communities. *Journal of Animal Ecology*, **44**, 497–520.

Sinclair, A.R.E. (1977). *The African Buffalo: A Study of Resource Limitation of Populations*. University of Chicago Press, Chicago.

Sinclair, A.R.E. (1979). The eruption of the ruminants. In *Serengeti: Dynamics of an Ecosystem* (Ed. by A.R.E. Sinclair & M. Norton-Griffiths), pp. 82–103. University of Chicago Press, Chicago.

Sinclair, A.R.E. (1985). Does interspecific competition or predation shape the African ungulate community? *Journal of Animal Ecology*, **54**, 899–918.

Sinclair, A.R.E. (1989). Population regulation in animals. In *Ecological Concepts* (Ed. by J.M. Cherrett), pp. 197–241. Blackwell Scientific Publications, Oxford.

Sinclair, A.R.E. (1995). Population limitation of resident herbivores. In *Serengeti II: Dynamics, Management, and Conservation of an Ecosystem* (Ed. by A.R.E. Sinclair & P. Arcese), pp. 194–219. University of Chicago Press, Chicago.

Sinclair, A.R.E. & Norton-Griffiths, M. (Eds) (1979). *Serengeti: Dynamics of an Ecosystem*. University of Chicago Press, Chicago.

Sinclair, A.R.E. & Norton-Griffiths, M. (1982). Does competition or facilitation regulate migrant ungulate populations in the Serengeti? A test of hypotheses. *Oecologia*, **53**, 364–369.

Sinclair, A.R.E., Dublin, H. & Borner, M. (1985). Population regulation of Serengeti Wildebeest: a test of the food hypothesis. *Oecologia*, **65**, 266–268.

Skogland, T. (1985). The effects of density-dependent resource limitations on the demography of wild reindeer. *Journal of Animal Ecology*, **54**, 359–374.

Sovell, J.R. (1993). *Attempt to determine the influence of parasitism on a snowshoe hare population during the peak and initial decline phases of the hare cycle.* MSc thesis, University of Alberta, Edmonton.

Stuart-Smith, K. & Boutin, S. (1995). Predation on red squirrels during a snowshoe hare decline. *Ecology*, **73**, 713–722.

Turkington, R., John, E., Krebs, C.J., Dale, M.R.T., Nams, V., Boonstra, R., Boutin, S., Martin, K., Sinclair, A.R.E. & Smith, J.N.M. (1998). The effects of fertilization on the vegetation of the boreal forest. *Journal of Vegetation Science* (in press).

Part 4

Effect of higher trophic levels on plant–herbivore interactions

Introductory remarks

L.E.M. Vet[1]

The *Science* issue of 10 January 1997 mentions the dramatic effect of the rabbit haemorrhagic disease (RHD) virus, introduced to Australia by humans in 1995. Apparently, the virus is doing its destructive work killing more than 95% of the rabbits in some areas. Native plants and animals are rapidly returning—plant species that have not been around for 70 years! Even animals thought to be unaffected by the presence of rabbits, such as western grey kangaroos, have increased their population significantly since the virus has spread. Perhaps we are witnessing a large-scale ecological experiment showing us what happens to interconnected species when the population of one dominating species abruptly changes: a strong top-down predator (pathogen) controlled force, leading to a trophic cascade. If so, it nicely introduces the present section that deals with the question: how do higher trophic levels influence the ecology and evolution of plant–animal interactions? What we see in Australia is an aberration imposed by humans first causing a pest, then successfully using biological control to solve the problem. The resulting cascading effects are certainly not typical for natural, species-rich terrestrial systems where reticulate and differentiated species interactions buffer such dramatic effects of top-down forces. But in these more complex systems, it is an even greater challenge to discover what the major forces are (bottom-up or top-down) that determine the abundance and distribution of organisms.

Plants are at the basis of all food webs. It is therefore a suitable choice that this section starts with a contribution that emphasizes the plant as mediator of both bottom-up and top-down effects in multitrophic systems. Dicke and Vet address the evolutionary aspects of indirect defence in plants. In response to herbivory, plants can release volatiles that are exploited by the third trophic level, the carnivores. Hence, plants can recruit the enemies of their enemies—a good example of integration of

1 Laboratory of Entomology, Wageningen Agricultural University, PO Box 8031, 6700 EH Wageningen, The Netherlands

bottom-up and top-down effects. Dicke and Vet discuss the ecological and evolutionary consequences of these plant–carnivore interactions for the plant, the herbivore and the carnivore.

The major issue of this section is about the effect of higher trophic levels on the ecology and evolution of lower level plant–hervibore interactions: ecology *and* evolution. And when we speak of evolution we think of individuals, variation, and costs and benefits of traits. Most mathematical models in ecology do not bother with individuals or their variation. In these models all individual organisms are considered equal and their characteristics can be caught in single parameters based on population averages. However, individuals differ and natural selection acts on these differences. Hence not all parameter values have the same degree of biological realism. There is a growing interest of population, and even system, ecologists for the level of the individual. Quite rightly, since there is increasing evidence that aspects of behaviour and physiology of individuals can strongly influence higher level processes such as the dynamics of whole populations. Examples of this are given in several contributions to this section (Murdoch *et al.*, Mooij and DeAngelis). Murdoch *et al.*, for example, take us from individual decision making to population dynamics and regulation with the use of stage-structured parasitoid–host models. Another promising solution to estimate population effects of variability at the individual level is through individual-based modelling. The computational capacity has now reached a point where the book-keeping of many different individuals is no longer a problem. Individual-based modelling has become a promising tool for expanding ecological theory. These models do not require the same types of simplifying assumptions as compared with state-variable models, and allow us to put individuals and their interactions where they belong: at the basis and therefore in the centre of all population and system processes. Some proponents of individual-based modelling predict we are witnessing the end of the state-variable models, which are still the principal modelling tools in ecological theory. Will they be replaced by individual-based modelling? Mooij and DeAngelis show us when individual-based models are useful, exemplified by their work on aquatic but also terrestrial food webs.

This final section of the symposium has the highest level of mutitrophy, it also has the greatest diversity of authors from different ecological levels, individual-level ecologists, population ecologists, community and ecosystem ecologists. Although their approaches may differ, their interests obviously have much in common. They are all searching for the functional and mechanistic understanding of species interactions in ecological systems. Some appreciate the small, and observe individuals, limiting themselves to a few interacting species. Others think large, and model whole systems. We should not focus on their differences but question where and how these different approaches can meet and complement each other. In my opinion this section has shown several promising developments as follows.

Combining ecology and evolution

Empirical research at the individual level can help us to determine the effects of individual variation at higher level processes such as population regulation and trophic interactions: the first step up. The tools are there (Murdoch *et al.*, Mooij and DeAngelis): stage-size and age-structured models, individual-based models. The work of Murdoch *et al.* shows us, for example, that subtle behavioural variation in how parasitoids respond to host size can have dramatic dynamic consequences. Murdoch also points to possible generalizations over different systems, such as the stabilizing effect of invulnerable host stages for parasitoid–host interactions and delayed feedback cycles as a result of the parasitoid differential response to hosts of different ages. Perhaps of even greater value for generalization is his message that different kinds of behavioural variation in individuals have little or similar effects on dynamics. These generalities are necessary for progressing to greater complexity. Furthermore, they may have a clear applied spin-off. Using these kinds of models may help us to bridge the remaining gap between ecological theory and the practice of biological control. Perhaps now we can predict which characteristics make carnivores good suppressors of herbivores.

For those of us who have great affinity for simple generalizing state-variable models it is a pleasure to hear that simple models can still be useful, and that the basic properties of simple Lotka–Volterra models often survive in many of the more complex stage-size or individual-based models. Several chapters in the present and other sections propagate a 'happy marriage' between different kinds of models for the future where some models can answer questions that others cannot, and similar outcomes of different models can strengthen conclusions. Perhaps we should also add here the potential of using evolutionary models such as evolutionarily stable strategy (ESS) approach models (similar to the work by Haccou *et al.* (Chapter 4) and Sabelis *et al.* (Chapter 5)) to understand the evolution of individual variation and species strategies. There is increasing integration of such evolutionary models with population-dynamic models.

Scale-up from single species interactions to systems?

Can unravelling the interactions between individuals and between a few species really help us to reach understanding of how complex systems work? To scale up from a few species to a food web we need to identify the subset of species that have greatest influence upon the community. Strong gives us the metaphor of the food web vignette. As he says, at least we are aware of our ignorance of the rest of the food web. We are aware that, when limiting ourselves to a subset of influential species, additional interactions at different time scales could substantially change the results of the interactions. But it is the best we can do, and in some clear cases it is perhaps all we need to do.

Several authors (e.g. Dicke and Vet) argue the need for comparative studies over systems to try to form generalizations on the ecological and evolutionary principles that are underlying species interactions. From there we can generate and test hypotheses on the importance of such general principles in other, perhaps more complex systems. For example, how important are plant–carnivore interactions for plant–herbivore inter-actions in different systems, e.g. early and late plant succession systems? Can we generalize about the costs and benefits of plant–carnivore interactions over different systems? Similarly, Strong shows us the value of comparing aquatic and terrestrial systems in order to predict trophic cascades. Not all aquatic multitrophic systems cascade. Some aquatic systems show similarity to more complex terrestrial systems where complete cascades still seem rare. At the same time his underground system of lupine–ghost moth–nematode shows that, indeed, not all trophic cascades are wet. It also illustrates how ignorant we still are of what happens in that soil habitat, as already stressed by some authors in other sections. Strong's work is highly encouraging of a stronger research effort on soil systems.

If judgement on which species are the most influential in the community is beyond reach, the alternative is to study whole communities. Mooij and DeAngelis show how individual-based modelling is a promising tool for describing even the full community of the endangered wetlands, the Everglades. In fact, the more complex, the better these models perform compared with other kinds of models. Let us wait a few years to see if they are right.

Bottom-up and top-down effects

As with so many controversies between ecologists, the earlier debate on whether the abundance and distribution of organisms in multitrophic systems is primarily determined by bottom-up or top-down forces seems to have smoothed out with a harmonious integration of both viewpoints. Consensus is reached that dominant forces can vary within and among systems. We are now challenged to answer the highly interesting questions of when and why we can expect bottom-up or top-down forces and how these interact and influence each other. The present section has significantly contributed to this need to emphasize heterogeneity and variability in our thinking of species interacting in multitrophic systems.

Chapter 16

Plant–carnivore interactions: evolutionary and ecological consequences for plant, herbivore and carnivore

M. Dicke[1] and L.E.M. Vet[1]

Summary

Many terrestrial food webs are dominated by direct and indirect inter-actions among higher plants and arthropods. Direct interactions occur between a consumer and its food. Indirect interactions do not involve consumption. They may occur when plants promote the effectiveness of carnivorous arthropods by emitting herbivore-induced volatiles in response to herbivory. These volatiles are an important solution to a foraging problem of carnivorous enemies. The volatiles are emitted in large amounts and the blend composition can be specific for the plant species, plant genotype or for the herbivore species or instar damaging the plant. The characteristics of the chemical information such as the degree of specificity and thus the information value for the responding insects differ largely among plant species. After their emission, the induced plant volatiles may also affect herbivores, other plants, or competitors or predators of the carnivores. Thus, an information web is superimposed upon the food web. This information web is more complex than the food web itself. The information-mediated interactions that result from the emission of herbivore-induced plant volatiles have consequences for the ecology and evolution of plants, herbivores and carnivores in a food web. For instance, other plants, neighbouring the signalling plants, may profit from the attracted carnivores, which may influence the strategy of the

1 Laboratory of Entomology, Wageningen Agricultural University, PO Box 8031, 6700 EH Wageningen, The Netherlands

signaller. Herbivores may exploit the information in selecting food plants to avoid competition or enemy-rich space and carnivores may use the information in developing a flexible foraging strategy with maximal reproductive success, e.g. through associative learning of the cues.

In this chapter we compare different simple multitrophic systems with the aim of reaching generalities with regards to the ecological and evolutionary consequences of the most important information flows.

Introduction

Multitrophic systems

Ecosystems consist of complex trophic interactions that vary in time and space. These trophic interactions constitute complex food webs mediated by direct and indirect interactions. For instance, members of the same trophic level may compete directly for resources, or they may compete indirectly through effects of a shared predator. With an increase in the number of species, the number of interactions in a food web increases exponentially (Gange & Brown 1997). Traditionally, most interactions in food webs have been investigated for two interacting species, e.g. a plant and a herbivore, or a herbivore and its predator. This may be necessary for pragmatic reasons. However, it has become clear that to understand food webs, it is essential to include more interactants (Hunter *et al.* 1992). For instance, to understand why many herbivores are specialists, one may need information on (i) plant–herbivore interactions and the role of secondary plant chemicals in plant defence and (ii) information on herbivore–carnivore interactions and the role of secondary plant chemicals in herbivore defence (Bernays & Graham 1988). By studying different bitrophic and tritrophic interactions we search for generalizations with regards to ecological and evolutionary principles and underlying mechanisms. It may then become possible to investigate food webs multitrophically.

In this chapter we restrict ourselves to higher plant–arthropod interactions. This may not constitute a severe restriction, as higher plants and arthropods comprise the majority of species known to date (Wilson 1992).

Direct defence vs. indirect defence of plants

Plants are fundamental to food webs, and may be attacked by a wealth of herbivorous insect species (about 50% of all insect species are herbivores; Strong *et al.* 1984). Plants have evolved two types of defence: (i) direct

484

defence, which acts on the herbivore directly and (ii) indirect defence, which promotes the effectiveness of natural enemies of the herbivore (Price *et al.* 1980; Dicke & Sabelis 1988; Dicke 1998). Indirect defence can be seen as a combination of bottom-up and top-down effects, which supports the assertion that these effects cannot be regarded in isolation (Hunter & Price 1992). Direct defence has been more thoroughly investigated than indirect defence, but both appear to be widespread.

Direct defence can be mediated by plant chemicals such as toxins, repellents or digestibility reducers, or by plant structures such as spines and thorns. Indirect defence can involve the provision of shelter, such as domatia (Walter 1996) or the provision of alternative food such as floral and extrafloral nectar (Koptur 1992). A third mode of indirect defence is the provision of information that can be used by carnivores during foraging. Chemical information on herbivore presence and identity may be essential for successful location of herbivores by carnivores, as discussed below.

Carnivores need plants

Most carnivores actively search for their herbivorous prey. However, herbivores have evolved to be inconspicuous to their natural enemies, and therefore the locational information they provide will be limited. Plants can provide a solution for this problem. The informational value of stimuli used by carnivores in the location of herbivore prey ultimately depends on two characteristics: (i) their reliability in indicating available and suitable prey and (ii) the detectability of the stimulus, i.e. the ease of stimulus reception. Unfortunately for the carnivores, reliability and detectability of stimuli are expected to be inversely correlated. Although herbivore-derived information is very reliable, it is difficult to detect. This will drive the evolution of 'indirect' search strategies, i.e. the use of information that is not derived directly from the herbivore itself but from other sources that predict its presence. This is where the plant comes in. But how useful is the plant for the carnivore? Although plants can provide more detectable cues because of their larger biomass, these cues are generally far less reliable for herbivore location. Smelling a host plant does not guarantee the presence of a suitable herbivore. Selection will favour animals that solve this reliability–detectability dilemma. Several solutions have been proposed (Vet *et al.* 1991; Vet & Dicke 1992). Firstly, carnivores can be more selective in the plant cues that they use, i.e. by preferring to respond to plant volatiles that are induced after herbivory (Dicke & Sabelis 1988, 1992; Turlings *et al.*

1990, 1995; Dicke 1998). This will be the major subject of this chapter. An example is the strong preference of the parasitoid *Cotesia glomerata* for the odours of host-damaged cabbage plants (from which the caterpillars have been removed) over odours from the caterpillars or their faeces (Fig. 16.1).

A second solution is associative learning by the foraging carnivore. Carnivores can associate plant cues with reliable herbivore-derived cues once the first suitable prey is encountered. Hence, learning can improve subsequent foraging efficiency, and plant (or other environmental) cues become the most important and useful in long-range herbivore location once learning has taken place (Vet *et al.* 1990, 1995; Tumlinson *et al.* 1993; Turlings *et al.* 1993). Plant cues may elicit innate responses in carnivores, the strength of which generally depends on their predictability as indicators of suitable herbivores over evolutionary time. However, the variable nature of plant cues (see below) and their low reliability constrain the evolution of innate responses to these cues. Furthermore, there will be a physiological

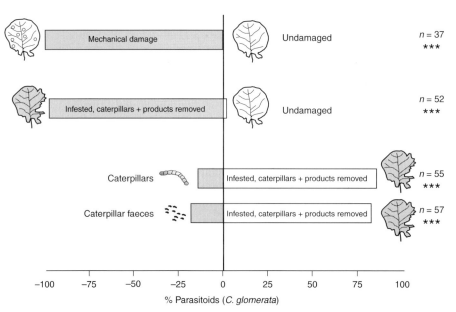

Figure 16.1 Response of adult female *Cotesia glomerata* parasitoids in a two-choice flight setup to volatiles from damaged plants, hosts (*Pieris brassicae* caterpillars) and host faeces. *n*, number of parasitoids, ***, $P < 0.001$ (Chi-squared test). For more details see Steinberg *et al.* (1993).

constraint on the width of innate responses (Vet *et al.* 1990). Associative learning allows for adaptive changes in foraging strategies of carnivores. In terms of Pavlov (1927), plant cues are the 'Conditioned Stimuli' being reinforced by a herbivore-derived cue (the 'Unconditioned Stimulus' or 'reward') that is recognized innately. Learning acts as a major optimization mechanism allowing carnivores to enhance their foraging efficiency in time and to deal adaptively with local and temporal variation in herbivore–food plant resources. Figure 16.2 illustrates a learning-induced change in preference of *C. glomerata* parasitoids for odours from different host-infested cabbage cultivars. The adaptive value of learning is expected to vary significantly between carnivore species, and to depend on factors such as predictability of herbivore presence, herbivore stage attacked and dietary specialization of the carnivore and the herbivore it is trying to locate (Vet *et al.* 1995).

A third solution is a bitrophic one. Carnivores can exploit cues that are released by their host/prey for unavoidable intraspecific communication

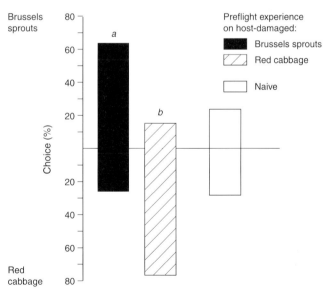

Figure 16.2 Effect of experience with host-damaged plants on preference of adult female *Cotesia glomerata* parasitoids for volatiles from either of two host-damaged cabbage cultivars. For more details see Geervliet *et al.* (1993). (a) and (b) indicate significant difference ($P < 0.05$, contingency table).

(Lewis *et al.* 1982; Noldus *et al.* 1991; Wiskerke *et al.* 1993; Hardie *et al.* 1994). An example is the sex pheromone communication system of moths, exploited by *Trichogramma* egg parasitoids (Noldus *et al.* 1991). Especially for carnivores that attack non-feeding herbivore stages such as eggs and pupae, this may be a major solution (Vet *et al.* 1995).

Outline of chapter

In this chapter we attempt to make an in-depth analysis of small components of a number of food webs. By comparing the specific characteristics of these, we aim to go beyond system specificity and to reach generality. Although aware of the restrictions of this approach, we feel that this is the only way of understanding general mechanisms underlying communication in multitrophic systems.

Both the production of plant volatiles and the responses of arthropods to these volatiles are subject to variation, and natural selection acts on both sides of the interaction. When plants produce carnivore-attracting volatiles in response to herbivory, this has costs and benefits for the different interactants. The costs and benefits are not restricted to the interaction between plant and carnivore. Herbivores may also exploit the volatiles. Furthermore, neighbouring plants, competing carnivores and second-order carnivores might also profit from plant-produced information. In this chapter we will consider costs and benefits of herbivore-induced plant volatiles for plants, carnivores and herbivores in a multitrophic context.

Firstly, we will discuss variation in plant-produced information and how carnivores deal with plant volatiles as a variable source of information. We subsequently review how herbivores respond to herbivore-induced plant volatiles. The possible effects of indirect defence on competition among plants are then considered. We will also discuss some likely implications of food web complexity for plant–carnivore interactions. Finally, we consider how herbivores may avoid falling victim to the plant–carnivore pact. We close the chapter by considering future directions.

Variation in plant odours

The existence of genetic variation in plant odours is common knowledge for plant breeders interested in natural plant-derived fragrances (e.g. Perry *et al.* 1996). Far less information is available on plant odours that are not of commercial interest. Flower odours may vary among genotypes

(Tollsten & Bergstrom 1993; Tollsten *et al.* 1994), and terpenoids emitted from foliar glands vary enormously among cotton cultivars (Elzen *et al.* 1985). Furthermore, the emission of plant volatiles varies with environmental factors, such as the growth condition of the plant (Visser 1986). For instance, Blaakmeer *et al.* (1994a) recorded a 10-fold increase in cabbage plant volatiles in greenhouses from winter to summer.

When herbivores feed on plants, plants can respond with an increase in volatile production. Hence, herbivores constitute an additional source of variation in plant volatiles. All this results in an impressive list of variation-inducing factors: plant volatiles may not only differ due to different abiotic factors, but also among plant species, plant genotypes, plant parts and according to different biotic factors, such as infestation by different herbivore species, stages and densities.

Carnivores that use plant volatiles in herbivore location are confronted with all these sources of variation. It is essential for them to distinguish between relevant and irrelevant variation in plant information. Relevant variation is correlated with herbivore presence, identity and quality. Irrelevant variation is noise that should be disregarded. How variation in plant odours is interpreted is basically determined by how insects perceive and process qualitative and quantitative variation in odour blends. Getz and Smith (1990) suggest that the quality of an odour is represented by the firing pattern of several different neurons, while quantity is coded by the overall absolute rate of firing. There are indications that qualitative differences in odour bouquets are more easily interpretable than quantitative differences (Vet *et al.* 1998).

When a parasitoid is specialized on a particular instar of a particular herbivore species, plant volatiles specifically indicating the presence of such a host are of great informational value. In general, we expect reliability of plant volatiles for the carnivore to be positively correlated with plant odour specificity. Plants can respond to herbivory in different ways. We distinguish three broad categories, based on changes in odour blend after herbivory (Fig. 16.3).

Let us first look at the effect of herbivory vs. mechanical damage. At one extreme there is no difference in the identity of volatiles produced in response to herbivory or mechanical damage (Fig. 16.3a). At the other extreme, novel information dominates the volatile blend: herbivore-damaged plants produce a large set of chemicals that are not produced when the plant is mechanically damaged and these novel compounds are major components of the odour blend (Fig. 16.3b,c).

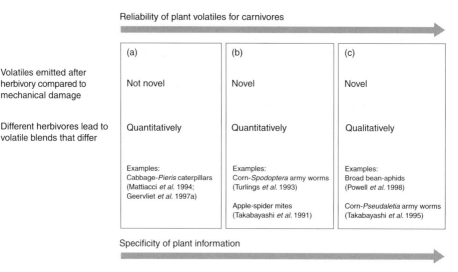

Reliability of plant volatiles for carnivores

	(a)	(b)	(c)
Volatiles emitted after herbivory compared to mechanical damage	Not novel	Novel	Novel
Different herbivores lead to volatile blends that differ	Quantitatively	Quantitatively	Qualitatively
	Examples: Cabbage-*Pieris* caterpillars (Mattiacci *et al.* 1994; Geervliet *et al.* 1997a)	Examples: Corn-*Spodoptera* army worms (Turlings *et al.* 1993) Apple-spider mites (Takabayashi *et al.* 1991)	Examples: Broad bean-aphids (Powell *et al.* 1998) Corn-*Pseudaletia* army worms (Takabayashi *et al.* 1995)

Specificity of plant information

Figure 16.3 Relative composition of herbivore-induced plant volatiles with regard to specificity and reliability.

Superimposed on these differences is variation in specificity of plant information related to the herbivore. Some plants emit qualitatively similar chemical blends irrespective of the herbivore that feeds on them. The informational value for the carnivore is restricted to quantitative differences, i.e. different ratios of the blend components (Fig. 16.3a,b). Other plants produce qualitatively specific blends, in response to different herbivore species or even different herbivore instars (Fig. 16.3c). Thus, from (a) to (c) in Fig. 16.3 we see a potential increase in the specificity of the plant information, and an increase in the reliability of this information for the foraging carnivore.

Table 16.1 summarizes studies for which chemical analyses of the volatiles emitted by herbivore-infested and mechanically damaged plants have been made. In 14 out of 18 different plant–herbivore complexes, herbivore-infested plants emit novel compounds (albeit that some plant species emit many novel compounds, while others emit only one). The explicit discrimination between plants that do or do not emit novel compounds is affected by the sensitivity of the analytical tools employed. Categorization into discrete classes is, of course, a simplification. In reality, there will be a continuum from one category to the next. Of 10 systems for which the emitted bouquets have been compared for infestations by different herbivores, four yielded qualitative differences (but here again the methodological aspects influence the conclusion). The greater the number of compounds emitted, the more likely it is that some qualitative differences will be found.

Table 16.1 shows that individual plants of one species, infested by different herbivores, may emit different odour bouquets. However, so far no studies have been carried out to elucidate how plants respond to simultaneous infestation by two or more herbivore species. This question is especially interesting in those cases where qualitative differences exist among bouquets from plants infested by different herbivore species.

How carnivores deal with plant information

Selection acts on the carnivore's ability to discriminate between relevant and irrelevant variation in plant information. The plants from categories (a), (b) and (c) in Fig. 16.3 present foraging carnivores with great differences in informational value. To be able to find their herbivorous prey on plants, carnivores may need different strategies, depending on the informational context.

Let us start with a carnivore that searches for herbivores on a plant species in category (a). All plants emit qualitatively the same blend of volatiles, albeit in different quantities. A damaged plant emits much larger quantities than an undamaged plant (Mattiacci *et al.* 1994). However, odour quantity will also decline with distance from the source, so a carnivore may experience different stimuli as a result of proximity as well as herbivore presence/absence. Furthermore, quantitative differences in compound ratios may not only reflect herbivore damage, but also mechanical damage. For example, in wind-tunnel bioassays, the parasitoid *Cotesia glomerata* is strongly attracted by mechanically damaged cabbage plants as well as by host-damaged plants from which the host and its products have been removed. Undamaged plants are not attractive at all (see Fig. 16.1).

The next problem carnivores face is to discriminate between plants infested by suitable and unsuitable herbivores. *C. glomerata* strongly prefers to fly to a cabbage plant infested by the non-host species *Plutella xylostella* or *Mamestra brassicae* when the alternative is a clean plant (Geervliet *et al.* 1996). Moreover, the parasitoids are equally attracted to plants infested by one of their hosts *Pieris brassicae* or the non-host *M. brassicae* (Geervliet *et al.* 1998b).

In conclusion, within category (a) (Fig. 16.3) any type of damage in the field will create a considerable amount of noise, i.e. plant volatiles not related to presence of suitable herbivores. How do carnivores deal with this? Carnivores may learn minor but consistent quantitative differences in odour blends or alternatively use other sensory modalities to discriminate noise from signal. Empirical data for *C. glomerata* show that during foraging this parasitoid does acquire the ability to distinguish even between plants infested by one of two host species that differ in suitability. It remains unknown whether this acquired ability is based on the use of minor quantitative differences in odour blends or on the use of visual cues. *C. glomerata* may use visual cues, such as the differences in size of the feeding holes, to distinguish between host plants infested with *P. brassicae* or *P. rapae*. In behavioural experiments with cabbage plants with artificial holes of different sizes, the parasitoids were shown to readily associate the size of the holes with the presence of hosts (J.M. Verdegaal, F.L. Wäckers & L.E.M. Vet, unpublished data).

The situation is quite different in categories (b) and (c) (Fig. 16.3) where plants emit novel compounds after herbivory (Dicke *et al.* 1990b; Turlings *et al.* 1990). Herbivore-infested plants stand out in the environment as a result of their unique odour blends. Their chemical profile is qualitatively and quantitatively different from that of artificially damaged plants (Dicke *et al.* 1990a; Turlings *et al.* 1990; Takabayashi *et al.* 1994a,b; Powell *et al.* 1998). Artificially damaged plants do not constitute noise, as they are not attractive to carnivores (Turlings *et al.* 1991; Shimoda *et al.* 1997; Powell *et al.* 1998; M. Dicke, unpublished data).

Categories (b) and (c) differ in the information available to carnivores with which to discriminate between plants infested by suitable and unsuitable herbivores. In category (c), the odour blends of such plants differ qualitatively, while in category (b) they only differ quantitatively.

We would expect that carnivores can more easily interpret qualitative than quantitative differences, and that learning is less important (or more easily achieved) in the case of qualitative differences in odour blends (Vet *et al.* 1998). Empirical data for category (c) show that an aphid parasitoid (*Aphidius ervi*) readily discriminates between broad bean plants infested with a suitable or a non-suitable aphid host. Experience is not needed for this discrimination. Plants infested with the non-host aphid are not even preferred over undamaged plants (Du *et al.* 1996; Powell *et al.* 1998).

Moreover, in category (c) there is evidence for discrimination among plants infested by herbivores of different larval instars. Without prior

Table 16.1 Tritrophic systems for which chemical analyses of herbivore-induced plant volatiles that attract carnivores have been made.

Tritrophic system

Plant species	Herbivore species	Carnivore species	Novel compounds emitted in response to herbivory?	Chemical differences in volatiles between plants with different herbivore types?	Carnivore discrimination between plants with different herbivore types?
Lima bean (*Phaseolus lunatus*) Fabaceae	Two-spotted spider mite (*Tetranychus urticae*) (Acari: Tetranychidae)	*Phytoseiulus persimilis* (Acari: Phytoseiidae)	Yes, several compounds (Dicke *et al.* 1990a)	Yes, quantitative: spider mite vs. thrips (*Frankliniella occidentalis*) (M. Dicke & M.A. Posthumus, unpublished data)	
Lima bean (*Phaseolus lunatus*) Fabaceae	Two-spotted spider mite (*Tetranychus urticae*) (Acari: Tetranychidae)	*Amblyseius andersoni* (= *A. potentillae*) (Acari: Phytoseiidae)	Yes, several compounds (Dicke *et al.* 1990a)		Yes: spider mite vs. leaf roller (*Adoxophyes orana*) (Dicke & Groeneveld 1986)
Lima bean (*Phaseolus lunatus*) Fabaceae	Two-spotted spider mite (*Tetranychus urticae*) (Acari: Tetranychidae)	*Scolothrips takahashii* (Thysanoptera: Thripidae)	Yes, several compounds (Dicke *et al.* 1990a; Shimoda *et al.* 1997)		
Bean (*Phaseolus vulgaris*) Fabaceae	*Liriomyza trifolii* (Diptera: Agromyzidae)	*Diglyphus isaea* (Hymenoptera: Eulophidae)	Yes, one compound, i.e. 1-octen-3-ol (Finidori-Logli *et al.* 1996)		
Cucumber (*Cucumis sativus*) Cucurbitaceae	*Tetranychus urticae* (Acari: Tetranychidae)	*Phytoseiulus persimilis* (Acari: Phytoseiidae)	Yes, several compounds (Dicke *et al.* 1990b; Takabayashi *et al.* 1994a)		
Tomato (*Lycopersicon esculentum*) Solanaceae	*Tetranychus urticae* (Acari: Tetranychidae)	*Phytoseiulus persimilis* (Acari: Phytoseiidae)	Yes, two minor compounds (J. Takabayashi & M. Dicke, unpublished data)		

Continued on p. 494

Table 16.1 (*Continued*)

Tritrophic system

Plant species	Herbivore species	Carnivore species	Novel compounds emitted in response to herbivory?	Chemical differences in volatiles between plants with different herbivore types?	Carnivore discrimination between plants with different herbivore types?
Gerbera (*Gerbera jamesonii*) Asteraceae	*Tetranychus urticae* (Acari: Tetranychidae)	*Phytoseiulus persimilis* (Acari: Phytoseiidae)	Yes, many compounds (O.E. Krips *et al.* unpublished data)		
Apple (*Malus domestica*) Rosaceae	*Tetranychus urticae* (Acari: Tetranychidae)	*Phytoseiulus persimilis* (Acari: Phytoseiidae)	Yes, a few compounds in minor quantities (Takabayashi *et al.* 1991)	Yes, quantitative differences: *T. urticae* vs. *P. ulmi* (Takabayashi *et al.* 1991)	Yes: *T. urticae* vs. *P. ulmi* (Sabelis & Van de Baan 1983)
Apple (*Malus domestica*) Rosaceae	European red mite *Panonychus ulmi* (Acari: Tetranychidae)	*Amblyseius andersoni* (=*A potentillae*), *Amblyseius finlandicus* (Acari: Phytoseiidae)	Yes, a few compounds in minor quantities (Takabayashi *et al.* 1991)	Yes, quantitative differences: *P. ulmi* vs. *T. urticae* (Takabayashi *et al.* 1991)	Yes: both predator species discriminate between *P. ulmi* and *T. urticae* (Sabelis & Van de Baan 1983; Sabelis & Dicke 1985)
Corn (*Zea mays*) Poaceae	Beet army worm (*Spodoptera exigua*) (Lepidoptera: Noctuidae)	*Cotesia marginiventris* (Hymenoptera: Braconidae)	Yes, several compounds (Turlings *et al.* 1990, 1991)	Yes, quantitative differences: beet army worm vs. fall army worm (Turlings *et al.* 1993)	
Corn (*Zea mays*) Poaceae	Common army worm (*Pseudaletia separata*) (Lepidoptera: Noctuidae)	*Cotesia kariyai* (Hymenoptera: Braconidae)	Yes, several compounds (Takabayashi *et al.* 1995)	Yes, qualitative differences: 1st–3rd vs. 6th larval instars *Pseudaletia separata* (Takabayashi *et al.* 1995)	Yes: 1st–3rd vs. 6th larval instars *P. separata* (Takabayashi *et al.* 1995)
Cotton (*Gossypium hirsutum*) Malvaceae	Corn earworm (*Helicoverpa zea*) (Lepidoptera: Noctuidae)	*Microplitis croceipes* (Hymenoptera: Braconidae)	Yes, minor differences in three fatty acid derived esters and indole (McCall *et al.* 1994)		

Plant	Herbivore	Natural enemy	Column 4	Column 5	Column 6
Cabbage (Brassica oleracea), various cultivars Brassicaceae	Large cabbage white Pieris brassicae, Small cabbage white Pieris rapae (Lepidoptera: Pieridae)	Cotesia glomerata (Hymenoptera: Braconidae)	No (Mattiacci et al. 1994; Agelopoulos & Keller 1994b; Geervliet et al. 1997a)	Yes, minor quantitative differences: P. brassicae and P. rapae (Blaakmeer et al. 1994a)	Yes: P. rapae vs. P. brassicae; experienced wasps (Geervliet et al. 1997b, 1998b)
Cabbage (Brassica oleracea, various cultivars Brassicaceae	Large cabbage white Pieris brassicae, Small cabbage white Pieris rapae (Lepidoptera: Pieridae)	Cotesia rubecula (Hymenoptera: Braconidae)	No (Mattiacci et al. 1994; Agelopoulos & Keller 1994b; Geervliet et al. 1997a)	Yes, minor quantitative differences: P. brassicae and P. rapae (Blaakmeer et al. 1994a)	No: P. rapae vs. P. brassicae; experienced wasps (Geervliet et al. 1997b)
Cabbage (Brassica oleracea) Brassicaceae	Plutella xylostella (Lepidoptera: Plutellidae)	Cotesia rubecula (Hymenoptera: Braconidae)	Yes, one compound, i.e. allyl isothiocyanate (Agelopoulos & Keller 1994b)	Yes, qualitative difference in one compound quantitative differences in several compounds. Pieris rapae vs. Plutella xylostella (Agelopoulos & Keller 1994b)	No: P. rapae vs. P. xylostella Yes: caterpillars vs. snails (Agelopoulos & Keller 1994a)
Yellow cress (Rorippa indica) Brassicaceae	Pieris rapae crucivora (Lepidoptera: Pieridae)	Cotesia glomerata (= Apanteles glomeratus) (Hymenoptera: Braconidae)	No (Takabayashi et al. 1998)		
Potato (Solanum tuberosum) Solanaceae	Colorado potato beetle Leptinotarsa decemlineata (Coleoptera: Chrysomelidae)	Perillus bioculatus (Hemiptera: Pentatomidae)	No (Bolter et al. 1997)		
Nasturtium (Tropaeolum majus) Tropaeolaceae	Pieris brassicae, P. rapae, P. napi (Lepidoptera: Pieridae)	Cotesia glomerata, Cotesia rubecula (Hymenoptera: Braconidae)	Yes, several compounds (Geervliet et al. 1997a)	Yes, qualitative and quantitative differences Pieris brassicae vs. P. rapae (Geervliet et al. 1997a)	No: Pieris brassicae vs. P. rapae; inexperienced wasps (Geervliet et al. 1996)
Broad bean (Vicia faba) Fabaceae	Pea aphid (Acyrthosiphon pisum) (Homoptera: Aphididae)	Aphidius ervi (Hymenoptera: Aphidiidae)	Yes, several compounds (Du et al. 1996; Powell et al. 1998)	Yes, qualitative and quantitative differences: Acyrthosiphum pisum vs. Aphis fabae (Powell et al. 1998)	Yes: Acyrthosiphum pisum vs. Aphis fabae; inexperienced wasps (Du et al. 1996)

oviposition experience, the parasitoid *Cotesia kariyai* is attracted to maize plants infested by first to fourth instar caterpillars, while plants infested by fifth to sixth instar larvae are not attractive. The same results are obtained when plants are treated with regurgitant of third instar larvae (attraction) or sixth instar larvae (no attraction) (Takabayashi *et al.* 1995).

In category (b), the opportunity for a carnivore to discriminate may be more restricted than in category (c) as the identity of the herbivore species is only indicated by quantitative differences. However, behavioural research shows that carnivores can discriminate among blends that differ quantitatively. In a choice situation, the predatory mites *Amblyseius andersoni* and *A. finlandicus* prefer volatiles from apple leaves infested by their preferred prey *Panonychus ulmi* over those from apple leaves infested by a non-preferred prey species, *Tetranychus urticae*. The predator *Phytoseiulus persimilis* shows the opposite prey preference, and the opposite response to plant volatiles (Sabelis & Van de Baan 1983; Sabelis & Dicke 1985). The parasitoid *Cotesia marginiventris* is attracted to host-infested maize plants on the basis of herbivore-induced terpenoids. Plants infested by different herbivores do not differ in these induced terpenoids, but consistent quantitative differences exist in green leaf volatiles emitted as a result of mechanical damage. These differences appear to be sufficient for the parasitoids to associatively learn to distinguish between plants infested by one of two host species (Turlings *et al.* 1993). Discrimination by carnivores has been observed in systems involving plants of categories (a), (b) and (c) (see Table 16.1). In eight out of 12 cases, carnivores discriminated between plants infested by different herbivores. In several cases the ability to discriminate is dependent on previous foraging experience.

Herbivore load of plants and reliability of plant information for carnivores

Degree of specialization of carnivores and the number of herbivore species that can potentially feed on a plant species will influence the usefulness of plant-derived chemical information to the carnivore. Consider a carnivore specialized on a specialist herbivore. Information from plant species that are outside the host range of its herbivore prey constitutes noise. This type of noise may be easy to interpret, as the variation in odour blends among plant species is qualitative in nature (Dicke *et al.* 1990b; Turlings *et al.* 1993; Takabayashi & Dicke 1996). In contrast, chemical profiles of a plant damaged by different herbivore species often differ only quantitatively (see Table 16.1: six out of 10 cases). Such variation may be more difficult to interpret than qualitative variation and discrimination seems to occur only after learning by the foraging carnivore (Turlings *et al.* 1993; Geervliet *et al.* 1998b).

With increasing numbers of non-target herbivore species feeding on the plant, the reliability of induced plant volatiles for a specialist carnivore decreases. Plants with a strong direct defence are expected to be used by small numbers of generalized herbivore species (Feeny 1976; Van der Meijden 1996). On such plants, a carnivore may encounter only a few specialist herbivores, and thus the reliability of herbivore-induced volatiles is relatively high. For plants with a weak direct defence, the situation is different: a specialist carnivore may encounter damage inflicted by many herbivore species (Feeny 1976; Van der Meijden 1996). A carnivore using herbivory-induced volatiles is here faced with much less reliable infor- mation. What we expect in such cases, however, is that carnivores will employ learning to locate only suitable herbivores, making use of additional sensory modes (as do more generalist parasitoids that attack a range of herbivores: Vet & Dicke 1992; Vet *et al.* 1995).

Qualitative vs. quantitative differences

Experiments by Vet *et al.* (1998) with *Leptopilina heterotoma*, a parasitoid of *Drosophila* larvae that feed in different substrates, showed that the wasps learned qualitative differences in odour cues more easily than quantitative differences. If we generalize the hypothesis that qualitative differences are important in odour recognition and discrimination learning in parasitoids, this has interesting implications for the study of plant–parasitoid inter- actions. Although many studies have shown that parasitoids can learn plant odours (see reviews by Turlings *et al.* 1993; Vet *et al.* 1995), how parasitoids deal behaviourally with natural variation in plant odour cues remains to be demonstrated. After having experienced plants infested with their host (caterpillars of *Pieris* spp.) the braconid parasitoid *C. glomerata* learns to discriminate between odours of different cabbage cultivars (see Fig. 16.2) (see also Geervliet *et al.* 1998b). The question remains, *what* do these parasitoids learn, and on what basis do they discriminate between these cabbage plants? Do they learn complete odour bouquets, or do they learn on the basis of qualitative differences in one (or a few) key components between plants, as perhaps suggested by studies with *L. heterotoma* (Vet *et al.* 1998)? Traditionally, the focus of research on the sensory physiology of plant–insect interactions is on how (mostly herbivorous) insects recognize and select plants on the basis of chemical information (Bernays

& Chapman 1994). Note that we ask a different question here, i.e. how do insects perceive and deal behaviourally (through learning) with *variation* in plant volatiles?

Assessment of profitability

Assessment of habitat and patch profitability is essential for all foraging animals to make optimal decisions on where to forage and how long to stay and exploit a particular food site. According to optimal foraging theory, patch residence times are predicted to vary with travel time between patches, and with average habitat profitability (Stephens & Krebs 1986). In many optimal foraging models, animals are assumed to be omniscient with regard to their foraging environment, i.e. they have complete knowledge about numbers and locations of resource units. This assumption has been criticized as being unrealistic. Is it not more likely that foraging animals need to sample, and that they have to learn from their experiences, assessing profitabilities as they forage? The latter may indeed be necessary when actual herbivore-encounter rates are needed for profitability assessment, or when carnivores use non-volatile, direct herbivore cues to estimate herbivore density and distribution. However, plants have great potential to provide carnivores with chemical information that is a powerful tool for profitability assessment from a distance. The production of induced volatiles may be positively correlated with the density of herbivores feeding on the plant; Geervliet *et al.* (1998a) showed a density-dependent response for *C. glomerata*. In wind-tunnel experiments, parasitoids could fly and choose between two Brussels sprout plants infested with different host (*Pieris brassicae*) densities. Percentages of responding females (that flew and chose) increased with increasing host density. Females preferred the plant with the highest density.

In conclusion, plant-derived information can help carnivores to optimize their arrival and leaving decisions, which is important for time-limited carnivores. With the help of the plant, carnivores do not necessarily have to land and sample to estimate plant profitabilities; they can make in-flight decisions on where best to go.

Effectiveness of individual carnivores

The link between plant and carnivore can be strong and clear when both partners significantly benefit from the interaction. For signals to evolve, they must be reliable for the receiver, and the receiver must respond in a way that is beneficial to the producer of the signal (Godfray 1995a).

In some tritrophic systems these conditions seem to be met. Clear signals from the plant are followed by a carnivore response, which significantly reduces the plant's herbivore load. The bean–spider mite–predatory mite system has these characteristics (Sabelis & Dicke 1985; Sabelis & Van der Meer 1986; Dicke *et al.* 1990b). In such cases, herbivore-induced plant volatiles have a clear mutualistic function, suggesting possible coevolution based on 1–3 level interactions.

Obviously, we do not expect or see such an intimate 1–3 link in all cases (see 'Food web complexity and plant–carnivore interactions' below). Due to a lack of reliability or availability of plant information, the value of the plant as an aid in herbivore location is limited for many species of carnivores. Similarly, many carnivore species that may respond to the plant's information may be of little value for enhancing the plant's fitness.

Non-feeding stages such as herbivore eggs generally induce little response from a plant (but see, for example, Blaakmeer *et al.* 1994b). Although egg parasitoids may be of great value to the plant — since they kill the herbivore before damage is done — attraction by herbivore-induced plant volatiles from a distance has, to our knowledge, not been reported for egg parasitoids.

Pupal parasitoids often lack plant aid in herbivore location. Larvae often leave the plant to pupate elsewhere. Larval dispersal away from highly detectable plants may, in fact, function to reduce location by carnivores that attack the pupal stage. Hence, the value of plant information is greatest for parasitoids and predators attacking actively feeding stages such as larvae or adults, and research on carnivore attraction by plant volatiles has focused on such systems.

From the plant's perspective, different carnivores may have different values. Predators can stop herbivore damage immediately by removing their prey. However, satiation of the individual predator may provide a constraint on the predator's benefit for the plant. Also of direct benefit to the plant are so-called idiobiont parasitoid species that paralyse or kill the host during oviposition. However, many parasitoids of larvae are koinobiont species, which allow their host to grow after oviposition, until an optimal stage or size has been reached, after which the host is killed and consumed. Some of these koinobiont species may even stimulate their host to increase its feeding rate compared with an unparasitized host (Slansky 1986), which is clearly disadvantageous to the plant.

Superimposed on the above variation in the strength of the 1–3 link is the effect of diet breadth of the carnivores. Overall, the importance of plant volatiles as information source and the way they are used (innate responses vs. learning) is related to the carnivore's diet breadth at both trophic levels, i.e. the number of herbivore species it attacks and the number of plant species that are used by these herbivores (Vet & Dicke 1992; Vet *et al.* 1995).

How herbivores respond to induced plant volatiles

To hypothesize on the evolution of plant–carnivore interactions, all costs and benefits of this interaction must be considered. Biosynthetic costs of the signals seem to be small, but ecological costs can be considerable (Dicke & Sabelis 1989). One of the ecological costs is the exploitation of the information by herbivores.

Until recently, research on the response of herbivores to plant volatiles was restricted to volatiles from undamaged plants. These responses have been studied with respect to phytochemistry, behaviour, sensory physiology and neurophysiology (Visser 1986; Bernays & Chapman 1994; Schoonhoven *et al.* 1998). Since the discovery of herbivore-induced plant volatiles that attract carnivores, the question arises how these induced volatiles affect herbivore foraging decisions. Exploiting the induced plant volatiles has advantages and risks for the herbivore. Increased amounts of volatiles after herbivory could enhance location of a (damaged) plant by herbivores, and may also provide information on the presence of intra- or interspecific competitors. They may even provide information on the density of these competitors, i.e. on the level of expected competition. Furthermore, these volatiles might inform the herbivore about the condition of the plant, e.g. induction or exhaustion of its defensive capabilities. In addition, there is the risk of entering an enemy-dense space, because of the simultaneous attraction of carnivores.

How can we relate this to the specificity of plant information as depicted in Fig. 16.3? At the left of the continuum, (a), plants do not provide specific information after herbivory. Chemically, plants just produce more of the same. As concentration of odour blends is distance dependent, the informational value of 'more of the same' is limited. Thus, herbivores remain uninformed with respect to whether something is feeding on the plant(s), and if so, who is feeding. A herbivore may simply be attracted by these odours, and use different cues at shorter distances or after arrival. At the opposite end of the continuum, (c), the specificity of the odour bouquet provides a herbivore with information on whether the plant is being fed on, and possibly by which species. Thus, the herbivore has more opportunities to adapt its searching strategy: attraction or avoidance, depending on the

Table 16.2 Responses of herbivores to plant volatiles of plants infested by conspecific or heterospecific herbivores.

System Plant	Herbivore	Behaviour to odour from plants infested by: Conspecifics	Heterospecifics	Novel compounds emitted in response to herbivory?	Reference
Potato	Colorado potato beetle (*Leptinotarsa decemlineata*) (Coleoptera: Chrysomelidae)	Attraction	*Spodoptera exigua* (Lepidoptera: Noctuidae) Attraction	No	Bolter *et al.* (1997)
Duranta repens	*Maladera matrida* (Coleoptera: Scarabeidae)	Attraction	*Schistocerca gregaria* (Orthoptera: Acrididae) Attraction	Not investigated	Harari *et al.* (1994)
Apple	Japanese beetle (*Popillia japonica*) (Coleoptera: Scarabeidae)	Attraction	*Hypantria cunea* (Lepidoptera: Arctiidae) Attraction	Yes	Loughrin *et al.* (1995b)
Cotton	Cabbage looper moth *Trichoplusia ni* (Lepidoptera: Noctuidae)	Attraction, but more oviposition on nearby undamaged plants	Not investigated	Not investigated	Landolt (1993)
Cabbage	Cabbage looper moth *Trichoplusia ni* (Lepidoptera: Noctuidae)	Avoidance	Not investigated	Not investigated	Landolt (1993)
Lima bean	Two-spotted spider mite (*Tetranychus urticae*) (Acari: Tetranychidae)	Attraction to low spider-mite densities Avoidance of high spider-mite densities	Not investigated	Yes	Dicke (1986) (behaviour), Dicke *et al.* (1990b) (chemistry)
Cucumber	Two-spotted spider mite (*Tetranychus urticae*) (Acari: Tetranychidae)	Attraction	*Frankliniella occidentalis* (Thysanoptera: Thripidae) Avoidance; heterospecific herbivore is facultative predator	Yes	Pallini *et al.* (1997) (behaviour) Dicke *et al.* (1990b) (chemistry)

costs of aggregation. Aggregation with heterospecifics may be costly due to food competition. For conspecific aggregation, food competition may be even more intense, given that individuals of the same species may compete for exactly the same resources. However, aggregation with conspecifics may be beneficial under certain circumstances, e.g. when aggregation is needed for mate finding, or when the *per capita* chance of being located by natural enemies in an aggregation decreases with the number of individuals in the aggregation. Whether living in an aggregation is advantageous will also depend on the relationship between the size of an aggregation and the chance that it is located by a carnivore. Questions on the function of aggregation of herbivores in response to herbivore-induced plant volatiles are very similar to functional questions on aggregation pheromones that are produced by herbivorous insects themselves.

Some empirical studies are now available on how herbivores respond to induced plant volatiles (Table 16.2). The majority of these studies shows that herbivores are attracted to odours from plants infested by conspecifics or even heterospecifics. It is too early to draw conclusions on the degree of differentiation of herbivore responses to plants with different herbivore densities, since all but one study tested only a single herbivore density. The two-spotted spider mite was attracted to bean leaves with a low density of conspecifics and deterred by bean leaves with a high density of conspecifics (Dicke 1986). One of the plant volatiles emitted from spider-mite-infested bean leaves, which is known to attract a predatory mite (Dicke *et al.* 1990a), has been reported elsewhere to deter the spider mite (Dabrowski & Rodriguez 1971).

The responses of herbivores to volatiles from infested plants comprise a recent research topic. Future studies should concentrate on the effect of herbivore density on herbivore response, to assess whether herbivores can discriminate between densities of competitors, and whether they are attracted or repelled.

Plant defence and competition among plants

To understand the evolution of plant defences, it is necessary to consider variation in plant defence between individuals and interactions among plant individuals that differ in defence strategies.

Plant genotypes vary in direct defence (Simms & Rausher 1989; Rausher 1992; Simms 1992; Vrieling *et al.* 1993), but also in indirect defence, with regard to herbivore-induced plant volatiles (Dicke *et al.* 1990b; Takabayashi *et al.* 1991; Loughrin *et al.* 1995a). The latter is a relatively little explored research area, although several reports on variation among genotypes and cultivars in the emission of volatiles, or the attraction of carnivores, are available. Apple cultivars differ qualitatively and quantitatively in the composition of the volatile blend induced by spider-mite feeding (Takabayashi *et al.* 1991), and naturalized cotton produces seven times more caterpillar-induced volatiles than agricultural cultivars (Loughrin *et al.* 1995a). This is an interesting observation, as it suggests that in natural systems herbivore-induced plant volatiles may be more abundant than in agricultural systems. It is likely that the trait has been lost in agricultural plants because there has been no selection for it. There is also behavioural evidence for differences among plant genotypes in the production of carnivore attractants. Two bean cultivars differed in the degree of attraction to predatory mites through spider-mite-induced volatiles (Dicke *et al.* 1990b). Thus, variation among plant genotypes in indirect defence has been documented.

Direct defence
Plant individuals compete with their neighbours for resources. However, defensive investments of an individual plant may profit its competing neighbours. Consequently, plants may have evolved to exploit the defence of their neighbours, instead of paying the costs of defence themselves. This could lead to polymorphism in defensive strategies. The question is, therefore, whether such polymorphism is evolutionarily stable. Augner (1995) explored this for direct plant defence strategies. Factors that are important in determining the predictions of the models are (i) the potential of herbivores to discriminate between plants that differ in defence intensity and (ii) the mobility of the herbivores, i.e. the chance that a herbivore moves from a plant to a neighbouring (i.e. competing) plant (Augner *et al.* 1991; Augner 1994; Tuomi *et al.* 1994).

Plants may signal their direct defence, e.g. through volatiles related to the defence. In doing so, they may envelope undefended neighbours with the signal, which can lead to associational defence because the presence of undefended plants is masked. Such associational defence of undefended plants, mediated by the defence-related signal of defended plants, can be essential for their stable coexistence (Augner 1994).

An important herbivore characteristic affecting the optimal strength of defence is mobility relative to the spatial structure of the plant population or community. A model study demonstrated that the greater the mobility of the herbivores, the weaker the optimal plant defence strength. The only situation that promoted lethal defence was when the herbivores were immobile (Tuomi *et al.* 1994). However, in the case of a mobile herbivore, a highly lethal defence will benefit the neighbours considerably.

Indirect defence

Also in the case of indirect defence, plants may benefit from the defensive activities of their neighbours. Carnivores attracted by herbivore-induced plant volatiles may also remove herbivores from neighbouring plants that do not signal. An evolutionarily stable strategy (ESS) approach demonstrated that there is a wide range of conditions for signalling plants to coexist with plants that spend their energy in alternative ways (Sabelis & De Jong 1988). The model used assumes (i) that the production of herbivore-induced volatiles bears costs due to biosynthesis of compounds that need to be replaced after emission and (ii) that neighbouring plants profit from the attracted carnivores (Sabelis & De Jong 1988). Godfray (1995b) also concludes that a condition for evolutionarily stable information transfer between plants and carnivores is that the volatiles are costly to produce. The cost level needed for evolutionary stability of plant–carnivore information transfer is dependent on the reliability of the volatiles for the carnivore and the benefit of carnivore attraction for the plant (Godfray 1995b).

What do we know about the biosynthetic costs of signalling? Calculations based upon emission rates yielded a value for biosynthetic costs of 0.001% of leaf production per day for spider-mite-induced volatile emissions by bean plants (Dicke & Sabelis 1989). This value may seem low, but it is likely to be an underestimate due to methods of determination of emission rate, and because costs of maintaining the biosynthetic pathway have not been included (Dicke & Sabelis 1989). Furthermore, other plant species emit much larger numbers and amounts of volatiles in response to herbivory, and so the cost value may differ significantly among plant species. Similar calculations for the volatiles emitted by cabbage plants infested by *Pieris brassicae* caterpillars (Mattiacci *et al.* 1994), or for the volatiles emitted by gerbera plants infested by *Tetranychus urticae* spider mites (O.E. Krips *et al.*, unpublished data), yield *c.* 100–150 times higher biosynthetic costs. Finally, even when biosynthetic costs are indeed low, this does not mean that they are evolutionarily insignificant, because of the effect of early costs on final biomass in exponentially growing plants (Gulmon & Mooney 1986), and because small but consistent fitness differences may still lead to gene replacement over evolutionarily short time scales (Dicke & Sabelis 1989). It is important to note that also in theoretical studies of the evolution of

504

direct defences, costs are a common and important assumption (Simms 1992; Augner 1995), although such costs have seldom been documented (Simms 1992). It is clear that biosynthetic costs of herbivore-induced plant volatiles are minor compared with ecological costs.

Do neighbours of the signalling plant benefit from carnivore attraction? This depends on the behavioural response of the carnivore. If plant volatiles attract carnivores to the particular site where the herbivore is feeding, the production of volatiles clearly benefits plants that can guide carnivores to their own damaged tissues, and selection will favour volatile production after attack. However, if plant volatiles are used by carnivores to locate a habitat, while herbivores are located in this habitat through other cues, such as odours of the herbivore's faeces, a plant may rely on its neighbours to attract carnivores to the habitat. Both types of carnivore responses have been recorded. For instance, the predatory mite *Phytoseiulus persimilis* locates prey patches through plant volatiles (Sabelis & Van de Baan 1983; Dicke *et al.* 1990a). Moreover, the predators do not emigrate from the prey patch as a result of their response to the plant odours (Sabelis *et al.* 1984) and this behaviour is essential for local prey extermination (Sabelis & Van der Meer 1986). In contrast, although the specialist parasitoid *Microplitis croceipes* is attracted to volatiles from host-damaged plants, within the habitat volatiles from its hosts' faeces are the major source of attractants (Eller *et al.* 1988).

Several factors may influence how carnivores respond to herbivore-induced plant volatiles. One factor is carnivore mobility. Carnivores that are highly mobile, such as parasitoids, can easily move between leaves and plants using host cues to locate individual herbivores. In contrast, wingless carnivores, such as predatory mites, are far less mobile and emigration to neighbouring plants is not common as long as prey remain available. Hence, for these mites herbivore-induced plant volatiles act as arrestants in addition to their function as attractants.

A second factor is the herbivore's spatial distribution. If the herbivore has a scattered distribution, herbivore-induced plant odours may be important to guide the carnivore from one individual herbivore to another. In contrast, if the herbivore has a clustered distribution, carnivores are likely to become arrested at the infested site, since subsequent prey can be found through area-restricted search after the first encounter.

A third factor is the degree of specialization of the carnivore. The more specialized the carnivore is (at the herbivore level), the more important herbivore products will be as foraging cues relative to plant volatiles (Vet &

Dicke 1992), decreasing the likelihood that neighbouring plants can exploit the signalling plant.

Food web complexity and plant–carnivore interactions

Simple linear relationships between the different trophic levels are an exception rather than a rule in nature (Polis & Strong 1996). In general, more than one herbivore will feed on a plant, and a single herbivore species is generally attacked by several carnivore species. Hence, information provided by plants can be used by several carnivores.

In a simple case of a linear 1–3-trophic-level interaction where the plant attracts the one carnivore that is regulating the abundance of the herbivore, the benefits for plants may be relatively straightforward. Attraction of the carnivore may have an immediate effect on herbivore abundance. When more than one carnivore species is attracted, such a regulating effect will be less straightforward, as complex interactions may positively or negatively affect herbivore densities (Polis et al. 1989; Polis & Holt 1992; Rosenheim et al. 1993, 1995). Plants may attract, for example, the natural enemies of carnivores, competing carnivores or intraguild predators (Janssen et al. 1998) (Fig. 16.4).

Whether and how carnivore interactions affect herbivore population densities is not only interesting from a fundamental ecological point of view. It has also received considerable attention from the applied field of biological control of insect pests. There have been several debates and theoretical (modelling) attempts to predict whether it would be better to introduce a single natural enemy species or several species to control a pest insect population, or which species would be the best in reducing pest densities (Kakehashi et al. 1984; Rosenheim et al. 1995; Murdoch & Briggs 1996, and references therein). For example, Briggs (1993) showed that a less effective parasitoid species can displace an existing parasitoid species, or that coexistence can reduce the effectiveness of the existing parasitoid. Murdoch and Briggs (1996) predict that when two parasitoid species attack different host stages, one parasitoid can gain competitive advantage by depressing the density of the stage attacked by its competitor. According to Murdoch and Briggs (1996) this does not necessarily imply that the superior competitor is the better one for pest control, since the pest life-stage whose density is minimized may not be the most economically damaging.

Another characteristic of intercarnivore interactions that can determine herbivore density is the degree of niche overlap between the competing

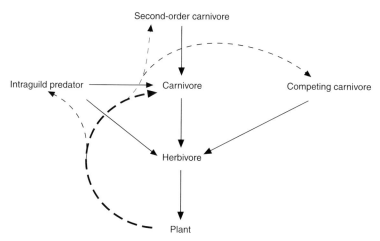

Figure 16.4 Herbivore-induced plant volatiles may attract carnivores at different positions in a food web. Solid lines indicate trophic interactions and broken lines indicate interactions mediated by herbivore-induced plant volatiles.

carnivore species. Models by Kakehashi *et al.* (1984) predict that a single species is better at reducing herbivore densities than two species when the niches of the two parasitoid species overlap to a great extent. Kakehashi *et al.* (1984) claim that the importance of competition was underestimated in earlier models that favoured multiple species introductions, such as those from May and Hassell (1981), because these models assume that the distribution of the attacks of each parasitoid species are independent of each other.

Differences in responses to information from the first trophic level may play a role in niche differentiation between carnivore species (Vet & Janse 1984). Firstly, carnivores may differ in their response level to quantities of plant odours through differences in sensory sensitivity. Bioassays show that parasitoid responses to infested host plants increase with host density (Geervliet *et al.* 1998a). Different parasitoid species may differ in this 'density response', and this may result in different subpopulations being found by two competitors. Secondly, carnivore species may have different preferences for odours from different plant parts or species, which can also lead to spatial segregation of attack of a shared herbivore prey. Comparative studies with ecologically related species are needed to assess to what extent the niches of (potentially) competing species are segregated through differential responses to plant cues.

Plant succession

The habitat in which a plant species normally occurs influences the complexity of the relevant food web. For instance, the number of (generalist) parasitoid species per herbivore species increases with succession (Price 1991). Furthermore, in later successional stages, plants are more predictably present and secondary plant chemicals in late successional plants are usually digestibility reducers rather than toxins. Consequently, a larger number of herbivore species is able to feed on late successional plants (Feeny 1976). Hence, the food web's complexity increases, which may reduce the importance of a specific plant–carnivore interaction. In early successional stages, carnivores have to track herbivores on ephemeral plants, while in late successional stages, spatial dynamics of the whole system are less fugitive. Hence, searching of herbivore-infested plants by carnivores is more important in early rather than late successional stages. Because in late successional stages carnivores are generalists, herbivore-induced plant volatiles are expected to be associatively learned rather than eliciting innate responses (Vet *et al.* 1995).

Early successional plants are characterized by defences through toxins, specialist herbivores that detoxify or sequester the plant toxins, and consequently associated with mostly specialist carnivores (Feeny 1976; Price 1991). Under precisely these ecological conditions, strong responses of carnivores to herbivore-induced plant volatiles are expected (Vet & Dicke 1992). Carnivores that function in early successional stages are highly dependent on reliable information on herbivore presence, as a result of the impermanent nature of herbivore distribution. Most research on herbivore-induced carnivore attraction is performed on cultivated plant species (Turlings *et al.* 1993; Dicke 1994, 1997; Takabayashi & Dicke 1996), which can be considered early successional plants. There is a need for comparison of the relative importance of herbivore-induced plant volatiles in different successional stages.

Can herbivores escape from plant–carnivore interactions?

The link between plant and carnivore is initiated by feeding activities of the herbivore itself. Herbivores must feed, but selection will act on their diet choices and feeding behaviour to reduce plant-produced information that reveals their presence. What can herbivores do to disconnect the plant–carnivore link?

Selection of a food plant also involves the selection of enemy-free or enemy-dense space. Therefore, plant selection by herbivores cannot be understood in a bitrophic context, as escape from carnivores can be a driving force (Price *et al.* 1980; Jeffries & Lawton 1984; Bernays & Graham 1988; Fox & Eisenbach 1992; Ohsaki & Sato 1994). Plant selection may involve, for example, the choice of plant species, plant genotypes, plants under different abiotic conditions, and plant parts. For instance, the butterfly *Pieris napi japonica* prefers to oviposit on *Arabis* spp. as these plants are of lower quality for larval development than alternative food plants. However, it was demonstrated that *Arabis* spp. were searched less readily by its major parasitoid, *Cotesia glomerata,* and thus *Arabis* spp. represent an enemy-free space (Ohsaki & Sato 1994).

The chances of a herbivore successfully escaping from carnivores through plant selection are best when there is a relative focus of the carnivores on plant stimuli over herbivore stimuli, i.e. for carnivores specialized at the plant level (Vet & Dicke 1992). A successful shift will then depend on the chemical relatedness of the induced odours of the two plant species, and the way carnivores use the odours. Insects such as bees are known to generalize between different classes of odours which they apparently interpret as chemically related (Getz & Smith 1990). However, as mentioned above ('Quantitative vs. qualitative differences'), little is known about how carnivorous arthropods perceive and interpret plant odour composition and its variation.

Plant species, genotypes and parts infested by the same herbivore differ in their attractiveness to carnivores. This, plus the fact that many carnivores learn to respond to plant volatiles during a successful attack, and subsequently focus on these learned cues, does create possibilities for a herbivore to find refuge on less attractive or less abundant host plants. In wind-tunnel experiments, naive *C. glomerata* parasitoids are much less attracted to *Tropaeolum majus* plants infested by *Pieris brassicae* caterpillars, than to *Brassica* spp. (Geervliet *et al.* 1996). However, this preference can easily be reversed when the parasitoids are given a preflight experience with host-infested *Tropaeolum* plants. In field experiments where a few host-infested *T. majus* plants were placed in a cabbage field, parasitization on *T. majus* plants was much lower than on host-infested cabbage plants placed in the same field (Geervliet *et al.* 1997c). A combination of low initial (innate) attraction to *T. majus* combined with a reinforcement of the preference for cabbage through learning creates a significant refuge for *P. brassicae* on *T. majus* when this plant is in the minority.

Herbivore density affects the amount of induced volatiles and thus the degree of carnivore attraction. If plants do not respond to a low herbivore load or if carnivores ignore the response of plants to a low herbivore density, such a density also reflects an enemy-free space. Hence, the evolution of herbivore clutch size may be influenced by plant–carnivore interactions.

The feeding behaviour of a herbivore will influence the induction of plant volatiles, e.g. a herbivore may chew the leaf petiole before consuming the detached leaf to sabotage induced defence (Weinstein 1990). Or herbivores may desert locations emitting volatiles (such as feeding holes) to continue feeding on a different part of the leaf (Mauricio & Bowers 1990). Some herbivore taxa may feed in a way that minimizes the damage inflicted on plant tissue. Aphids penetrate their stylets between parenchymal cells and feed from the phloem (Tjallingii & Hogen Esch 1993). However, aphids can still induce volatiles that attract their natural enemies (Du *et al.* 1996; Powell *et al.* 1998), although this has not been found in another plant–aphid system (Turlings *et al.* 1998). The elicitor of plant volatiles has been identified as a component of herbivore oral secretions (Mattiacci *et al.* 1995; Alborn *et al.* 1997). The quality and quantity of oral secretions will differ among herbivore species. Comparison of herbivores with different feeding mechanisms such as biting and sucking on the same plant species will elucidate its role in the induction of plant–carnivore interactions (cf. Turlings *et al.* 1998).

Conclusion

Food webs consist of many interrelated interactions among species. Information conveyance through chemicals is an important mediator of species interactions. So, superimposed on a food web, there is an information web, which is more complex than the food web itself, because information conveyance occurs irrespective of trophic relationships. To get a functional and mechanistic understanding of food webs and their information networks it is essential to start unravelling interactions among individual species. From here the interrelated species interactions should be investigated, but this is not an easy task. Because of the reticulate nature of most webs, integration cannot be reached by simply adding up individual components. Many new interactions appear by adding only one trophic interaction.

For pragmatic reasons, limitation of the research to key interactions is a necessity. But which criteria should be used to determine which interactions drive the system? This is more easy to assess in simple natural or agricultural systems, where species diversity is relatively low and regulating forces are more clear, e.g. in the case of a pest outbreak and its control by a natural enemy. By comparing different simple multitrophic systems, one may reach generalities with regards to the most important information flows. These generalities can propagate hypotheses that can be tested in more complex systems, where we expect the reticulate nature of the interactions to obscure regulating forces.

Acknowledgements

We thank Joop J.A. van Loon, Oliver Cheesman and an anonymous reviewer for comments that significantly improved the manuscript. MD is supported in part by the Uyttenboogaart-Eliasen Foundation, Amsterdam.

References

Agelopoulos, N.G. & Keller, M.A. (1994a). Plant–natural enemy association in the tritrophic system, *Cotesia rubecula–Pieris rapae*–Brassicaceae (Crucifera): I. Sources of infochemicals. *Journal of Chemical Ecology*, **20**, 1725–1734.

Agelopoulos, N.G. & Keller, M.A. (1994b). Plant–natural enemy association in the tritrophic system, *Cotesia rubecula–Pieris rapae*–Brassicaceae (Crucifera): III. Collection and identification of plant and frass volatiles. *Journal of Chemical Ecology*, **20**, 1955–1967.

Alborn, H.T., Turlings, T.C.J., Jones, T.H., Stenhagen, G., Loughrin, J.H. & Tumlinson, J.H. (1997). An elicitor of plant volatiles identified from beet armyworm oral secretion. *Science*, **276**, 945–949.

Augner, M. (1994). Should a plant always signal its defence against herbivores? *Oikos*, **70**, 322–332.

Augner, M. (1995). *Plant–plant interactions and the evolution of defences against herbivores.* PhD thesis, University of Lund, Sweden.

Augner, M., Fagerström, T. & Tuomi, J. (1991). Competition, defence and games between plants. *Behavioural Ecology and Sociobiology*, **29**, 231–234.

Bernays, E.A. & Chapman, R.F. (1994). *Host–Plant Selection by Phytophagous Insects.* Chapman & Hall, New York.

Bernays, E.A. & Graham, M. (1988). On the evolution of host specificity in phytophagous arthropods. *Ecology*, **69**, 886–892.

Blaakmeer, A., Geervliet, J.B.F., Van Loon, J.J.A., Posthumus, M.A., Van

Beek, T.A. & De Groot, A.E. (1994a). Comparative headspace analysis of cabbage plants damaged by two species of *Pieris* caterpillars: consequences for in-flight host location by *Cotesia* parasitoids. *Entomologia Experimentalis et Applicata*, 73, 175–182.

Blaakmeer, A., Hagenbeek, D., Van Beek, T.A., De Groot, A.E., Schoonhoven, L.M. & Van Loon, J.J.A. (1994b). Plant response to eggs vs. host marking pheromone as factors inhibiting oviposition by *Pieris brassicae*. *Journal of Chemical Ecology*, 20, 1657–1665.

Bolter, C.J., Dicke, M., Van Loon, J.J.A., Visser, J.H. & Posthumus, M.A. (1997). Attraction of Colorado potato beetle to herbivore-damaged plants during herbivory and after its termination. *Journal of Chemical Ecology*, 23, 1003–1023.

Briggs, C.J. (1993). Competition among parasitoid species on an age-structured host, and its effect on host suppression. *American Naturalist*, 141, 372–397.

Dabrowski, Z.T. & Rodriguez, J.G. (1971). Studies on resistance of strawberries to mites. 3. Preference and nonpreference responses of *Tetranychus urticae* and *T. turkestani* to essential oils of foliage. *Journal of Economic Entomology*, 64, 387–391.

Dicke, M. (1986). Volatile spider-mite pheromone and host-plant kairomone, involved in spaced-out gregariousness in the spider mite *Tetranychus urticae*. *Physiological Entomology*, 11, 251–262.

Dicke, M. (1994). Local and systemic production of volatile herbivore-induced terpenoids: their role in plant-carnivore mutualism. *Journal of Plant Physiology*, 143, 465–472.

Dicke, M. (1998). Evolution of indirect defence of plants. In *The Ecology and Evolution of Inducible Defences* (Ed. by C.D Harvell & R. Tollrian), pp. 62–88. Princeton University Press, Princeton.

Dicke, M. & Groeneveld, A. (1986). Hierarchical structure in kairomone preference of the predatory mite *Amblyseius potentillae*: dietary component indispensable for diapause induction affects prey location behaviour. *Ecological Entomology*, 11, 131–138.

Dicke, M. & Sabelis, M.W. (1988). How plants obtain predatory mites as bodyguards. *Netherlands Journal of Zoology*, 38, 148–165.

Dicke, M. & Sabelis, M.W. (1989). Does it pay plants to advertize for bodyguards? Towards a cost–benefit analysis of induced synomone production. In *Causes and Consequences of Variation in Growth Rate and Productivity of Higher Plants* (Ed. by H. Lambers, M.L. Cambridge, H. Konings & T.L. Pons), pp. 341–358. SPB Academic Publishing, The Hague.

Dicke, M. & Sabelis, M.W. (1992). Costs and benefits of chemical information conveyance: proximate and ultimate factors. In *Insect*

Chemical Ecology: An Evolutionary Approach (Ed. by B.D. Roitberg & M.B. Isman), pp. 122–155. Chapman & Hall, New York.

Dicke, M., Van Beek, T.A., Posthumus, M.A., Ben Dom, N., Van Bokhoven, H. & De Groot, A.E. (1990a). Isolation and identification of volatile kairomone that affects acarine predator–prey interactions. Involvement of host plant in its production. *Journal of Chemical Ecology*, **16**, 381–396.

Dicke, M., Sabelis, M.W., Takabayashi, J., Bruin, J. & Posthumus, M.A. (1990b). Plant strategies of manipulating predator–prey interactions through allelochemicals: prospects for application in pest control. *Journal of Chemical Ecology*, **16**, 3091–3118.

Du, Y.-J., Poppy, G.M. & Powell, W. (1996). Relative importance of semiochemicals from first and second trophic levels in host foraging behavior of *Aphidius ervi*. *Journal of Chemical Ecology*, **22**, 1591–1605.

Eller, F.J., Tumlinson, J.H. & Lewis, W.J. (1988). Beneficial arthropod behavior mediated by airborne semiochemicals. VII. Source of volatiles mediating the host-location flight behavior of *Microplitis croceipes* (Cresson) (Hymenoptera: Braconidae), a parasitoid of *Heliothis zea* (Boddie) (Lepidoptera: Noctuidae). *Environmental Entomology*, **17**, 745–753.

Elzen, G.W., Williams, H.J., Bell, A.A., Stipanovic, R.D. & Vinson, S.B. (1985). Quantification of volatile terpenes of glanded and glandless *Gossypium hirsutum* L. cultivars and lines by gas chromatography. *Journal of Agricultural and Food Chemistry*, **33**, 1079–1082.

Feeny, P. (1976). Plant apparency and chemical defense. *Recent Advances in Phytochemistry*, **10**, 1–40.

Finidori-Logli, V., Bagneres, A.G. & Clement, J.L. (1996). Role of plant volatiles in the search for a host by parasitoid *Diglyphus isaea* (Hymenoptera: Eulophidae). *Journal of Chemical Ecology*, **22**, 541–558.

Fox, L.R. & Eisenbach, J. (1992). Contrary choices: possible exploitation of enemy-free space by herbivorous insects in cultivated vs. wild crucifers. *Oecologia*, **89**, 574–579.

Gange, A.C. & V.K. Brown (Eds) (1997). *Multitrophic Interactions in Terrestrial Systems*. Blackwell Science Ltd, Oxford.

Geervliet, J.B.F., Van Aaken, R., Savelkoul, C., ter Smitte, S.M., Brodeur, J., Vet, L.E.M. & Dicke, M. (1993). Comparative approach to infochemical use by parasitoids for the case of *Cotesia glomerata* and *Cotesia rubecula*. *Proceedings of Experimental and Applied Entomology*, **4**, 33–38.

Geervliet, J.B.F., Vet, L.E.M. & Dicke, M. (1996). Innate responses of the parasitoids *Cotesia glomerata* and *C. rubecula* (Hymenoptera: Braconidae) to volatiles from different plant–herbivore complexes. *Journal of Insect Behavior*, **9**, 525–538.

Geervliet, J.B.F., Posthumus, M.A., Vet, L.E.M. & Dicke, M. (1997a). Comparative analysis of headspace from different caterpillar-infested or uninfested food plants of *Pieris* species. *Journal of Chemical Ecology*, 23, 2935–2954.

Geervliet, J.B.F., Snellen, H., Vet, L.E.M. & Dicke, M. (1997b). Host-location behaviour of the larval parasitoids *Cotesia glomerata* and *C. rubecula* (Hymenoptera: Braconidae): the effect of host and plant species. In *Infochemical Use by Insect Parasitoids in a Tritrophic Context: Comparison of a Generalist and a Specialist* (Ed. by J.B.F. Geervliet), pp. 95–110. PhD thesis, Wageningen Agricultural University, The Netherlands.

Geervliet, J.B.F., Verdel, M.S.W., Schaub, J., Snellen, H., Dicke, M. & Vet, L.E.M. (1997c). Coexistence and niche segregation by field populations of the parasitoids. *Cotesia glomerata* and *C. rubecula* in the Netherlands: analysis of a parasitization and parasitoid behaviour. In *Infochemical Use by Insect Parasitoids in a Tritrophic Context: Comparison of a Generalist and a Specialist* (Ed. by J.B.F. Geervliet), pp. 141–158. PhD thesis, Wageningen Agricultural University, The Netherlands.

Geervliet, J.B.F., Ariëns, S.J., Dicke, M. & Vet, L.E.M. (1998a). Long-distance assessment of patch profitability through volatile infochemicals by the parasitoids *Cotesia glomerata* and *C. rubecula* (Hymenoptera: Braconidae). *Biological Control*, 11, 113–121

Geervliet, J.B.F., Vreugdenhil, A.I., Dicke, M. & Vet, L.E.M. (1998b). Learning to discriminate between infochemicals from different plant–host complexes by the parasitoids *Cotesia glomerata* and *C. rubecula* (Hymenoptera: Braconidae). *Entomologia Experimentalis et Applicata*, 86, 241–252.

Getz, W.M. & Smith, K.B. (1990). Odorant moiety and odor mixture perception in free-flying honey bees (*Apis mellifera*). *Chemical Senses*, 15, 111–128.

Godfray, H.C.J. (1995a). Signaling of need between parents and young: parent–offspring conflict and sibling rivalry. *American Naturalist*, 146, 1–24.

Godfray, H.C.J. (1995b). Communication between the first and third trophic levels: an analysis using biological signalling theory. *Oikos*, 72, 367–374.

Gulmon, S.L. & Mooney, H.A. (1986). Costs of defense and their effects on plant productivity. In *On the Economy of Plant Form and Function* (Ed. by T.J. Givnish), pp. 681–699. Cambridge University Press, Cambridge.

Harari, A.R., Ben-Yakir, D. & Rosen, D. (1994). Mechanism of aggregation behavior in *Maladera matrida* Argaman (Coleoptera: Scarabaeidae). *Journal of Chemical Ecology*, 20, 361–371.

Hardie, J., Hick, A.J., Holler, C., Mann, J., Merritt, L., Nottingham, S.F., Powell, W., Wadhams, L.J., Witthinrich, J. & Wright, A.F. (1994). The responses of *Praon* spp. parasitoids to aphid sex pheromone components in the field. *Entomologia Experimentalis et Applicata*, **71**, 95–99.

Hunter, M.D. & Price, P.W. (1992). Playing chutes and ladders: heterogeneity and the relative roles of bottom-up and top-down forces in natural communities. *Ecology*, **73**, 724–732.

Hunter, M.D., Ohgushi T. & Price, P.W. (Eds) (1992). *Effects of Resource Distribution on Animal–Plant Interactions*. Academic Press, New York.

Janssen, A., Pallini, A., Venzon, M. & Sabelis, M.W. (1998). Behaviour and indirect food web interactions among plant inhabiting mites. *Experimental and Applied Acarology* (in press).

Jeffries, M.J. & Lawton, J.H. (1984). Enemy free space and the structure of ecological communities. *Biological Journal of the Linnean Society*, **23**, 269–286.

Kakehashi, M., Suzuki, Y. & Iwasa, Y. (1984). Niche overlap of parasitoids in host–parasitoid systems: its consequence to single versus multiple introduction controversy in biological control. *Journal of Applied Ecology*, **21**, 115–131.

Koptur, S. (1992). Extrafloral nectary-mediated interactions between insects and plants. In *Insect–Plant Interactions*, Vol. IV (Ed. by E.A. Bernays), pp. 81–129. CRC Press, Boca Raton.

Landolt, P.J. (1993). Effects of host plant leaf damage on cabbage looper moth attraction and oviposition. *Entomologia Experimentalis et Applicata*, **67**, 79–85.

Lewis, W.J., Nordlund, D.A., Gueldner, R.C., Teal, P.E.A. & Tumlinson, J.H. (1982). Kairomones and their use for management of entomophagous insects. XIII. Kairomonal activity for *Trichogramma* spp. of abdominal tips, excretion, and a synthetic sex pheromone blend of *Heliothis zea* (Boddie) moths. *Journal of Chemical Ecology*, **8**, 1323–1331.

Loughrin, J.H., Manukian, A., Heath, R.R. & Tumlinson, J.H. (1995a). Volatiles emitted by different cotton varieties damaged by feeding beet armyworm larvae. *Journal of Chemical Ecology*, **21**, 1217–1227.

Loughrin, J.H., Potter, D.A. & Hamilton-Kemp, T.R. (1995b). Volatile compounds induced by herbivory act as aggregation kairomones for the Japanese beetle (*Popillia japonica* Newman). *Journal of Chemical Ecology*, **21**, 1457–1467.

Mattiacci, L., Dicke, M. & Posthumus, M.A. (1994). Induction of parasitoid attracting synomone in brussels sprouts plants by feeding of *Pieris brassicae* larvae: role of mechanical damage and herbivore elicitor. *Journal of Chemical Ecology*, **20**, 2229–2247.

Mattiacci, L., Dicke, M. & Posthumus, M.A. (1995). β-Glucosidase: an elicitor of herbivore-induced plant odor that attracts host-searching parasitic wasps. *Proceedings of the National Academy of Sciences, USA*, **92**, 2036–2040.

Mauricio, R. & Bowers, M.D. (1990). Do caterpillars disperse their damage?: larval foraging behaviour of two specialist herbivores, *Euphydras phaeton* (Nymphalidae) and *Pieris rapae* (Pieridae). *Ecological Entomology*, 15, 153–161.

May, R.M. & Hassell, M.P. (1981). The dynamics of multiparasitoid–host interactions. *American Naturalist*, 117, 234–261.

McCall, P.J., Turlings, T.C.J., Loughrin, J., Proveaux, A.T. & Tumlinson, J.H. (1994). Herbivore-induced volatile emissions from cotton (*Gossypium hirsutum* L.) seedlings. *Journal of Chemical Ecology*, 20, 3039–3050.

Murdoch, W.W. & Briggs, C.J. (1996). Theory for biological control: recent developments. *Ecology*, 77, 2001–2013.

Noldus, L.P.J.J., Potting, R.P.J. & Barendregt, H.E. (1991). Moth sex pheromone adsorption to leaf surface: bridge in time for chemical spies. *Physiological Entomology*, 16, 329–344.

Ohsaki, N. & Sato, Y. (1994). Food plant choice of *Pieris* butterflies as a trade-off between parasitoid avoidance and quality of plants. *Ecology*, 75, 59–68.

Pallini, A., Janssen, A. & Sabelis, M.W. (1997). Odour-mediated responses of phytophagous mites to conspecific and heterospecific competitors. *Oecologia*, 100, 179–185.

Pavlov, I.P. (1927). *Conditioned Reflexes*. Oxford University Press, Oxford.

Perry, N.B., Baxter, A.J., Brennan, N.J., van Klink, J.W., McGimpsey, J.A., Douglas, M.H. & Joulain, D. (1996). Dalmatian sage. Part 1. Differing oil yields and compositions from flowering and non-flowering accessions. *Flavor and Fragrances Journal*, 11, 231–238.

Polis, G.A. & Holt, R.D. (1992). Intraguild predation: The dynamics of complex trophic interactions. *Trends in Ecology and Evolution*, 7, 151–154.

Polis, G.A. & Strong, D.R. (1996). Food web complexity and community dynamics. *American Naturalist*, 147, 813–846.

Polis, G.A., Myers, C.A. & Holt, R.D. (1989). The ecology and evolution of intraguild predation: potential competitors that eat each other. *Annual Review of Ecology and Systematics*, 20, 297–330.

Powell, W., Pennacchio, F., Poppy, G.M. & Tremblay, E. (1998). Strategies involved in the location of hosts by the parasitoid *Aphidius ervi* Haliday (Hymenoptera: Braconidae, Aphidiinae). *Biological Control*, 11, 104–112.

Price, P.W. (1991). Evolutionary theory of host and parasitoid interactions. *Biological Control*, 1, 83–93.

Price, P.W., Bouton, C.E., Gross, P., McPheron, B.A., Thompson, J.N. & Weis, A.E. (1980). Interactions among three trophic levels: influence of plant on interactions between insect herbivores and natural enemies. *Annual Review of Ecology and Systematics*, 11, 41–65.

Rausher, M.D. (1992). Natural selection and the evolution of plant–insect interactions. In *Insect Chemical Ecology. An Evolutionary Approach* (Ed. by B.D. Roitberg & M.B. Isman), pp. 20–88. Chapman & Hall, New York.

Rosenheim, J.A., Wilhoit, L.R. & Armer, C.A. (1993). Influence of intraguild predation among generalist insect predators on the suppression of an herbivore population. *Oecologia*, **96**, 439–449.

Rosenheim, J.A., Kaya, H.K., Ehler, L.E., Marois, J.J. & Jaffee, B.A. (1995). Intraguild predation among biological-control agents: theory and evidence. *Biological Control*, **5**, 303–335.

Sabelis, M.W. & De Jong, M.C.M. (1988). Should all plants recruit bodyguards? Conditions for a polymorphic ESS of synomone production in plants. *Oikos*, **53**, 247–252.

Sabelis, M.W. & Dicke, M. (1985). Long-range dispersal and searching behaviour. In *Spider Mites. Their Biology, Natural Enemies and Control* (Ed. by W. Helle & M.W. Sabelis), pp. 141–160. World Crop Pests, Vol. 1B. Elsevier, Amsterdam.

Sabelis, M.W. & Van de Baan, H.E. (1983). Location of distant spider mite colonies by phytoseiid predators: demonstration of specific kairomones emitted by *Tetranychus urticae* and *Panonychus ulmi*. *Entomologia Experimentalis et Applicata*, **33**, 303–314.

Sabelis, M.W. & Van der Meer, J. (1986). Local dynamics of the interaction between predatory mites and two-spotted spider mites. In *Dynamics of Physiologically Structured Populations* (Ed. by J.A.J. Metz & O. Diekman), pp. 322–343. Springer Lecture Notes in Biomathematics, Vol. 68. Springer, New York.

Sabelis, M.W., Vermaat, J.E. & Groeneveld, A. (1984). Arrestment responses of the predatory mite, *Phytoseiulus persimilis*, to steep odour gradients of a kairomone. *Physiological Entomology*, **9**, 437–446.

Schoonhoven, L.M., Jermy, T. & Van Loon, J.J.A. (1998). *Insect–Plant Biology. From Physiology to Evolution*. Chapman & Hall, London.

Shimoda, T., Takabayashi, J., Ashihara, W. & Takafuji, A. (1997). Response of the predatory insect *Scolothrips takahashii* toward herbivore-induced plant synomone under both laboratory and field conditions. *Journal of Chemical Ecology*, **23**, 2033–2048.

Simms, E.L. (1992). Costs of plant resistance to herbivory. In *Plant Resistance to Herbivores and Pathogens. Ecology, Evolution and Genetics* (Ed. by R.S. Fritz & E.L. Simms), pp. 392–425. University of Chicago Press, Chicago.

Simms, E.L. & Rausher, M.D. (1989). The evolution of resistance to herbivory in *Ipomoea purpurea*. II. Natural selection by insects and costs of resistance. *Evolution*, **43**, 573–585.

Slansky, F., Jr. (1986). Nutritional ecology of endoparasitic insects and their hosts: an overview. *Journal of Insect Physiology*, **32**, 255–261.

Steinberg, S., Dicke, M. & Vet, L.E.M. (1993). Relative importance of infochemicals from first and second trophic level in long-range host location by the larval parasitoid *Cotesia glomerata*. *Journal of Chemical Ecology*, **19**, 47–59.

Stephens, D.W. & Krebs, J.R. (1986). *Foraging Theory*. Princeton University Press, Princeton.

Strong, D.R., Lawton, J.H. & Southwood, T.R.E. (1984). *Insects on Plants: Community Patterns and Mechanisms*. Harvard University Press, Cambridge.

Takabayashi, J. & Dicke, M. (1996). Plant–carnivore mutualism through herbivore-induced carnivore attractants. *Trends in Plant Science*, **1**, 109–113.

Takabayashi, J., Dicke, M. & Posthumus, M.A. (1991). Variation in composition of predator-attracting allelochemicals emitted by herbivore-infested plants: relative influence of plant and herbivore. *Chemoecology*, **2**, 1–6.

Takabayashi, J., Dicke, M., Takahashi, S., Posthumus, M.A. & van Beek, T.A. (1994a). Leaf age affects composition of herbivore-induced synomones and attraction of predatory mites. *Journal of Chemical Ecology*, **20**, 373–386.

Takabayashi, J., Dicke, M. & Posthumus, M.A. (1994b). Volatile herbivore-induced terpenoids in plant–mite interactions: variation caused by biotic and abiotic factors. *Journal of Chemical Ecology*, **20**, 1329–1354.

Takabayashi, J., Takahashi, S., Dicke, M. & Posthumus, M.A. (1995). Developmental stage of herbivore *Pseudaletia separata* affects production of herbivore-induced synomone by corn plants. *Journal of Chemical Ecology*, **21**, 273–287.

Takabayashi, J., Sato, Y., Horikoshi, M., Yamaoka, R., Yano, S., Ohsaki, N. & Dicke, M. (1998). New aspects on the solution of reliability–detectability problems by parasitic wasps with volatiles from plants infested by their hosts. *Biological Control*, **11**, 97–103.

Tjallingii, W.F. & Hogen Esch, T. (1993). Fine structure of aphid stylet routes in plant tissues in correlation with EPG signals. *Physiological Entomology*, **18**, 317–328.

Tollsten, L. & Bergstrom, L.G. (1993). Fragrance chemotypes of *Platanthera* (Orchidaceae) — the result of adaptation to pollinating moths? *Nordic Journal of Botany*, **13**, 607–613.

Tollsten, L., Knudsen, J.T. & Bergstrom, L.G. (1994). Floral scent in generalistic *Angelica* (Apiaceae) — an adaptive character? *Biochemical and Systematic Ecology*, **22**, 161–169.

Tumlinson, J.H., Lewis, W.J. & Vet, L.E.M. (1993). How parasitic wasps find their hosts. *Scientific American*, **268**, 100–106.

Tuomi, J., Augner, M. & Nilsson, P. (1994). A dilemma of plant defences: is it really worth killing the herbivore? *Journal of Theoretical Biology*, **170**, 427–430.

Turlings, T.C.J., Tumlinson, J.H. & Lewis, W.J. (1990). Exploitation of herbivore-induced plant odors by host-seeking parasitic wasps. *Science*, **250**, 1251–1253.

Turlings, T.C.J., Tumlinson, J.H., Eller, F.J. & Lewis, W.J. (1991). Larval-damaged plants: source of volatile synomones that guide the parasitoid *Cotesia marginiventris* to the micro-habitat of its hosts. *Entomologia Experimentalis et Applicata*, **58**, 75–82.

Turlings, T.C.J., Wäckers, F.L., Vet, L.E.M., Lewis, W.J. & Tumlinson, J.H. (1993). Learning of host-finding cues by Hymenopterous parasitoids. In *Insect Learning* (Ed. by D.R. Papaj & A.C. Lewis), pp. 51–78. Chapman & Hall, New York.

Turlings, T.C.J., Loughrin, J.H., McCall, P.J., Röse, U.S.R., Lewis, W.J. & Tumlinson. J.H. (1995). How caterpillar-damaged plants protect themselves by attracting parasitic wasps. *Proceedings of the National Academy of Sciences, USA*, **92**, 4169–4174.

Turlings, T.C.J., Bernasconi, M., Bertossa, R. & Caloz, G. (1998). The induction of volatile emissions in maize by three herbivore species with different feeding habits: possible consequences for their natural enemies. *Biological Control*, **11**, 122–129.

Van der Meijden, E. (1996). Plant defence, an evolutionary dilemma: contrasting effects of (specialist and generalist) herbivores and natural enemies. *Entomologia Experimentalis et Applicata*, **80**, 307–310.

Vet, L.E.M. & Dicke, M. (1992). Ecology of infochemical use by natural enemies in a tritrophic context. *Annual Review of Entomology*, **37**, 141–172.

Vet, L.E.M. & Janse, C.J. (1984). Fitness of two sibling species of *Asobara* (Braconidae: Alysiinae), larval parasitoids of Drosophilidae in different microhabitats. *Ecological Entomology*, **9**, 345–354.

Vet, L.E.M., Lewis, W.J., Papaj, D.R. & Van Lenteren, J.C. (1990). A variable-response model for parasitoid foraging behavior. *Journal of Insect Behavior*, **3**, 471–490.

Vet, L.E.M., Wäckers, F.L. & Dicke, M. (1991). How to hunt for hiding hosts: the reliability–detectability problem in foraging parasitoids. *Netherlands Journal of Zoology*, **41**, 202–213.

Vet, L.E.M., Lewis, W.J. & Carde, R.T. (1995). Parasitoid foraging and learning. In *Chemical Ecology of Insects 2* (Ed. by R.T. Carde & W.J. Bell), pp. 65–101. Chapman & Hall, New York.

Vet, L.E.M., De Jong, A.G., Franchi, E. & Papaj, D.R. (1998). The effect of complete vs. incomplete information on odour discrimination in a parasitic wasp. *Animal Behaviour*, **55**, 1271–1279.

Visser, J.H. (1986). Host odor perception in phytophagous insects. *Annual Review of Entomology*, **31**, 121–144.

Vrieling, K., De Vos, H. & Van Wijk, C.A.M. (1993). Genetic analysis of the concentration of pyrrolizidine alkaloids in *Senecio jacobaea*. *Phytochemistry*, **32**, 1141–1144.

Walter, D.E. (1996). Living on leaves: mites, tomenta, and leaf domatia. *Annual Review of Entomology*, **41**, 101–114.

Weinstein, P. (1990). Leaf petiole chewing and the sabotage of induced defences. *Oikos*, **58**, 231–233.

Wilson, E.O. (1992). *The Diversity of Life*. Harvard University Press.

Wiskerke, J.S.C., Dicke, M. & Vet, L.E.M. (1993). Larval parasitoid uses aggregation pheromone of adult hosts in foraging behaviour: a solution to the reliability–detectability problem. *Oecologia*, **93**, 145–148.

Chapter 17

Dynamics of consumer–resource interactions: importance of individual attributes

W.W. Murdoch,[1] C.J. Briggs[2] and R.M. Nisbet[1]

Summary

We focus on models of interacting populations of consumers and resources, particularly insect parasites (parasitoids) and hosts that incorporate individual attributes. Differences among individuals can have profound effects on population dynamics. In parasitoid–host models, invulnerable classes are stabilizing, and a short host adult stage can induce single-generation cycles. In many parasitoids the gain to the future female parasitoid population increases with the age of the encountered host, and this can both stabilize the inherent 'predator–prey' cycles and induce delayed feedback cycles. Models of predators and prey also show invulnerable stages are stabilizing, but little has been done on other age-related differences. Age structure in the consumer population is typically a source of time lags and hence is destabilizing. We know little, however, about the dynamical effects of other age-related properties, especially of predators. Juvenile and adult predators often feed on different resources and this has hardly been investigated.

We also investigate dynamical effects of individual properties of the resource population that are unrelated to age and that affect its vulnerability to attack by the consumer, e.g. vulnerability related to spatial position. Insight is gained by considering whether the effect on vulnerability is transient or persistent relative to the system's time scale. Persistent, but not transient, differences tend to have substantial dynamical effects.

1 Department of Ecology, Evolution and Marine Biology, University of California, Santa Barbara, CA 93106, USA
2 Department of Integrative Biology, University of California, Berkeley, CA 94720, USA

Finally, although much can be learned from models that recognize differences among individuals, we argue that simple models remain useful.

Introduction

In this chapter we explore differences among individuals and their potential effects on the dynamics of interacting populations of consumers and resources. The symposium organizers asked that we compare parasitoids and predators and this is one of our themes, although we make no claim to an exhaustive survey. We also discuss one type of herbivore–plant interaction.

We separate our discussion of individual properties into two broad classes. In the first section we consider differences that arise as a result of the individual's ageing and hence passing through different development stages and/or getting larger. In the second section we consider differences that arise from other sources, including spatial location and physiological state. In each section we deal first with properties of the resource population, host or prey, and then with those of the consumer, predator or parasitoid. Although our focus is individual attributes, which usually implies complex models, we also argue for the importance of simple models in ecology.

Population dynamics theory should be able to account for the range of dynamics seen in real populations, e.g. stability, and cycles of various amplitudes and periods. Virtually every species is either a consumer, a resource or both, and theory suggests that much of the observed dynamics results from consumer–resource interactions. The theory discussed here represents attempts to provide at least plausible explanations for this range of dynamics.

For the most part, our concern is not to compare model output with the dynamics of particular populations. We concentrate instead on qualitative dynamics–mainly whether the model has a stable equilibrium or limit cycles of one sort or another. We aim to give an overview of the effects of different classes of individual attributes on these aspects of dynamics.

Effects of age/stage/size

The effects of age that matter to dynamics are frequently associated with developmental stage or size. All three are typically correlated, even if in some cases quite imperfectly, and until we need to distinguish among them we will refer only to age. We begin with age-related differences among hosts and prey, then consider predator and parasitoid age. However, since, the effects of age have been much more thoroughly analysed for insect host–parasitoid than in prey–predator systems, we deal separately with host and prey age.

Effects of host age/stage/size in insect host–parasitoid systems

Variation among host individuals of different ages has been shown in models to have potentially profound effects on dynamics, and we summarize these findings below. First, we provide some essential natural history.

Adult female parasitoids lay an egg or eggs on or in a host individual, and the immatures develop by feeding on that single host individual. The adult parasitoid typically feeds on nectar or some other source of energy. Pro-ovigenic parasitoid species emerge as adults with their lifetime complement of eggs. Other species feed as adults on some host individuals (usually of the species that is parasitized but not usually on the same individual) and use the nutrients to mature new eggs and possibly also for maintenance (Heimpel & Collier 1996).

Host properties typically change with age. Individuals usually increase in size as they age and may also change their form and habits. Thus, age tends to be correlated with size and/or developmental stage. In the models discussed in this section hosts usually are classified by stage, which is formally a discrete age class but is typically considered to be a developmental stage or size range that is treated by the parasitoid in a distinctive way. Age and size or stage will be most nearly synonymous in circumstances where the parasitoid suppresses the host well below the density at which host resources limit its development or growth rate.

Searching adult female parasitoids often respond differently to the different host stages. Depending on the stage encountered, the parasitoid may: (i) ignore or be unable to attack the host; (ii) lay a fixed or variable number of eggs with a fixed or variable sex ratio; or (iii) (in some species) may host feed. These responses may be fixed or they may be flexible; in particular, the propensity to attack or to oviposit rather than host feed, and the number of eggs laid, may depend on the number of mature eggs carried by the searching female (Minkenberg *et al.* 1992). Thus, parasitoid response may depend on the host state, the parasitoid state or both.

Invulnerable host classes

The most basic age-related difference is between immatures and adults,

523

and in parasitoid–host interactions the adult host is typically invulnerable. Surprisingly, a model with only these differences between hosts, plus an immature parasitoid stage (Fig. 17.1a), contains most of the dynamic

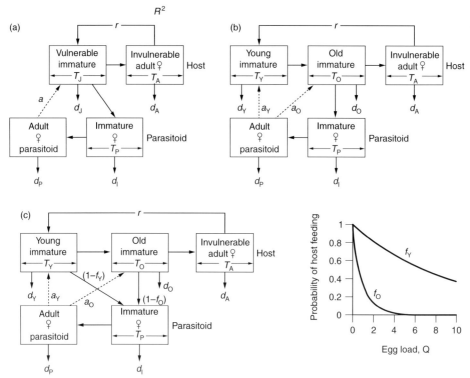

Figure 17.1 Structure of models described in text. (a) Basic model (from Murdoch *et al.* 1987): single immature host stage on which the parasitoid oviposits only. (b) Host-size-dependent sex-allocation and host-feeding model (from Murdoch *et al.* 1992): two immature host stages, small (young) stage is host fed or receives a male egg, large (old) stage receives a female egg. (c) Parasitoid state- and host-size-dependent sex-allocation and host-feeding model (from Murdoch *et al.* 1997): the probability of host-feeding or ovipositing on each of the two immature host stages depends on the parasitoid egg load in the manner shown in the graph on the right. In all models, pre-adult host and parasitoid stages have fixed durations, and *per capita* death rates that are independent of age. Adult hosts and parasitoids have constant, age-independent *per capita* death rates, d_P and d_A, respectively, and have no upper age limit. Thus, the mean durations for these stages (shown in the figures) are $T_P = d_P^{-1}$ and $T_A = d_A^{-1}$. Attack rates on host stage i is denoted by a_i. All symbols introduced in (a), (b) or (c) are defined in Table 17.1.

range seen in more complex models. We refer to this as the 'basic model' hereafter.

The model describes insects in which births and deaths are continuous (Table 17.1), i.e. there is no discrete breeding season, and there is the potential for all life stages to overlap. The juvenile development period of the host, T_J, is the basic time scale, and all other time periods and instantaneous rates are scaled to it. Parasitoids have a type 1 functional response (Table 17.1).

The stability properties of the basic model are plotted in Fig. 17.2 as a function of the duration of the invulnerable adult stage and the length of the parasitoid's developmental delay, both in relation to the duration of the immature host stage. We can think of any particular parasitoid–host

Table 17.1 Basic model with invulnerable adult stage.

Variables

$J(t)$	Juvenile host density
$A(t)$	Adult host density
$P(t)$	Adult parasitoid density
$M_J(t)$	Maturation rate out of juvenile host stage

Parameters

T_J	Duration of juvenile host stage
T_P	Duration of juvenile parasitoid stage
T_A	Average duration of adult host stage (= $1/d_A$)
r	Host rate of increase
a	Parasitoid attack rate
d_J	Juvenile host death rate
d_A	Adult host death rate
d_P	Adult parasitoid death rate
S_I	Juvenile parasitoid through-stage survival probability

Equations

$dJ(t)/dt = (r/T_A) A(t) - M_J(t) - aP(t) J(t) - d_J J(t)$	Juvenile hosts
$dA(t)/dt = M_J(t) - d_A A(t)$	Adult hosts
$dP(t)/dt = aJ(t - T_P) P(t - T_P)s_I - d_P P(t)$	Adult parasitoids

$$M_J(t) = (r/T_A) A(t - T_J)\exp\left\{-\int_{t-T_J}^{t} aP(x) + d_J dx\right\}$$ Host maturation rate

Equilibria

$J^* = d_P/(as_I)$	Juvenile hosts
$P^* = [\ln(r) - d_J T_J]/(aT_J)$	Adult parasitoids
$A^* = J^*(aP^* + d_J)/[d_A(r-1)]$	Adult hosts

interaction as a point on the graph. For example, interactions in which the host has a long-lived adult, such as beetles or locusts, would fall to the right of the graph, while those with hosts like mayflies and moths would be close to the *y* axis. At each point, the figure tells us whether the equilibrium is stable or unstable for a defined set of parameter values.

A nice property of the basic model is that the equilibrium at the origin in Fig. 17.2a, where there is no maturation delay in the parasitoid and the infinitesimally short adult host stage produces all its young in a single pulse, is neutrally stable like that of the simplest neutrally stable

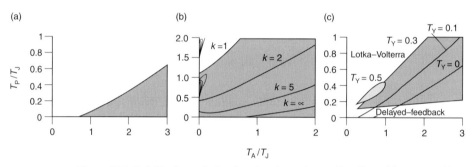

Figure 17.2 Stability boundaries showing the regions of locally stable or unstable equilibria in terms of the duration of the adult host stage (T_A) and the duration of the juvenile parasitoid stage (T_P), relative to the host development time, T_J. In each case, the equilibria are locally stable in the shaded region or to the right of the boundary. Boundaries are shown for: (a) the basic model (from Murdoch *et al.* 1987), $T_J = 1$, $r = 33$, $d_P = 8$, $a = 1$, $s_J = 1$, and $d_J = 0$; (b) the basic model with density dependence in the parasitoid attack rate, i.e. the *per capita* parasitoid attack rate decreases with increasing parasitoid density. k is a measure of the strength of density dependence in the parasitoid attack rate (with low k representing strong density dependence, and $k = \infty$ representing no density dependence). For each value of k, the equilibrium is locally stable to the right and Lotka–Volterra-like cycles occur in most of the region to the left of the labelled line. In the small enclosed regions near the *y* axis, single-generation cycles of the type reported by Godfray and Hassell (1989) occur, with the regions decreasing in size as k increases. Thus, the main effect of density dependence in the parasitoid attack rate is to stabilize the Lotka–Volterra-like cycles so that the generation cycles can be observed. The shaded area indicates the region of stable equilibria for $k = 1$, $d_P = 5$, all other parameters as in (a). (c) The host-size-dependent sex-allocation and host-feeding (SSH) model (from Murdoch *et al.* 1992), for different values of T_Y (with $T_J = T_Y + T_O$) (all other parameters as in (a)).

Lotka–Volterra predator–prey model. So, at its simplest, the basic model behaves like the basic Lotka–Volterra model.

The model shows that a sufficiently long-lasting invulnerable adult host stage stabilizes the interaction, and that stability is more likely the longer the invulnerable stage persists (Fig. 17.2a). Stability can also be induced by an invulnerable immature stage (Murdoch *et al.* 1987; for a related model see Hastings 1984). The unstable region is largely characterized by stable limit cycles ('Lotka–Volterra-like' or 'predator–prey' cycles) with a period of a few to many host generations.

An unexpected, although possibly biological unimportant, result is that the invulnerable adult stage is actually destabilizing when it is short. This is seen in Fig. 17.2a where, as we move along the *x* axis from the neutrally stable equilibrium at the origin, the equilibria are initially unstable. This result contrasts with an earlier result of Smith and Mead (1974) who showed that an invulnerable adult stage always stabilizes the neutrally stable equilibrium in the basic Lotka–Volterra model, if the juvenile stage duration is exponentially distributed. In general, however, we will see that insight from simple Lotka–Volterra models often survives, with modification of the details, through to more realistic models.

Short-lived adult hosts and single-generation cycles

Godfray and Hassell (1989) showed the existence of single-generation cycles in a structurally more complex model that distinguishes between vulnerable larvae and invulnerable eggs, pupae and adults. The cycles occur only if the adult host stage is short-lived, and are more likely if the hosts are vulnerable for only a small fraction of their development time. The key to exposing the single-generation cycles is density dependence in the *per capita* parasitoid attack rate, which suppresses the Lotka–Volterra-like cycles, thus allowing the single-generation cycles to be expressed, although if too strong, density dependence suppresses the single-generation cycles as well. These authors suggested that the model can explain the single-generation cycles seen in some tropical insect pests.

In fact, these single-generation cycles are lurking in the basic model. There they occur in a tiny area along the *y* axis, with the parameter values as in Fig. 17.2a, but they are obscured by Lotka–Volterra cycles, which are unstable in this parameter region. Their period is close to the host development time, T_J, which is itself very close to the host generation time when the adult stage is short-lived. Again, they can be exposed by adding density dependence to the parasitism rate, thus suppressing the

Lotka–Volterra-like cycles (Fig. 17.2b).

Each peak of density in a single-generation cycle represents a strong cohort. Such cohorts are, of course, the rule in species such as annuals that breed at discrete intervals. But they are remarkable in a continuously breeding population in a constant environment, where we would expect indistinct and strongly overlapping cohorts to be the rule. They arise because the brevity of the adult life allows the host population to develop dominant cohorts. In the absence of parasitism, these cohorts would eventually become desynchronized, but with appropriate timing the parasitoids may reinforce them. If the development time of the parasitoid is a fraction (or $1 + a$ fraction) of the host development time, then attacks on the dominant cohorts of hosts will produce peaks in the density of searching adult parasitoids (and therefore peaks in host mortality) at times when other host cohorts are in the vulnerable stage. Thus, hosts produced outside of the main cohorts suffer higher mortality, and host reproduction comes into synchrony.

This mechanism is analogous to that producing discrete cohorts in continuously breeding single-species populations (Gurney & Nisbet 1985). A short-lived adult stage is again crucial. In single-species populations, however, density-dependent survival of the young substitutes for the mechanism in the parasitoid–host models, namely feedback of host density on host survival via the production of new parasitoids. An example in a single-species population is provided by laboratory populations of the Indian flour moth, *Plodia* (Gurney *et al.* 1983).

We later distinguish these cycles, whose period is determined by the generation time of the resource (host) population, from those with a period equal to a single generation of the consumer (predator) population.

Older hosts produce a larger gain to the parasitoid

In this section we discuss various ways in which parasitoids treat older (often larger) vulnerable hosts differently from younger vulnerable hosts. Although these various responses are quite different, and might be expected to have different dynamical effects, they are in fact just different manifestations of a widespread phenomenon in parasitoids: the gain to the future female parasitoid population typically is greater when an older host is attacked (Murdoch *et al.* 1997) (we concentrate on female parasitoids because only they kill hosts). The host–parasitoid characteristics that produce this change in yield of female parasitoid offspring with host age include those listed in Table 17.2.

For parasitoids that kill or paralyse their host upon attack (idiobionts), older and therefore larger host individuals represent a larger packet of resources. Murdoch *et al.* (1992) explored the case where smaller hosts yield male offspring (or are fed upon), and so the attack does not result in a female parasitoid offspring, while larger hosts yield a female offspring. (The model ignores potential gains from meals and males, a point we discuss below.) This differential response to different hosts appears to induce a kind of delayed density dependence in the parasitoid *per capita* recruitment rate. Suppose, for example, we increase now the number of searching females above the equilibrium density. This will cause an increase in the death rate of young hosts, and hence a lower density of older hosts at a future time. In turn this will lead to a lower per parasitoid production of female offspring than would have been the case if the parasitoid population had remained at equilibrium (see Fig. 17.1b). If the density dependence is not too strong or too delayed, it tends to stabilize the Lotka–Volterra cycles (moving the stabilizing boundary to the left in Fig. 17.2c), but if it is stronger or the delay is longer it may also induce a different type of cycle: delayed feedback cycles (Fig. 17.2c).

The above analysis ignores firstly the potential value of male production. Female eggs require the mother to be fertilized (male eggs are haploid), and an additional adult male parasitoid will contribute additional future

Table 17.2 Older (usually larger) hosts provide a larger gain to the future parasitoid population: size- and state-dependent patterns of attack in parasitoid–host interactions that have the same dynamical consequences. See text and Murdoch *et al.* (1997) for further explanation. Victims attacked by idiobiotic parasitoids cease growth and development when they are attacked; victims of koinobiotic parasitoids continue to grow for some time after being parasitized.

Idiobiont parasitoids
1 Parasitoid feeds on young hosts and parasitizes old hosts
2 Sex ratio of parasitoid offspring becomes increasingly female with host size/age
3 Clutch size increases with host size/age
4 Decisions in 1 or 3 depend on parasitoid egg load
5 Combinations of 1–4
6 Older hosts yield larger and more fecund female parasitoids
7 Juvenile parasitoids survive better in older hosts

Koinobiont parasitoids
8 Juvenile parasitoids often survive better in older hosts
9 Juvenile parasitoids often develop faster in older hosts

searching female parasitoids if it fertilizes a female that would otherwise remain unmated. The analysis also ignores the contribution of host meals, which can add to the future production of female eggs. Taking these contributions into account reduces the difference in gain between attacks on young and old hosts. C.J. Briggs (unpublished observations) shows that these factors tend to suppress the amount of delayed density dependence, and hence reduce the dynamical effects described above. Male production is less important the more females each male can fertilize, and host feeding is less important the larger the initial egg complement of an emerging adult female. Clearly, we need more quantitative information on these individual properties and processes in real systems before we can evaluate their likely relative importance and their propensity to suppress Lotka–Volterra cycles and induce delayed feedback cycles.

Murdoch *et al.* (1997) show that the various parasitoid size- and state-dependent behaviours listed in Table 17.2 (with the structure of one example illustrated in Fig. 17.1c) all have the same dynamical effect. The table includes a kind of parasitoid not yet discussed, namely koinobionts, whose victims keep growing after they have been parasitized. In species in which the juvenile parasitoids grow faster in older (larger) hosts, the dynamical result parallels an increase in gain, even though the mechanism is different (Murdoch *et al.* 1997). The gain from older hosts is not greater (unless faster growth implies lower mortality); instead, the shorter development time in older hosts causes the same dynamical effect.

There are some cases in which the female-offspring yield to the parasitoid does not increase in older hosts (e.g. in some cases when the pupal stage is parasitized). This opposite pattern fits easily into our framework. As might be expected, the dynamical effect is opposite to that above: Lotka–Volterra-like limit cycles occupy a larger region of parameter space (Murdoch *et al.* 1997).

The models in which these results were developed have discrete host stages. The effect of this formulation is that yield of future female offspring increases as a step function with host age. However, in many real systems the increase in yield to the parasitoid, especially if associated with increasing host size, tends to change more or less continuously with host age. C.J. Briggs *et al.* (unpublished observations) show that the above results still hold if the yield the function increases continuously with host age, but delayed-feedback cycles are more likely if the yield is an accelerating function of host age (as is found for the parasitoid *Aphytis melinus*).

Effects of prey age/stage/size in predator–prey systems

The effect of an invulnerable stage (see above) has also been investigated in models of predator–prey interactions, including cases where the predator has a destabilizing type 2 functional response. These models also show that an invulnerable adult stage is stabilizing (Hastings 1983; van den Bosch & Diekmann 1986).

However, Hastings (1983) obtained less transparent results in a model with vulnerable adults and invulnerable juveniles. Here the juvenile stage is implicit since it acts simply as a developmental delay of duration, T. Adult prey (A) are converted to predators (P) at efficiency, c, to give:

$$
\begin{aligned}
dA/dt &= aA\ (t-T)-d_A A-Pf(A)\\
dP/dt &= cPf(A)-d_p P
\end{aligned}
\tag{1}
$$

where $f(A)$ is the functional response. This and other variants produce results that are not straightforward and that we cannot analyse in detail here. Briefly, Hastings found that certain values for the juvenile stage duration, T, could compensate for a destabilizing type 2 functional response and lead to stability, whereas others could not. As T was increased, the equilibrium moved in and out of stability. We suspect this occurs because there are multiple regions of parameter space that show single-generation cycles (see Briggs & Godfray 1995, for a related model of insect pathogens).

Just as in parasitoids (see above), smaller prey individuals, when eaten, probably contribute less to current and future predator population growth than do larger prey. But the transfer of meals to offspring in true predators is less direct than in host-feeding parasitoids.

Effects of parasitoid and predator age/stage/size

The properties of both parasitoids and predators typically change with age, especially between immatures and adults, or with size, and these changes have potentially substantial dynamical effects. Most obviously, only adults reproduce. In parasitoids the resulting developmental lag is destabilizing, as illustrated in Fig. 17.2. This simple result is seen also in predator–prey models (e.g. van den Bosch & Diekmann 1986).

Some of the differences between adult and juvenile predator individuals, however, give rise to more complications than do those between juvenile and adult parasitoids, and the models referred to so far do not deal with these complications. In the models mentioned above, prey density affects only the reproductive rate of adult predators. But in real predators, prey density also affects juvenile performance: juveniles that eat more may survive better, mature faster, or be larger and hence more fecund at maturation. Growth and an attendant increase in fecundity may also continue in the adult class. Models that incorporate such processes are necessarily more complicated and less amenable to analysis. They range from relatively simple stage-structured models to simulations of ensembles of individuals and in all cases become more narrowly focused as they increase in realism.

A single prey species is attacked

To illustrate how new dynamical features arise as we take into account differences among predator individuals related to age and size, we use three structured models that portray the interaction between the zooplankter *Daphnia* and its edible algal prey. All three models include:

1 a hypothesis about how an individual *Daphnia* allocates assimilated energy between growth and 'maturation' (in juveniles) or reproduction (in adults);

2 a hypothesis specifying the factor that determines when *Daphnia* mature to become adults;

3 relationships between *Daphnia*'s size and its feeding and growth rates at a given algal density; and finally

4 rules relating *Daphnia*'s death rate to its age and feeding rate.

The unstructured algae grow logistically and *Daphnia*'s (size-dependent) functional response is type 2.

In the simplest model, feeding and maintenance rates are assumed to scale in direct proportion to a daphnid's body weight, and death rates are independent of age or size. With this assumption, the total *Daphnia* and algal *biomasses* can be shown to obey two simple ordinary differential equations (Nisbet *et al.* 1991, 1997):

$$dE/dt = rE(1 - E/K) - I_{max}ED/(E + E_h)$$
$$dD/dt = eI_{max}ED/(E + E_h) - (m + b)D \tag{2}$$

This is a classic 'paradox of enrichment' model (Rosenzweig 1971); it has a stable equilibrium in nutrient-poor environments that gives way to large-amplitude cycles as the environment is enriched, as shown in Fig. 17.3a.

Two other models confirm the robustness of the prediction of large-amplitude cycles in enriched systems (Fig. 17.3b,c). The first, studied by de Roos *et al.* (1990; see also de Roos 1997) makes different assumptions about the size dependence of feeding rate, and assumes a set of priorities for energy utilization. The other is a stage-structured model that recognizes

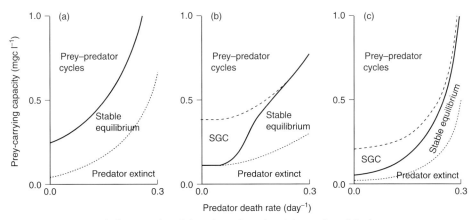

Figure 17.3 Dynamical properties of three 'paradox of enrichment' models that vary in complexity, in relation to algal carrying capacity (K) in mgc l^{-1}, and *Daphnia* death rate (m) in units day^{-1}. Case (a) is the simplest model (Eq. 2), case (b) is a fully individual-based model by de Roos *et al.* (1990) and case (c) is a stage-structured model of intermediate complexity (Nisbet *et al.* 1989; Murdoch *et al.* 1998). SGC, single-generation cycles.

only juveniles and adults, and has an empirically based representation of juvenile development (Nisbet *et al.* 1989; McCauley *et al.* 1996). These models also add a new kind of dynamics — single-generation cycles — not found in the simple model of Fig. 17.3a. Note that the period of the cycle is approximately the generation time of the *consumer*, whereas in the single-generation cycles in the parasitoid–host models discussed above the period corresponds with the generation time of the *resource*. They are thus quite different types of dynamics.

A complication in predators that does not occur in parasitoids is that larger predator individuals may eat larger prey individuals. For example, while all instars of the backswimmer, *Notonecta*, will eat mosquito larvae, or zooplankton, in each case larger *Notonecta* prefer prey from larger size classes (Scott & Murdoch 1983). We know of no modelling work that has been done on the dynamical consequences of such behaviour.

Different ages (stages) attack different prey species
Consider now parasitoids or predators in which the species attacked changes with the attacker's age, and especially where the adult obtains its nutrients from a different species than does the immature. These diet shifts

are sufficiently different in parasitoids and predators that we discuss them separately. A consistent theme in our comments, however, is the potential for new dynamical effects to appear as a result of coupling between different parts of the community resulting from age-related differences in diet.

Adults of parasitoids and of some predatory insects typically obtain energy from nectar and other sugar sources, and the availability of such energy sources may affect the adult 'background' death rate. However, this will have only a quantitative effect on stability (an increase in death rate is destabilizing, Murdoch *et al.* 1987), unless the adult population actually suppresses the energy source, in which case we need to describe the coupled interaction, or the adults need to divide a fixed amount of energy. The latter could induce density-dependent adult mortality.

The adults of host-feeding parasitoids require protein as well as sugar. The protein is used to mature new eggs but may also be used for maintenance, in which case it may prevent egg resorption and/or reduce mortality (Collier 1995). The protein typically comes from host individuals that are attacked and killed but not parasitized, with the potential dynamical implications discussed on p. 529.

Some species, however, may also obtain protein from other sources including pollen or by feeding on a prey species other than the host (Kidd & Jervis 1989), and this can be strongly stabilizing (Briggs *et al.* 1995). It is well known that models of ecological systems in general can be stabilized if there is an external source of immigrants, and the Briggs *et al.* model suggests an external source of nutrients that are used for reproduction can act in an analogous way. The key, as in the case of an energy source, is that the external source be a constant flux, i.e. that its dynamics be independent of (uncoupled from) the dynamics of the system under investigation.

Among predators, it is common for older (larger) predators to eat different prey species from younger (smaller) individuals. This is especially so in freshwater communities and is probably typical of fish. Indeed, in fish the two sets of prey may exist in different habitats. In bluegill and small-mouth bass, for example, young fish feed on plankton and older fish feed on benthos (e.g. DeAngelis *et al.* 1991). In such food webs there are two distinct predator–prey interactions, which are linked through predator reproduction and prey maturation.

We have not been able to discover any models of this situation. Preliminary investigations suggest the dynamics may be complicated when both the adult's and the juvenile's prey dynamics are strongly coupled to the dynamics of, respectively, adult and juvenile predator. However, it seems possible, for reasons similar to those discussed above, that the system might be stabilized if the dynamics of either the juvenile's, or the adult's prey are uncoupled from those of the predator. Weak coupling might arise in general predators where either the adult or the juvenile attacks several to many prey species and has only a weak effect on the dynamics of any one of these populations (Murdoch 1992a).

Here again, theory for predator–prey interactions has fallen behind that for parasitoid–host systems, for the obvious reason that the former are more complex.

Other sources of individual variation

Differences among individual hosts and prey

We know of no studies in this area showing that insight may differ markedly between parasitoids and predators, and we do not distinguish between them. The initial set of models all refer, however, to parasitoid–host interactions.

Persistent vs. transient differences in vulnerability

In this section we suggest a way of interpreting the models discussed that may provide some broad insight. The models can be divided into two groups. Models in the first group are consistent with assuming that each host (or prey) individual is assigned a value measuring its relative vulnerability to attack at the start of the vulnerable period, and retains this value throughout the period. Usually this is not the only possible assumption, but it is a natural one, and other assumptions are more difficult to imagine being true in real systems. Models in the second group are not consistent with such an interpretation. Instead, they require us to assume that the relative vulnerability of a particular individual is changing throughout the vulnerable period more or less rapidly with respect to other rates in the model. We refer, for brevity, to individuals in the two classes of models as having, respectively, persistent properties or transient properties, while recognizing that this is but a short-hand.

Bailey *et al.* (1962) were probably the first to explore the dynamical effects of variation among individual hosts in degree of vulnerability to parasitism. They modelled the situation where there is a distribution of relative vulnerability, i.e. in the relative risk of parasitism (Chesson & Murdoch 1986).

Bailey *et al.* (1962) imagined that a host individual 'presents' to the searching parasitoid an 'area of discovery', *a*, within which it is successfully attacked if a parasitoid enters the area; *a* thus measures the relative vulnerability of that host to parasitism. Different host individuals present different areas, so there is a distribution of relative vulnerability, $f(a)$, across the host population. Bailey *et al.* assumed the distribution is continuous and, in some cases, is defined by a gamma distribution whose shape is determined by the parameter *k*. Then the number surviving to the end of the generation, when the number of hosts at the start is λH_t and $f(a)$ defines the distribution of their relative vulnerabilities, is:

$$H_{t+1} = \lambda H_t \int_0^\infty \exp(-aP) f(a) \, da \tag{3}$$

where *P* is the number of searching parasitoids, which is constant throughout the generation. Notice that there is no explicit mechanism causing *a* to take on different values for different host individuals, and the mechanism can be any process consistent with the model's assumptions.

Bailey *et al.* (1962) showed that the otherwise unstable Nicholson–Bailey model can be stabilized if the distribution of relative vulnerabilities among hosts is highly skewed (see also Chesson & Murdoch 1986), so that some hosts are extremely vulnerable but the majority have low vulnerability. In the case of the gamma distribution, with coefficient of variation $\sqrt{(1/k)}$, stability requires *k* to be small. The skew implies that the parasitoid concentrates its attacks on a relatively small fraction of the initial host population.

This model attempts to describe in a discrete-time framework a process that is actually taking place continuously throughout the generation. In reality, parasitoids are attacking a declining number of hosts as the generation progresses, but this is not portrayed in Eq. 2. In line with the classification suggested earlier, a reasonable interpretation of the model's approximation, therefore, is that individuals are assigned a relative vulnerability at the start of the generation and retain this value thereafter. When $f(a)$ is very skewed, as required for stability, parasitoids are thus essentially continually re-attacking the same individuals. This induces inefficiency in the parasitoid, which causes the host equilibrium to be higher than it would be otherwise (May 1978; Murdoch 1992b). The inefficiency increases with parasitoid density, and it is this parasitoid density dependence that is stabilizing.

Spatial location: persistent relative vulnerability

Bailey *et al.* (1962) proposed spatial location as one mechanism that could induce differences in relative vulnerability among hosts. The differences could arise, firstly because some sites give more access to parasitoids than others, or secondly because parasitoids spend more time at certain sites. May (1978) envisaged parasitoids distributed among patches according to a gamma distribution and attacking prey at random within the patch, so the numbers of hosts and parasitoids in generation $t+1$, averaged over patches, is then

$$H_{t+1} = \lambda H_t (1 + aP_t/k)^{-k}$$
$$P_{t+1} = H_t[1 - (1 + aP_t/k)^{-k}]$$

(4)

where k is now also the clumping parameter of the negative binomial distribution. These models assume that host relative vulnerability, or equivalently the fraction of parasitoids visiting the patch, is uncorrelated with the number or fraction of hosts in the patch (Chesson & Murdoch 1986), i.e. parasitoid aggregation independent of local host density. These models lead to the well-known criterion that, under certain circumstances, the equilibrium is stable if $cv^2 > 1$, where cv is the coefficient of variation of the frequency of parasitism among hosts (Pacala & Hassell 1991).

Models of variation in vulnerability among hosts owing to location in space, beginning with Bailey *et al.* (1962), nicely demonstrate how apparent persistence of individual differences affects dynamics. Once again a consistent interpretation is that hosts are assigned their position in space, and hence their degree of relative vulnerability, at the start of the generation and retain it thereafter; indeed, this fits the biology of many insect host species which, as immatures, are relatively immobile. The stabilizing effect is thus associated with the persistence of individual differences over a long time scale—a host generation. These persistent differences in vulnerability induce inefficiency in the parasitoid, which can be thought of as persistently attacking hosts (or patches) that have already been most heavily attacked.

A fixed assignment of relative vulnerability through the whole vulnerable period was explicitly modelled by Reeve *et al.* (1994). Here, the vulnerable host stage, the egg, is distributed by the adults among 10 different microhabitats each with a fixed relative vulnerability drawn from a (discrete) gamma distribution. Adults lay eggs continuously and distribute them evenly among the 10 habitats, and each habitat is visited by the same number of parasitoids. The resulting dynamics are obtained simply by summing across patches. In this model stability ensues if the shape parameter $k \ll 1$. As in the Godfray and Hassell (1989) model discussed above, somewhat larger values of k still suppress the Lotka–Volterra-like cycles but now also expose the stable single-generation cycles inherent in this stage-structured model.

Rohani *et al.* (1994) studied patchy Nicholson–Bailey models in which parasitoids aggregate at some patches more than at others, but now the number of hosts in a patch changes within the season as hosts are killed. They looked at aggregation independent of local host density. At the start of the season, hosts are distributed uniformly across five patches: hosts are implicitly assumed not to move thereafter, although the number in a patch changes as a result of parasitism. Parasitoids are distributed unevenly across patches at the start of the season, but independently of how many hosts are present. The key assumption is that the fraction of searching parasitoids visiting each patch thereafter does not change: parasitoids move among patches, but this movement has no effect on dynamics since the *distribution* among patches remains fixed—a parasitoid that has left a patch is simply replaced by another, and hosts in any given patch are visited by a fixed number of parasitoids throughout the vulnerable period. Thus, the model is consistent with the interpretation that the relative vulnerability of the hosts in each patch is fixed over the vulnerable period, as in Eq. 4.

Unsurprisingly, this model retains the same stability properties as the model in Eq. 4: if the parasitoid distribution, and hence the distribution of host relative vulnerabilities, are sufficiently skewed to yield a stable equilibrium in the absence of parasitoid movement, they remain sufficient for stability when the nominal parasitoid movement is added.

An interpretation in terms of apparent persistence of the vulnerability of individual hosts may also be useful in understanding some recent spatially explicit individual-based models. McCauley *et al.* (1993) examined predator–prey populations in which individuals are distributed on a spatial grid. There is local prey density dependence and the predator has a type 2 functional response. They showed that, when individuals can move about the grid without restriction (i.e. the populations are completely mixed), the expected 'paradox-of-enrichment' type of instability occurs. However, when individuals can move only a relatively short distance the system tends to stabilize, in association with out-of-phase dynamics developing in different parts of the system. Thus, a stabilizing effect was found only when persistent differences in vulnerability among individuals was maintained through limited mobility.

McCauley *et al.* (1993) showed that stability of this system is more sensitive to restrictions in prey than in predator movement (see also McCauley *et al.* 1996). Even when predators can move about rapidly, restricted prey movement can stabilize the system, and de Roos *et al.* (1998) show that this occurs even when the predators are so mobile as to be evenly distributed across the space.

Spatial location: transient relative vulnerability
We now turn to models that assume relative vulnerability of an individual's changes through the vulnerable period. We begin with a model at the other extreme from those discussed above.

Murdoch and Stewart-Oaten (1989) investigated a continuous-time model in the Lotka–Volterra framework. They looked at the effects of both aggregation to local host density and aggregated attacks independent of local host density, which is the version relevant to our discussion here. They noted that the dynamics of the basic neutrally stable Lotka–Volterra model are unaffected if the host and parasitoid are independently distributed, regardless of the actual distribution.

If H and P are the numbers on a randomly chosen patch and parasitoids search randomly within patches, then on that patch

$$\begin{aligned} dH/dt &= rH - bHP \\ dP/dt &= cbHP - dP \end{aligned} \tag{5}$$

If we average over all patches we get

$$\begin{aligned} dh/dt &= rh - bE(HP) \\ dp/dt &= cdE(HP) - dp, \end{aligned} \tag{6}$$

where the average numbers of hosts and parasitoids per patch are, respectively, $h = E(H)$ and $p = E(P)$, and the expected number of encounters between hosts and parasitoids per unit time is proportional to expectation of HP, namely $E(HP)$. If parasitoids and hosts are distributed independently of each other $E(HP) = E(H)E(P)$ and we get

$$\begin{aligned} dh/dt &= rh - bhp \\ dp/dt &= cbhp - dp, \end{aligned} \tag{7}$$

which is just the original Lotka–Volterra model. Thus, this type of aggregation has no effect on stability.

In this model, the distribution of relative vulnerability among hosts is embodied in the joint distribution of H and P. The model assumes that, whatever the parasitoid and host spatial distributions are, they are constant through time. This can only be achieved, as hosts and parasitoids on patches die and recruit, if both species move around continuously. It is not clear just how fast this movement has to be for the model to be a reasonable approximation in the face of continuous births and deaths and continuously changing densities, but it presumably needs to be fast relative to the time scale set by the vital rates: the location (and hence relative vulnerability) of any particular host must change fast relative to, for example, the expected host life span. In the framework of our short-hand, such transient differences among prey individuals thus have no effect on dynamics.

Rohani *et al.* (1994) also studied density-dependent aggregation, which Hassell and May (1973) showed can in some circumstances induce stability in a Nicholson–Bailey model. Hosts are again distributed among patches, but now, at each time during the vulnerable stage, parasitoids preferentially visit patches that *currently* have more hosts, rather than those that had more hosts at the start of the generation, as in the initial formulation. As a result, especially when the hosts are immobile, the relative vulnerability of hosts in any particular patch changes throughout the vulnerable period. When the parasitoid responds quickly to the changes in relative density in patches, i.e. host relative vulnerability changes relatively rapidly, the stabilizing effect of parasitoid aggregation found by Hassell and May (1973) is lost. This is consistent with a loss of dynamical effect as the differences among individuals become more transient. However, in this model the amount of difference in density, and hence the amount of difference in relative vulnerability among hosts, are also reduced through the season, and this also results in a loss of the stabilizing effect. We are currently investigating models in which relative vulnerability of hosts is always independent of local density, but its distribution among hosts changes through the generation.

A spatial refuge

Physical refuges represent perhaps the most extreme case of spatial variation in individual vulnerability. When a constant fraction, s, is safe in the refuge, and a fraction $1-s$ is vulnerable, we have (Bailey *et al.* 1962; Hassell 1978):

$$H_{t+1} = \lambda s H_t + \lambda(1-s)H_t \exp(-aP_t)$$
$$P_{t+1} = (1-s)H_t[1-\exp(-aP_t)]$$

(8)

This type of refuge is weakly stabilizing. The model is consistent with the interpretation that a fraction of the *initial* host population, sH_t, goes into the refuge and these individuals remain there for the entire generation, so the relative vulnerabilities in the refuge and non-refuge classes can be thought of as persisting over the generation.

The contrast with a Lotka–Volterra fractional refuge is illuminating. When a fraction of the host population, s, is safe and a fraction $1-s$ is vulnerable, we have:

$$dH/dt = r[sH + (1-s)H] - b(1-s)HP$$
$$dP/dt = cb(1-s)HP - dP \tag{9}$$

But this collapses to a model whose dynamical properties are effectively the same as the basic neutrally stable Lotka–Volterra model with $b(1-s) = a$:

$$dH/dt = rH - aHP$$
$$dP/dt = caHP - dP \tag{10}$$

Thus, stability is unaffected when a constant fraction is in the refuge. In such a Lotka–Volterra population, with continuously changing densities and continuously occurring births and deaths, a constant fraction can be maintained in the refuge only if individuals are continuously moving in and out of it, i.e. in our short-hand, individual properties are transient. By contrast, a constant *number* in the refuge does stabilize the equilibrium in a Lotka–Volterra model. Where R is the number in the refuge we have

$$dH/dt = rH - bP(H-R)$$
$$dP/dt = cbP(H-R) - dP \tag{11}$$

so the R individuals never take part in the interaction with the parasitoid, although they reproduce at rate r. This time, it is possible to interpret the model as saying, as in Eq. 7, that the refuge is populated by a distinct set of individuals that remain there for their lifetimes and leak offspring into the true predator–prey interaction.

Factors other than space

In the Nicholson–Bailey models discussed above, and those of Reeve *et al.* (1994) and Murdoch and Stewart-Oaten (1989), spatial location is but one interpretation of many that might be imagined for the differences in relative vulnerability among host individuals. Bailey *et al.* (1962) recognized that space was only one possible or plausible cause of differences in vulnerability. In fact, their first proposed mechanism involved host individuals

541

having a range of 'intrinsic defensive properties' (although some intrinsic differences might be heritable and this would add a so far unexamined complication). Other types of differences in vulnerability that might persist for much of the host's life include those arising from early accidents whose persistent effects make the individual more vulnerable, or initial random differences in, say, size, that become reinforced if initially larger prey can eat faster, and larger prey are less vulnerable. We would expect these differences to have substantial dynamical effects (see discussion of DeAngelis *et al.* (1991), below).

Transient individual differences could also arise in a variety of ways. For example, such differences could be created by a concatenation of many small randomly acting processes: how active the host is, how well its camouflage operates at any given time, how the light filtering through a forest canopy makes it more or less visible at any given time, etc. Such combinations of environmental and spatial variation should lead to rapidly changing vulnerabilities in a given individual, which we would expect to have little dynamical effect.

Finally, Abrams and Walters (1996) explored models directly related to the degree of persistence of individual properties. Theirs have the basic structure of a paradox of enrichment model, with logistic prey and type 2 functional response. The prey, however, are of two types: vulnerable and invulnerable (which in principle might or might not relate to spatial location), and prey move between the two classes at a constant rate. The great bulk of their results are consistent with our notion that dynamical effects are more likely the more apparently persistent are differences in relative vulnerability among individuals. In this case, equilibria are more likely to be stable when the transition rate between prey classes is low, i.e. when the differences between individuals are more persistent. As the transition rate increases, and individual differences are more transient, we recover the paradox of enrichment instability of the unstructured model.

Differences among individual parasitoids and predators

We explore here differences among parasitoids or predators induced by their physiological state.

Among parasitoids, egg load (the number of eggs that a searching female parasitoid is carrying) is a key physiological state that affects patterns of attacks (Minkenberg *et al.* 1992). For example, it has been observed in a number of species that a female parasitoid is more likely to parasitize an encountered host when her egg load is high, and more likely to host feed to gain nutrients when her egg load is low. We might expect substantial dynamical effects to flow from such decisions dependent on egg load, since egg load presumably reflects recent host encounters and should therefore respond to host (and indirectly parasitoid) density. Dependence on egg load might therefore effectively lead to density–dependent behaviour by the parasitoid. This expectation turns out, however, to be incorrect, as dependence on egg load, *per se*, has little effect on stability. Kidd and Jervis (1989) simulated an interaction in which the probability that an encountered host is host fed rather than parasitized, declines with egg load (see Collier *et al.* 1994 for an empirical relationship); they found no effect on stability. Briggs *et al.* (1995) confirmed this result analytically in a model lacking age structure in either population. They also showed the dynamical mechanism that leads to the result; the fraction of parasitoids in each egg-load class reaches a stable distribution, even though the parasitoid population itself may be cycling.

In a model in which the probability of host feeding declines with egg load, but faster on the larger class of two vulnerable immature host classes (since the latter are preferred for parasitism) (see Fig. 17.1c), exactly the same results are obtained as would be the case if there were simply a preference for ovipositing in the larger host (see Fig. 17.2c). That is, once again dependence on egg load *per se* has no effect beyond that caused by the difference in yield from the two host classes, discussed on p. 529 (Murdoch *et al.* 1997).

A second type of parasitoid behaviour that commonly has been found to depend on egg load is clutch size (Courtney *et al.* 1989; Courtney & Hard 1990; Odendaal & Rausher 1990). Parasitoids are more likely to lay more than one egg in an encountered host when their egg load is high, but lay a single egg at low egg loads. As with the case of the egg-load-dependent probability of host feeding, egg-load-dependent parasitoid clutch size has no effect on dynamics (C.J. Briggs, unpublished observations).

The absence of a significant dynamical effect related to egg load in these models is due to two distinctive features, both of which disconnect mortality from physiological state. Firstly, egg load does not affect parasitoid mortality. Secondly, the decision to host feed or oviposit also does not affect the host death rate. By contrast, Briggs *et al.* (1995) showed that an adult parasitoid death rate that increases as egg load decreases is stabilizing. For example, T. Collier (personal communication, 1995) has observed that *Aphytis melinus* females typically have no eggs when they die in the laboratory.

In general, we expect that dependence of vital rates on physiological state *will* have significant effects on dynamics. An example is provided by models motivated by microbial experiments in continuous culture (Kooijman 1993). As in the *Daphnia* example discussed earlier, greater stability is observed than is predicted by simple models (Cunningham & Nisbet 1983; Nisbet *et al.* 1983). Kooijman introduces models in which the fecundity of an organism depends on the level of its internal energy reserves, and this enhances stability.

Environmentally induced differences in physiological state are typical in detailed simulations of fish cohort models. DeAngelis *et al.* (1991) described an individual-based model of the first-year dynamics of small-mouthed bass with dynamic prey. In a stochastic environment, initial random differences among individuals in the smallest size class, caused by differences in laying date, are reinforced by subsequent experience. Thus, earlier-born individuals in the cohort are, at any time, larger and feed and grow faster; hence the initial size differences become reinforced. These differences allow some individuals to 'jump ahead' and achieve a size that permits them to survive through the winter. As a consequence, the relationship between the number recruiting to the year 1 class, as a function of the number born (the year 0 class) tends to be an increasing under-compensating function (i.e. Beverton–Holt curve) and hence stabilizing, whereas if all individuals are the same, the relationship tends to be over-compensating (a Ricker curve), leading to potentially unstable dynamics. Size-dependent predation on the cohort tends to reinforce this difference.

Discussion

We have shown how differences among individuals, be they in the consumer or resource population, can have profound effects on population dynamics. Models containing such differences, especially differences related to age or size among individuals in the resource population, are most developed for parasitoid–host interactions. A wide range of these age-related properties, including various responses by parasitoids, have been studied, and some generalities have emerged including the stabilizing effect of invulnerable host stages, the emergence of cycles with a period of a single host generation and of delayed feedback cycles as a result of the parasitoid responding in different ways to hosts of different ages. We also saw many apparently different parasitoid responses to hosts of different age fall nicely under a single rubric—the gain to the future female parasitoid population increases with host age. They also have the same dynamical effect — they suppress Lotka–Volterra-like oscillations and under some circumstances generate delayed-feedback cycles.

Some processes explored in parasitoid–host interactions have analogies in predator–prey systems and have the same kind of dynamical effects in the latter. Most obvious among these are the tendency for an invulnerable class of prey and a dynamically uncoupled food supply for one predator stage to be stabilizing, and for developmental delays to be destabilizing. Work by Hastings (1983, 1984), however, suggests that the interaction between delays and an invulnerable stage may be more complex in some predator–prey models than in parasitoid–host models. We return to the theme of greater complexity in predators below.

When we explored the dynamical effects of variation among host/prey individuals that arise from sources other than age, we suggested that some insight can be gained by answering the following question: does the model allow the interpretation that an individual's degree of vulnerability to attack persists for a long period relative to, say, its expected lifespan? If the answer is yes, large differences in vulnerability among individuals are likely to have substantial dynamical effects, but not otherwise. Such differences might arise from spatial location or some other source of a heterogeneity.

We emphasize our relative ignorance about the effect of individual differences in true predator–prey systems. Predator–prey models aspiring to even modest realism are by necessity much more complicated than equivalent models for parasitoid–host interactions. Complications peculiar to predator–prey systems include at least the following.

1 Juvenile predators actively search for and attack prey.

2 Juvenile maturation rate, death rate or size at maturation can all depend on juvenile feeding history.

3 Some adults can continue to grow, which can affect their feeding, fecundity and possibly death rates.

4 Juveniles may eat smaller individuals of the prey species than do adults, or they may eat different prey species, which may be in a different part of the environment.

We noted above that, in spite of these differences, some insights from parasitoid–host models apply to predator–prey systems. But we know remarkably little about the dynamical implications of the numbered list above. Most existing models for predator–prey systems are too simple to include these features. A few, for example de Roos *et al.* (1990), include some factors — in this case 1, 2 and 3 from the above list. Some individual-based simulation models also include 1–3, and aspects of 4 (e.g. DeAngelis *et al.* 1991). But they do not have multigenerational coupled predator–prey dynamics.

Theory for predator–prey systems has thus fallen behind that for parasitoid–host interactions, mainly because of the greater structural complexity of the former. Such complexity poses particularly acutely the usual challenge of ecological modelling: to create moderately realistic representations of individual properties that still yield useful insight into dynamical effects.

Finally, with this last point in mind, we stress that simple models remain useful despite the new insights gained from complex models that incorporate the attributes of individual organisms. Indeed, Murdoch and Nisbet (1996) argue that complex models are most useful when their connections to simpler, more analysable and more understandable models are explicit and fully understood. This is the case for several classes of models discussed above. For example, the basic properties of simple Lotka–Volterra models survive in many of the more complex stage-, size- and age-structured parasitoid–host models. Table 17.1 shows that the equilibrium densities of hosts and parasitoids in the basic model are exact analogues of those for the simple Lotka–Volterra predator–prey model (see also Murdoch *et al.* 1987, 1992).

The point is reinforced by models for the interaction between *Daphnia* and its algal prey (p. 532), which also illustrate the use of simple models to test hypotheses in the field. Although both the stage-structured model and the fully individual-based age- and size-structured model for this interaction expose single-generation cycles not seen in the simplest 'paradox-of-enrichment' model, their most striking feature is that they show standard Lotka–Volterra instability in parameter regions analogous to those found in the simple model (Fig. 17.3). As in the parasitoid–host models, equilibrial values also respond to changes in parameter values in fundamentally the same way in the complex and simple models. Murdoch *et al.* (1998) were able to use these features to test hypotheses seeking to explain the stability of this interaction in lakes: in each of the hypotheses, the mechanism maintaining stability in the face of enrichment also causes the predicted equilibrium algal density to increase to levels above those seen in real lakes.

Thus, in most cases it appears that from the added structure of more complex models we mainly gain insight into potential *additional* dynamic behaviour, such as additional sources of density dependence, or the single-generation cycles and delayed-feedback cycles arising from resource and/ or consumer age structure. Many basic properties, however, appear to persist across levels of complexity. Progress is thus likely to continue to rely on a judicious blend of simple and complex models.

Acknowledgements

This work was supported by grants DEB94-20286, DEB96-29136 and DEB93-19301 from NSF, and No. 96-35302-3753 from the NRI Competitive Grants Program, USDA. We are grateful to Priyanga Amarasekare, Bill Gurney, Bas Kooijman, Ed McCauley, Andre de Roos and Will Wilson for useful discussions. Louise Vet and Jef Huisman provided very useful reviews.

References

Abrams, P.A. & Walters, C.J. (1996). Invulnerable prey and the paradox of enrichment. *Ecology*, **77**, 1125–1133.

Bailey, V.A., Nicholson, A.J. & Williams, E. (1962). Interaction between hosts and parasites when some host individuals are more difficult to find than others. *Journal of Theoretical Biology*, **3**, 1–18.

Briggs, C.J. & Godfray, H.C.J. (1995). The dynamics of insect–pathogen interactions in stage-structured populations. *American Naturalist*, **145**, 855–887.

Briggs, C.J., Nisbet, R.M., Murdoch, W.W., Collier, T.R. & Metz, J.A.J. (1995). Dynamical effects of host-feeding in parasitoids. *Journal of Animal Ecology*, **64**, 403–416.

Chesson, P.L. & Murdoch, W.W. (1986). Aggregation of risk: relationships among host–parasitoid models. *American Naturalist*, **127**, 696–715.

Collier, T.R. (1995). Host feeding, egg maturation, resorption, and longevity in the parasitoid *Aphytis melinus* (Hymenoptera: Aphelinidae). *Annals of the Entomological Society of America*, **88**, 206–214.

Collier, T.R., Murdoch, W.W. & Nisbet, R.M. (1994). Egg load and the decision to host-feed in the parasitoid, *Aphytis melinus. Journal of Animal Ecology*, **63**, 299–306.

Courtney, S.P. & Hard, J.J. (1990). Host acceptance and life-history traits in *Drosophila buskii*: tests of the hierarchy-threshold model. *Heredity*, **64**, 371–375.

Courtney, S.P., Chen, G.K. & Gardner, A. (1989). A general model for individual host selection. *Oikos*, **55**, 55–65.

Cunningham, A. & Nisbet, R.M. (1983). Transients and oscillations in continuous culture. In *Mathematics in Microbiology* (Ed. by M. Bazin), pp. 77–104. Academic Press, London.

DeAngelis, D.L., Godbout, L. & Shuter, B.J. (1991). An individual-based approach to predicting density-dependent dynamics in smallmouth bass populations. *Ecological Modelling*, 57, 91–115.

de Roos, A.M. (1997). A gentle introduction to physiologically structured population models. In *Structured-Population Models in Marine, Terrestrial, and Freshwater Systems* (Ed. by S. Tuljapurkar & H. Caswell), pp. 119–204. Population and Community Biology Series, 18. Chapman & Hall, New York.

de Roos, A.M., Metz, J.A.J., Evers, E. & Leipoldt, A. (1990). A size dependent predator–prey interaction: who pursues whom? *Journal of Mathematical Biology*, 28, 609–643.

de Roos, A.M., McCauley, E. & Wilson, W.G. (1998). Pattern formation and the spatial scale of interaction between predators and their prey. *Theoretical Population Biology*, 53, 108–130.

Godfray, H.C.J. & Hassell, M.P. (1989). Discrete and continuous insect populations in tropical environments. *Journal of Animal Ecology*, 58, 153–174.

Gurney, W.S.C. & Nisbet, R.M. (1985). Fluctuation periodicity, generation separation, and the expression of larval competition. *Theoretical Population Biology*, 28, 150–180.

Gurney, W.S.C., Nisbet, R.M. & Lawton, J.H. (1983). The systematic formulation of tractable single-species population models incorporating age structure. *Journal of Animal Ecology*, 52, 479–496.

Hassell, M.P. (1978). *The Dynamics of Arthropod Predator–Prey Systems*. Princeton University Press, Princeton, NJ.

Hassell, M.P. & May, R.M. (1973). Stability in insect host–parasite models. *Journal of Animal Ecology*, 42, 693–736.

Hastings, A. (1983). Age dependent predation is not a simple process. I. Continuous time models. *Theoretical Population Biology*, 23, 347–362.

Hastings, A. (1984). Delays in recruitment at different trophic levels: effects on stability. *Journal of Mathematical Biology*, 21, 33–44.

Heimpel, G.E & Collier, T.R. (1996). The evolution of host-feeding behaviour in insect parasitoids. *Biological Reviews*, 71, 373–400.

Kidd, N.A.C. & Jervis, M.A. (1989). The effects of host-feeding behaviour on the dynamics of parasitoids–host interactions, and the implications for biological control. *Researches on Population Ecology*, 31, 235–274.

Kooijman, S.A.L.M. (1993). *Dynamic Energy Budgets in Biological Systems: Theory and Applications in Ecotoxicology*. Cambridge University Press, Cambridge.

May, R.M. (1978). Host–parasitoid systems in a patchy environment: a phenomenological model. *Journal of Animal Ecology*, **47**, 833–844.

McCauley, E. & Murdoch, W.W. (1990). Predator–prey dynamics in rich and poor environments. *Nature*, **343**, 455–457.

McCauley, E., Wilson, W.G. & de Roos, A.M. (1993). Dynamics of age-structured and spatially structured predator–prey interactions individual-based models and population-level formulations. *American Naturalist*, **142**, 412–442.

McCauley, E., Wilson, W.G. & de Roos, A.M. (1996). Dynamics of age-structured predator–prey populations in space: asymmetrical effects of mobility in juvenile and adult predators. *Oikos*, **76**, 485–497.

Minkenberg, O.P.J.M., Tatar, M. & Rosenheim, J.A. (1992). Egg load as a major source of variability in insect foraging and oviposition behavior. *Oikos*, **65**, 134–142.

Murdoch, W.W. (1992a). Individual-based models for predicting effects of global change. In *Biotic Interaction and Global Change* (Ed. by P. Kareiva, J. Kingsolver & R. Huey), pp. 147–162. Sinauer Associates, Sunderland, MA.

Murdoch, W.W. (1992b). Ecological theory and biological control. In *Applied Population Biology* (Ed. by L. Botsford & S. Jain), pp. 197–221. Dr S. Junk Publishers, The Netherlands.

Murdoch, W.W. & Nisbet, R.M. (1996). Frontiers of population ecology. In *Frontiers of Population Ecology* (Ed. by R.B. Floyd & A.W. Sheppard), pp. 31–43. CSIRO Press, Melbourne.

Murdoch, W.W. & Stewart-Oaten, A. (1989). Aggregation by parasitoids and predators: effects on equilibrium and stability. *American Naturalist*, **134**, 288–310.

Murdoch, W.W., Nisbet, R.M., Blythe, S.P. & Gurney, W.S.C. (1987). An invulnerable age class and stability in delay-differential parasitoid–host models. *American Naturalist*, **129**, 263–282.

Murdoch, W.W., Nisbet, R.M., Luck, R.F., Godfray, H.C.J. & W.S.C. Gurney (1992). Size-selective sex-allocation and host feeding in a parasitoid–host model. *Journal of Animal Ecology*, **61**, 533–541.

Murdoch, W.W., Briggs, C.J. & Nisbet, R.M. (1997). Dynamical effects of parasitoid attacks that depend on host size and parasitoid state. *Journal of Animal Ecology*, **66**, 542–556.

Murdoch, W.W., Nisbet, R.M., McCauley, E., de Roos, A.M. & Gurney, W.S.C. (1998). Plankton abundance and dynamics across nutrient levels: tests of hypotheses. *Ecology*, **79**, 1339–1356.

Nisbet, R.M., Cunningham A. & Gurney W.S.C. (1983). Endogenous metabolism and the stability of microbial predator–prey systems. *Biotechnology and Bioengineering*, **25**, 301–306.

Nisbet, R.M., Gurney, W.S.C., Murdoch, W.W. & McCauley, E. (1989). Structured population models: a tool for linking effects at individual and population level. *Biology Journal of the Linnean Society*, **37**, 79–99.

Nisbet, R.M., McCauley, E., de Roos, A.M., Murdoch, W.W. & Gurney, W.S.C. (1991). Population dynamics and element recycling in an aquatic plant–herbivore system. *Theoretical Population Biology*, **40**, 125–147.

Nisbet, R.M., McCauley, E., Gurney, W.S.C., Murdoch, W.W. & de Roos, A.M. (1997). Simple representations of biomass dynamics in structured populations. In *Case Studies in Mathematical Modeling: Ecology, Physiology, and Cell Biology* (Ed. by H.G. Othmer, F.R. Adler, M.A. Lewis & J.C. Dillon), pp. 61–79. Prentice-Hall, NJ.

Odendaal, F.J. & Rausher, M.D. (1990). Egg load influences search intensity, host selectivity, and clutch size in *Battus philenor* butterflies. *Journal of Insect Behavior*, **3**, 183–194.

Pacala, S.W. & Hassell, M.P. (1991). The persistence of host–parasitoid associations in patchy environments. II. Evaluation of field data. *American Naturalist*, **138**, 584–605.

Reeve, J.D., Cronin, J.T. & Strong, D.R. (1994). Parasitism and generation cycles in a salt-marsh planthopper. *Journal of Animal Ecology*, **63**, 912–920.

Rohani, P., Godfray, H.C.J. & Hassell, M.P. (1994). Aggregation and the dynamics of host–parasitoid systems: a discrete-generation model with within-generation redistribution. *American Naturalist*, **144**, 491–509.

Rosenzweig, M.L. (1971). The paradox of enrichment: destabilization of exploitation ecosystems in ecological time. *Science*, **171**, 385–387.

Scott, M.A. & Murdoch, W.W. (1983). Selective predation by the backswimmer, *Notonecta*. *Limnology and Oceanography*, **28**, 352–366.

Smith, R.H. & Mead, R. (1974). Age structure and stability in models of predator–prey systems. *Theoretical Population Biology*, **6**, 308–322.

van den Bosch, F. & Diekmann, O. (1986). Interactions between egg-eating predator and prey: the effect of the functional response and of age structure. *IMA Journal of Mathematics Applied in Medicine and Biology*, **3**, 53–69.

Chapter 18

Individual-based modelling as an integrative approach in theoretical and applied population dynamics and food web studies[1]

W.M. Mooij[2] and D.L. DeAngelis[3]

Summary

Traditional modelling approaches in population dynamics and food web studies do not incorporate the basic biological principles that each individual is different and that interactions between individuals are inherently local. This notion has led to the development of so-called individual-based models of populations and communities. A population or community is described as a database of individuals, whereby each individual is described in a number of traits and a number of processes that change these traits over time. This approach allows for incorporating significantly more detail compared with unstructured and physiologically structured models. Nowadays, the individual-based approach is widely applied in both animal and plant ecology. These applications, of which some representative examples are briefly described in this chapter, show that individual-based models can be used to bridge the gap between theoretical studies that focus on general features of ecosystems and applied studies that are designed to predict the dynamics of ecosystems under management scenarios. Individual-based models can contribute to theoretical studies because they accommodate small but biologically relevant detail, which cannot be included in analytical models. In applied studies, they have proven to be

1 Publication 2337 of the Netherlands Institute of Ecology, Centre for Limnology
2 Netherlands Institute of Ecology, Centre for Limnology, Rijksstraatweg 6, 3631 AC Nieuwersluis, The Netherlands
3 US Geological Survey, Biological Resources Division and University of Miami, Department of Biology, PO Box 249118, Coral Gables, FL 33124, USA

551

quite flexible in using those data that are typically available from field studies. The advantages of individual-based modelling come at the cost of technical and numerical complexity, necessary to cope with the numerous book-keeping operations that individual-based models essentially come down to. Various frameworks are now available to enhance the development of individual-based models. We expect individual-based models to become established as a standard tool in ecological research within the near future because the individual-based modelling paradigm provides a general structure of the formalization and analysis of community and food web dynamics.

Introduction

In a paper with the provocative title 'New computer models unify ecological theory', Huston *et al.* (1988) stress that traditional modelling approaches in population dynamics and food web studies, such as Lotka–Volterra equations, do not incorporate two of the basic principles of biology. Firstly, each individual is different in its physiology and behaviour, as a result of its unique combination of genetic and environmental influences. Secondly, interactions between individuals are inherently local, i.e. organisms are only affected in a mechanistic way by those other organisms with which they locally interact, irrespective of whether these other organisms serve as food, competitors or predators. Physiologically structured population dynamical models (Metz & Diekmann 1986), which take only major traits as age or size into account but ignore many other components of variation among individuals, have been proposed as one solution to incorporate these features. However, the validity of focusing only on major traits in these models is often not tested but merely assumed.

Interest in testing the role of individual variation and local interactions has led to the development of so-called individual-based models of populations and communities (DeAngelis & Gross 1992). These models allow one to take more traits into account than just age or size. Conceptually this is extremely simple (Fig. 18.1). A population or community is described as a database of individuals, whereby each individual is described in a number of traits. Biological processes are also implemented at the level of the individual and modify the physiological, behavioural or life history traits of the individuals in time. The essential demographic processes of birth and death are implemented by adding records to, or deleting records from, the database, respectively. Spatially explicit individual-based models

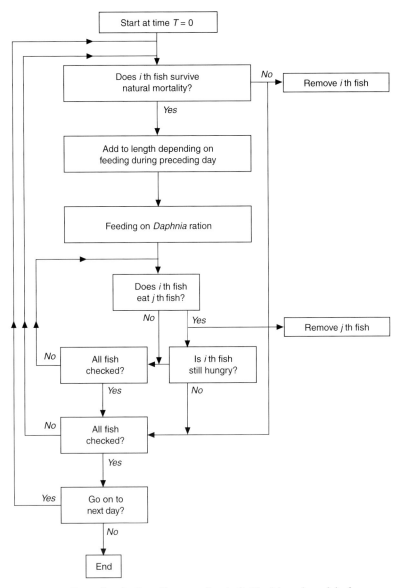

Figure 18.1 Example of a flow diagram of an individual-based model of cannibalism in the first-year cohort of large-mouth bass. Fish are given a *Daphnia* ration as a supplement to their cannibalistic feeding. The outcome of the model was tested against tank experiments and the results were in good agreement. Redrawn from DeAngelis *et al.* (1979).

keep track of the spatial position of each individual in time. By adding information about the local physico-chemical environment the full dynamics of an ecosystem can be covered. This conceptual simplicity, however, is obtained at the cost of the technical and numerical complexity, necessary to stay on top of the numerous book-keeping operations that individual-based modelling essentially comes down to. Moreover, many analytical techniques (e.g. stability analysis) cannot be applied to these models and numerical techniques should be used instead.

Explorations with individual-based population models probably began at least as early as the 1960s, with early applications in animal ecology including models of arthropods (Hilborn 1975; Myers 1976), fish (DeAngelis *et al.* 1979), zooplankton (Hogeweg & Richter 1982) and mammals (David *et al.* 1982). From the early 1990s onwards, the individual-based modelling approach has taken off (DeAngelis & Gross 1992), and has now been applied to organisms as large as ungulates (Turner *et al.* 1994) and as small as bacteria (Jaworska *et al.* 1996). In plant ecology an analogous development has taken place as in animal ecology, with pioneers in the 1970s (Botkin *et al.* 1972; Shugart & West 1977) and a broader application in recent years (Pacala & Deutschman 1995; Pacala *et al.* 1996; Humphries *et al.* 1996; Post & Pastor 1996).

Several authors have presented their perspectives on the future development of the individual-based approach (Hogeweg & Hesper 1990; Murdoch 1993; Judson 1994; Uchmanski & Grimm 1996), but these papers mention relatively little of the actual applications. We therefore feel that there is need for a paper that concentrates on what has actually been achieved by applying the individual-based approach. Specifically we want to focus on how the individual-based approach can be used to bridge the gap between theoretical studies that focus on general features of communities and ecosystems (e.g. diversity, stability, importance of initial conditions) and applied studies, designed to predict the dynamics of whole ecosystems under a variety of management scenarios. As a trade-off, we will pay relatively little attention to the use of individual-based models in experimental science. Also, we will limit ourselves to applications of the individual-based approach in animal ecology.

Before discussing some of the recent applications of individual-based models in theoretical and applied studies, using both aquatic and terrestrial examples, we will give a brief overview and some definitions of contemporary modelling approaches in population dynamics and food web studies. We often sense confusion about what principles are referred to when a model is called unstructured, physiologically structured or individual-based. We also include a section that focuses on the techniques of individual-based modelling, because we realize that those who are interested in applying this relatively new approach may be limited in its use by mere technical difficulties. We end this chapter with some general conclusions about the state of the art in individual-based modelling and some perspectives that we see for the near future.

Unstructured, physiologically structured and individual-based models

The simplest models employed in population dynamics and food web studies are unstructured models. These models describe the state of a population in a single state variable, representing the total number of individuals or biomass of a given population. The dynamics of that population are described by a differential equation that links the rate of change to the state of the population. The well-known Lotka–Volterra predation and competition equations are examples of this approach (e.g. Yodzis 1975). When an analytical solution of this differential equation exists, the state of the system at a given moment in time can be directly calculated given an initial state. Otherwise, a numerical solution is used. Models of interacting populations can be built by linking several differential equations. Also, this format allows for the explicit modelling of resource dynamics next to population dynamics (Tilman 1977; DeAngelis 1992). Although quite abstract, unstructured models have the advantage that their dynamics can be analysed and understood for the full parameter range. Isocline analysis can reveal regions of model behaviour that may reflect some of the essentials of the dynamics of natural systems, such as the existence of two stable states at intermediate levels of eutrophication in aquatic ecosystems (Scheffer *et al.* 1993) and in terrestrial ecosystems (see Chapter 8; Van de Koppel *et al.* 1996).

Physiologically structured models (Metz & Diekmann 1986) describe the state of a population in more detail. Only the trait of the individual that is assumed to play a key role in the dynamics of the population is selected as the structuring variable: often age or size (Ebenman & Persson 1988). In addition to birth and death rates that change the total number of individuals, processes that change the size, age or stage structure of the population, such as ageing or somatic growth, have to be formulated in these models. Matrix models (Caswell 1989) provide a format for physiologically structured population models that are discontinuous for both the structuring trait and in time. Partial differential equations give a continuous description along both axis (Metz & Diekmann 1986). As with unstructured models, physiologically structured population models have the advantage of a formal notation and a well-understood mathematical behaviour. Analytical solutions for partial differential equations are mostly not available, however, and numerical techniques have to be applied to calculate the state of the system at a given moment in time (de Roos *et al.* 1992). Extending a line of research initiated by Sinko and Streifer (1967) McCauley *et al.* (1996) showed the applicability of the partial differential approach to understand the dynamics of herbivorous zooplankton.

Unstructured and physiologically structured population models have been extended to spatial descriptions of populations through two general approaches. The first method is to consider space to consist of a number of discrete areas. Then the dynamics with each region is modelled as well as continuous migration between the areas (e.g. Levin 1974). An alternative approach is to model space as a continous variable. In this case partial differential equations, with variables representing time and space, are used (e.g. Okubo 1980).

At first sight there seems to be no limit to the number of traits that can be entered in physiologically structured population models. However, a matrix model that takes six traits into account with a resolution of 10 classes per trait would already need a six-dimensional matrix with 10^6 cells to describe the state of the population at a given moment in time. Moreover, most of these cells will contain zeros, due to the inherent covariance in physiological, behavioural and life history traits. This would lead to a lot of unnecessary computational overhead in evaluating these matrices. Entering six traits in the partial differential approach may easily lead to such complex equations that the mathematical aspects of the study will dominate over the biological aspects.

In studies, in which one aims to evaluate the dynamics of several traits in a coherent analysis, the sampling approach underlying individual-based models may be more appropriate than the matrix or partial differential approach. We argue that even for some tens of traits, a sample of 1000–10000 model individuals would give a fair numerical representation of the density distribution of a population in multidimensional trait space. This numerical representation contains implicitly all the information about the covariances among traits, and these relations can be extracted from it by standard statistical techniques. The dynamics of each trait can be entered in an individual-based model as a function of the current state of the individual itself and of its perception of, and interaction with, other individuals and the local physico-chemical environment. The parameters of these functions can be described at the population level or at the individual level. The latter approach allows for the evaluation of clonal differences with a population, or, given some genetic inheritance mechanism, even microevolutionary processes.

Individual-based models that are spatially structured and in which the individuals can move over the spatial grid are called spatially explicit individual-based models (Tyler & Rose 1994). Rather than focusing on variation among individuals, spatially explicit individual-based models allow for focusing on the effects of the spatial configuration of the interactions between individuals and their resources (Huston *et al.* 1988). When one scans the literature, models of terrestrial systems generally focus on spatial patterns (Turner *et al.* 1994; Pacala *et al.* 1996; Comiskey *et al.* 1997), whereas models for the pelagic zones of aquatic systems, which tend to have less spatial structure, focus on the variations among individuals in size (DeAngelis *et al.* 1979; Madenjian 1991; Mooij *et al.* 1997).

There is no general answer to whether the unstructured, the physiologically structured, or the individual-based modelling approach is the most appropriate for a given problem or ecosystem. Besides the obvious application for modelling the demographic stochasticity of small populations (Ruxton 1996), we argue that individual-based models are useful when individuals have complex phenotypes with multiple relevant traits, when there is large variation among individuals due to genetic differences or adaptations to the local environment or when the environment is spatially complex. Individual-based models are often assumed to be best suited for describing organisms at higher trophic levels, which are mostly larger, have bigger home ranges and longer lifespans, and which occur in smaller numbers. However, smaller organisms perceive their environment at a much finer scale, and relative to their size they may encounter the same amount of heterogeneity as do larger organisms. This notion has led to the development of individual-based models for zooplankton (Hogeweg & Richter 1982; Mooij & Boersma 1996; Mooij *et al.* 1997) and even bacteria (Jaworska *et al.* 1996).

DeAngelis and Rose (1992), Lomnicki (1992) and Murdoch *et al.* (1992) give some more criteria for choosing between the physiologically structured and the individual-based approach, but we argue that if possible, using several approaches concurrently may be the better than choosing among them. After all, a model is only a vehicle to derive insights in the inherent dynamics of the system under study, and these insights themselves should not be dependent on the choice for a certain modelling formalism. For complex, applied problems, however, spatially explicit individual-based modelling may be the only possible format.

Implementation of individual-based models

At first sight, focusing on the implementation of a concept, such as individual-based modelling, into algorithms seems of limited relevance. However, it is important to realize that there often is a tight relationship between the development of a theoretical concept and the tools to represent it. For instance, the development of mathematics would have been almost impossible without a mathematical notation. There is general agreement about the need for such a notation to document the structure of individual-based models (Gross *et al.* 1992). Technically, this structure tends to be rather complex because of the large number of book-keeping operations required to keep track of all individuals, their traits, their positions and their interactions. This complexity has two sides. On the one hand, the number of state variables of the model may easily be several orders of magnitude higher than in unstructured or physiologically structured population models. On the other hand, the number of process formulations may be considerably higher and dependent on more state variables compared with traditional models. This makes implementation of individual-based models in procedural programming languages such as FORTRAN 77 or ANSI C a feasible but tedious task.

There is general agreement that the object-oriented programming paradigm (Meijer 1988) is well-suited for the implementation of individual-based models (Baveco & Lingeman 1992; Maley & Caswell 1993; Mooij & Boersma 1996). This style of programming is based on a hierarchy of structured variables. More concrete and complex variables can be derived from more abstract and simple ones by a mechanism called inheritance. Each structured variable is accompanied by a set of functions that acts on its contents. By performing data exchange between variables only via this set of so-called member functions is the risk of unwanted side-effects of specific algorithms limited, the structure of the programme becomes more clear and maintenance more simple. Object-oriented programming is available in languages such as C++ (Stroustrup 1991), SMALLTALK (Pinson & Wiener 1988) and some dialects of PASCAL.

The complex variables of object-oriented programming can directly represent ecological hierarchies such as individual, population and community (O'Neill *et al.* 1996) and ecological operational units such as condition, resource and habitat (Begon *et al.* 1990). The advantage of using these basic units as building blocks for the model is that they have fixed mutual relationships. The statement that an ecosystem is a network of habitats, each with a local community of individuals, interacting with each other and with the local resources under control of the local conditions, will hold for any system (Fig. 18.2). Due to these fixed relationships, this structure can be embedded in a framework for the implementation of individual-based models (Baveco & Lingeman 1992; Maley & Caswell 1993; Mooij & Boersma 1996).

Different mechanisms can be used to adopt the framework for a specific application. Baveco and Lingeman (1992) and Maley and Caswell (1993) propose to use the inheritance mechanism of object-oriented programming for the insertion of application-specific detail in the model. Although logical at first sight, this solution implies redefinition of the building blocks of the framework (i.e. individual, population, habitat) for each application. Mooij and Boersma (1996) propose a different solution by giving each of the building blocks of their framework OSIRIS (Fig. 18.2) an arbitrary number of each of the three low-level data types of contemporary computers (floating point, integer and string) as a standard feature. The user of the framework only has to specify how many of each of these low-level variables are needed for a certain application to store all relevant properties and traits of the individuals, populations, etc. This approach allows for a better separation between framework and application and considerably reduces the size and complexity of the application specific code.

An obvious problem with the implementation of individual-based models is that the number of individuals in the natural system under study is often larger than what can be handled by contemporary computers in terms of memory and computational requirements (Haefner 1992). Only in studies of small populations of organisms is a one to one representation possible (e.g. Nolet & Baveco 1996). In such studies, the use of individual-based modelling has the advantage that the effects of demographic stochasticity, which may be important for such small populations, can be explored directly with the model. In most cases, however, the model will only represent a sample of individuals, each representing a much larger number of individuals in the natural system (Scheffer *et al.* 1995). In such

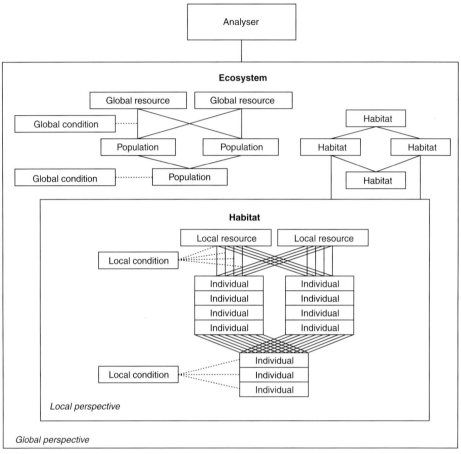

Figure 18.2 Structure of the OSIRIS framework for the implementation of spatially explicit individual-based models (Mooij & Boersma 1996). This framework recognizes nine basic entities: analyser, ecosystem, habitat, global condition, global resource, population, local condition, local resource and individual. The classic food web is implemented at the ecosystem level as interactions between global resources and populations, with global conditions as modifiers. The spatial structure of the ecosystem is implemented as a network of habitats. The actual interactions take place at the habitat level between local resources and individuals, with local conditions as modifiers. The global food web acts as a template for these local interactions and is used to store information that is not habitat specific. The simulation is controlled by the analyser object, representing the ecologist.

cases the demographic stochasticity exhibited by the model population is an artefact, and its effect should be reduced to an acceptable level by taking a sufficient number of individuals into account (Mooij & Boersma 1996). However, the number of individuals should not be too large because of the unnecessary reduction of performance due to evaluating redundant information.

The question of how many individuals should be represented in the model is closely related to the question of how many individuals should be sampled in a field programme or followed in an experimental design, and the same rules of thumb can be applied. The answer to this question is also linked to the spatial extent of the model and the choice for a specific spatial extent implies the choice for a specific number of model individuals. Populations that undergo rapid changes in numbers raise another problem, which can be handled properly by random cloning or deleting of model individuals while keeping the number of model individuals constant (Rose *et al.* 1993; Mooij & Boersma 1996). Finally, it should be noted that while the model may take only a fraction of the total population into account, the number of interactions that an individual has with other individuals may still be represented with a reasonable degree of realism in the individual-based model.

In contrast with unstructured and physiologically structured models, individual-based models are stochastic by nature because birth and death are modelled as discrete events and not as continuous rates (see Fig. 18.1). Moreover, several stochastic components may be added to the model, representing natural variation in certain processes. This stochasticity entered at the level of the individual, in conjunction with environmental stochasticity that is imposed on the system, will result in a stochastic behaviour of the model at the highest, i.e. system, level. This simply means that two runs with the same model and the same parameters will give different results. Because modelling has traditionally focused on deterministic processes, this may be perceived as an unwanted situation. One way to handle this inherent stochasticity is to rerun the model as often as required to produce an approximation to the whole set of possible outcomes for a given parameter setting. Mooij and Boersma (1996) suggest making the analysis of the model a formal part of it by introducing a dedicated operational unit in their framework. A discussion of whether the distribution of model behaviours gives a realistic description of the possible range of system behaviours, of which we see only one realization in the natural system under study, is far beyond the scope of this chapter.

At least, however, having a distribution for each output variable enables a formal falsification of the model against the natural situation for community and system characteristics (i.e. diversity and stability indices) for which no replications in the natural situation can be obtained.

Individual-based models in theoretical studies

In the next two sections of this chapter we want to briefly discuss some examples that we think highlight the possibilities of individual-based models in theoretical and applied studies. Although the individual-based approach is often associated with empirical studies we want to start with some examples that illustrate how individual-based modelling can contribute to theoretical insights. In these studies the individual-based paradigm is typically used to study the effect of initial variation in size structure or the inherent stochasticity in processes such as somatic growth rate and reproduction rates. Another line of models explores the effects of the spatial extent of ecosystems. Because generating insight is the target in these knowledge-orientated studies, one is actually looking for differences rather than similarities between the outcome of comparative models. These differences may then lead to the falsification of the hypothesis underlying each model. In applied, prediction-orientated models, which are discussed after the theoretical models, similarity between the model and the natural system is obviously the target, but a formal analysis that could lead to falsification of the model is often lacking in these studies.

Mooij and Boersma (1996) compared the estimate of the population growth rate r of an individual-based model of a *Daphnia* population with the estimate of r obtained from a cohort life table (Stearns 1992). The latter approach is based on the Euler–Lotka equation. Both models had the same number of parameters for the deterministic components, derived from individual-based experiments. In the individual-based model only, however, experimentally derived stochastic components were added to the duration of the moulting stages and the moulting stage-specific fecundities. Nevertheless, both models gave similar estimates of r, thereby validating the assumption that ignoring the natural variation in moulting stage duration and fecundity does not lead to a biased estimate of r.

In an extended analysis, Mooij *et al.* (1997) studied the effect of size-selective harvesting on the *per capita* birth rate of individuals of a *Daphnia* population. The population was kept in steady state by adjusting the harvesting rate so that it equalled the birth rate. Under random selection, this model resembled the results of the cohort life table, but size-selective harvesting had a strong effect on the *per capita* birth rate of the individuals in the culture (Fig. 18.3). Moreover, by adding a simple length–weight relationship to the model, the effects of size-selective harvesting on the production to biomass (*P*/*B*) ratio of the population can be studied in parallel (Fig. 18.3). The *P*/*B* ratio has the same dimension as the *per capita* birth rate of individuals (1/*d*), and by definition both parameters have equal values under random selection from a population that is in steady state. Under size-selective harvesting, however, the *P*/*B* ratio is much

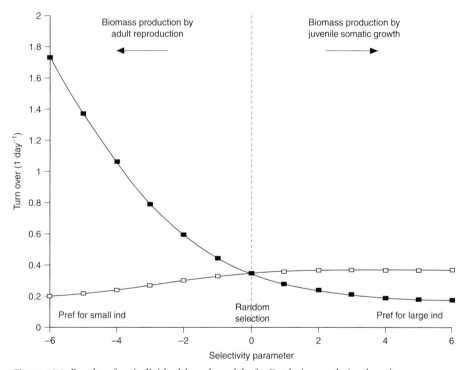

Figure 18.3 Results of an individual-based model of a *Daphnia* population kept in steady state by size-selection harvesting. This example shows the ability of the individual-based model to study the *per capita* birth rate (closed squares) and the production to biomass (*P*/*B*) ratio (open squares) of the population concurrently. Under random harvesting, both measures of turnover have equal values by definition. However, the *per capita* birth rate is highly sensitive for size-selective harvesting, whereas the *P*/*B* ratio is rather stable. This result validates the use of constant *P*/*B* ratios in simple food web models. Redrawn from Mooij *et al.* (1997).

more constant than the *per capita* birth rate of individuals (Fig. 18.3). This result is interesting because it gives some credit to the use of constant turnover rates of zooplankton as a functional group (or trophic species) in food web studies (Christensen & Pauly 1993). These results could not have been obtained from a cohort life table because under size-selective harvesting the relative mortality rates of each size class are a function of the a priori unknown size structure of the population.

Another analysis of complementary modelling approaches was performed by DeAngelis *et al.* (1993) in a study on juvenile fish cohort dynamics. Analytical solutions were obtained for three partial differential equation models of the early life history of the fish cohorts and were compared with the results of equivalent individual-based models (Fig. 18.4). The individual-based models gave similar results as the partial differential equation models, although as asymptotic approximations. As with the *Daphnia* example, these results suggested that one can use both approaches interchangeably (Fig. 18.4). Simple uncorrelated stochasticity in daily growth was then added to the individual-based models, and again it was shown that this does not result in a significant difference from the deterministic model (Fig. 18.4). However, when the stochasticity was correlated in time such that fish growing faster on one day also tends to grow faster the next day, strong effects on the outcomes of the simulations were observed. This temporal correlation, which has a biological basis, could not be incorporated in the standard partial differential equation model.

A more complex exploratory use of an individual-based model was made by Basset *et al.* (1997). Using an abstract spatially explicit individual-based model of a population of grazers moving through a landscape while foraging on grass, they analysed the effect of the shape of the functional response of the grazers on the stability of this consumer–resource interaction. From simple non-spatially explicit mathematical models it is known that a Holling type 2 (Holling 1959) functional response of the grazers tends to give unstable results for a wide range of realistic parameters. Functional responses such as Holling type 3 tend to stabilize the consumer–resource interaction (see also Chapter 12). For the spatially explicit individual-based model, however, the effect of these two functional responses on the stability of the system was reversed, given certain assumptions concerning the movement of grazers on the landscape. Basset *et al.* (1997) concluded that this apparently paradoxical result is caused by the fact that, in line with conventional wisdom, the Holling type 2 functional response allows grazers to graze the grass locally to a lower level than with

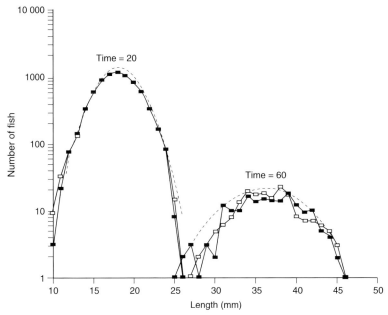

Figure 18.4 Results of a fish growth model that was concurrently implemented as a deterministic partial differential equation (dotted line), as a deterministic individual-based model (closed squares) and a stochastic (random variation in growth increments of up to ± 0.15 mm day^{-1}) individual-based model (open squares). All three models give similar results. This validates the assumption that underlies the PDE model, namely that ignoring the variation in growth among individuals does not lead to a bias in the predicted length frequencies. Redrawn from DeAngelis *et al.* (1993).

type 3. However, this local overexploitation leads the grazers to be slower in reaching areas with higher resource densities. This prevents the system from being overexploited globally, thereby enhancing the stability of the coexistence of consumers and their resource.

The above examples show that for any of the analytical models, an individual-based counterpart could be created, often with the same number of parameters. At first hand this may seem useless, because the individual-based model will only give asymptotic approximations of the analytical solutions, at the cost of a considerable computational overhead. However, taking these comparative models as a starting point, the individual-based models can accommodate small but biologically relevant details, which cannot, or only with complex mathematics, be included in the analytical models. Sometimes these differences will not affect the results, thereby providing formal evidence for the validity of the assumptions underlying the analytical model. The example of the grazer–grass interaction, however, shows that these explorations may lead to fundamentally different conclusions about the stability of the system under study.

Individual-based models in applied studies

Individual-based models can contribute significantly to prediction-orientated, applied ecology because of their capability of incorporating the details of specific systems. In contrast to classical population models, there are virtually no limits to types of physiological, behavioural and life history information that can be integrated into an individual-based model. Consequently, much of the information on individuals collected by field ecologists can be accommodated in these models. The question is, what information is relevant for addressing a particular question and what information is available, rather than what type of information can the model handle. Similarly, landscapes of almost any degree of complexity can be simulated in these models. Once the rules for the physiology and behaviour of the organisms have been formulated, it is technically no more difficult to model them on a complex landscape than on a simple one.

As early as the late 1970s DeAngelis *et al.* (1979) developed an individual-based model to predict the outcome of a series of experiments on the cohort dynamics of juvenile fish (see Fig. 18.1). The populations in question were cohorts of first-year largemouth bass (*Micropterus salmoides*) in 50-gallon aquaria. Model simulations using the experimental starting conditions and variations on these showed the extreme sensitivity of the dynamics of these cohorts to the precise initial size distribution of the bass. A change of a few millimetres in the sizes of the largest few bass in a cohort could make a large difference in the distribution of sizes after 2 or 3 months. A sufficiently large enough bass had the opportunity to consume many of the small bass and to grow at a rate allowing it to eventually consume most of the others. However, if there was initially no sufficiently large bass present, a large fraction of the cohort would survive until the end of the simulation. The results of the simulations agreed well with the outcome of the experiments.

In line with this application, Madenjian *et al.* (1991) developed an individual-based fish model in a study on the stocking strategies for fingerling walleyes (*Stizostedion vitreum vitreum*). They simulated fish stocking in order to investigate the effects of the size of stocked fingerlings and the timing of stocking on subsequent young-of-the-year walleye growth. From these simulations it was concluded that if walleye fingerlings were stocked on 28 June, at an average length of 60 mm rather than 50 mm, the proportion of large fish in the young-of-the-year walleye population at the end of the growing season would have increased threefold over the observed proportion. Moreover, economic cost per large walleye was minimized when average total length at stocking was 62 mm. Finally, the model of Madenjian *et al.* (1991) showed that stocking 50-mm walleye fingerlings on 14 June instead of 28 June also resulted in a tripling of the percentage of large walleyes at the end of the growing season.

Whereas applications of individual-based models for aquatic systems tend to focus on size structure, models for terrestrial systems tend to focus on the effects of the spatial extent. Turner *et al.* (1994) developed a spatially explicit individual-based model to simulate winter interactions among ungulates, vegetation and fire in northern Yellowstone Park (WY, USA). Their model was developed to predict the effects of scale and pattern of fire on the winter foraging dynamics and survival of free-ranging elk and bison. The model simulated the individual search, movement and foraging activities of 18 000 elk and 600 bison at daily intervals for a duration of 180 days. The 77 020-ha landscape of the park was represented as an irregular polygon with a spatial resolution of 1 ha. The model of Turner *et al.* (1994) showed that winter severity played a dominant role in ungulate survival in relation to fire patterns. It was concluded that the effects of fire on ungulate survival became important when winter conditions were average to severe. Moreover, the spatial patterning of fire influenced ungulate survival when fires covered small to moderate proportions of the landscape and when winter snow conditions were moderate to severe. Finally, ungulate survival was higher with a clumped than with a fragmented fire pattern, suggesting that a single, large fire is not equivalent to a group of smaller disconnected fires.

Due to their ability to take demographic stochasticity into account, in combination with environmental stochasticity, individual-based models can also be used to perform a population viability analysis. Nolet and Baveco (1996) monitored survival, reproduction and emigration of a translocated beaver population in the Biesbosch (The Netherlands) and used an individual-based model to assess its viability. The observed mortality was initially high but gradually fell to normal rates while the observed breeding success remained low during the 5 years of observation. They hypothesized that this was either a temporary phenomenon (the translocation

567

hypothesis) or a permanent feature (the poor habitat hypothesis). According to the results of the individual-based model, the isolated population was viable under the first but not under the second hypothesis. In the latter case, the prospects generally improved by the foundation of another population in a nearby wetland. However, under the poor habitat hypothesis, this second habitat should be optimal for beavers in order to reduce the extinction probability of the first population to below 10% in 100 years. Nolet and Baveco (1996) therefore concluded that the beaver population in the Biesbosch was not viable unless reproductive success would increase, either in the Biesbosch itself or in a nearby population.

Within ATLSS (Across Trophic Level System Simulation), a modelling effort within the ongoing Florida Everglades restoration project, a set of individual-based models (Fleming et al. 1994; Wolff 1994; Comiskey et al. 1997) are currently being developed and integrated with GIS (Geographical Information System) maps, hydrological models and physiologically structured models of lower trophic levels to describe the full community of this endangered wetland (ATLSS 1994; Davis & Ogden 1994). This approach applied on such a large scale is relatively new, so it is not easy to say how successful the coupling of different models will be in describing some of the cascading effects (see Chapter 10; Carpenter & Kitchell 1993) in the biological community of the Everglades under different hydrological scenarios. Because making the best possible predictions is the target in this project, coupling of the models is not a goal in itself and should be avoided where it creates instabilities in the model that are more likely to be an artefact than an inherent property of the system. In line with our plea for using concurrent modelling approaches, analysis of the system with more abstract models may play a role in making these judgements.

Recently, Levin et al. (1997) commented on individual-based modelling by stating that the amount of detail in such models cannot be supported in terms of what we can measure and parameterize. We argue that models like the those of Turner et al. (1994) and Nolet and Baveco (1996) are quite flexible in using those data that are typically available from field studies, rather than requiring data that are difficult to obtain. At the process level, complex mechanisms such as foraging strategies (Stevens & Krebs 1986) and even learning (Krebs & Inman 1992) can be included. Where essential data on the mechanistic details are absent, a correlative approach (i.e. habitat quality) can be used instead and integrated with the more mechanistically described components of the model.

Conclusions and perspectives

The goal of this paper has been to illustrate some of the recent progress that has been made in individual-based modelling. In order to limit the extent of the chapter, the examples chosen have been restricted to a few areas of ecology. However, the example models span the range from quite abstract to highly detailed. This illustrates that individual-based models have been formulated for a variety of reasons, from general theoretical questions to applied problems pertaining to specific systems. We hope to have shown that individual-based models can contribute to both of these areas. Because theoretical and applied individual-based models differ only in their scope and level of detail, theory and application can be joined in a more seamless way than with traditional modelling approaches. Field ecologists are continuously aware of the great complexity of natural systems that the models employed by theoretical ecologists often ignore. Individual-based models may help to bridge the gap between these two groups of ecologists.

The conclusion that individual-based models can be made to be either simple theoretical or complex applied models does not guarantee their success. One can make an analogy with the use of laboratory-scale eco-systems (i.e. microcosms, ecotrons) in ecological research. Those who favour this experimental approach say that laboratory-scale ecosystems combine the best of both worlds: the realism of natural systems with the rigid control of experiments. Those who criticize the approach argue that these laboratory ecosystems are, in fact, a bad combination of the lack of control of natural systems with the artefacts of experiments. The same holds for individual-based models. It is easy to make a model that is too complex to be fully understood from an analytical point of view, while being far too unrealistic to serve any management purpose. As Starfield and Bleloch (1986) stress, any modelling effort can only be successful if the objective of the model is clearly stated a priori. Only with this objective in mind can judicious decisions be made for the scope and level of detail of the model.

It is not easy to say how rapidly individual-based modelling will become established as a standard tool in ecological research, although we do not doubt that it will be established in the long term. The absence of a general notation and suitable software now limits the use of this approach to those who fully concentrate on developing such models. In analogy with the increasing use of complex statistical techniques by ecologists due to the increasing availability of user-friendly software, one can expect an increasing use of individual-based models once such software becomes available. Of course, as with statistics, this will enable both use and abuse of this technique as an exploratory and predictive tool. The frameworks that are currently available (Baveco & Lingeman 1992; Mooij & Boersma 1996) can be seen as a first step in this process. These frameworks can only be refined by applying them to a wide variety of problems, while checking each time whether the given problem can be represented without too much distortion. Unfortunately, the development of a solid-proof user-friendly interface around these frameworks takes an enormous programming effort for which it is hard to raise money from scientific funding agencies. As with statistical software, development of these interfaces will probably have to wait until it becomes commercially attractive.

For those who hope for the discovery of general laws in ecology as they exist in physics, the development of the individual-based modelling paradigm may not seem a big step forwards. Instead of generally applicable laws, individual-based modelling provides only a general structure for the formalization and analysis of the dynamics of ecological systems. It does so by taking the individual as the elementary particle of ecology, governed internally by its own genetic, physiological and behavioural laws. We argue that at the level of ecosystem dynamics, having a general approach is probably the furthest we can get in the foreseeable future. Using this approach, however, abstract knowledge-orientated individual-based models can help to generate new insights in the general, qualitative behaviour of these systems, picking up where models based on (partial) differential equations are limited by what can be handled by analytical mathematics. With exactly the same format, prediction-orientated individual-based models can be generated, which, although sometimes not fully understood in their mathematical behaviour, can play a role in the conservation of endangered species and in ecosystem management in general.

Acknowledgements

W.M. Mooij thanks his colleagues at the Netherlands Institute of Ecology, Centre for Limnology for the stimulating discussions on individual-based modelling and their comments on the manuscript.

References

ATLSS (1994). *Across-Trophic-Level System Simulation for the Everglades and*

Big Cypress Swamp. Report of the ATLSS workshop, Homestead, Florida, June 21–22, 1994.

Basset, A., DeAngelis, D.L. & Diffendorfer, J.E. (1997). The effect of functional response on stability of a grazer population on a landscape. *Ecological Modelling*, **101**, 153–162.

Baveco, J.M. & Lingeman, R. (1992). An object-oriented tool for individual-oriented simulation: host–parasitoid system application. *Ecological Modelling*, **61**, 267–286.

Begon, M., Harper, J.R. & Townsend, C.R. (1990). *Ecology: Individuals, Populations, Communities*, 2nd edn. Blackwell Scientific Publications, Oxford.

Botkin, D.B., Janak, J.F. & Wallis, J.R. (1972). Some ecological consequences of a computer model for forest growth. *Journal of Ecology*, **60**, 849–873.

Carpenter, S.R. & Kitchell, J.F. (1993). *The Trophic Cascade in Lakes*. Cambridge University Press, Cambridge.

Caswell, H. (1989). *Matrix Population Models: Construction, Analysis and Interpretation*. Sinauer Associates, Sunderland, MA.

Christensen, V. & Pauly, D. (1993). *Trophic Models of Aquatic Ecosystems*. ICLARM conference proceedings 26. ICLARM, Manila.

Comiskey, E.J., Gross, L.J., Fleming D.M., Huston, M.A., Bass, O.L., Luh, H.K. & Wu, Y. (1997). A spatially-explicit individual-based simulation model for Florida panther and white-tailed deer in the Everglades and Big Cypress landscapes. In *Proceedings of the Florida Panther Conference, Fort Myers, FL, November 1994* (Ed. by D. Jordan). US Fish and Wildlife Service, Gainesville, FL.

David, J.M., Andral, L. & Artois, M. (1982). Computer simulation of the epi-zootic disease of vulpine rabies. *Ecological Modelling*, **15**, 107–125.

Davis, S.M. & Ogden, J.C. (1994). *Everglades. The Ecosystem and its Restoration*. St Lucie Press, Delray Beach, FL.

DeAngelis, D.L. (1992). *Dynamics of Nutrient Cycling and Food Webs*. Chapman & Hall, London.

DeAngelis, D.L. & Gross, L.J. (Eds) (1992). *Individual-Based Models and Approaches in Ecology. Populations, Communities and Ecosystems*. Chapman & Hall, New York.

DeAngelis, D.L. & Rose, K.A. (1992). Which individual-based approach is most appropriate for a given problem? In *Individual-Based Models and Approaches in Ecology* (Ed. by D.L. DeAngelis & L.J. Gross), pp. 67–87. Chapman & Hall, New York.

DeAngelis, D.L., Cox, D.K. & Coutant, C.C. (1979). Cannibalism and size dispersal in young-of-the-year largemouth bass: experiment and model. *Ecological Modelling*, **8**, 133–148.

DeAngelis, D.L., Rose, K.A., Crowder, L.B., Marschall, E.A. & Lika, D. (1993). Fish cohort dynamics–application of complementary modeling approaches. *American Naturalist*, **142**, 604–622.

de Roos, A.M., Diekmann, O. & Metz, J.A.J. (1992). Studying the dynamics of structured population models—a versatile technique and its application to *Daphnia*. *American Naturalist*, **139**, 123–147.

Ebenman, B. & Persson, L. (Eds) (1988). *Size-Structured Populations: Ecology and Evolution*. Springer-Verlag, Berlin.

Fleming, D.M., Wolff, W.F. & DeAngelis, D.L. (1994). Importance of landscape heterogeneity to wood storks in Florida Everglades. *Environmental Management*, **18**, 743–757.

Gross, L.J., Rose, K.A., Rykiel, E.J., VanWinkle, W. & Werner, E.E. (1992). Individual-based modeling: summary of a workshop. In *Individual-Based Models and Approaches in Ecology* (Ed. by D.L. DeAngelis & L.J. Gross), pp. 511–522. Chapman & Hall, New York.

Haefner, J.W. (1992). Parallel computers and individual-based models: an overview. In *Individual-Based Models and Approaches in Ecology* (Ed. by D.L. DeAngelis & L.J. Gross), pp. 126–164. Chapman & Hall, New York.

Hilborn, R. (1975). The effect of spatial heterogeneity on the persistence of predator–prey interactions. *Theoretical Population Biology*, **8**, 346–355.

Hogeweg, P. & Hesper, B. (1990). Individual-oriented modelling in ecology. *Mathematical and Computer Modelling*, **13**, 83–90.

Hogeweg, P. & Richter, A.F. (1982). INSTAR, a discrete event model for simulating zooplankton population dynamics. *Hydrobiologia*, **95**, 275–285.

Holling, C.S. (1959). Some characteristics of simple types of predation and parasitism. *Canadian Entomologist*, **91**, 385–398.

Humphries, H.C., Coffin, D.P. & Lauenroth, W.K. (1996). An individual-based model of alpine plant distributions. *Ecological Modelling*, **84**, 99–126.

Huston, M.A., DeAngelis, D.L. & Post, W.M. (1988). New computer models unify ecological theory. Computer simulations show that many ecological patterns can be explained by interactions among individual organisms. *Bioscience*, **38**, 682–691.

Jaworska, J.S., Hallam, T.G. & Schultz, T.W. (1996). A community model of ciliate tetrahymena and bacteria *E. coli*: Part I. Individual-based models of tetrahymena and *E. coli* populations. *Bulletin of Mathematical Biology*, **58**, 247–264.

Judson, O.P. (1994). The rise of the individual-based model in ecology. *Trends in Ecology and Evolution*, **9**, 9–14.

Krebs, J.R. & Inman, A.J. (1992). Learning and foraging—individuals, groups, and populations. *American Naturalist*, **140**, S63–S84.

Levin, S.A. (1974). Dispersion and population interactions. *American Naturalist*, **108**, 207–228.

Levin, S.A., Grenfell, B., Hastings, A. & Perelson, A.S. (1997). Mathematical and computational challenges in population biology and ecosystem science. *Science*, **275**, 334–343.

Lomnicki, A. (1992). Population ecology from the individual perspective. In *Individual-Based Models and Approaches in Ecology* (Ed. by D.L. DeAngelis & L.J. Gross), pp. 3–17. Chapman & Hall, New York.

Madenjian, C.P., Johnson, B.M. & Carpenter, S.R. (1991). Stocking strategies for fingerling walleyes — an individual-based model approach. *Ecological Applications*, **1**, 280–288.

Maley, C.C. & Caswell, H. (1993). Implementing i-state configuration models for population dynamics — an object-oriented programming approach. *Ecological Modelling*, **68**, 75–89.

McCauley, E., Nisbet, R.M., De Roos, A.M., Murdoch, W.W. & Gurney, W.S.C. (1996). Structured population models of herbivorous zooplankton. *Ecological Monographs*, **66**, 479–501.

Meijer, B. (1988). *Object-oriented Software Construction*. Prentice-Hall, New York.

Metz, J.A.J. & Diekmann, O. (Eds) (1986). *The Dynamics of Physiologically Structured Populations*. Springer Lecture Notes in Biomathematics, Vol. 68. Springer-Verlag, Berlin.

Mooij, W.M. & Boersma, M. (1996). An object-oriented simulation framework for individual-based simulations (OSIRIS): *Daphnia* population dynamics as an example. *Ecological Modelling*, **93**, 139–153.

Mooij, W.M., Boersma, M. & Vijverberg, J. (1997). The effect of size-selective predation on the population growth rate, the production to biomass ratio and the population structure of *Daphnia galeata*: a modelling approach. *Archive für Hydrobiologie (Special Issues Advances in Limnology)*, **49**, 87–97.

Murdoch, W.W. (1993). Individual-based models for predicting effects of global change. In *Biotic Interactions and Global Change* (Ed. by P.M. Kareiva, J.G. Kingsolver & R.B. Huey), pp. 147–162. Sinauer Associates, Sunderland, MA.

Murdoch, W.W., McCauley, E., Nisbet, R.M., Gurney, W.S.C. & De Roos, A.M. (1992). Individual-based models: combining testability and generality. *Individual-Based Models and Approaches in Ecology* (Ed. by D.L. DeAngelis & L.J. Gross), pp. 18–35. Chapman & Hall, New York, London.

Myers, J.H. (1976). Distribution and dispersal in populations capable of resource depletion. *Oecologia*, **23**, 255–269.

Nolet, B.A. & Baveco, J.M. (1996). Development and viability of a translocated beaver castor fiber. *Biological Conservation*, **75**, 125–137.

Okubo, A. (1980). *Diffusion and Ecological Problems: Mathematical Models.* Springer-Verlag, Berlin.

O'Neill, R.V., DeAngelis, D.L. & Allen, T.F.H. (1996). *A Hierarchical Concept of the Ecosystem.* Princeton University Press, Princeton, NJ.

Pacala, S.W. & Deutschman, D.H. (1995). Details that matter: the spatial distribution of individual trees maintains forest ecosystem function. *Oikos*, **74**, 357–365.

Pacala, S.W., Canham, C.D., Saponara, J., Silander, J.A., Jr., Kobe, R.K. & Ribbens, E. (1996). Forest models defined by field measurements: estimation, error analysis and dynamics. *Ecological Monographs*, **66**, 1–43.

Pinson, L.J. & Wiener, R.S. (1988). *An Introduction to Object-oriented Programming and Smalltalk.* Addison Wesley Publishing Company, Reading, MA.

Post, W.M. & Pastor, J. (1996). LINKAGES — an individual-based forest ecosystem model. *Climatic Change*, **34**, 253–261.

Rose, K.A., Christensen, S.W. & DeAngelis, D.L. (1993). Individual-based modeling of populations with high mortality — a new method based on following a fixed number of model individuals. *Ecological Modelling*, **68**, 273–292.

Ruxton, G.D. (1996). Dispersal and chaos in spatially structured models: an individual-level approach. *Journal of Animal Ecology*, **65**, 161–169.

Scheffer, M., Hosper, S.H., Meijer, M.-L., Moss, B. & Jeppesen, E. (1993). Alternative equilibria in shallow lakes. *Trends in Ecology and Evolution*, **8**, 275–279.

Scheffer, M., Baveco, J.M., DeAngelis, D.L., Rose, K.A. & Van Nes, E.H. (1995). Super-individuals: a simple solution for modelling large populations on an individual basis. *Ecological Modelling*, **80**, 161–170.

Shugart, H.H. & West, D.C. (1977). Development of an Appalachian deciduous forest succession model and its application to assessment of the impact of the chestnut blight. *Journal of Environmental Management*, **5**, 161–179.

Sinko, J.W. & Streifer, W. (1967). A new model for age-size structure of a population. *Ecology*, **48**, 910–918.

Starfield, A.M. & Bleloch, A.L. (1986). *Building Models for Conservation and Wildlife Management.* Macmillan Publishing Company, New York.

Stearns, S.C. (1992). *The Evolution of Life-Histories.* Oxford University Press, Oxford.

Stevens, D.W. & Krebs, J.R. (1986). *Foraging Theory.* Princeton University Press, Princeton, NJ.

Stroustrup, B. (1991). *The C++ Programming Language.* Addison-Wesley Publishing Company, Reading, MA.

Tilman, D. (1977). Resource competition between planktonic algae: an experimental and theoretical approach. *Ecology*, **58**, 338–348.

Turner, M.G., Wu, Y.G., Wallace, L.L., Romme, W.H. & Brenkert, A. (1994). Simulating winter interactions among ungulates, vegetation, and fire in northern Yellowstone Park. *Ecological Applications*, **4**, 472–486.

Tyler, J.A. & Rose, K.A. (1994). Individual variability and spatial heterogeneity in fish population models. *Reviews in Fish Biology and Fisheries*, **4**, 91–123.

Uchmanski, J. & Grimm, V. (1996). Individual-based modelling in ecology: what makes the difference? *Trends in Ecology and Evolution*, **11**, 437–441.

Van de Koppel, J., Huisman, J., Van der Wal, R. & Olff, H. (1996). Patterns of herbivory along a gradient of primary productivity: an empirical and theoretical investigation. *Ecology*, **77**, 736–745.

Wolff, W.F. (1994). An individual-oriented model of a wading bird nesting colony. *Ecological Modelling*, **72**, 75–114.

Yodzis, P. (1975). *Introduction to Theoretical Ecology*. Harper & Row, New York.

Chapter 19

Predator control in terrestrial ecosystems: the underground food chain of bush lupine

D.R. Strong[1]

Summary

Predator control in terrestrial ecosystems is one of the great ecological enigmas of our time. How general is the trophic cascade? Do plants flourish primarily because carnivores suppress herbivores? Certainly, trophic cascades are strong in some streams, lakes and marine systems. In natural terrestrial ecosystems, however, the evidence is not so convincing; potent carnivores are not known to be linked to potent carnivores in isolation of other forces. Part of the challenge is our current ignorance of how herbivores and carnivores are influenced by other elements of food webs on land. I present an example of a subterranean food chain that bears the trappings of a trophic cascade. An entomopathogenic nematode kills root-feeding caterpillars, which can kill seedling and mature bush lupine. However, the nematode has a very high mortality rate in the soil that appears to be the product of other organisms. Nematophagous fungi are the most prominent suspects. These fungi have dual-mode nutrition and are probably sustained by detritus as well as by nematodes, which suggests that the web involves saprotrophism. The root-feeding caterpillars also experience competition from other herbivores of bush lupine, each with their own natural enemies. Thus, an imbroglio of interactions much more complex than the simple chain of Hairston *et al.* (1960; HSS) is suggested by this evidence. From the evidence at hand, we cannot infer that HSS-like chains are the reason that the terrestrial world is green.

1 University of California, Davis and Bodega Marine Laboratory, Box 247 Bodega Bay, CA 94923, USA

Introduction

Food Webs

Ecology is the most complex of sciences (Holt & Polis 1997), and the topic of the 1997 BES Symposium is a large slice of this complexity. Most multitrophic systems in nature are a tangle of interspecific interactions. Food webs, the 'ecologically flexible scaffolding around which communities are assembled and structured' (Paine 1996), are the conceptual tool that ecology has used most effectively to make sense of this farrago. Scaling down to manageable size, most analyses are based upon a subset of species that are most obvious or otherwise known well. These 'interaction webs' (Menge & Sutherland 1976) have also been termed 'community modules' (Holt & Polis 1997). For the purposes of reflecting upon my general topic of predator control in multitrophic systems, which is one of the least settled topics in food web studies, I find useful the additional metaphor of a food web vignette. As in interaction webs and modules, well-resolved interactions among a few familiar species are the focus of vignettes; however, vignettes fade out of focus as the inevitable, lesser-known species and environmental influences are considered. Being explicit about ignorance plays an important role in science (Dupre 1993), and asserting that a module is a vignette is a way of taking account of our ignorance. As larger communities and spatial scales, longer time spans, and externalities are taken into account in food web studies, the outcome of central sets of interactions can change.

The HSS model

Indirect interactions in food webs change direct interspecific influences between pairs of species (Menge 1995). The trophic cascade of Hairston *et al.* (1960, 'HSS') is a particular indirect interaction that has had unusually great influence upon ecological thought. It is germane to topics concerning predator control and to this symposium. The HSS trophic cascade works from the top down by means of a predator, which changes the outcome of the direct interaction between a herbivore and plant. By suppressing the numbers of the herbivore population, the predator indirectly allows the plant to escape suppression by its consumer; thus, in HSS, the world is green because predators suppress herbivores. The important antecedents to the discovery of the trophic cascade include the idea of the food web itself, termed the 'food cycle', by Elton (1927), which generalized producer–consumer relationships in nature. A subsequent, very important deduction was 'bottom-up' trophic ecology, or trophodynamics, which stressed the thermodynamic implications of indirect interactions in chains of energy transfer from sunlight through green plants, to herbivores and to carnivores (Lindeman 1942). Slobodkin (1960) brought trophodynamics to a head by emphasizing the inefficiency of energy transfer through the food chain and by proposing that this inefficiency limited the length of food chains; '... there are not fiercer dragons on earth ... because the energy supply will not stretch to the support of super dragons' (Colinvaux 1978). Thus, 1960 was a big year for trophic ecology. The yin of bottom-up trophodynamics met the yang of the top-down trophic cascade.

The HSS trophic cascade is a special case of a two-linked chain in which green plants are the basal resource species (Menge 1995). Longer cascades have a third link in which a secondary predator suppresses the primary predator. Thus, HSS cascades can be reversed by a secondary predator that suppresses the primary predator; this leads to a flourishing of the herbivore population and suppression of the plant population. Scores of trophic cascades have been demonstrated in lakes (Carpenter & Kitchell 1993), in flowing waters (Power 1990) and on sea shores (Paine 1992; Estes & Duggins 1995). These archetypal trophic cascades are aquatic, and algae are the basal, plant species. Where the primary carnivore is absent, or suppressed by a secondary carnivore, algal biomass is substantially reduced by the herbivore; where the primary carnivore is abundant, the herbivore is suppressed and plant biomass is high. In lakes, two-linked trophic cascades often have as the carnivore a planktivorous fish that consumes zooplankton, and this allows phytoplankton to flourish and renders the water turbid. A piscivorous species can reverse this effect and clarify the water.

The demonstrations of so many of these linear chains of predator control recommends the trophic cascade as a touchstone. While bottom-up control in plant–herbivore–carnivore food chains is the proper null hypothesis of trophic ecology (Hunter & Price 1992), the trophic cascade is an alternative that ecologists have found particularly compelling. Yet, while some authors advocate HSS (indirect plant protection by carnivores) as a unifying element of terrestrial ecosystems (Hairston & Hairston 1993, 1997), others have disagreed (Crawley 1989). On land, we lack the collection of clear-cut examples of simple linear control by predators that has accumulated for aquatic systems. Evidence to date suggests that terrestrial food webs are more diverse than the exemplary aquatic trophic cascades (Strong 1992; Polis & Strong 1996). Certainly, there is no lack of potent herbivores and carnivores on land. What seems to be rare are the dominant chains — two or three links together — that are so well resolved in aquatic systems. Some common themes are that trophic levels are poorly resolved, and potent herbivores are not necessarily controlled by predators (Bazely & Jefferies 1997). Nutrient cycling and plant competition, rather than predator control, can dominate vegetation patterns (Pastor & Naiman 1992). An important new idea is that simple linear trophic cascades, based upon more edible plants early in succession, evolve toward reticulate food webs as more lignified, less edible plants come to dominate in later successional communities on land (see Chapter 10).

The centrepiece of my contribution is a recently discovered two-linked trophic cascade of native species in a natural terrestrial system; the central message is that this cascade is real, but that our knowledge of it is a vignette. The interactions of the central species of this vignette are well resolved, much observation and a recent field experiment support the inference that the carnivore indirectly affects biomass and population dynamics of the plant. However, interactions with other species that could be important to the system are less well resolved, and the role of the cascade in the larger arena of vegetation dynamics — which after all is the fascination of this topic for ecologists — remains to be understood. Thus, I demur (contrary to Hairston & Hairston 1997) that our discovery in this particular system is some verification of the general importance of predator suppression of herbivores in promoting the greenness of the terrestrial world.

The lupine–ghost moth–nematode cascade

First, the central clear area of the vignette. Field observations suggested that an entomopathogenic nematode, *Heterorhabditis hepialus*, could indirectly protect bush lupine, *Lupinus arboreus*, by killing root-feeding ghost moth caterpillars, which appeared to kill unprotected lupines. Bush lupine is a rapidly growing, nitrogen-fixing, perennial shrub (Maron & Connors 1996). Ghost moth caterpillars of *Hepialus californicus* (Lepidoptera; Hepialidae) feed upon lupine roots, are univoltine and largely monophagous at the study site at the Bodega Marine Reserve, Sonoma Co., California. Ghost moths, also known as 'swifts' or 'swift moths', are strong flyers and disperse widely. On the wing, a single moth broadcasts upwards of 2000 small (0.5 mm diameter) eggs around and beneath *L. arboreus* plants (Tobi *et al.* 1993). Hatching within a few weeks, the tiny larvae burrow into the soil and feed upon the exterior of lupine roots, and by early summer caterpillars have grown large, and have bored inside the root. In the autumn, prepupal caterpillars bore upwards into the shoot, where they make an exit hole through which the imago will pass in winter or spring.

Inspired by observations of dead bush lupine and of root damage caused by these caterpillars, we found a spatial correlation among four sites between death rates of this woody plant and the caterpillar densities inside the roots (Strong *et al.* 1995). In 1993, a year of particularly high ghost moth densities, about 40% of mature lupine bushes, aged 3–7 years old, died in an area where there was an average of about 38 and a maximum of 62 large caterpillars per root. At the site with the opposite, low extreme of caterpillar densities (*c.* six per root), only about 2% of the plants died. This spatial correlation was consistent with a 40-year history among the sites in die-off of patches of lupine, inferred from a series of aerial photographs taken since 1955 (Fig. 19.1). The large round canopies of bush lupine are distinctive against the background of low-lying grasses and forbs in these photographs. No trees grow on the California coastal dunes and headlands where this study took place, and other species of shrubs that would make interpretation of the photographs difficult are extremely rare in this habitat. The photographs showed that bush lupine on sites with high plant mor-

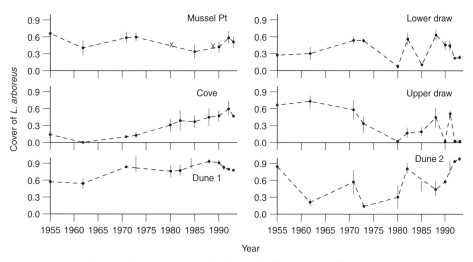

Figure 19.1 Spatial cover (means and standard errors) of six stands of *Lupinus arboreus* at the Bodega Marine Reserve from 1955 to present. Data are from aerial photographs in Strong *et al.* (1995).

581

tality in 1993 had repeatedly died off and regenerated, while over the same period of 40 years, no die-off was seen in the sites that had low mortality in 1993. Experimental exclusion of ghost moth caterpillars, by means of systemic insecticides, demonstrated that even low numbers of caterpillars caused increased mortality of bush lupine. Large bush lupine can withstand the feeding of a few ghost moth caterpillars each year, but seedlings can be killed by a single caterpillar (Maron 1998).

Entomopathogenic nematode as a potent natural enemy

The main predator of ghost moth caterpillars is remarkable. After a great deal of sampling, we must conclude that insect parasitoids are all but missing from our food web. Instead, we have found that a soil-dwelling, entomopathogenic nematode, *Heterorhabditis hepialus*, is the major source of biotic mortality to caterpillars of *Hepialus californicus*. This nematode caused very high mortality to ghost moth caterpillars that were mining into lupine roots (Fig. 19.2). Consistent with a trophic cascade, the prevalence of the nematode in lupine rhizospheres was negatively correlated in space with the pattern of lupine die-off (Fig. 19.3). Although these insect-killing

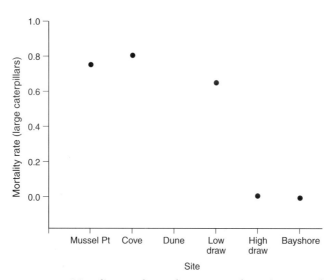

Figure 19.2 Mortality rate due to the entomopathogenic nematode *Heterorhabditis hepialus* of large caterpillars of the ghost moth *Hepialus californicus*. From Strong *et al.* (1996, Table 1).

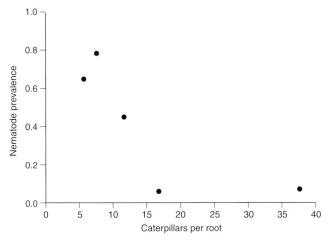

Figure 19.3 Prevalence in rhizospheres of the entomopathogenic nematode *Heterorhabditis hepialus* and the number of ghost moth caterpillars per root of bush lupine. Points represent different areas of the Bodega Marine Reserve. From Strong *et al.* (1996, Table 2).

nematodes are diverse (Poinar 1979), widely distributed in the soils of the world (Hominick & Reid 1990) and believed to act as generally important enemies of root-feeding insects (Kline 1990), little ecological work has addressed entomopathogenic nematodes in natural settings (Gaugler *et al.* 1997).

The biology, if not the ecology, of entomopathogenic nematodes is well known (Gaugler & Kaya 1990). They are strictly carnivorous, with insects and a few other arthropods as the only hosts in nature. The non-feeding 'infective juveniles' or third-instar, dauer larvae, are about 0.5 mm in length, live in the soil and search for hosts (Kaya 1990). Attracted by carbon dioxide and other waste gasses, they move up to 6 cm day^{-1} through moist soil towards prospective hosts (Strong *et al.* 1996). An infective juvenile enters the host through a spiracle or other orifice, punctures a membrane, then regurgitates the mutualistic bacterium, *Photorhabdus luminescens*, which kills the host within 48 hours (Forst & Nealson 1996). This bacterium occurs only in symbiotic association with entomopathogenic nematodes in the genus *Heterorhabditis*. The host cadaver is preserved in the soil by antibiotics produced by the bacterial mutualists, which also lyse the interior of the host and produce the food for the nematodes. These nematodes reproduce only inside host cadavers. One or a few infective juveniles enter the insect, then mature. After several generations within 6 weeks or so, when resources are exhausted within the host, as many as 410 000 descendants stream out of the cadaver, each searching for a host in the soil.

A field experiment with plant, herbivore and natural enemy

The first of our experiments has dealt with lupine seedlings, which are germane to the HSS idea because of the lack of self-thinning. Without herbivory, high densities of germinating seedlings persist into the adult phase to set seed themselves (Maron & Simms 1997). The lack of self-thinning means that herbivory and the entomopathogenic nematodes can have (opposite) large effects upon adult densities and the population dynamics of bush lupine. Early events, during late winter and spring, are crucial. The tiny ghost moth caterpillars can sever young tap roots within the first few months after germination, and the entomopathogenic nematode kills many small ghost moth caterpillars in the soil in March, April and May. We placed hatchling ghost moth caterpillars upon the soil near the stems of seedlings growing in deep pots that were sunk into the soil, in the field. The hatchling caterpillars were very small. Emerging from the egg, they were about 50% larger than a comma on this page. The lupine seedlings were growing rapidly and began the experiment at about 15 cm in height with tap roots about 15 cm long. Roots were about 5 mm in diameter just beneath the soil surface, where the tiny ghost moth caterpillars begin to feed. The experiment was a fully random, factorial design with four levels of hatchling ghost moth caterpillars (caterpillars: 0, 8, 16 and 32) crossed with the entomopathogenic nematode (nematodes: present or absent) in soil with a single lupine seedling. Each treatment combination had 15 replicate seedlings (Strong *et al.*, unpublished). The caterpillars burrowed into the soil quickly, and damage to the seedlings was almost entirely below ground. Neither the activities of the herbivore nor those of its natural enemy were visible above ground.

The results of this experiment demonstrated a powerful, but subtle, subterranean trophic cascade (Fig. 19.4). Two months after the beginning of the experiment, 46% (21/45) of seedlings had died in treatments with ghost moth caterpillars and no entomopathogenic nematodes, while only 11% (5/45) of seedlings with the nematode *Heterorhabditis hepialus* and caterpillars had died. More ghost moth caterpillars meant fewer survivors of the lupine seedlings, in the absence of the nematode. To interpret the experimental results, we explored a series of 'one-hit dose response' models for lupine survival. With maximum likelihood estimates of parameters, we

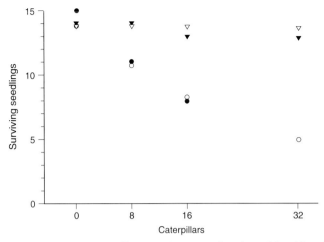

Figure 19.4 Lupine seedlings surviving as a function of densities of hatchling ghost moth caterpillars placed upon the soil surface at the beginning of the experiment. Treatment combinations with no entomopathogenic nematodes are indicated by triangles, those with entomopathogenic nematodes in the rhizospheres are indicated by circles. Solid symbols indicate the data and open symbols indicate expected values (see text). Each treatment combination had a sample size of 15. From Strong *et al.* (unpublished).

selected the best-fitting model using the Schwartz Information Criterion (SIC), which is a more scientifically powerful statistical approach than pairwise hypothesis testing. In the absence of nematodes, the best fitting model was one of exponentially decreasing seedling survival as a function of numbers of ghost moth caterpillar hatchlings. At the highest experimental density of 32 hatchling ghost moth caterpillars, the model had 4.9 lupine survivors and we observed five of 15 surviving seedlings. In the presence of the nematodes, caterpillar density had no influence upon seedling survival; 13.8 survivors were expected from the model, and we observed 13 of 15 seedling survivors at a density of 32 hatchling caterpillars. Interestingly, the standard procedure of pairwise testing of the logistic regression model failed to identify the most ecologically salient feature of the experimental outcome, the statistical interaction of the two nematode treatments with the caterpillar density treatment. The interaction means that the nematode virtually cancels the effect of this potent herbivore. The message from a statistical point of view was that model selection among an ecologically realistic assortment of possibilities can be a more fruitful avenue than the standard procedure of pairwise hypothesis testing for understanding complex ecological interactions. The ecological message was an unambiguous indication of a terrestrial trophic cascade, at least at the centre of the lupine–ghost moth–entomopathogenic nematode vignette.

Moving to the edges of the vignette

Several sets of potentially important species frame the central, focused part of the lupine–ghost moth–entomopathogenic nematode vignette. The systematics and biology of soil-dwelling enemies of root-feeding arthropods are familiar to specialists in biological control and insect pathology (Tanada & Kaya 1993), but little is known of either the population biology or food web influences of these organisms. While plant–insect food chains above ground have been vigorously investigated over several decades (Strong *et al.* 1984; Hawkins & Sheehan 1994; Price 1997), the complementary below-ground chains and webs have been all but ignored. As I will argue in the discussion, root-feeding insects often kill or seriously damage both seedlings and adult plants, and ignorance of the biotrophic food web below ground is a serious deficiency in ecological knowledge.

Predators of the entomopathogenic nematode?

Could the entomopathogenic nematode in this food chain be controlled by one or more secondary predators? *Heterorhabditis hepialus* suffers extremely high rates of mortality in the soils of the study site, and local extinctions and recolonizations at the scale of individual bush lupine rhizospheres are consistent with biogenic mortality (Strong *et al.* 1996).

Nematophagous fungi

Fungi are a major group of nematode predators that can cause high mortality to entomopathogenic nematodes (Timper *et al.* 1991). These fungi do not attack insects or live plants, and thus do not have the potential to be food chain omnivores. For this reason, they raise the possibility of a three-linked trophic cascade (at least a vignette of a trophic cascade) like that with piscivorous fish in lakes or streams. We have concentrated preliminary work on these owing to their abundance and diversity in lupine rhizospheres and to the complimentary paucity of other enemies of these nematodes (nematophagous insects, mites, Collembola, tardigrades and carnivorous nematodes; Coleman & Crossley 1996). So far, we have identified 13 species of nematophagous fungi in the soils of our study site (Table 19.1).

Table 19.1 Nematophagous fungi observed and isolated from soil in stands of *Lupinus arboreus* at the Bodega Marine Reserve, California. From Jaffee *et al.* (1996).

Fungus	Trophic structures or mode
Arthrobotrys brochopaga	Constricting rings
Arthrobotrys musciformis	Adhesive networks
Arthrobotrys oligospora	Adhesive networks
Arthrobotrys superba	Adhesive networks
Geniculifera paucispora	Adhesive networks
Hirsutella rhossiliensis	Adhesive conidia
Monacrosporium cionopagum	Adhesive branches
Monacrosporium deodycoides	Constricting rings
Monacrosporium eudermatum	Adhesive networks
Monacrosporium parvicollis	Adhesive knobs
Nematoctonus concurrens	Adhesive, glandular cells
Stylopage sp.	Adhesive hyphae
Pleurotis ostreatus	Nematoxin from secretory cells

Although the available techniques (soil placed upon nematode-baited agar) probably deliver many false negatives, because the jostling of soil during sampling can destroy the adhesiveness of the trap structures (as with 'bubble gum in sand', Jaffee 1996), we detected as many as 695 propagules of these fungi per gram of soil from the study area. One species, *Arthrobotrys oligospora*, was found at all sites and in 92% (44/48) of the samples. Another species, *Pleurotis ostreatus*, the oyster mushroom, is common in the sporocarp (mushroom) stage on dead lupines at the study site, but its propagules were not detected by our procedures. Soil inoculated with hyphae of *P. ostreatus* causes extremely high mortality to *Heterorhabditis hepialus* (D.R. Strong & J. Johnston, unpublished observations), suggesting that the nematicidal toxins produced by specialized hyphal cells of this fungus (Kwok *et al.* 1992) could play a role in the lupine–ghost moth food web. Five additional species of these fungi from the field site that we tested in the laboratory all readily killed the entomopathogenic nematode on agar (Fig. 19.5). In soil from the field site, *Arthrobotrys oligospora* (Ao), *Monacrosporium cionopagum* (Mc) and *Nematoctonus concurrens* (Nc) had no statistically significant effect upon the mortality of the nematode when added to pasteurized soils. This indicates that the mortalities caused by species in raw (unpasteurized) soils was probably caused by other unidentified microbes. For *Geniculifera paucispora* (Gp) and *M. eudermatum* (Me), significant numbers of nematodes were killed in pasteurized soils to which the fungi were added, indicating that these species probably caused the mortalities.

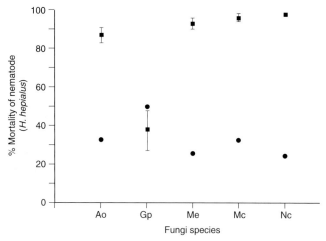

Figure 19.5 Mean mortality of the entomopathogenic nematode *Heterorhabditis hepialus* caused by cultures of five species of nematophagous fungi from the soils of bush lupine stands at the Bodega Marine Reserve. Data are from laboratory experiments. All points are statistically greater than zero. Error bars indicate one standard deviation. Squares represent experiments on agar, after 36 h. Circles represent experiments in unpasteurized (raw) soils, after 48 h, from the study site to which each fungus species was added. In the cases of *Arthrobotrys oligospora* (Ao), *Monacrosporium cionopagum* (Mc), and *Nematoctonus concurrens* (Nc), no statistically significant effect upon the mortality of the nematode was obtained from the addition of the fungi to pasteurized soils. This indicates that the mortalities shown for these species were probably caused by unidentified microbes in the raw soil. For *Geniculifera paucispora* (Gp) and *M. eudermatum* (Me), significant numbers of nematodes were killed in pasteurized soils to which the fungi were added, indicating that these species probably caused the deaths of the nematodes. Data from Koppenhoffer *et al.* (1996, Table 3).

How nematophagous fungi might be involved in the lupine–ghost moth food chain

Dual nutrition: saprophytism and parasitism. Recent thought about nematophagous fungi has emphasized their dual modes of nutrition. Many species probably consume dead plant material as well as live nematodes (Cooke 1977). The nematologists, mycologists and plant pathologists who study these fungi term the dual modes 'saprophytism' and 'parasitism'. Conventional wisdom is that nematodes provide the nitrogen that is in short supply in the plant detritus upon which hyphae proliferate (Barron

1992). The hyphae of the oyster mushroom, which is conspicuous at the study site, are particularly compelling candidates for this sort of decomposer–predator (Barron & Thorn 1987). A spectrum from more detritivorous to more carnivorous is proposed (Stirling 1991); while most species have some saprophytic dependence upon detritus, at the extreme are a few completely carnivorous ('endoparasitic') species such as *Hirsutella rhossiliensis* (Jaffe & Zehr 1985) and perhaps *Monacrosporium ellipsosporum* (Jaffe & Muldoon 1995). Most nematophagous fungi are potentially polyphagous in terms of the nematode species that they can trap and consume, but variation in behaviour and other features among nematode species lead to differences in vulnerabilities to particular fungi (Jaffee 1996; Van den Boogert *et al.* 1994).

The oldest idea of the food web functioning of these fungi ignores saprophytic nutrition, and advocates that nematophagous fungi function as omnivorous carnivores (Linford 1937); subsidized by microbivorous nematodes, these fungi increase and suppress populations of other kinds of nematodes. Decreases in plant-parasitic nematodes of pineapple were seen to be the result of numerical increases in nematophagous fungi that resulted from increased numbers of microbivorous nematodes caused by additions of plant detritus to the soil.

Indirect and direct interactions. The dual modes of nutrition mean that nematophagous fungi could come into play in the lupine–ghost moth food web in a variety of ways (Fig. 19.6). Two indirect possibilities involve plant detritus, which would lead to an increase in biomass and density of hyphae. The density of trapping structures, and thus the threat to the entomopathogenic nematode, would be a function of hyphal density. In the Cooke–Barron hypothesis (Fig. 19.6A), hyphae proliferate by direct cellulolytic digestion of plant detritus. In the Linford hypothesis (Fig. 19.6B), fungal proliferation is supported by microbivorous nematodes, which increase by consuming the microbes that digest plant detritus. Direct interaction between entomopathogenic nematodes and nematophagous fungi is also possible (Fig. 19.6C). The very high local abundances of infective juvenile entomopathogenic nematodes escaping cadavers of the host insect could support substantial fungal growth.

In the lupine rhizospheres, 86% of the average of *c.* 30 000 nematodes per kilogram of this soil were bactivorous, 9% were fungivorous, 9% were plant parasites and about 0.01% were predacious dorylaim nematodes (bottom four bars of Fig. 19.7). In the vast majority of samples and most of the year,

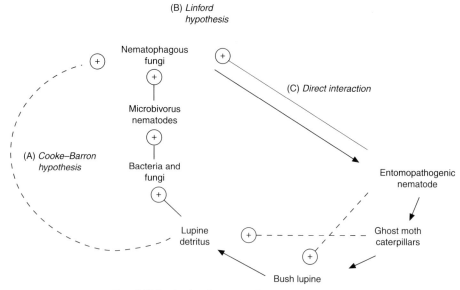

Figure 19.6 Possibilities for involvement of nematophagous fungi in the lupine–ghost moth food chain. The dual modes of nutrition of these fungi make possible two general indirect avenues (A and B) and one direct avenue (C) for suppression of the entomopathogenic nematode *Heterorhabditis hepialus*. In the Cooke–Barron hypothesis (A), the cellulolytic activities of the fungi themselves leads to a proliferation of hyphae, which form trapping structures that kill this insect's natural enemy. In the Linford hypothesis (B), microbivorous nematodes sustained by fungi and bacteria (which perform the cellulolysis of lupine litter) lead to proliferation of the hyphae. In the direct interaction (C), the abundance of entomopathogenic nematodes that escape from cadavers of the host insect support a numerical response in the fungi.

entomopathogenic nematodes were very rare in the rhizospheres. However, when a large ghost moth caterpillar is killed by *Heterorhabditis hepialus*, up to 410 000 individual nematodes of this entomopathogenic nematode are produced by the cadaver and introduced to a small volume of soil near the root (Strong *et al.* 1996). Even when dispersed through the approximately 1 kg of soil in the upper 10 cm of the lupine rhizosphere, these hundreds of thousands of entomopathogenic nematodes numerically overwhelm the tens of thousands of other kinds of nematodes (see Fig. 19.5). Frequently, *H. hepialus* killed several large ghost moth caterpillars on a single root, magnifying the predominance of this entomopathogenic nematode in lupine

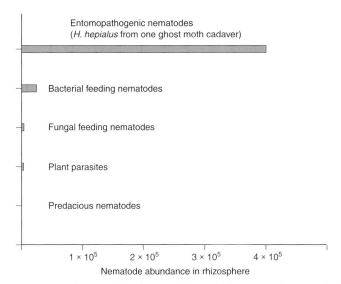

Figure 19.7 Average abundances of the trophic groups of nematodes in lupine rhizospheres of the study site in which a single large caterpillar has been killed by *Heterorhabditis hepialus*, per kilogram soil. Determinations of bacterial feeders, fungal feeders, plant parasites and predacious nematodes by D. Freckman and R. Niles.

rhizospheres several times that shown in Fig. 19.5. A candidate predator for this direct interaction with *H. hepialus* is *Arthrobotrys oligospora*. Propagules of this fungus are common and sometimes abundant in soils of the study site (Jaffee *et al.* 1996), and it proliferates upon infective juveniles of *H. hepialus* (D.R. Strong, personal observation). Thus, the resources in entomopathogenic nematodes exist for substantial numerical response and population growth of nematophagous fungi. This is a direct interaction, and the three-linked linear chain that would result is analogous to the aquatic trophic cascade in which piscivorous fish suppress planktivorous fish (Carpenter & Kitchell 1993).

Detrital subsidy? Detrital subsidy of nematophagous fungi, i.e. both the Linford (1937) and the Cooke (1977)–Barron and Thorn (1987) hypotheses, involve circuitous indirect interactions through the saprotrophic channels of the food web (Wiegert & Owen 1971; Polis & Strong 1996), while the direct population interaction of these fungi with *H. hepialus* would result in a strictly linear biotrophic chain extending from the living resources of the lupine, through ghost moth caterpillars and the entomopathogenic nematode (see Fig. 19.6). Neither of these scenarios are consistent with HSS, nor is intraguild predation (IGP) among nematophagous fungi. We examined the potential for IGP in terms of interspecific competition, which may be important among these fungi (Quinn 1987). In laboratory studies with cultures from the field site, *Arthrobotrys oligospora* had a slightly negative affect upon the abilities of *Geniculifera paucispora* and *Monacrosporium eudermatum* to suppress the entomopathogenic nematode. Population densities of *M. eudermatum* and *G. paucispora* were less in the presence than in the absence of *A. oligospora*. In contrast, *A. oligospora* population density was unaffected or possibly enhanced by the other fungi (Koppenhoffer *et al.* 1996, 1997). These small effects do rot suggest a large role for intraguild predation among nematophagous fungi. However, it is possible that saprophytism and field conditions, which we did not study, could change these relationships.

Granivores

Four other known sets of interactions affect bush lupine greatly and are relevant to the factors involved in the green world of *Lupinus arboreus* on the Central California coast (Fig. 19.8). The first concerns the heavy seed predation upon this plant by the deer mouse, *Peromyscus maniculatus* (Maron & Simms 1997). In dunes, where the seed bank is sparse, rodents consumed or otherwise removed 65% and 86% of seeds in 2 successive years. This significantly reduced emergence of seedlings by 18% and 19%, respectively, in these 2 years. In contrast, the lupine seed bank is dense in coastal prairie (where the ghost moth work was carried out), and consumption by rodents of almost 60% of seeds did not significantly affect recruitment of the plant. Thus, this granivorous rodent strongly suppressed lupine seedling numbers in one, but not another habitat. The habitat effect was related to the propensity of seed to be lost by the mice much more readily in prairie soil than in the sands of the dunes, leading to the denser seed bank in the prairie. This kind of difference between the habitats is not one consistent with HSS, although predators of the deer mouse were not studied.

Flower and foliage herbivores

The second interaction concerns flower-feeding and seed-eating insects (Fig. 19.8), which reduced output of seed from the bush lupine by as much

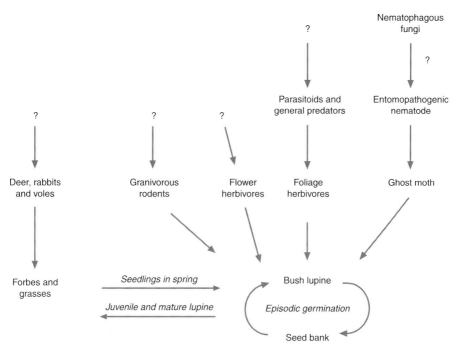

Figure 19.8 The known sets of food web interactions involving bush lupine on the central California coast. See text for details.

as 78% (Maron 1998). Because of the seed bank mentioned above, the effects of these substantial reductions need to be factored into the full life history equation of the lupine. No information on the effects of predators and parasites of these insects on lupine density or dynamics is available, so the role of the HSS cascade is unknown in this instance. The third additional interaction is based upon foliage feeding by tussock moth, *Orgyia vetusta*. These caterpillars periodically defoliate lupine (Harrison & Maron 1995). Generalist predators and parasitoids can suppress populations of the moth that are on the periphery of outbreaks (Maron & Harrison 1997). If these natural enemies are not much affected by yet other species (via intraguild predation, apparent competition, etc.), this part of the lupine food web would be consistent to the HSS model.

Apparent competition between plant species

The final additional interaction is competition between lupine seedlings and grasses and forbs (Fig. 19.8) (Maron 1997). Deer, voles and rabbits graze these grasses, and it is a reasonable hypothesis that the grazing reduces the competitive impact of these competitors on lupine seedlings. No information exists about the influence of the natural enemies of these vertebrate herbivores upon the food web.

Discussion

Things to learn

Very few natural enemies of underground insects are known ecologically. Especially in wild settings, even many underground insect herbivores are probably yet to be found. Even though underground insect herbivores have been 'out of sight, out of mind', recent work has shown that the diversity of herbivorous insects attacking plant roots represents all of the orders of insect herbivores found above ground (Brown & Gange 1990). Plants are quite sensitive to root herbivory (Parker 1985; Reichman & Smith 1991; this study), and subterranean herbivores can substantially influence plant performance, competition and succession (Seastedt *et al.* 1988; Masters & Brown 1997). Thus, I would speculate that natural enemies of root-feeding insects will prove to be as important to plants as are the enemies of insects that feed upon leaves, stems and other above-ground parts of plants.

A matrix of interesting questions about the lupine–ghost moth–entomopathogenic nematode food web vignette concern the possibility that nematophagous fungi function in a top-down fashion, by suppressing the entomopathogenic nematode and thus cancelling the protection that this natural enemy gives to bush lupine. Two possibilities in the matrix are not mutually exclusive: suppression could derive from species of these fungi that proliferate upon detritus, and it could derive from direct carnivorous proliferation upon nematodes. Indirect interactions could result from microbivorous nematodes and they could result from fungal proliferation upon detritus directly (leaving out other microbes that would be consumed by microbivorous nematodes). Potential direct interactions are proliferation of fungi on the very high local densities of *H. hepialus* that issue from cadavers of the ghost moth infected with this nematode. The complexities of these interactions are multiplied by the number of different fungal species in the system, and they are reduced by the fact that these fungi are not food chain omnivores; they do not attack the insect or the plant. The protocol for these experiments is logically simple (manipulate fungal species, detritus and nematodes; Stirling 1991), while being practically quite sophisticated (e.g. Jaffee 1996). Identifying effective natural enemies of the nematode would contribute to the solution of the enigma of their patchy distribution across apparently homogenous stretches of lupine (Strong *et al.* 1996). Of course, it is possible that as far as entomopathogenic nematodes are concerned, nematophagous fungi are donor controlled.

HSS—Loyal opposition

The attitude of loyal opposition to HSS, as an explanation of the greenness of the terrestrial world, is a combination of respect and scepticism. While HSS established a basis for ecologists to appreciate indirect interactions, evidence is not compelling that predator suppression of herbivores is the major factor in the greenness of the terrestrial world. We know what HSS looks like in nature because it operates so perfectly in the two-linked aquatic trophic cascades in lakes (Carpenter & Kitchell 1993), in flowing waters (Power 1990; Wootton *et al.* 1996) and shallow marine communities (Estes & Duggins 1995). The strength and clarity of the top-down effects are remarkable; runaway consumption by planktivorous fish greatly reduces the population of herbivores, and this propagates to great influence upon the algae that are the basal organisms in the chain. The singular feature to both basic and applied ecology of the aquatic trophic cascades is the remarkable biomass suppression of the algae in the absence of the primary carnivore or when this species is suppressed by a secondary carnivore.

Trophic trickles?

But even these purest of cascades are not free of richer ecological stews of the sort that I have argued are likely to be brewing in the soils of the lupine–ghost moth–entomopathogenic nematode food chain. The simplest trophic cascades are strictly linear and have rates of consumption per predator that are a linear function of number or biomass of the prey. This prey dependence (Oksanen *et al.* 1981) predicts that increases in primary productivity get passed up the chain proportionally, and this leads to the famous steps of density increase for the species with increasing primary productivity (see Fig. 5.5 in DeAngelis 1992). For a lake, this means that when productivity exceeds a certain low level, a herbivorous zooplankton is supported that holds the density of phytoplankton constant with increasing productivity; all increases in productivity go into density increase in the herbivore. A switch occurs when further increases in productivity exceed the minimum necessary to support a carnivore (planktivorous fish). At this point, the carnivore holds the density of the herbivore constant with increasing productivity.

Using meta-analysis, Brett and Goldman (1996, 1997) found evidence that the simple prey-dependent model did not apply generally. Phytoplankton increased little, if any, in roughly two-thirds of the 54 different planktivorous fish addition experiments that they analysed; these were more 'trophic trickles' (Strong 1992) than strong trophic cascades. Brett and Goldman (1997) identified two general patterns in these pelagic food webs: (i) in two-level systems of zooplankton and phytoplankton (without zooplanktivorous fish), nutrient addition experiments (i.e. increased phytoplankton productivity) caused large increases in phytoplankton with modest, if any, increases in zooplankton densities; (ii) phytoplankton densities in three-level systems (with zooplanktivorous fish) increased less than in two-level systems, with equivalent increases in nutrients. We have argued (McCann *et al.* 1998) that a slightly more complex model that includes trophic interference between species of zooplankton is more consistent with Brett and Goldman's findings than the prey-dependent model. Species interactions would be more reticulate than in the linear trophic cascade.

As good a model as is the aquatic trophic cascade, we need to remember that not all chains of fish, zooplankton and phytoplankton cascade. The archetypal trophic cascade occurs in small north temperate dimictic lakes from glaciated regions. Subtropical dimictic lakes give a well-documented example of lakes with a different kind of food web, one that lacks a prominent trophic cascade (Porter 1996). In these dimictic lakes, the role of large Cladocera, such as *Daphnia* spp., in linking photosynthetic productivity to fish is limited to the winter. In summer, these lakes stratify thermally, and most primary productivity is by smaller taxa of algae and cyanobacteria, which are consumed by flagellates, rotifers and ciliates. Larval fish consume these. In other lakes, the lack of a simple linear cascade is correlated with detritus being a major source of food for the ichthyofauna (Darnell 1964; Vadas 1990). As noted by Liebold and Wilbur (1992), species specificity leads to an assortment of indirect interactions richer than the trophic cascade.

Coda

A schema for food web research that goes beyond the HSS model has been proposed by Polis and Strong (1996). Substantial evidence urges approaches to food webs that appreciate the reticulation and inter-connection of webs in nature. Although vignettes of parts of webs, like that addressed in this chapter, are composed of discrete trophic levels, this will be the exception in nature. Omnivory is the rule, and a trophic spectrum of foods will be consumed by many species during their ontogeny. An important fact is that most primary productivity becomes detritus directly, without being consumed by a herbivore. This detritus forms the basis of the saprotrophic channels of the food web. Detritus plays a large role in subterranean food webs in general (De Ruiter *et al.* 1995) and it could play a very large role in the lupine–ghost moth–entomopathogenic nematode web discussed in this chapter. Most progress will be made by studies that accommodate the potential richness of food web interactions.

References

Barron, G.L. (1992). Lignolytic and cellulolytic fungi as predators and parasites. In *The Fungal Community* (Ed. by G.C. Carroll & D.T. Wicklow), pp. 311–326. Marcel Dekker, New York.

Barron, G.L. & Thorn, R.G. (1987). Destruction of nematodes by species of *Pleurotis*. *Canadian Journal of Botany*, **65**, 774–778.

Bazely, D.R. & Jefferies, R.L. (1997). Trophic interactions in arctic ecosystems and the occurrence of a terrestrial trophic cascade, In *Ecology of Arctic Environments* (Ed. by S.J. Woodin & M. Marquiss), pp. 183–208. Blackwell Science Ltd, Oxford.

Brett, M.T. & Goldman, C.R. (1996). A meta-analysis of the fresh-water trophic cascade. *Proceedings of the National Academy of Sciences, USA*, **15**, 7723–7726.

Brett, M.T. & Goldman, C.R. (1997). Consumers versus resource control in freshwater pelagic food webs. *Science*, **275**, 304–306.

Brown, V.K. & Gange, A.C. (1990). Insect herbivory below ground. *Advances in Ecological Research*, **20**, 1–58.

Carpenter, S.R. & Kitchell, J.F. (1993). *The Trophic Cascade in Lakes*. Cambridge University Press, New York.

Coleman, D.A. & Crossley, D.A. (1996). *Fundamentals of Soil Ecology*. Academic Press, New York.

Colinvaux, P. (1978). *Why Big Fierce Animals are Rare*. Princeton University Press, Princeton.

Cooke, R.C. (1977). *The Biology of Symbiotic Fungi*. John Wiley & Sons, New York.

Crawley, M.J. (1989). Insect herbivores and plant population dynamics. *Annual Review of Entomology*, **34**, 531–564.

Darnell, R.M. (1964). Organic detritus in relation to secondary production in aquatic communities. *Verhandlungen Internationale Vereninigung für theoretishe und angewandte Limnologie*, 15, 462–470.

De Ruiter, P., Neutel, A.M. & Moore, J.C. (1995). Energetics, patterns of interaction strengths, and stability in real ecosystems. *Science*, 269, 1257–1260.

DeAngelis, D.L. (1992). *Dynamics of Nutrient Cycling and Food Webs*. Chapman & Hall, New York.

Dupre, J. (1993). *The Disorder of Things*. Harvard University Press, Cambridge, MA.

Elton, C. (1927). *Animal Ecology*. Sidgwick & Jackson, London.

Estes, J.A. & Duggins, D.O. (1995). Sea otters and kelp forests in Alaska: generality and variation in a community ecological paradigm. *Ecological Monographs*, 65, 75–100.

Forst, S. & Nealson, K. (1996). Molecular biology of the symbiotic-pathogenic bacteria *Xenorhabdus* spp. and *Photorhabdus* spp. *Microbiological Reviews*, 60, 21–43.

Gaugler, R. & Kaya, H.K. (Eds.) (1990). *Entomopathogenic Nematodes in Biological Control*. CRC Press, Boca Raton.

Gaugler, R., Lewis, E. & Stuart, R. (1997). Ecology in the service of biological control. *Oecologia,* 109, 483–489.

Hairston, N.G., Smith, F.E. & Slobodkin, L.B. (1960). Community structure, population control, and competition. *American Naturalist*, 94, 421–425.

Hairston, N.G. & Hairston, N.G., Jr. (1993). Cause–effect relationships in energy flow, trophic structure and interspecific interactions. *American Naturalist*, 142, 379–411.

Hairston, N.G., Jr. & Hairston, N.G., Sr. (1997). Does food web complexity eliminate trophic-level dynamics? *American Naturalist*, 149, 1001–1007.

Harrison, S. & Maron, J.L. (1995). Impacts of defoliation by tussock moths (*Orgyia vetusta*) on the growth and reproduction of bush lupine (*Lupinus arboreus*). *Ecological Entomology*, 20, 223–229.

Hawkins, B.A. & Sheehan, W. (Eds) (1994). *Parasitoid Community Ecology*. Oxford University Press, Oxford.

Holt, R.D. & Polis, G.A. (1997). A theoretical framework for intraguild predation. *American Naturalist*, 149, 745–764.

Hominick, W.M. & Reid, A.P. (1990). Perspectives on entomopathogenic entomology. In *Entomopathogenic Nematodes in Biological Control* (Ed. by R. Gaugler & H.K. Kaya), pp. 327–349. CRC Press, Boca Raton.

Hunter, M.D. & Price, P.W. (1992). Playing chutes and ladders: heterogeneity and the relative roles of bottom-up versus top-down forces in natural communities. *Ecology*, **73**, 724–732.

Jaffee, B.A. (1996). Soil microcosms and the population biology of nematophagous fungi. *Ecology*, **77**, 690–693.

Jaffee, B.A. & Zehr, E.I. (1985). Parasitic and saprophytic ability of the fungus *Hirsutella rhossiliensis* on the nematode *Criconemella xenoplax*. *Journal of Nematology*, **17**, 341–345.

Jaffee, B.A. & Muldoon, A.E. (1995). Numerical responses of the nematophagous fungi *Hirsutella rhossiliensis, Monacrosporium cionopagum*, and *M. ellipsosporum. Mycologia*, **87**, 643-650.

Jaffee, B.A., Strong, D.R. & Muldoon, A.E. (1996). Nematode-trapping fungi in a natural shrubland: tests for foodchain involvement. *Mycologia*, **88**, 554–564.

Kaya, H.K. (1990). Soil ecology. In *Entomopathogenic Nematodes in Biological Control* (Ed. by R. Gaugler & H.K. Kaya), pp. 93–116. CRC Press, Boca Raton.

Kline, M.G. (1990). Efficacy against soil-inhabiting insect pests. In *Entomopathogenic Nematodes in Biological Control* (Ed. by R. Gaugler & H.K. Kaya), pp. 195–210. CRC Press, Boca Raton.

Koppenhoffer, A.M., Jaffee, B.A., Muldoon, A.E., Strong, D.R. & Kaya, H.K. (1996). Effect of nematode-trapping fungi on an entomopathogenic nematode originating from the same field site in California. *Journal of Invertebrate Pathology*, **68**, 246–252.

Koppenhoffer, A.M., Jaffee, B.A., Muldoon, A.E. & Strong, D.R. (1997). Suppression of an entomopathogenic nematode by the nematode-trapping fungi *Geniculifera paucispora* and *Monacrosporium eudermatum* as affected by the fungus *Arthrobotrys oligospora. Mycologia*, **89**, 220–227.

Kwok, O.C.H., Plattner, D., Weisleder, D. & Wicklow, D.T. (1992). A nematicidal toxin from *Pleurotis ostreatus* NRRL 3526. *Journal of Chemical Ecology*, **18**, 127–136.

Liebold, M.A. & Wilbur, H.M. (1992). Interactions between food-web structure and nutrients on pond organisms. *Nature*, **360**, 341–343.

Lindeman, R. (1942). The trophic–dynamic aspect of ecology. *Ecology*, **23**, 399–418.

Linford, M.B. (1937). Stimulated activity of natural enemies of nematodes. *Science*, **85**, 123–124.

McCann, K.S., Hastings, A. & Strong, D.R. (1998). Trophic cascades and trophic trickles in pelagic foodwebs. *Proceedings of the Royal Society, London B*, **265**, 1–5.

Maron, J.L. (1997). Interspecific competition and insect herbivory reduce bush lupine (*Lupinus arboreus*) seedling survival. *Oecologia*, **110**, 284–290.

Maron, J.L. (1998). Insect herbivory above and below ground: individual and joint effects on plant fitness. *Ecology*, **79**, 1281–1293.

Maron, J.L. & Connors, P.G. (1996). A native nitrogen-fixing shrub facilitates weed invasion. *Oecologia*, **105**, 302–312.

Maron, J.L. & Harrison, S. (1997). Spatial pattern formation in an insect host–parasitoid system. *Science*, **278**, 1619–1622.

Maron, J.L. & Simms, E.L. (1997). Effect of seed predation on seed bank size and seedling recruitment of bush lupine (*Lupinus arboreus*). *Oecologia*, **111**, 76–83.

Masters, G.J. & Brown, V.K. (1997). Host-plant mediated interactions between spatially separated herbivores: effects on community structure. In *Multitrophic Interactions in Terrestrial Systems* (Ed. by A.C. Gange & V.K. Brown), pp. 217–237. Blackwell Science Ltd, Oxford.

Menge, B. & Sutherland, J. (1976). Species diversity gradients: synthesis of the roles of predation, competition and temporal heterogeneity. *American Naturalist*, **110**, 351–369.

Menge, B.A. (1995). Indirect effects in marine rocky intertidal interaction webs: patterns and importance. *Ecological Monographs*, **65**, 21–74.

Oksanen, L., Fretwell, S.D., Arruda, J. & Niemela, P. (1981). Exploitation ecosystems in gradients of primary productivity. *American Naturalist*, **118**, 240–261.

Paine, R.T. (1992). Food web analysis: field measurements of per capita interaction strength. *Nature*, **355**, 73–75.

Paine, R.T. (1996). Preface. In *Food Webs: Integration of Patterns and Dynamics* (Ed. by G.A. Polis & K.O. Winemiller), pp. ix–x. Chapman & Hall, New York.

Parker, M. (1985). Size-dependent herbivore attack and the demography of an arid grassland shrub. *Ecology*, **66**, 850–860.

Pastor, J. & Naiman, R.J. (1992). Selective foraging and ecosystem processes in boreal forests. *American Naturalist*, **139**, 690–705.

Poinar, G.O., Jr. (1979). *Nematodes for Biological Control of Insects*. CRC Press, Boca Raton.

Polis, G.A. & Strong, D.R. (1996). Food web complexity and community dynamics. *American Naturalist*, **147**, 813–846.

Porter, K.G. (1996). Integrating the microbial loop and the classic food chain into a realistic planktonic foodchain. In *Food Webs: Integration of Patterns and Dynamics* (Ed. by G.A. Polis & K.O. Winemiller), pp. 51–59. Chapman & Hall, New York.

Power, M.E. (1990). Effects of fish in river food webs. *Science*, **250**, 811–814.

Price, P.W. (1997). *Insect Ecology*. John Wiley & Sons, New York.

Quinn M.A. (1987). The influence of saprophytic competition on nematode predation by nematode-trapping fungi. *Journal of Invertebrate Pathology*, **49**, 170–174.

Reichman, O.J. & Smith, S.C. (1991). Responses to simulated leaf and root herbivory by a biennial, *Tragopogon dubius*. *Ecology*, **72**, 116–124.

Seastedt, T.R., Ramundo, R.A. & Hayes, D.C. (1988). Maximization of densities of soil animals by foliage herbivory: empirical evidence, graphical, and conceptual models. *Oikos*, **51**, 243–248.

Slobodkin, L.B. (1960). Ecological energy relationships at the population level. *American Naturalist*, **94**, 213–236.

Stirling, G.R. (1991). *Biological Control of Plant Parasitic Nematodes*. CAB International, Wallingford.

Strong, D.R. (1992). Are trophic cascades all wet? The differentiation and donor-control in speciose ecosystems. *Ecology*, **73**, 747–754.

Strong, D.R., Lawton, J.H. & Southwood, T.R.E. (1984). *Insects on Plants*. Harvard University Press, Cambridge, MA.

Strong, D.R., Maron, J.L., Connors, P.G., Whipple, A.V., Harrison, S. & Jefferies, R.L. (1995). High mortality, fluctuation in numbers, and heavy subterranean herbivory in bush lupine, *Lupinus arboreus*. *Oecologia*, **104**, 85–92.

Strong, D.R., Kaya, H.K. Whipple, A.V., Child, A.L., Kraig, S., Bondonno, M., Dyer, K. & Maron, J.L. (1996). Entomopathogenic nematodes: natural enemies of root-feeding caterpillars on bush lupine. *Oecologia*, **108**, 167–173.

Tanada, Y. & Kaya, H.K. (1993). *Insect Pathology*. Academic Press, New York.

Timper, P., Kaya, H.K. & Jaffee, B.A. (1991). Survival of entomogenous nematodes in soil infested with the nematode-parasitic fungus *Hirsutella rhossiliensis* (Deuteromycotina: Hyphomucetes). *Biological Control*, **1**, 42–50.

Tobi, D.R., Grehan, J.R. & Parker, B.L. (1993). Review of the ecological and economic significance of forest Hepialidae (Insecta: Lepidoptera). *Forest Ecology and Management*, **56**, 1–12.

Vadas, R.L. (1990). The importance of omnivory and predator regulation of prey in freshwater fish assemblages of North America. *Environmental Biology of Fish*, **27**, 285–302.

Van den Boogert, P.H.J.F., Velvus, H., Ettema, C.H. & Bouwman, L.A. (1994). The role of organic matter in the population dynamics of the endoparasitic nematophagous fungus *Drechmeria coniospora* in microcosms. *Nematologica*, **40**, 249–257.

Wiegert, R.G. & Owen, D.F. (1971). Trophic structure, available resources and population density in terrestrial vs. aquatic ecosystems. *Journal of Theoretical Biology*, **30**, 69–81.

Wootton, J.T., Parker, M.S. & Power, M.E. (1996). Effects of disturbance on river food webs. *Science*, **273**, 1558–1561.

Author Index

603

Subject Index

Subject index